The Nutrient
Requirements
of Ruminant Livestock

The Nutrient Requirements of Ruminant Livestock

TECHNICAL REVIEW BY AN AGRICULTURAL
RESEARCH COUNCIL WORKING PARTY

PUBLISHED ON BEHALF OF THE
AGRICULTURAL RESEARCH COUNCIL BY THE
COMMONWEALTH AGRICULTURAL BUREAUX

CAB

COMMONWEALTH AGRICULTURAL BUREAUX

Published on behalf of the Agricultural Research Council by:

Commonwealth Agricultural Bureaux
Farnham Royal
Slough SL2 3BN
England
Tel. Farnham Common 2281
Telex 847964

ISBN 0 85198 459 2

Printed by Unwin Brothers
The Gresham Press,
Old Woking,
Surrey

Working Party on Nutrient Requirements of Ruminants

Chairman	SIR KENNETH BLAXTER, Ph.D., D.Sc., F.I.Biol., F.R.S.E., F.R.S.	Rowett Research Institute
	G. ALDERMAN, B.Sc. (Hons.)	Agricultural Development and Advisory Service, Ministry of Agriculture, Fisheries and Food
	PROFESSOR D. G. ARMSTRONG, Ph.D., D.Sc., F.R.I.C., F.R.S.E.	University of Newcastle upon Tyne
	C. C. BALCH, Ph.D., D.Sc.	National Institute for Research in Dairying
	A. C. FIELD, Ph.D., M.V.Sc., M.R.C.V.S.	Animal Diseases Research Association
	A. S. FOOT, M.Sc.	National Institute for Research in Dairying
	I. McDONALD, M.A.	Rowett Research Institute
	E. L. MILLER, Ph.D.	Department of Applied Biology, Cambridge University
	C. F. MILLS, Ph.D., F.R.I.C., F.R.S.E.	Rowett Research Institute
	E. R. ØRSKOV, Ph.D., D.Sc.	Rowett Research Institute
	PROFESSOR J. W. G. PORTER, M.A., Ph.D., F.I.Biol.	National Institute for Research in Dairying
	PROFESSOR SIR ALEXANDER ROBERTSON, C.B.E., LL.D., D.V.Sc., Ph.D., F.R.C.V.S., F.R.I.C., F.R.S.H., F.R.S.E.	Royal (Dick) School of Veterinary Studies
	PROFESSOR J. A. F. ROOK, Ph.D., D.Sc., F.I.Biol., F.R.I.C., F.R.S.E.	Hannah Research Institute
	J. H. B. ROY, M.A., Ph.D., D.Sc.	National Institute for Research in Dairying
	R. H. SMITH, Ph.D., D.Sc.	National Institute for Research in Dairying
	N. F. SUTTLE, Ph.D.	Animal Diseases Research Association
	S. Y. THOMPSON, Ph.D., D.Sc.	National Institute for Research in Dairying
	PROFESSOR P. N. WILSON, M.Sc., Ph.D.	BOCM Silcock Limited

Technical Secretary   *J. F. D. GREENHALGH, Ph.D.   Rowett Research Institute

D. A. GRUBB, L.I.Biol. (Assistant to   Rowett Research Institute
  Technical Secretary)

*Now Professor of Animal Production and Health, University of Aberdeen.

Preface

This review of the nutrient requirements of cattle and sheep is much larger than its predecessor, published in 1965. The reason is an obvious one; a very considerable amount of investigation relevant to the requirements of ruminants has taken place since that time. Much of that investigation was perhaps engendered by the fact that serious gaps in our knowledge were evident in the first edition and it is hoped that the obvious gaps and uncertainties in this review will similarly be filled through new experiments in the coming years.

When work commenced on the review it seemed possible that a simple expansion of the previous edition might suffice. It soon became evident that the new information was so great that a complete rewriting was necessary to accommodate it. The opportunity was then taken to deal separately with the composition of the animal body and its products, so as to avoid unnecessary repetition in subsequent chapters which deal with energy and specific nutrient metabolism and requirements.

The careful analysis and consideration of published and unpublished experiments has involved all members of the Working Party in considerable labour. While groups within the Working Party were charged to consider particular nutrients or groups of nutrients and to provide firstly discussion papers and later quantitative estimates of requirements, all members took part in the discussions which led to the final conclusions and share responsibility for them.

It will be obvious from the text that some parts of the review were completed before others and that there has been no attempt to ensure a synchrony. The chapter dealing with the composition of the body was completed first and none of the reviews contains work published after the summer of 1977. The final chapter in which nutrient requirements are expressed in terms of dietary concentrations was completed early in 1978 and draws only on the information in the preceding chapters.

Throughout the five years of work all members of the Working Party have received much help from colleagues both in this country and abroad. Unpublished detail of published work was made freely available to us and on occasion the results of studies which had not been published at all. Those with expert knowledge about particular facets of ruminant nutritional physiology gave us help where needed and on occasion attended our meetings. In addition, we received considerable technical assistance from closer colleagues in the massive tasks of collation of numerical information and its mathematical and statistical analysis. As in the previous edition, Mr. Brian F. Bone, Librarian at the National Institute for Research in Dairying, checked the references to the literature for their accuracy. Dr. Dorothy Duncan, ex-Director of the Commonwealth Bureau of Nutrition, assisted in the final editing, and compiled the index. To all these helpers we are most grateful.

All members of the Working Party wish to express our sincere thanks to our Technical Secretary, Dr. J. F. D. Greenhalgh. Besides taking a very active part in the more detailed work of analysis of experimental studies relevant to three chapters he has been responsible for the final assembly of material, the standardization of referencing and of presentation of tabulated data, and for the organization of all our meetings.

Kenneth Blaxter
Chairman, Working Party
on the Nutrient Requirements
of Ruminants

Contents

Chapter 1

Composition of the Ruminant's Body and its Products   1

Introduction   1

  1. Data   1
  2. Methods of Analysis   1
  3. Energy Value of Body Components   2

Body Composition of Sheep   3

  1. Composition of the Foetus and Associated Structures   3
  2. Protein, Fat and Energy in Growing Sheep   9
  3. Mineral Content of Growing Sheep   17
  4. Composition of Body Weight Changes in Adult Sheep   22

Body Composition of Cattle   24

  1. Composition of the Foetus and Associated Structures   24
  2. Protein, Fat and Energy in Growing Cattle   29
  3. Mineral Content of Growing Cattle   34
  4. Composition of Body Weight Changes in Adult Cattle   37

Gut Contents of Ruminants   38

Composition of Animal Products   45

  1. Composition of Milk   45
  2. Composition of the Fleece   49

Appendix Tables   51

Chapter 2

Feed Intake   59

Introduction   59

Feed Intake of Growing Ruminants   59

  1. The Data and Their Analysis   59
  2. Conclusions   64

Feed Intake of Lactating Ruminants   66

   1. Lactating cows   66
   2. Lactating Ewes   70

Intake of Milk and Milk Substitutes   71

Chapter 3

Requirements for Energy

Requirements for Energy   73

Introduction   73

   1. Energy Units   73
   2. Primary Measurements and Simple Derived Quantities   74
   3. Feed Intake and Energy Retention   75
   4. Metabolizability and Feeding Level   76

The Efficiency of Utilization of Metabolizable Energy   78

   1. Efficiency for Maintenance   78
   2. Metabolizable Energy for Maintenance   80
   3. Problems in Meeting Maintenance Requirements   81
   4. Efficiency for Muscular Work   82
   5. Efficiency for Growth and Fattening   83
   6. Efficiency during Pregnancy   87
   7. Efficiency in Lactation   89

Estimates of Requirements   95

   1. Fasting Metabolism   95
   2. Muscular Work   100
   3. Correction for Feeding Level   103
   4. The Energy Value of Gains in Weight   105

Energy Requirements for Production   107

   1. Cattle   107
   2. Sheep   114

Appendix 3.I   117

Appendix 3.II   118

Chapter 4

Requirements for Protein 121

Methods of Expressing Requirements and the Value of Feeds 121

The New Approach 122

1. Factors Involved in the New Approach 123
2. Proportion of Apparently Digested Organic Matter that is Apparently Digested in the Rumen 125
3. Microbial Nitrogen Yield in the Rumen 126
4. Efficiency of Conversion of Degraded Dietary Nitrogen into Microbial Nitrogen 127
5. Proportion of Total Microbial Nitrogen Present as Amino Acid Nitrogen 129
6. Apparent Absorbability in the Small Intestine of Amino Acids Derived from Microbial and Dietary Protein 129
7. Efficiency of Utilization of Absorbed Amino Acid Nitrogen 130
8. Endogenous Urinary Nitrogen Excretion and Dermal Losses of Nitrogen by Cattle 131
9. Extent of Degradation of Dietary Protein in the Rumen 134
10. Summary of the Calculation of Nitrogen Requirement 136
11. Formulation of a Ration to Meet the Protein Requirements of the Animal 138
12. The Protein Requirements of the Preruminant Animal 145
13. Summary of Rumen Degradable (RDP) and Undegraded (UDP) Protein Requirements for Various Productive Processes and Classes of Ruminant 148
14. Comparison of the Protein Requirements Calculated by the New System with the Results of Practical Trials and with Earlier Estimates of Protein Requirements 159

Appendix 4.I   The Sulphur Requirements of Ruminants 166

1. The Relationship Between Sulphur and Nitrogen Requirements 166
2. Recycling of Sulphur 167
3. Formulation of Sulphur Requirement 167
4. Comparison of Recommendations with Results of Practical Trials 168

Appendix 4.II.   Toxicity of Nitrogenous Compounds 168

Chapter 5

Requirements for the Major Mineral Elements: Calcium, Phosphorus, Magnesium, Potassium, Sodium and Chlorine 183

Introduction 183

1. Estimation of Requirements 183

2. The Validation of Factorial Estimates   184

3. The Presentation of Data   184

Calcium   184

1. The Endogenous Losses of Calcium from the Body   184
2. Efficiency of Absorption of Dietary Calcium   186
3. Requirements for Calcium   188
4. Comparison of the Estimates of Requirement with Those of the First Edition   190
   and of the National Research Council
5. Comparison of the Estimates of Calcium Requirements with the Results of   192
   Practical Trials

Phosphorus   192

1. The Endogenous Losses of Phosphorus from the Body   192
2. Efficiency of Absorption of Dietary Phosphorus   195
3. Requirements for Phosphorus   196
4. Comparisons of the Estimates of Requirement with Those of the First Edition   197
   and of the National Research Council
5. Comparison of the Estimates of Phosphorus Requirements with the Results of   199
   Practical Trials

The Ratio of Calcium to Phosphorus in Ruminant Diets   201

Magnesium   201

1. The Endogenous Losses of Magnesium from the Body   201
2. The Coefficient of Absorption of Dietary Magnesium   203
3. Requirements and Recommended Dietary Allowances for Magnesium   206
4. Comparisons of the Present Estimates with Those of the First Edition and of   208
   the National Research Council
5. Comparison of the Estimates of Magnesium Requirements with the Results of   208
   Practical Trials
6. Supplementation   210

Potassium   211

1. The Loss of Potassium Through the Skin of Cattle and in Saliva   211
2. Inevitable Losses of Potassium in the Faeces and Urine   211
3. Requirements for Potassium and Their Agreement with Other Estimates and   212
   the Results of Practical Feeding Trials

Sodium and Chlorine   213

1. Losses of Sodium and Chlorine Through the Skin of Cattle and in Dribbled   213
   Saliva

2. Inevitable Losses of Sodium and Chlorine in the Faeces and Urine   213

3. Estimates of Dietary Requirements for Sodium and Chlorine   215

4. Comparison of the Estimates of Requirement with Those of the First Edition and Other Standards   216

5. Comparison of Estimates with Results of Practical Trials   216

6. Salt Tolerance   216

Appendix Tables   217

Chapter 6

Trace Elements

Trace Elements   221

Introduction   221

Copper   221

1. Factorial Computation of Net Requirements of Sheep for Copper   222

2. Factorial Computation of Net Requirements of Cattle for Copper   223

3. Absorption Coefficients for Converting Net to Gross Requirements for Copper   226

4. Derivation and Validation of Factorially-derived Statements of Copper Requirement   230

5. Copper Toxicity   233

Iron   234

1. Changes in Tissue Iron Content During the Development of Iron Deficiency   234

2. Influence of Iron Supply on Growth   235

3. The Influence of Iron Supply on Blood Haemoglobin and Muscle Myoglobin   235

4. Availability of Dietary Iron   236

5. Iron Requirements   237

6. Iron Toxicity   239

Cobalt   240

1. Utilization of Dietary Cobalt for Vitamin B_{12} Synthesis   240

2. Relationship of Cobalt to Weight Gain and Tissue Vitamin B_{12} Content   241

3. Cobalt Requirements of Ruminants   242

4. Cobalt Toxicity   242

Selenium and Vitamin E   243

1. Terminology and Units of Vitamin E Activity   243

2. Selenium and Vitamin E in Ruminants   243

3. Vitamin E and Selenium in Relation to Diseases in Ruminants   244
4. Dietary Composition and the Incidence of Selenium/Vitamin E Deficiency   244
   Syndromes
5. Absorption and Utilization of Selenium   248
6. Absorption and Utilization of Vitamin E   248
7. Production of Vitamin E-Responsive Disorders by Polyunsaturated Fatty   249
   Acids of Dietary Origin
8. Summary of Selenium and Vitamin E Requirements   249
9. Selenium Toxicity   250

Iodine   251

1. Effects of Iodine Deficiency   251
2. Methods of Assessing Adequacy of Iodine Intake   251
3. Estimates of Requirement and Factors Affecting Dietary Requirements for   252
   Iodine
4. Iodine Requirements   255
5. Methods of Supplementation   255
6. Toxicity of Iodine   256

Zinc   256

1. Effects of Zinc Deficiency   256
2. Experimental Studies of Zinc Deficiency and Requirements   256
3. Factorial Estimates of Zinc Requirement   258
4. Comparison of Factorial Estimates of Zinc Requirement with Experimental   260
   and Field Observations
5. Conclusions   262
6. Zinc Toxicity   262

Manganese   263

1. Manganese Metabolism and Effects of Deficiency   263
2. Influence of Dietary Manganese on Development of Skeletal Lesions   264
3. Manganese Requirements for Growth   264
4. Manganese and Fertility in Ruminants   265
5. Summary of Requirements for Manganese   265
6. Manganese Toxicity   265

Fluorine Toxicity   265

Lead Toxicity   266

Chapter 7

Requirements for Fat-Soluble Vitamins

Requirements for Fat-Soluble Vitamins   269

Vitamin A   269

1. Basis of Assessment   269
2. Factors Affecting Utilization   271
3. Hypervitaminosis A   273
4. Requirements of Cattle   274
5. Requirements of Sheep   276
6. Use of Recommended Requirements in Practice   277

Vitamin D   278

1. The Importance of Dietary Calcium, Phosphorus and Energy in Relation to Vitamin D Requirements   278
2. Storage of Vitamin D in the Body   280
3. Avitaminosis D in Farm Livestock in Practical Conditions   280
4. The Minimum Requirement for Vitamin D to Prevent Rickets in Calves   281
5. Vitamin D Requirement of Growing Cattle   283
6. The Requirement of Vitamin D for Pregnancy and Lactation in Dairy Cows   283
7. The Requirement of Vitamin D for Sheep   284
8. Vitamin D and Bovine Fertility   285
9. Vitamin D and the Prevention of Milk Fever   286
10. Toxicity of Vitamin D   286
11. Sunlight and Vitamin D   287
12. Comparison of the Requirements Suggested Here with Requirements Given in Literature   288

Essential Fatty Acids   289

Chapter 8

The Vitamin B Complex

The Vitamin B Complex   291

1. Introduction   291
2. Requirements of the Milk-fed Ruminant   291
3. Requirements of the Ruminating Animal   292

Chapter 9

Requirements for Water

Requirements for Water   295

Introduction   295

Factors Affecting Water Intake   296

   1. Physiological Condition and Stage of Growth of the Animal   296
   2. Ambient Temperature   300
   3. Relative Humidity, Wind Velocity and Rainfall   301
   4. Other Factors   302
   5. Quantity of Dry Matter Consumed   302
   6. Composition of the Diet   302
   7. Variation   304
   8. Temperature of Drinking Water   304
   9. Frequency and Periodicity of Watering   304
  10. pH and Toxicity of Water   305

Conclusions   305

Chapter 10

Nutrient Requirements of Ruminants Expressed as Dietary Concentrations

Nutrient Requirements of Ruminants Expressed as Dietary Concentrations   307

Introduction to the Tables   307

Dry Matter Intake   308

Protein   308

   1. Rumen-degradable Protein   308
   2. Requirements for Undegraded Protein: Optimal Degradability and Concentration of Crude Protein   309

Minerals   309

Fat-Soluble Vitamins   310

References

References   315

Index

Index   349

Chapter 1

Composition of the Ruminant's Body and its Products

Introduction

Many of the estimates of requirement given in this volume are calculated by the factorial method, in which requirements for maintenance and for growth, pregnancy and lactation are estimated separately, and then summed to give the total net requirement. The quantities of nutrients retained by growing animals are estimated from data on the chemical composition of the body. In the first edition (Agricultural Research Council 1965) data for body composition were reviewed for each nutrient in the appropriate section. Information on body composition has now increased to such an extent that it is more convenient to review it in a separate section.

1. Data

The body components considered in this chapter are protein, fat and energy, and the major mineral elements, calcium, phosphorus, magnesium, potassium, sodium and chlorine. Trace elements have been excluded because there are few reliable data on their concentrations in the whole animal. The data used in the analyses are generally those for the chemical composition of the whole empty body, although the data on sheep do not always include wool, and those for prenatal growth have been extended to include the adnexa, i.e., the uterine wall, membranes and fluids. Indirect estimates of body composition, e.g., by marker dilution techniques, have been excluded.

Nearly all the data analysed were for individual animals. The primary selection of suitable material was made from published experiments, but in many cases the actual values were obtained directly from authors. Much unpublished information was also provided.

2. Methods of analysis

In estimating rates of accretion of body components from slaughter data, it is necessary to assume that the differences between values obtained from different animals killed at successive ages or weights are equivalent to the changes that would take place in the course of time in an individual animal. This assumption may be reasonable for a group of animals of the same sex and genotype, reared in the same environmental conditions and on the same feed and plane of nutrition, but not otherwise. The data were therefore subdivided before analysis, according to known differences in these factors, and were recombined only when statistical analysis indicated them to be homogeneous. The term "data source" as used in this section may therefore refer to the results of a single experiment, or to a group of animals within an experiment.

The data sources were first grouped by two species (sheep or cattle) and three age classes (the developing foetus, the newborn animal and the growing animal), the first two age classes being used to derive estimates for prenatal growth. For each source of foetal data, the total weight of the foetus or of the weight of each of its chemical components (y) was regressed on foetal age (x) by equations of Gompertz type, $\log y = a - be^{-ex}$. This form of analysis was used also for the total or component weights of the structures associated with the foetus (the adnexa). Laird (1966) has given reasons why the Gompertz equation is to be preferred to the large variety of other forms that have been used to describe foetal growth, and has argued that the existence of allometric relationships between components may be a consequence of underlying Gompertzian growth patterns. For each data source for newborn or growing animals the weights of chemical components (y) were regressed on empty-body weight (x) by allometric equations of the type, $\log y = a + b \log x$.

For both the pre- and the post-natal phases of growth, the next stage of the analysis was to draw the regression lines derived from each data source and to assess the differences between sources in the light of factors known to affect body composition. The primary determinant of composition before birth was assumed to be age, and of composition at and after birth, empty-body weight. Although differences in composition between breeds and sexes were expected, there were often insufficient sources to permit clear distinctions to be made. In sheep, the number of foetuses being carried (birth type) may influence their composition.

An additional source of variation in composition may be growth rate or, alternatively, the age at which a given weight is achieved. When growth rate is constant these two variables are equivalent to one another, and age is preferred on account of its precise definition. However, if growth rate varies, perhaps because a period of feed scarcity is followed by one of feed abundance with the possibility of compensatory growth, it becomes a more complex variable than age at a given weight.

A final factor to be considered is nutrition, which may affect body composition indirectly, through its influence on growth rate, or directly as a consequence of specific dietary deficiencies. As the purpose of the analyses has been to define feeding standards for animals fed on balanced diets, data derived from animals given grossly unbalanced diets have been excluded as far as possible.

Data sources which formed a homogeneous group were combined to give representative values for the composition of the animal, or for the composition of each unit of empty-body gain, at various weights or ages. It was frequently impossible to combine data sources by any systematic mathematical procedure, because they varied considerably with respect to numbers of animals and range in body weight or age. The procedure commonly adopted was to select representative values from graphs and then to fit a Gompertz or an allometric equation.

3. Energy value of body components

Few data sources included directly determined values for the energy concentration in the animal, i.e., heat of combustion of samples representative of the complete empty body. In some instances, energy concentration had been calculated from chemical composition by the use of factors which had been derived from the experiment or taken from the literature. In the interests of uniformity we have selected standard factors for estimating energy concentration from chemical composition and applied these to all sources of data, whether or not they included values for energy concentration.

Factors in common use are intended to apply to ether-extracted material (subsequently referred to as fat) and to either crude protein or fat-free organic matter. The last-named fraction is calculated by difference and is numerically equal to crude protein only when it contains nitrogen at exactly 160 g/kg. The nitrogen content of the fat-free organic matter of the complete ruminant body is commonly, although not invariably, less than 160 g/kg; this means that energy concentration per unit of fat-free organic matter is generally lower than energy concentration per unit of crude protein.

The conventional factors of Armsby (1917), 9.5 kcal/g fat and 5.7 kcal/g protein, were based mainly on German work that has since been summarized by Blaxter & Rook (1953) and Franke & Weniger (1958). This work involved the analysis of selected tissues rather than of samples representative of the whole body. More recent factors have been derived from whole body samples by one of two methods. In one, the samples are fractionated into fat and fat-free organic matter, and each fraction is subjected to bomb calorimetry (e.g., Jagusch et al 1970). In the other, a range of samples is analysed both for heat of combustion and for chemical components, and energy value is regressed on chemical composition; the energy value of each component is calculated by extrapolation of the regression equations (e.g., Paladines et al 1964). The two methods give similar results. For sheep six sets of factors were collated (Garrett 1958, Reid et al 1968, Langlands & Sutherland 1969, Bull 1969, Jagusch et al 1970, Drew 1971); they gave mean values of 39.1 kJ (9.348 kcal)/g fat and 23.2 kJ (5.542 kcal)/g fat-free organic matter. For cattle, the means of two sets (Reid et al 1968, Garrett & Hinman 1969) were 39.5 kJ (9.442 kcal)/g fat and 23.0 kJ (5.493 kcal)/g fat-free organic matter. The small differences between sheep and cattle were considered to be of little significance, and mean values of 39.3 kJ (9.40 kcal)/g fat and 23.1 kJ (5.52 kcal)/g fat-free organic matter were selected.

The choice of a factor for crude protein is made difficult by the scarcity of data for the nitrogen concentration of fat-free organic matter. Two values for sheep of 155.8 g/kg (Paladines et al 1964) and 158.0 g/kg (Jagusch et al 1970) agree reasonably well with one for cattle, of 155.0 g/kg (Moulton et al 1922), but not with a second value for cattle, of 163.3 g/kg (Garrett & Hinman 1969). The mean of the first three (156.3 g/kg) gives a factor for crude protein of $23.0 \times 160.0/156.3 = 23.6$ kJ(5.62 kcal)/g.

Body Composition of Sheep

1. Composition of the foetus and associated structures

To calculate the quantities of nutrients deposited in the gravid uterus at successive stages of pregnancy it is necessary to know:

(a) the total quantities present in the foetus at birth;
(b) the additional quantities deposited in the foetal membranes and fluids and in the uterine wall;
(c) the quantities of nutrients deposited at different stages of gestation (the time courses).

(a) Composition of the lamb at birth

The data sources are given in Appendix Table 1.1 and identified by code numbers. They comprised 68 lambs killed soon after birth, and 47 killed up to 5 days later; most lambs had been analysed for major minerals as well as for protein, fat and ash. In addition fat or protein concentrations or both were available for 70 foetuses taken from ewes killed 140 to 143 days after conception and thus shortly before full term (147 days). For four further experiments in which ewes had been killed at various stages of pregnancy and the foetuses analysed, Gompertz equations were fitted to the data and extrapolated to provide estimates of component weights at 147 days.

Most of these sources included single and twin lambs, but equations derived separately for each birth type were generally found to be combinable. The main exceptions were for calcium and phosphorus in newborn lambs (Sykes & Field 1972b) where singles and twins had to be treated as separate sources, and for foetuses (Rattray et al 1974a) taken from ewes given different feed allowances during the second half of pregnancy; feed allowance affected the weights of the foetuses but not their composition.

Table 1.1 Composition of perinatal lambs (g/kg empty-body weight)

		Foetuses at 140–143 days			Foetuses at 147 days (predicted values)				Newborn lambs						Lambs up to 5 days old	
Source*: No. of lambs:	Birth weight (kg)	11 (16)	12 (36)	15 (18)	5 (49)	10 (47)	13 (143)	14 (25)	1 (4)	2 (4)	3 (4)	6 (14)	8 (24)	9 (18)	4 (43)	7 (4)
Protein	2															
	3															
	4											141			192	
	5			125					157	152		155	168	151	190	
	6			128					154	158	178	164	169	163	188	178
	7			131	139		134	159	162	164	182	172	170	171	187	170
Fat	2															
	3															
	4											26.7			13.1	
	5	25.9	15.0	6.7					25.9	19.5		33.6	14.8	11.7	22.2	
	6	27.9	16.4	8.1					27.4	22.9	3.02	39.6	16.4	13.6	33.3	43.5
	7	29.7	17.7	9.3	39.0	35.2	22.3	23.8	28.7	26.0	43.0	45.0	17.7	15.1	46.5	43.1
Ash	2															
	3															
	4											40			46	
	5								37	40		41			43	
	6								33	40	37	41			41	42
	7				41		31		31	40	36	41			39	42

		Foetuses at 147 days (predicted values)		Newborn lambs				Lambs up to 5 days old	
Source*: No. of lambs:	Birth weight (kg)	5 (49)	10 (47)	2 (4)	6 (14)	8 (24)	9 (18)	4 (43)	7 (4)
Calcium	2				12.9				
	3			10.7	13.2	12.9	15.2	11.3	
	4			10.5	13.3	12.8	12.8	10.6	12.5
	5	12.8	10.7	10.3	13.4	12.8	11.4	10.0	11.8
Phosphorus	2				7.3				
	3			6.3	7.6	6.5	7.7	8.4	
	4			6.0	7.9	6.7	6.9	7.4	6.0
	5	6.1	5.7	5.7	8.1	6.9	6.4	6.7	5.8
Magnesium	2				0.59				
	3			0.56	0.53	0.51	0.61		
	4			0.45	0.49	0.50	0.53		0.28
	5		0.29	0.38	0.46	0.50	0.47		0.29
Potassium	2				1.46				
	3				1.59	1.44	1.53		
	4				1.69	1.66	1.68		1.96
	5	1.91	1.67		1.77	1.83	1.80		1.77
Sodium	2				1.31				
	3				1.33	2.33	2.62		
	4				1.33	2.45	2.58	1.89	
	5	2.33	2.43		1.34	2.54	2.54	1.77	

* For Source see Appendix Table 1.1.

Table 1.2. Representative values for the composition of the lamb at birth (g/kg body weight)†

	Body weight (kg)				Overall value (where appropriate)
Component	2	3	4	5	
Protein	150	160	168	175	—
Fat	—	25	30	35	30
Ash	40	40	40	40	40
Calcium	13	13	13	12	13
Phosphorus	7	7	7	6	7
Magnesium	0.60	0.55	0.45	—	0.5
Potassium	1.5	1.6	1.8	1.8	1.7
Sodium	2.5	2.5	2.4	2.4	2.4

† based on values given in Table 1.1

Estimates of the composition of lambs obtained from each source are given for an appropriate range of birth weights in Table 1.1. Representative values are shown in Table 1.2; these are based mainly on estimates for newborn lambs and foetuses, because there is an appreciable change in composition of foetuses in the last week of gestation. Sources for data in the text Tables are shown in Appendix Table 1.1.

Protein and fat

For protein, equations derived from different sources for newborn lambs were combinable, but lower values appeared to be obtained from foetuses and higher values from slightly older lambs. Heavy lambs generally had a higher concentration of protein and also of fat. Overall, the data suggested that the water concentration in the lamb fell appreciably about the time of birth, and possibly that heavy lambs had a lower water concentration at birth. In slightly older lambs (source 4, Appendix Table 1.1) the normal inverse relationship between protein and fat concentration occurred.

For fat, the differences between sources were obviously very large. Merinos (sources 3, 5, 6 and 7) appeared to be fatter than Scottish Blackfaces (8, 9, 11 and 12). Suffolk×Targhee (13) and Rambouillet×Columbia foetuses (14), although large, had low fat concentrations. The fat concentration in the lamb may be partly determined by maternal nutrition (Alexander 1962b); this was supported by the finding that the foetuses of Scottish Blackface ewes fed *ad libitum* on good roughage (11) were considerably fatter than the other lambs of that breed. Moreover, the foetuses with the lowest fat concentration (15) were from ewes on a low plane of nutrition.

The effect of large relative differences in fat concentration on the energy content of newborn lambs is smaller than might be expected, because the values for fat concentration are low. Nevertheless, a 5-kg lamb with 45 g/kg of fat contains about 43% more energy than one with only 15 g/kg.

Ash

In general, the concentration of ash varied less than that of fat or protein. The main exception was the value of 31 g/kg from source 13 for foetuses which also had low concentrations of protein and fat and an unusually high water concentration.

Calcium and phosphorus

The relationship of calcium concentration to birth weight (Table 1.1) was generally negative, although it was positive for source 6. For both calcium and phosphorus, the slopes of the regressions differed between single (8) and twin lambs (9) from the same experiment. However, the differences between sources and between birth weights appeared small enough to justify the selection of a single representative concentration for each element.

Magnesium

The data for newborn lambs were in reasonable agreement. The values for foetuses (10) and for lambs a few days old (7) were low, and were excluded in the selection of the representative values.

Potassium

There was reasonable agreement between different sources.

Sodium

The range for this element was remarkable. The four highest estimates (2.3–2.6 g/kg) were in good agreement, and were comparable to values for the newborn calf. Much lower values occurred in newborn lambs from source 6, which appeared to have been shorn before slaughter and the wool discarded, but they agree with data for older sheep and cattle. High values for foetal and newborn lambs may possibly be due to contamination with amniotic fluid, which contains sodium 2.4 g/litre (Phillips & Sundaram 1966). Slightly older lambs (7) had an intermediate sodium concentration. The higher range of values appear to provide the best estimate of sodium accretion in pregnancy.

(b) Deposition of nutrients in the adnexa

Three data sources (5, 13, 14) included values for the weights and gross chemical composition of fluids, membranes and uterus at successive stages of pregnancy, and source 5 included values for major minerals. Gompertz equations fitted to the data were used to predict the total weights and weights of components for those structures at full term, and each weight so calculated was then expressed as a fraction of the corresponding weight of the term foetus (Table 1.3). The "total" values in this table may be used to calculate the weight of a nutrient deposited in the complete gravid uterus. For example, a 4-kg lamb containing $4 \times 168 = 672$ g protein at birth (Table 1.2) would require that the ewe store a total of $672 \times (1.0 + 0.22) = 820$ g protein in the uterus during pregnancy. The sources used for Table 1.3 differed considerably with respect to lamb birthweight and birth type, but gave very similar estimates of the factor required to calculate total deposition in the gravid uterus from the amount present in the foetus. Additional data for foetuses before full term (sources 11, 12 and 15; Wallace 1948) confirmed that factors for single and twin lambs are in close agreement. However the distribution of nutrients between fluids, membranes and uterus differed between sources. The data for mineral deposition were restricted to one source, which included only single foetuses. No data were available for magnesium; in the previous edition the quantity deposited in the gravid uterus was assumed to be 12% greater than that in the foetus.

Table 1.3.   Deposition of nutrients in structures associated with the sheep foetus (quantity at full term as a fraction of the corresponding quantity for the foetus at full term)

Component	Fluids	Membranes	Uterus*	Total	Sources†
Total weight	0.23	0.08	0.13	0.44	5, 13, 14
Protein	0.029	0.071	0.121	0.22	5, 13, 14
Fat	0.004	0.041	0.054	0.10	5, 13, 14
Ash	0.061	0.028	0.031	0.12	5, 13, 14
Energy	0.027	0.081	0.089	0.20	5, 13, 14
Calcium	0.004		0.007	0.01	5
Phosphorus	0.023		0.028	0.05	5
Potassium	0.305		0.168	0.47	5
Sodium	0.129		0.121	0.25	5

* excluding the uterus of the nonpregnant ewe, which is estimated to weigh 76 g and to contain 13 g protein, 0.8 g fat, 0.9 g ash, 0.33 MJ energy, 0.008 g Ca, 0.11 g P, 0.16 g K and 0.17 g Na (from sources 5 and 13)
† For sources see Appendix Table 1.1

(c) The time course of nutrient deposition

Several of the Gompertz equations referred to above were used to estimate the proportion of the quantity of each component at full term that was deposited by a specific age (Table 1.4). The ages chosen span the last 12 weeks of gestation; for earlier ages the equations are unreliable. Source 13 included values for single and twin foetuses, and separate equations indicated that they differed in growth pattern; e.g., at 17 weeks, twin foetuses had reached 43.1% of their final weight, whereas a single foetus had reached only 40.1%. Sources 5 and 14 gave estimates of the time course of protein and fat deposition which agreed well with those from source 13, although the gravid uteri from sources 5 and 14 matured earlier with respect to total weight. For minerals the only estimates available were for the foetus alone (10) and for the gravid uterus (5).

Table 1.4.   Estimates of the fraction of final weight reached by specified ages for the total and component weights of the sheep foetus and gravid uterus

		Age (days from conception)*			
	Component	63	91	119	Source†
Foetus	Total weight	0.010	0.102	0.416	13
	Protein	0.004	0.064	0.340	
	Fat	0.002	0.052	0.356	
	Ash	0.006	0.073	0.356	
	Energy	0.004	0.063	0.340	
	Calcium	—	0.066	0.380	10
	Phosphorus	—	0.090	0.444	
	Magnesium	—	0.086	0.428	
	Potassium	—	0.196	0.641	
	Sodium	—	0.122	0.452	
Gravid uterus‡	Total weight	0.102	0.247	0.534	13
	Protein	0.048	0.156	0.428	
	Fat	0.023	0.101	0.399	
	Ash	0.038	0.132	0.408	
	Energy	0.042	0.144	0.417	
	Calcium	0.009	0.093	0.398	5
	Phosphorus	0.029	0.176	0.521	
	Potassium	0.093	0.363	0.715	
	Sodium	0.223	0.560	0.838	

* Age for final weight is 147 days
† For sources see Appendix Table 1.1
‡ less values for non-pregnant uterus

(d) Composition of the foetus and gravid uterus at various stages of gestation

The estimates obtained for the components of the lamb at birth, for the components of the gravid uterus and for the time courses of nutrient deposition have been combined in Table 1.5. The lamb at birth is assumed to have an empty-body weight of 4 kg and to have the composition shown by the representative values in Table 1.2. The weights of the foetus and its components at term have been increased to the corresponding weights for the gravid uterus by using the values given in Table 1.3. Finally, the component weights of the foetus alone or of the gravid uterus at term have been

Table 1.5. Estimated weights for the sheep foetus and gravid uterus and their components at various ages (g)

| | | Age (days from conception) | | | |
	Component	63	91	119	147
Foetus alone	Total	40	408	1664	4000
	Protein	2.7	43	228	672
	Fat	0.2	6.2	43	120
	Ash	1.0	11.7	57	160
	Energy (MJ)	0.08	1.3	6.9	20.2
	Calcium	—	3.4	20	52
	Phosphorus	—	2.5	12	28
	Magnesium	—	0.17	0.86	2.0
	Potassium	—	1.3	4.4	6.8
	Sodium	—	1.2	4.3	9.6
Gravid uterus‡	Total	588	1423	3076	5760
	Protein	39	128	351	820
	Fat	3.0	13.3	53	132
	Ash	6.8	23.6	73	179
	Energy (MJ)	1.0	3.5	10.1	24.2
	Calcium	0.48	4.9	21	53
	Phosphorus	0.84	5.1	15	29
	Magnesium	(0.02)*	(0.20)*	(0.88)*	2.2†
	Potassium	0.93	3.6	7.2	10.0
	Sodium	2.68	6.7	10.1	12.0

* tentative values calculated on the assumption that the time course of deposition of magnesium is the same as that for calcium.
† see text.
‡ less values for uterus of nonpregnant sheep.

Table 1.6. Parameters for estimating weight (Y) of components of the sheep foetus and gravid uterus at ages between 9 and 21 weeks from conception (X) by substitution in the equation:
$$\log_{10} Y = A - Be^{-CX},$$
(X = time from conception (days), Y = weight in g)

| | Foetus* | | | Gravid uterus† | | |
Component	A	B	C	A	B	C
Total	4.199	7.819	0.01750	5.250	3.641	0.00608
Protein	3.579	9.203	0.01704	4.928	4.873	0.00601
Fat	2.638	12.761	0.02129	3.766	5.553	0.00828
Ash	3.023	8.049	0.01554	4.540	5.334	0.00576
Energy (MJ)	2.039	9.310	0.01728	3.322	4.979	0.00643
Calcium	2.333	11.739	0.02124	2.499	7.406	0.01535
Phosphorus	1.813	12.692	0.02412	1.981	5.862	0.01650
Magnesium	0.715	11.679	0.02271	(1.044	7.644	0.01626)‡
Potassium	0.949	19.778	0.03492	1.143	5.255	0.02450
Sodium	1.517	7.316	0.01780	1.130	4.142	0.02987

* All derived estimates are relative to a lamb weighing 4 kg at 147 days of gestation.
† All estimates are for the gravid less the nonpregnant uterus.
‡ Tentative (see Table 1.5).

projected back, by means of the growth factors given in Table 1.4, to produce weights at earlier ages. For lambs of empty-body weight greater or less than 4 kg at birth, and for twin lambs, the values in Table 1.5 can be increased or decreased in direct proportionality to total lamb weight with little error. For more precise calculations, the tendency for composition to change with birth weight (Table 1.2) should be taken into account. Donald & Russell (1970) give equations derived from 15 breeds of sheep for estimating total lamb weight at birth from the weight of the ewe at mating.

Estimates of weights of constituents of the foetus and of the gravid uterus (less those of the nonpregnant uterus) at ages intermediate to those used in Table 1.5 may be made by using Gompertz equations which have been fitted to the values in that table. Parameters for these equations are given in Table 1.6. Estimates from the equations are again all relative to a foetus weighing 4 kg at 147 days of gestation, and would require proportionate changes to fit other birth weights.

(e) Comparison with earlier estimates

In the previous edition, few data sources were available for the composition of the foetus or newborn lamb, and for some nutrients values for cattle were employed. For protein, the present estimates agree well with the earlier estimate of 162 g/kg. For minerals, the present estimates are higher than those used earlier; e.g., the value accepted for calcium concentration is 13 g/kg, compared with the earlier value of 10.5 g/kg, and for magnesium the corresponding values are 0.50 and 0.36 g/kg. In the previous edition, no estimates were made of fat or energy storage in the sheep foetus.

The conversion of weights of nutrients in the foetus to those in the complete gravid uterus was based mainly on the data of Wallace (1948) for gross weights of foetus and adnexa, together with some isolated values for the chemical composition of adnexa. Estimates for the gravid uterus were not given beyond 140 days from conception and comparison with present values is therefore difficult. Nevertheless, the earlier estimate of the quantity of phosphorus present in a gravid uterus containing a foetus of 5.9 kg at 140 days was 38 g, whereas the present estimate is 37 g; but, for potassium the present estimate of 13.9 g is much lower than the previous value of 20 g.

2. Protein, fat and energy in growing sheep

(a) Classification of the data sources

In the first edition, estimates of the protein and energy concentrations in gains made by sheep were based on data for 63 animals. During the past 15 years the composition of the growing sheep has been studied much more extensively, and 67 data sources comprising in total about 1360 sheep (Appendix Table 1.2) were available for analysis. These included many breeds and crosses, and some subdivision of the data was thought desirable. About half the data were for Merinos or Merino crosses, and these were analysed separately. A further subdivision was by sex (males, castrated males and females) but there were relatively few Merino males or females. Most of the data were for shorn sheep, and those which included the fleece were analysed separately and are identified in Appendix Table 1.2; one source was analysed with and without the fleece (20 and 21).

(b) Analysis of the data

Allometric equations for each source of data are given in Appendix Table 1.3. The overall range in empty-body weight for which there was a sufficient number of data sources was 10 to 45 kg. Equations from most sources showed a consistent pattern of protein concentration falling and fat concentration rising with increasing body weight. However, at any one body weight there were large differences between sources in predicted composition, and the ranges in fat and protein concentration are illustrated in Fig. 1.1(a). For Fig. 1.1(b) the equations were differentiated to provide estimates of the composition of gain at each empty-body weight.

Some of the variation between sources was due to breed type and sex, and this is discussed below. Differences were introduced also by the inclusion of the fleece. A comparison of sources showed that the mean protein concentration in the empty body

Fig. 1.1   Protein and fat in the empty body and in empty body gain of sheep; values predicted from the equations of Table 1.7.

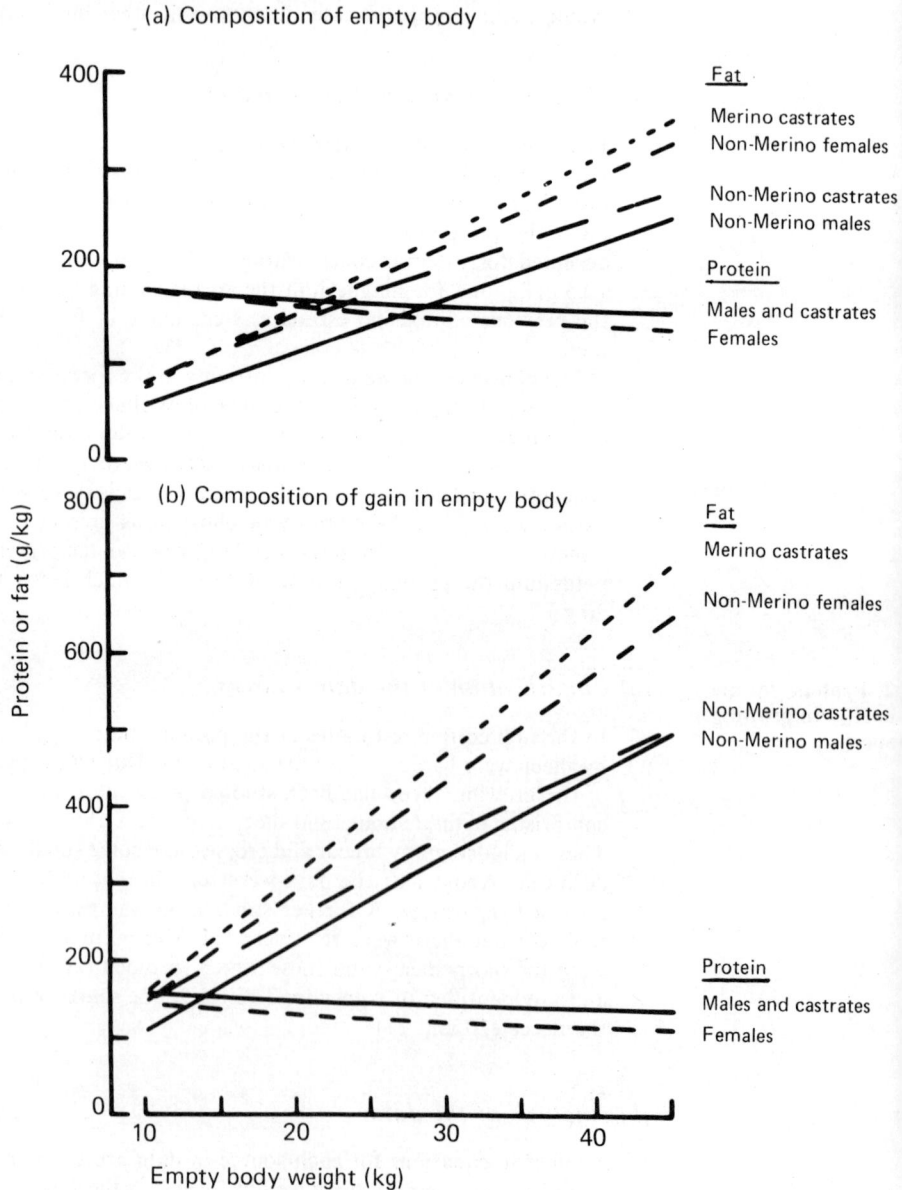

(a) Composition of empty body

Fat
Merino castrates
Non-Merino females

Non-Merino castrates
Non-Merino males

Protein
Males and castrates
Females

(b) Composition of gain in empty body

Fat
Merino castrates

Non-Merino females

Non-Merino castrates
Non-Merino males

Protein
Males and castrates
Females

Protein or fat (g/kg)

Empty body weight (kg)

was about 14 g/kg greater when the fleece was included, but fat concentration was not affected. These differences appeared to be independent of body weight and were in agreement with those found from sources 20 and 21, which were analysed with and without fleece, respectively.

The proportion of the variation in body composition due to age as well as weight of the animal was examined for selected sets of data covering wide ranges in weight (x_1) and age (x_2). Multiple regression equations of the form

$$\text{Log } y = a + b \log x_1 + c\ x_2$$

were calculated, and the residual standard deviations of these equations are compared in Appendix Table 1.4 with those of the original allometric equations. In general, age had little effect on the precision of the regressions and was not considered in subsequent analyses.

Table 1.7. Representative equations for sheep. Allometric equations of the form $\log_{10}$ (component wt., kg) $= a + b \times \log_{10}$ (empty-body wt., kg)

		Class of data	b	a
Protein	All data:	upper limit	0.9589	−0.6915
	,, ,,	lower ,,	0.7487	−0.5183
	All breeds:	males and castrates	0.8955	−0.6451
	,, ,,	females	0.8164	−0.5660
Fat	All data:	upper limit	1.868	−1.789
	,, ,,	lower ,,	2.251	−2.774
	Non-Merino:	males	1.987	−2.239
	,, ,,	castrates	1.821	−1.918
	,, ,,	females	1.975	−2.100
	Merino:	castrates	2.024	−2.149

The first general equations fitted to the data were those describing the upper and lower limits of the data used for Fig 1.1; these equations are included in Table 1.7. Equations were then fitted to representative values for the composition of Merinos and non-Merinos of each sex. In deriving the representative values for protein, a deduction of 14 g protein/kg was made for data that included the fleece. For protein, the equations for the two breed types were found to be combinable, but there remained a distinction between males and castrates on the one hand and females on the other; the two equations required are included in Table 1.7 and illustrated in Fig. 1.1. For fat, the situation was more complex. For the non-Merinos, separate equations were required for males, castrates and females. The Merino castrates also formed a distinct group, but there were insufficient data to derive additional equations for Merino males or females. The equations for fat are included in Table 1.7 and illustrated in Fig. 1.1. Table 1.8 gives values predicted from all equations except those describing the upper and lower limits. Also included in Table 1.8 are estimates of the energy concentration in the empty body and in gains in empty-body weight.

There are some apparent inconsistencies in these estimates, which probably arise from differences between breeds and sexes being confounded with other differences. While it is reasonable to expect that the gains of castrates would contain more fat than those of males, it is surprising that this difference between the sexes should be manifested so early in life. The probable explanation is that most of the young non-Merino males were early-weaned lambs and the castrates were suckled lambs.

A possible test for the internal consistency of the data is to calculate the protein concentration in the fat-free empty body, which is commonly assumed to be constant. Reid et al (1968) quote a mean value of 203.5 (± 8.4) g protein per kg fat-free empty-

Table 1.8. The protein, fat and energy concentration in the fleece-free empty body of sheep, and of gains in empty-body weight, as predicted by the equations in Table 1.7

| | Protein (g/kg) | | Fat (g/kg) | | | | Energy (MJ/kg)† | | | |
| | Males and castrates | Females | Non-Merino | | | Merino | Non-Merino | | | Merino |
Empty-body weight (kg)			Males	Castrates	Females	Castrates	Males	Castrates	Females	Castrates
Empty-body										
10	178	178	56	80	75	75	6.4	7.3	7.2	7.2
15	171	165	84	112	111	114	7.3	8.3	8.3	8.4
20	166	157	111	141	147	152	8.3	9.5	9.5	9.7
25	162	150	139	170	183	192	9.3	10.5	10.7	11.1
30	159	146	166	197	219	231	10.3	11.5	12.1	12.5
35	156	141	193	224	254	270	11.3	12.5	13.3	13.9
40	154	138	220	250	290	310	12.3	13.4	14.7	15.4
45	152	135	248	275	325	349	13.3	14.4	16.0	16.9
Empty-body gain										
10	159	145	111	146	148	152	8.1	9.5	9.2	9.7
15	153	135	166	203	220	230	10.1	11.6	11.8	12.6
20	148	128	221	257	291	308	12.2	13.6	14.5	15.6
25	145	123	275	309	362	387	14.2	15.6	17.1	18.6
30	142	119	330	359	432	467	16.3	17.5	19.8	21.7
35	140	115	384	407	502	546	18.4	19.3	22.4	24.8
40	138	113	438	455	572	626	20.5	21.1	25.1	27.9
45	136	110	492	501	642	707	22.6	22.9	27.8	31.0

† 23.6 MJ/kg protein and 39.3 MJ/kg fat

body weight, although calculations from their results suggest that this value rises as the sheep increases in weight, from about 198 g/kg at 20 kg empty-body weight to 211 g/kg at 50 kg. From Table 1.8 it may be calculated that at 20 kg empty-body weight, non-Merino males, castrates and females contain 187, 193 and 184 g protein per kg fat-free empty body, and at 45 kg the comparable values are 202, 210 and 200 g/kg.

(c) The milk-fed lamb

The interval from birth to 10 kg empty-body weight is not covered by the values in Table 1.8. Norton et al (1979) have made a special study of this period, and their estimates are given in Table 1.9. Approximate values can also be calculated from the differences in composition between the lamb at birth (Table 1.2) and the 10 kg lamb (Table 1.8); these are 200 g protein, 100 g fat and 8.6 MJ/kg empty body gain (including wool).

Table 1.9.   Composition of gains in milk-fed lambs (Norton et al 1970)

| Base | Rate of gain (g/day) | Composition of weight gain | | |
		Protein (g/kg)	Fat (g/kg)	Energy (MJ/kg)
Wool-free empty body	50	164	70	6.3
	100	166	91	7.2
	200	165	123	8.5
	300	160	153	9.6
Empty body including wool	50	195	70	7.2
	100	189	91	7.9
	200	183	122	9.0
	300	178	151	10.0

(d) Comparison with other estimates

The values given in Table 1.8 and the conclusions about the influence of breed, sex and age (or growth rate) on the relationship between body weight and body composition need to be compared with the findings of others.

In Tables 1.10 and 1.11, estimates of the protein and energy concentrations in weight gain are compared. Those in the first edition were based on data for 63 sheep,

Table 1.10.   Estimates of the protein concentration in gains made by sheep in fleece-free empty-body weight (g/kg)

| Source: | Present estimates | | ARC (1965) | Reid et al (1968) | Langlands & Sutherland (1969) | |
Class of sheep:	Males and castrates	Females	*	Castrates	Merino castrates and females	
Gain (g/day):	*	*	†	*	50	200
Empty-body weight (kg)						
10	159	145	156†	—	169	158
20	148	128	156†	139	160	152
30	142	119	156†	128	150	146
40	138	113	150†	121	144	144
50	—	—	150†	—	138	144

* unspecified.
† including protein in wool.

Table 1.11.   Estimates of the energy concentration of gains made by sheep in fleece-free empty-body weight (MJ/kg)

Source:	Present estimates				ARC (1965)		Reid et al (1968)	Langlands & Sutherland (1969)		Rattray et al (1973a)				Rattray et al (1973b)	
	Non-Merino			Merino				Merino castrates and females		"White-faced" lambs; castrates and females		"White-faced" yearling castrates		"White-faced" females	
Class of sheep:	Males	Castrates	Females	Castrates	*		Castrates								
Gain (g/day)†:	*	*	*	*	50	200	*	50	200	50	200	50	200	50	200
Empty-body weight (kg)															
10	8.1	9.5	9.2	9.7	—	—	—	9.4	11.8	—	—	—	—	—	—
20	12.2	13.6	14.5	15.6	13.9	15.9	14.3	11.4	12.9	—	—	—	—	16.8	16.8
30	16.3	17.5	19.8	21.7	15.5	17.7	18.2	13.5	14.2	20.6	20.6	—	—	23.0	23.0
40	20.5	21.1	25.1	27.9	17.2	19.2	21.5	14.9	14.7	25.5	25.5	31.4	20.7	28.4	28.4
50	—							16.1	15.0	30.3	30.3	37.2	24.5	33.6	33.6

* unspecified
† 50 and 200 g/day have been selected as being representative of low and high rates of gain; the original sources include estimates for other rates

most of which are included in the analysis of non-Merino castrates. Some of the data of Reid et al (1968) are also included in the present analyses. Langlands & Sutherland (1969) analysed 107 Merino sheep aged up to 9 years, and derived estimates for two age series and three rates of gain within each series. Only the younger age series is included here, as the second series appears to apply to sheep unlikely to be found in Britain (e.g., weighing 25 kg at 40 months of age). Rattray et al (1973a, b) estimated the net energy requirements for weight gain in sheep, which are presumably equal to the energy content of the gain. These were based partly on direct estimates of body composition which are included in the present analysis, and partly on data derived from the specific gravity of sheep carcasses.

For protein, the estimates for males and castrates fall between those of Reid et al (1968) and Langlands & Sutherland (1969), but the present estimates for females are lower. For energy the comparisons are complicated by differences due to rate of gain, which are discussed below. In general, the estimates for non-Merino castrates agree well with those of Reid et al (1968), but for heavier sheep are higher than those in the first edition. The values of Langlands & Sutherland (1969) appear low (especially for Merinos), whereas those of Rattray et al (1973a, b) are generally higher than the estimates for non-Merino males and females.

Some estimates not included in the tables are those of Searle et al (1972), who used tritiated water to obtain several estimates of body composition in Border Leicester×Merino lambs growing at 50 or 100 g/day to 27 months of age. Four growth phases were distinguished: suckling, weaning, prepubertal (to 25–35 kg) and fattening. In the prepubertal phase the protein concentration in gain in fleece-free live weight was 131 and 125 g/kg for sheep gaining 50 and 100 g/day, respectively, the corresponding values for energy being 12.0 and 13.7 MJ/kg. These values would probably be increased by 10 to 15% if recalculated for empty-body weight gain; their protein values would then agree with the present estimates but their energy values would appear to be lower. In fattening sheep gaining 50 or 100 g/day, the protein concentration in gain was 97 or 87 g/kg and energy concentration 22.8 or 26.0 MJ/kg. In this phase, the energy values agree with the present estimates for sheep of 40 kg, but the protein values of Searle et al appear to be lower.

Breed effects

Reid et al (1968) found the composition of six breeds or crosses to differ significantly, although each breed was apparently studied in a separate experiment. Searle & Graham (1972) found the Merino to be significantly fatter than the Border Leicester×Merino at the same weight, though Gharaybeh et al (1969) found no significant difference between those breeds. Similarly, there was no significant difference in fat or protein concentration between the Merino and Dorset Horn×Merino (Allden 1970) or between the Merino and Dorset Horn×(Border Leicester×Merino) (Kellaway 1973). Thus the arbitrary division that has been adopted between Merinos and their crosses on the one hand and all other sheep on the other hand may not be fully justified. Rattray et al (1973c) found relatively small differences in composition between unselected Targhee sheep and a line selected for heavy weaning weight.

Sex

In two comparisons of males and females, the latter were significantly fatter at the same empty-body weight (Bull 1969, Andrews & Ørskov 1970a). Differences between castrates and females have been less clear-cut and not always significant (Gharaybeh et al 1969, Kellaway 1973, Rattray et al 1973a, b).

Growth rate

The effects of growth rate on composition have been studied directly, by rearing animals to the same weight at different growth rates, or indirectly by multiple regression analysis of composition on weight and age. In four out of seven studies of the first type, there were significant differences in the composition of gain. Morton et al (1970) reported that as the weight gain of milk-fed lambs increased from 50 to 300 g/day, the protein concentration in the gain decreased from 195 to 178 g/kg and the fat concentration increased from 70 to 151 g/kg. Searle et al (1972) found that an increase in weight gain of lambs from about 50 to 100 g/day reduced the protein concentration in gains by about 10% and increased the fat concentration by about 20%. Andrews & Ørskov (1970a) found opposite effects, faster growth being associated with higher protein and lower fat contents, especially when a high-protein diet was given. In further trials Ørskov et al (1971a) found faster growth associated with higher protein and lower fat deposition when lambs were fed *ad libitum* on diets varying in protein content. In the remaining three experiments, gains ranged from 70 to 176 g/day and there was no significant difference in body composition (Burton & Reid 1969, Searle & Graham 1972, Kellaway 1973). In addition, the estimates of Rattray et al (1973a b) for lambs (see Table 1.11) suggest that the energy concentration in empty-body gain is independent of rate of gain. In yearlings, however, the energy concentration in gain apparently *decreased* as rate of gain increased.

Langlands & Sutherland (1969) found age to have significant effect in multiple regressions of composition on weight and age. But their age series was extreme in that it included animals 9 years old, and their results suggest that age had an appreciable effect on the composition of gain only at weights less than 25 kg. Rate of gain had a relatively small and inconsistent effect on the composition of gain (see Tables 1.10 and 1.11). Reid et al (1968) concluded, from their analysis of data for 221 sheep, that age *per se* had little effect on body composition. On balance, there seems to be little justification for providing separate estimates of the composition of gain for different rates of gain, or for different ages of animal, except for those of the milk-fed lamb.

(e) *Weight loss and compensatory growth*

Searle & Graham (1972) considered that the composition of weight loss in sheep was similar to that of weight gain, but this seems unlikely to be true when weight losses are large. When sheep were allowed to lose weight and then return to their starting weight, in three out of four experiments (Reid et al 1968, Keenan et al 1969, Burton et al 1974) their protein content was increased and their fat content reduced, but in the fourth (Kellaway 1973) there was no significant effect. In three further experiments, sheep were grown to a target weight by continuous growth or by routes which included a phase of weight loss. In one (McManus et al 1972) the sheep which had experienced interrupted growth contained less fat, in the second (Allden 1970) they did not differ from sheep which had gained continuously and in the third (Meyer & Clawson 1964) the sheep were fatter after interrupted growth. Thus, insufficient reliable evidence exists to justify separate estimates of the composition of gains made by sheep with interrupted growth.

(f) *Wool growth in relation to growth in empty-body weight*

Wool growth has been considered separately from the growth of the empty body. Requirements for wool growth could be estimated from the anticipated weight of the fleece at shearing, and in doing so, the known differences between breeds could be taken into account. Alternatively, the relationship between wool growth and growth in other tissues might be used, as in the first edition where it was assumed that in mature sheep wool growth accounted for 20% of total N retention. In growing lambs, however, the rate of N deposition in tissues other than wool is greater than that in mature sheep, and the proportion of N stored as wool, as indicated earlier, is only about 10%. At the other extreme, appreciable wool growth occurs when N retention in other tissues is low or negative. Several of the data sources listed in Appendix Table 1.2 include information on wool growth. If protein retained in wool (y g/day) is regressed against protein retained in other tissues (x g/day) it appears that for British breeds of sheep, i.e., excluding Merinos and their first crosses, the approximate relationship is

$$y = 3 + 0.1x.$$
[1.1]

(g) Conclusions

The variation in the composition of growing sheep and of their gains is such that mean estimates cannot be accurate in every situation. But the separate estimates for different sexes and, in some cases, for different breed types allow for the main sources of variation in composition. As information accumulates it may be possible to cater for other sources of variation. In the meantime the estimates given in Tables 1.8 and 1.9 provide an approximate indication of the protein and energy requirements of sheep for tissue growth.

3. Mineral content of growing sheep

(a) Sources of data

Ten data sources are included in Appendix Table 1.2. The total numbers of sheep analysed for each element were for calcium 419; phosphorus 389; magnesium 144; potassium and sodium 213. The range of sources was insufficient to justify the subdivision by breed type and sex, which was made for data on protein and fat concentration in sheep. Source 21 was unusual in that some of the diets given to the sheep contained high concentrations of calcium in the form of calcium salts of volatile fatty acids.

In some data sources wool was included in the empty body. Its inclusion would be expected to reduce the concentrations of all of the elements considered here, except potassium, by up to 5%. Potassium is present in high concentration in suint, but all data for the potassium concentration in the body were from sources from which the fleece had been excluded.

(b) Calcium

The calcium concentration in individual body tissues varies widely. Fat depots contain virtually no calcium, soft tissues about 0.1 g Ca/kg, and bone 110 to 200 g/kg. In growing animals, the main factor determining the calcium concentration in weight gain will be the relative growth rate of the skeleton, which contains 99% of total body calcium. In older animals, bone growth and mineralisation proceed slowly, if at all, and weight gains, consisting mainly of fat, will contain little calcium.

Allometric equations and estimates obtained from them for the calcium concentration in the empty body of sheep are given in Table 1.12. There appears to be no universally applicable relationship between calcium concentration and empty-body

Table 1.12.   Estimates of the calcium concentration in the empty body of sheep

Source*	Regression of $\log_{10}$ Ca (g) on $\log_{10}$ empty-body weight (kg)			Concentration (g/kg) estimated from regression at empty-body weight (kg) of:							
	Coefficient	Intercept	Residual SD	10	15	20	25	30	35	40	45
16	0.804	1.246	0.063	11.2	10.4	9.8	9.4	9.0	8.8		
17, 18	0.934	1.064	0.041		9.7	9.5	9.4	9.3	9.2	9.1	9.0
20‡	1.049	0.967	0.035				10.9	10.9	11.0	11.1	11.2
20§	1.265	0.696	—					12.2	12.7	13.2	13.6
28	—	—				9.4†	11.6	(13.4†)(10.0)†	9.4†		
2	1.063	1.005	0.066			12.2	12.4	12.5	12.7	12.8	
61	0.722	1.243	0.067	9.2	8.2						
70	1.159	0.959	0.050	13.1	14.0	14.7	15.2	15.6			
82	0.375	2.112	0.070						14.3	13.1	12.2
71	1.015	1.111	0.037	13.4	13.4	13.5	13.6	13.6	13.6		
72	1.334	0.852	0.059	15.3							
30, 31	1.093	0.944	0.058		11.3	11.6	11.9	12.1	12.2	12.4	12.5
	1.058	0.957	0.056		10.6	10.8	10.9	11.0	11.1	11.2	11.3
33	1.042	0.970	0.046	10.3	10.5	10.6	10.7	10.8	10.8	10.9	11.0

* For sources see Appendix Tables 1.1 and 1.2
† Mean values, not estimates from regression equations
‡ Control diet
§ High Ca diet

Table 1.13.   Estimates of the phosphorus concentration in the empty body of sheep

Source*	Regression of $\log_{10}$ P (g) on $\log_{10}$ empty-body weight (kg)			Concentration (g/kg) estimated from regression at empty-body weight (kg) of:							
	Coefficient	Intercept	Residual SD	10	15	20	25	30	35	40	45
16	0.850	0.959	0.054	6.5	6.1	5.8	5.6	5.5	5.3		
17, 18	0.889	0.893	0.040		5.8	5.6	5.5	5.3	5.3	5.2	5.1
20‡	0.926	0.936	0.030				6.8	6.7	6.6	6.6	6.5
20§	1.068	0.757	0.030				7.1	7.2	7.3	7.3	7.4
2	0.994	0.857	0.054			7.1	7.1	7.0	7.0	7.0	
61	0.830	0.968	0.057	6.3	5.9						
70	1.027	0.880	0.053	8.1	8.2	8.2	8.3	8.3			
82	0.238	2.089	0.073						8.2	7.4	6.8
71	0.892	1.015	0.050	8.1	7.7	7.5	7.3	7.2	7.1		
72	1.135	0.676	0.041	6.5							
30, 31	0.980	0.770	0.040		5.6	5.5	5.5	5.5	5.5	5.5	5.5
	1.005	0.699	0.048		5.1	5.1	5.1	5.1	5.1	5.1	5.1
33	0.974	0.807	0.045	6.0	6.0	5.9	5.9	5.9	5.9	5.8	5.8

* For sources see Appendix Tables 1.1 and 1.2
‡ Control diet
§ High-Ca diet

weight. In young non-Merino lambs fed on concentrate diets (16, 17, 18, 19), calcium concentration declined with increasing body weight, and estimates derived from these sources were generally lower than the rest. For most of the remaining sources, calcium concentration increased as the animals grew, but the increase was generally quite slow. In slow-growing Merino castrates (70) the increase in calcium concentration was more marked, but in a group of older Merino females (82) calcium concentration declined with increasing weight. For source 21, the diet richest in calcium (30 g/kg) resulted in body concentrations which were higher than those produced by the control diet (13 g/kg) and exceeded the values from all other sources at higher body weights.

A suitable representative value for the calcium concentration in growing sheep is 11 g/kg. Since this value is assumed to apply at all body weights, the estimate of the calcium concentration in empty-body gain is also 11 g/kg. The application of this value would probably overestimate calcium accretion in very young lambs, since the calcium concentration in the newborn lamb is estimated to be 13 g/kg, and also overestimate that of concentrate-fed lambs or mature, fattening sheep. Conversely, calcium deposition in slowly-growing lambs would probably be underestimated. However, exact quantification of these differences cannot at present be justified.

(c) Phosphorus

The distribution of phosphorus in the body differs from that of calcium. About 75 to 80% of the phosphorus present in the adult body is in bone, in concentrations of 50 to 100 g/kg. Muscles contain 2 to 3 g/kg and brain and nervous tissue, 4 g/kg. The triglycerides of depot fats contain negligible amounts of phosphorus, but lecithins and other phospholipids are phosphoric acid esters.

Allometric equations and estimates derived from them are given in Table 1.13. Most sources show a slight fall or little change in phosphorus concentration with increasing body weight. The concentration of phosphorus, unlike that of calcium, is not lower for intensively fed lambs (16–19), but values from source 70 are, like those of calcium, higher than the rest. A representative value is 6 g phosphorus per kg empty body or empty-body gain, in comparison with a value of 7 g/kg for the newborn lamb. The representative value may underestimate the phosphorus accretion of slowly-growing lambs.

(d) Magnesium

About 70% of body magnesium occurs in bone, at a concentration of about 2 g/kg. Soft tissues contain 0.1 to 0.2 g Mg/kg, or about 140 mg/g N.

The data sources for this element provide estimates in reasonable agreement with one another (Table 1.14). There is little evidence that magnesium concentration changes as body weight increases; a representative value is 0.41 g Mg/kg empty-body weight or empty-body gain, which may be compared with the estimate of 0.5 g/kg for the newborn lamb.

(e) Potassium

The highest concentrations of potassium in body tissues are found in muscle (c. 4 g/kg) and nervous and secretory tissues (c. 3.5 g/kg). Tissue fluids and serum contain about 0.2 g/kg, and bone, less than 0.05 g/kg.

Table 1.14.   Estimates of the magnesium, potassium and sodium concentrations in the empty body of sheep

Element	Source*	Regression of $\log_{10}$ Mg, K or Na (g) on $\log_{10}$ empty-body weight (kg)			Concentration (g/kg) estimated from regression at empty-body weight (kg) of:							
		Coefficient	Intercept	Residual SD	10	15	20	25	30	35	40	45
Magnesium	71	1.093	−0.460	0.068	0.43							
	72	0.992	−0.340	0.079		0.45	0.45	0.45	0.45	0.44	0.44	0.44
	30, 31	0.949	−0.319	0.085		0.42	0.41	0.41	0.40	0.40	0.40	0.40
	33	0.925	−0.289	0.041	0.43	0.42	0.41	0.40	0.40	0.39	0.39	0.39
Potassium	70	0.826	0.546	0.067	2.45	2.28	2.17	2.09	2.02			
	70	0.298	0.135	0.049						1.85	1.69	1.56
	82	0.777	0.630	0.065	2.56	2.34	2.19	2.08	2.00	1.93		
	71	1.030	0.238	0.061	1.85							
	72	1.043	0.223	0.068		1.88	1.90	1.92	1.93	1.95	1.96	1.97
	30, 31	0.809	0.544	0.058		2.08	1.97	1.89	1.82	1.77	1.73	1.69
	33	1.320	−0.014	0.049	2.02	2.30	2.53					
Sodium	70	1.089	0.017	0.073	1.28	1.32	1.36	1.38	1.41			
	70	−0.329	2.237	0.040						1.53	1.28	1.10
	82	1.085	−0.002	0.090	1.21	1.25	1.28	1.31	1.33	1.35		
	71	0.862	0.348	0.036	1.62							
	72	0.451	0.866	0.075		1.66	1.42	1.25	1.13	1.04	0.97	0.91
	30, 31	0.857	0.328	0.064		1.45	1.39	1.34	1.31	1.28	1.26	1.24
	33	0.704	0.554	0.041	1.81	1.61	1.48	1.38	1.31	1.25	1.20	1.16

* For sources see Appendix Table 1.2

Allometric equations and estimates for potassium are included in Table 1.14. For three of the five major data sources, potassium concentration declined with increasing body weight, and for the remaining two sources it increased either slightly or considerably. As the weight of potassium in the body is commonly found to be related to lean-body mass a declining concentration would be expected.

From the value of 1.7 g K/kg for the newborn lamb, and from an assumed value of 1.8 g/kg for the 45-kg sheep, the potassium content of empty-body gain in sheep is calculated to be 1.8 g/kg.

(f) Sodium (and chlorine)

The distribution of sodium differs markedly from that of potassium. The sodium ion enters the mineral lattice of bone, which contains about 4 g Na/kg. In soft tissues, sodium is mainly an extracellular ion and therefore occurs in low concentrations (c. 0.75 g/kg). Body fluids, however, contain about 3.5 g Na/kg. Chlorine distribution is similar to that of sodium, although there is proportionately less chlorine in bone.

Table 1.14 includes allometric equations and estimates for sodium. Some sources showed an increase and some a decline in sodium concentration with increasing body weight but, as for potassium, a declining concentration is theoretically more likely. In the group of heavy castrates (source 70), the decline in sodium concentration was so great that 45-kg animals apparently contained a smaller weight of sodium than those weighing 35 kg.

The sodium concentration of the newborn lamb was 2.4 g/kg. If a concentration of 1.2 g Na/kg is accepted for the 45-kg sheep, the sodium concentration in empty-body gain is calculated to be about 1.1 g/kg.

No data are available for the chlorine concentration in growing sheep. However, if the ratio of sodium to chlorine is the same in sheep as it is in cattle (see p. 36), the chlorine concentration in weight gain of sheep will be about 0.8 g/kg.

(g) Conclusions

The estimates of the mineral composition of weight gains made by growing sheep are summarized in Table 1.15, together with the estimates for the newborn lamb, and the estimates for growing sheep from the first edition. Except for sodium, the present estimates agree well with those given earlier. For sodium and potassium the earlier estimates were not based on analytical data, since none were available for sheep, but

Table 1.15.   Estimates of the mineral concentrations in gains in empty-body weight of sheep, together with those present in the empty-body weight of the newborn lamb (from Table 1.2)

Component	Newborn lamb (g/kg)	Growing sheep (g/kg weight gain)	
		Present estimate	ARC (1965)*
Calcium	13	11	10
Phosphorus	7.0	6.0	5.5
Magnesium	0.50	0.41	0.40
Potassium	1.7	1.8	1.8
Sodium	2.4	1.1	1.5
Chlorine	—	0.8	1.0

* Values increased by 10% to convert from live weight to empty-body weight basis.

were derived by analogy from values for cattle. The data now available suggest that for both newborn and growing animals, the sodium concentration in sheep is lower than that in cattle (cf. Tables 1.15 and 1.26).

As fat depots contain virtually no minerals, variations in the fat concentration in weight gain should be reflected in their mineral concentrations. Thus the mineral concentration in weight gain would be expected to decline with increasing body weight, to vary according to sex and breed type, and possibly to vary according to rate of gain. Differences of these kinds were not detected, possibly because of the heterogeneity of the data sources examined. Langlands & Sutherland (1969) concluded that the mineral concentration in weight gains declined with increasing weight, or age, and that in sheep up to 25 kg live weight, faster-growing animals had lower concentrations of minerals per unit of weight gain. Kellaway (1973) found that growth rate and sex had relatively small effects on the concentrations of minerals in the empty body of sheep, but Merinos had concentrations of calcium, phosphorus, magnesium and sodium that were 5 to 10% higher than those of Border Leicester×Dorset Horn crossbreds.

4. Composition of body weight changes in adult sheep

Breeding ewes in the UK usually show a net loss of weight from early pregnancy to immediately after parturition. In lactation an early loss of weight is gradually changed to a gain as milk yield falls and feed intake increases. There is a steady gain of weight between weaning and remating, culminating in a period of accelerated growth if flushing is practised.

(a) Pregnancy

Rattray et al (1974b) studied the composition of 106 Targhee ewes slaughtered at between 70 and 140 days of pregnancy. All the ewes had been fed at a maintenance rate up to day 70. Thereafter, some animals received 1.5 times their maintenance requirement but gained little in maternal body weight; the remainder received 2.0 times their maintenance requirement, and gained about 0.1 kg/day. Additional compositional data were obtained from nonpregnant ewes fed to appetite on the experimental diet and from ewes killed at conception. Allometric equations fitted to all the data for pregnant animals gave the following estimates for the composition of empty-body weight gain (or loss):

Empty-body weight (kg)	Protein (g/kg gain)	Fat (g/kg gain)	Energy (MJ/kg gain)
50	94	503	22.0
60	88	596	25.5
70	83	688	29.0

The weight gains of the nonpregnant animals (mean empty-body weight 60 kg) contained less protein (68 g/kg) but more fat (770 g/kg) and energy (32 MJ/kg). Foot (1969) measured the fat content of Scottish Blackface ewes that generally gained in maternal empty-body weight during pregnancy. Allometric equations fitted to her data for 16 ewes killed at full term (mean weight 54 kg) estimate the fat concentration of weight gain to be about 700 g/kg. Again, nonpregnant ewes and those in early pregnancy had gains containing more fat.

In Britain, hill ewes lose much weight in pregnancy; Russel et al (1968) found that pregnant Scottish Blackface ewes, weighing 33.9 kg in November, lost 7.4 kg in maternal empty-body weight (fleece excluded) by April. Of this, 424 g/kg was fat and 191 g/kg was fat-free dry matter; if the latter contained 800 g protein per kg, the energy content of the loss would have been 20.3 MJ/kg. Half the loss of fat occurred in the last month of pregnancy. Field et al (1968) recorded that similar ewes weighing 32.2 kg in October had lost 9.4 kg in maternal empty-body weight (skin excluded) by early May, when they were lactating. The loss contained 600 g/kg of fat, from which its protein content may be estimated as approximately 100 g/kg, and its energy value as 26 MJ/kg. From October to January there were also large losses of Ca, P, Mg and K; these were partially replaced between January and April by means of a dietary supplement but further losses were experienced in early lactation. Over the period October to May, mineral losses were equivalent to about 9 g calcium, 4 g phosphorus, 0.3 g magnesium, 1.5 g potassium, all per kg maternal empty-body weight loss; there was no net loss of sodium. In a later experiment (Sykes & Field 1972a), ewes fed indoors on a high-protein diet lost 4.0 kg in maternal empty-body weight during pregnancy. This contained 800 g fat per kg, and thus about 50 g protein and 33 MJ/kg. A loss of K equivalent to 4 g/kg body-weight loss was experienced, but there were gains in Ca, P, Mg and Na. On a low-protein diet, ewes lost more weight (6.6 kg), containing a lower concentration of fat (500 g/kg) and energy (estimated at 23 MJ/kg). Mineral losses were about 5 g calcium, 3 g phosphorus, 0.1 g magnesium, 4 g potassium and 0.4 g sodium per kg body-weight loss.

In Canada, Lodge & Heaney (1973) showed that the maternal body-weight loss of Rambouillet×Columbia ewes (excluding head, feet and fleece) contained about 130 g protein, 570 g fat and 25.4 MJ per kg.

(b) Lactation

Between May and July the lactating ewes of Field et al (1968) gained 10 kg empty-body weight, and thus returned to their October weight. The fat content of the gain was apparently much lower than that of the loss (170 against 600 g/kg). The gain was associated with a further loss of Ca, little change in P, Mg or Na and a considerable increase in K, equivalent to 3 g/kg gain.

(c) Nonpregnant, nonlactating ewes

Mature Dorper ewes were fed to lose 5 kg or gain up to 9 kg from an initial empty-body weight (less skin) of 25 kg (O'Donovan & Elliott 1971b). Their weight gains or losses were calculated by the authors to contain 111 g protein and 859 g fat per kg dry matter (i.e., 78 g protein, 607 g fat and 25.7 MJ/kg fresh weight). These values were almost identical to those found for lambs of the same breed weighing initially 15 kg (O'Donovan & Elliott 1971a), but are higher for fat and energy, and lower for protein, than those calculated for growing sheep in general (see Table 1.8).

Panaretto (1964) measured by tritium dilution the composition of weight losses in two groups of nonpregnant ewes, 6 moderately fat and 4 very fat animals. The protein concentration of their weight loss was 111 and 90 g/kg, fat concentration was 464 and 602 g/kg and energy, 21.1 and 26.0 MJ/kg, respectively. In a similar experiment Farrell et al (1972a) found comparable concentrations of fat and energy in body-weight loss, but a higher concentration of protein (154 g/kg).

(d) Conclusions

Overall, the data suggest a representative value of 26 MJ/kg for the energy concentration of empty-body weight gain or loss in adult ewes. Animals that are neither pregnant nor lactating may have gains rather higher in energy concentration and, on the basis of limited evidence, lactating ewes that are restoring weight lost in pregnancy may have gains containing less than 26 MJ/kg.

A representative value for protein concentration is more difficult to select, since directly determined values for ewes mainly maintaining or gaining weight, of 83 to 94 g protein per kg (Rattray et al 1974b) and 78 g/kg (O'Donovan & Elliott 1971b), are considerably lower than that for ewes losing weight, of 130 g/kg (Lodge & Heaney 1973). The results of Russel et al (1968) for fat-free dry matter suggest a protein concentration of about 150 g per kg empty-body loss. If the difference is real it may be due to the fact that when gaining weight adult sheep store much protein in wool but when losing weight they are unable to draw on wool protein. It is suggested that tentative values for the protein concentration of empty-body weight gain and loss in ewes should be 90 and 130 g/kg, respectively.

Evidence on the mineral components of gains and losses in weight in adult sheep is insufficient to justify the selection of representative values.

Body Composition of Cattle

1. Composition of the foetus and associated structures

The procedure used for cattle to estimate the composition of the foetus and its associated structures (adnexa) was the same as that employed for sheep (see p. 3).

(a) Composition of the calf at birth

The data sources are listed in Appendix Table 1.5. Of the four data sources for calves 0–3 days old, only one (103) was sufficiently large and detailed to justify the fitting of allometric equations and hence allow the prediction of composition at various birth weights. The remaining three sources for calves 0–3 days old were each used to provide single estimates of composition at their mean birth weight, and single estimates were obtained from Gompertz equations fitted to the two data sources for cattle foetuses.

The range in birth weight in source 103 was due partly to breed and partly to plane of nutrition of the cow. On average, Jersey calves weighed less than Herefords (23.3 against 33.8 kg), but did not differ to any marked extent in composition. Any effects of birth weight on composition (see below) are therefore largely due to maternal nutrition.

The only comprehensive data for mineral components are those of source 103. Samples from these animals were analysed for phosphorus by Haigh et al (1920) and for other elements by Hogan & Nierman (1927), who also reanalysed for phosphorus. The later values for phosphorus (P_2) were generally lower than the earlier analyses (P_1), possibly because of differences in ashing procedures, of sample deterioration and of missing samples. Values for all elements determined by Hogan & Nierman (1927) have therefore been multiplied by the factor P_1/P_2.

Estimates of the composition of the newborn calf are shown for each source in Table 1.16, which also includes representative values.

Table 1.16. The composition of calves at birth.

Composition (g/kg)	Foetuses			Newborn calves						Representative value
Source No.* (No. of calves)	101 (24)	102 (21)	136 (16)	103 (18)			104 (7)	105 (3)		
Empty-body weight (kg)	54.2	35.2	40.2	20	30	40	49.3	40.2		
Water	716.4	(752.9)†	747.0	736.7	729.6	724.6	741.9	722.1		730
Protein	—	163.9	166.7	176.6	182.9	187.5	(188.9)†	200.3		185
Fat	42.9	45.0	37.4	34.9	35.3	35.6	28.0	36.2		40
Ash	—	38.2	41.2	43.4	44.4	45.2	41.2	45.7		43
Calcium	13.8	—	—	13.6	14.3	14.7	13.7	—		14
Phosphorus	7.6	—	—	7.5	8.0	8.3	7.6	—		8
Magnesium	—	—	—	0.51	0.51	0.58	—	—		0.55
Potassium	—	—	—	1.70	2.04	2.33	—	—		2.1
Sodium	—	—	—	2.13	2.36	2.55	—	—		2.4
Chlorine	—	—	—	1.44	1.69	1.89	—	—		1.7

* For sources see Appendix Tables 1.5 and 1.6
† calculated by difference

Protein and ash

Like lambs, calves heavier at birth have higher concentrations of protein and ash, and a lower concentration of water; there is therefore an appreciable change in the ratio of protein to water with increasing birth weight.

Fat

Estimates derived from foetuses are higher than those from newborn calves. The difference could be a consequence of the errors involved in extrapolating the Gompertz equations, but could also be due to metabolism of fat during the neonatal period; a relatively high representative value of 40 g/kg has therefore been chosen.

Mineral elements

For calcium and phosphorus there is reasonable agreement between sources. For other elements the estimates come only from source 103; for magnesium and sodium there is good agreement with values for the lamb (Table 1.2), but for potassium the estimate for the calf (2.1 g/kg) is higher than that for the lamb (1.7 g/kg).

(b) Deposition of nutrients in the adnexa

For the quantities of nutrients present at full term in the foetal fluids and membranes, and in the uterus, the main data source is 106. Additional estimates can be obtained from Gompertz equations fitted to data source 102. The estimates for the uterus are for the weights over and above the following weights (kg) of the components in the uterus of nonpregnant cows: total weight, 0.700; protein, 0.119; fat, 0.007; ash, 0.008. The weight of the uterus in the nonpregnant cow is taken from Hammond (1927), and the components have been calculated from values for the ovine uterus (Langlands & Sutherland 1968, Rattray et al 1972a), since none was available for the cow.

Table 1.17.   Deposition of nutrients in the adnexa of cattle (quantity at full term as a fraction of corresponding quantity for the foetus at full term)

Component	Fluids	Membranes	Uterus*	Total
Total weight	0.50	0.12	0.20	0.82
Protein	0.08	0.06	0.14	0.28
Fat	—	0.01	0.11	0.12
Ash	0.08	0.03	0.06	0.17
Energy	0.05	0.05	0.13	0.23

* excluding amounts in uterus of nonpregnant cow (see text).

Factors for estimating the quantities of nutrients in the adnexa are given in Table 1.17. As an example of their use, a calf weighing 40 kg at birth and containing $40 \times 0.185 = 7.40$ kg protein (Table 1.16) would necessitate the cow's storing a total of $7.40 \times 1.28 = 9.47$ kg protein in its uterus during pregnancy. The weight of fluids differs between source 106, where the weight was equivalent to 61% of the weight of the foetus, and source 102, where the corresponding value was 30%.

The former value is supported by the estimates of Swett et al (1948) and Becker et al (1950) (see also Arthur 1969), hence the value of 50% given in Table 1.17. The fractions for the total adnexa in this table are similar to the corresponding values for sheep, except that for total weight which is nearly twice the sheep value.

No comprehensive data are available for the mineral contents of the bovine uterus, membranes and foetal fluids. However, by recourse to the review of Needham (1931) and to Forbes et al (1935), Blaxter & Rook (1957) and Bitman et al (1961), the quantities of minerals in these tissues at full term can be estimated, and the total uterine depositions can be calculated as fractions of the quantities in the foetus. These approximate fractions are: Ca, 0.01; P. 0.07; Mg, 0.06; K. 0.40; Na, 0.57; Cl, 0.84. With the exception of that for sodium, these values are close to those calculated for sheep.

(c) The time course of nutrient deposition

Data sources used were 101 for the foetus alone and 102 for the foetus and adnexa. Estimates of the proportions of the total quantities at full term that are deposited by specified ages are given in Table 1.18.

Table 1.18. Estimates of the fraction of final weight reached by specified ages, for the total and component weights of the cattle foetus and gravid uterus

	Component	Age (days from conception)†					Source*
		141	169	197	225	253	
Foetus	Total	0.044	0.104	0.214	0.391	0.651	101 and 102
	Protein	0.023	0.062	0.147	0.307	0.579	102
	Fat	0.007	0.032	0.094	0.244	0.529	101 and 102
	Ash	0.023	0.063	0.148	0.310	0.582	102
	Energy	0.021	0.052	0.122	0.264	0.531	102
	Calcium	0.013	0.041	0.112	0.262	0.540	101
	Phosphorus	0.017	0.048	0.123	0.274	0.549	101
Gravid uterus‡	Total	0.151	0.244	0.367	0.531	0.741	102
	Protein	0.062	0.116	0.212	0.373	0.637	
	Fat	0.048	0.079	0.146	0.285	0.549	
	Ash	0.051	0.103	0.195	0.354	0.612	
	Energy	0.064	0.107	0.187	0.330	0.581	

† Age for final weight is 281 days.
* For sources see Appendix Table 1.5.
‡ less uterus of nonpregnant cow.

(d) Composition of the foetus and gravid uterus at various stages of gestation

Estimates for the components of the newborn calf, for the components of the gravid bovine uterus and for the time courses of nutrient deposition have been combined in Table 1.19. The calf is assumed to have an empty-body weight of 40 kg at birth, and its composition has been calculated from the representative values in Table 1.16. The weights of components of the foetus at earlier ages have been calculated by using the proportions of final weight that are achieved at those ages, given in Table 1.18. The values given for K and Na in the developing foetus are based on only two samples (Blaxter & Rook 1957). It is assumed that the K concentration of the foetus increases linearly from 1.6 g/kg at 141 days to 2.1 g/kg at full term. For Na, a constant concentration of 2.4 g/kg has been assumed. No data are available for Mg or Cl in the developing foetus. Tentative estimates involve the assumption that throughout pregnancy the ratios of Mg to Ca and of Cl to Na are the same as at full term, 0.04 and 0.7, respectively.

Table 1.19.   Estimated weights for the cattle foetus and gravid uterus and their components at various ages

Component		141	169	197	225	253	281
		Age (days from conception)					

	Component	141	169	197	225	253	281
Foetus alone	Total (kg)	1.76	4.16	8.56	15.64	26.04	40.00
	Protein (kg)	0.17	0.46	1.09	2.27	4.28	7.40
	Fat (kg)	0.011	0.051	0.150	0.390	0.846	1.600
	Ash (kg)	0.040	0.108	0.255	0.533	1.001	1.720
	Energy (MJ)	4.9	12.2	28.5	61.7	124.1	233.8
	Calcium (g)	7.3	23.0	62.7	147	302	560
	Phosphorus (g)	5.4	15.4	39.4	88	176	320
	Magnesium (g)	(0.3)*	(0.9)	(2.5)	(5.9)	(12)	22
	Potassium (g)†	3	7	15	30	52	84
	Sodium (g)†	4	10	21	38	62	96
	Chlorine (g)	(3)	(7)	(15)	(27)	(43)	68†
Gravid‡ uterus	Total (kg)	11.0	17.8	26.7	38.7	53.9	72.8
	Protein (kg)	0.59	1.10	2.01	3.53	5.94	9.47
	Fat (kg)	0.086	0.14	0.26	0.51	0.98	1.79
	Ash (kg)	0.10	0.21	0.39	0.71	1.23	2.01
	Energy (MJ)	18	31	54	95	167	288
	Calcium (g)	(29)	(58)	(110)	(200)	(346)	566
	Phosphorus (g)	(17)	(35)	(67)	(121)	(209)	342
	Magnesium (g)	(1.2)	(2.4)	(4.5)	(8.1)	(14)	23
	Potassium (g)	(6)	(12)	(23)	(42)	(72)	118
	Sodium (g)	(8)	(16)	(29)	(53)	(92)	151
	Chlorine (g)	(6)	(13)	(24)	(44)	(76)	125

* values in parentheses are tentative (see text).
† see text.
‡ less values for uterus of nonpregnant cattle.

Data for the foetus at full term have been converted to values for the gravid uterus by means of the factors given in Table 1.17 and in the text. For components of the gravid uterus the values have then been calculated for earlier ages by means of the factors of Table 1.18. For minerals this was not possible, and the tentative values included in Table 1.19 were calculated by assuming that the proportion of the final amount of any element that is deposited by a particular age is the same as for ash (see Table 1.18). At earlier ages this method of calculation is likely to have overestimated the quantities of Ca, P and Mg in the gravid uterus, and underestimated the quantities of other minerals.

Table 1.20.   Parameters for estimating weights (Y) of components of the cattle foetus and gravid uterus at ages between 141 and 281 days from conception (X) by substitution in the equation
$$\log_{10} Y = A - Be^{-CX}$$
(X = time from conception (days), Y = weight (kg))

Component	Foetus*			Gravid uterus†		
	A	B	C	A	B	C
Total (kg)	2.580	5.606	0.00621	2.932	3.347	0.00406
Protein (kg)	2.327	6.614	0.00538	3.707	5.698	0.00262
Fat (kg)	1.361	9.527	0.00750	378.000	380.474	0.0000256
Ash (kg)	1.674	6.599	0.00542	2,530	5.603	0.00328
Energy (MJ)	5.188	7.207	0.00334	151.665	151.640	0.0000576
Calcium (g)	4.326	7.651	0.00562	(5.190	5.715	0.00303)
Phosphorus (g)	4.205	7.141	0.00511	(4.661	5.538	0.00340)
Magnesium (g)	(2.929	7.585	0.00557)‡	(3.988	5.826	0.00285)
Potassium (g)	3.275	5.848	0.00522	(4.441	5.687	0.00312)
Sodium (g)	2.827	5.882	0.00691	(5.203	6.153	0.00253)
Chlorine (g]	(2.773	5.642	0.00637)	(4.150	5.540	0.00353)

* All derived estimates are relative to a calf weighing 40 kg at 281 days of gestation.
† All estimates are for the gravid less the nonpregnant uterus.
‡ Parameters in parentheses are tentative (see text relating to Table 1.19).

Gompertz equations, fitted to the values shown in Table 1.19, are given in Table 1.20, and may be used to predict the weights of components of the foetus and gravid uterus at ages intermediate to those given in Table 1.19. The equations give a good fit, but not an exact fit, and hence should not be expected to agree exactly with Table 1.19.

(e) Comparison with previous estimates

In the previous edition, estimates of the weights of nutrients in the foetus alone or in the gravid uterus were compiled for a 45-kg calf plus 33 kg of adnexa. The current estimates are for a 40-kg calf, again with 33 kg of adnexa. Comparisons of the present and earlier estimates are therefore made in terms of nutrient concentrations rather than of absolute quantities.

The current estimates of the concentrations of nutrients in the bovine foetus at full term (Table 1.16) are generally rather higher than the earlier ones. The differences, which are greatest for protein (185 g/kg compared with 162 g/kg) and for sodium (2.4 against 1.44 g/kg), probably reflect our inclusion of more data sources, particularly those for newborn calves. The previous estimate for sodium was based on two calves analysed by Forbes et al (1935) and included some of the structures associated with the foetus.

Current and previous estimates differ less for the gravid uterus at full term than for the foetus alone. The reasons for this are complex, and include the differences in the ratio of foetus to adnexa mentioned above. New and old values for the gravid uterus differ most with respect to energy (4.0 and 3.0 MJ/kg, respectively) and potassium (1.6 and 2.0 g/kg). The earlier value for energy was based entirely on an equation of Jakobsen (1957) which appears not to fit very well at ages approaching term. The new estimate of the total quantity of energy stored in the gravid uterus is therefore 25% higher than the earlier value (288 against 231 MJ).

The difference in potassium concentrations is no greater than is to be expected when grossly inadequate data are replaced by values that still leave much to be desired. Indeed, it must be emphasized that all estimates of the mineral contents of calf foetuses and their adnexa are based on inadequate information.

2. Protein, fat and energy in growing cattle

(a) The data and their analysis

Sources of data are summarized in Appendix Table 1.6. For the first edition, estimates of the composition of gains made by cattle were based on data for bullocks of British beef breeds (sources 107–112). These data have now been extended to include about 600 animals covering wide ranges in breed type, sex, diet and rate of growth. Sources were subdivided according to the above criteria and were recombined only when statistical analysis indicated homogeneity. Data for calves frequently showed a discontinuity at weaning, particularly with respect to fat content. At the other extreme, in sources 110–112, four very fat animals of more than 500 kg empty-body weight were excluded from analysis because they were inconsistent with any relationship fitted to other animals from these sources. Equations fitted to the sources are given in Appendix Table 1.7, and those from representative sources are illustrated in Fig. 1.2.

In the first edition, the main source of variation among the six sets of data analysed was growth rate. In the present larger collection of data, two additional sources of variation were recognized. When compared with castrates of the same body weight,

Fig. 1.2   Composition of the empty body of cattle (see Appendix Tables 1.6 and 1.7 for details of the individual data sources and regression equations illustrated; the heavy lines illustrate the representative equations of Table 1.21).

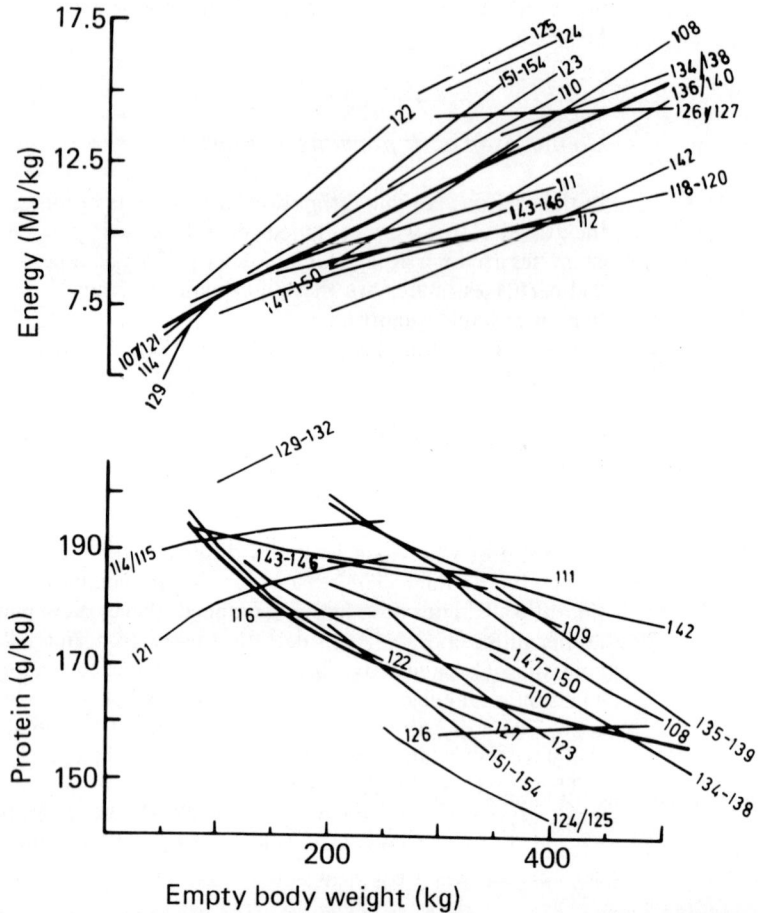

females contained more energy and less protein (sources 147–150 and 151–154) and conversely, intact males contained less energy and more protein (sources 147–150 and 143–146). Composition was affected also by mature size; when compared with medium-sized breeds (e.g., Friesian) at the same weight, the smaller breeds contained more energy and less protein, whereas the larger breeds contained less energy and more protein. The comparison is illustrated by Aberdeen Angus crosses and Charollais crosses from the same experiment (sources 124 and 125 compared with 126 and 127).

It was difficult to make quantitative estimates of the effects on composition of gain produced by differences in growth rate, sex and breed type. There was only scattered information within experiments, mostly covering no more than one factor at a time. Comparisons between experiments were complicated by confounding of the various factors, and by differences in the weight ranges that were covered. At a late stage of the analyses, data were obtained from a single experiment (sources 143–154; Ayala 1974) which included two breed types (Holstein and Aberdeen Angus), three sexes, and two rates of growth on the same diet (about 0.6 and 1.0 kg live weight per day). For such a complex experiment the animals were limited in number, as was the range in slaughter weight. Nevertheless, the data obtained provided a valuable framework on which to superimpose data from other sources.

(b) Composition of empty-body gain

Estimates of the composition of gain were reached by two stages. For the first stage, allometric equations were calculated so as to agree as well as possible with the mean values of protein and of fat concentration both in empty-body weight and in empty-body weight gain. These "mean" equations were assumed to describe the composition of castrate males of an average-sized breed gaining about 0.6 kg empty-body weight per day; they are given in Table 1.21 and included in Fig. 1.2. The mean equations were differentiated to provide estimates of the composition of gains at various empty-body weights, which are also included in Table 1.21, as are derived estimates of the energy concentration of the empty body and of empty-body weight gain.

Table 1.21. Representative regression equations for cattle of $\log_{10}$ protein or fat on $\log_{10}$ empty-body weight and derived estimates of the protein, fat and energy concentrations of the empty body and empty-body weight gain

Equations:		Protein	Fat	Energy*
	Coefficient	0.8893	1.788	
	Intercept	−0.5037	−2.657	
	Empty-body weight (kg)			
Composition of empty body (g/kg or MJ/kg)	50	203	48	6.68
	75	194	66	7.17
	100	188	83	7.70
	150	180	114	8.73
	200	174	143	9.73
	300	167	197	11.68
	400	162	247	13.53
	500	158	295	15.32
Composition of empty-body weight gain (g/kg or MJ/kg)	50	181	86	7.65
	75	173	118	8.72
	100	167	148	9.76
	150	160	204	11.80
	200	155	256	13.72
	300	148	353	17.36
	400	144	442	20.77
	500	140	527	24.01

* 23.6 MJ/kg protein and 39.3 MJ/kg fat.

Table 1.22. Correction factors for the composition of gains made by cattle

		Percentage addition or subtraction to values in Table 1.21	
		Protein	Energy
Breed:	small	−10	+15
	large	+10	−15
Sex:	female	−10	+15
	intact male	+10	−15
Gain:	for each 0.1 kg/day more than 0.6 kg/day	− 1.3	+ 2
	for each 0.1 kg/day less than 0.6 kg/day	+ 1.3	− 2

Secondly, for other types of cattle, growing at rates other than 0.6 kg/day, the correction factors shown in Table 1.22 were calculated. These factors are intended to be additive, and in practice the maximum corrections are likely to be about ± 30% for protein and ± 40% for energy (e.g., a heifer of a small breed growing at 1.2 kg/day or a bull of a large breed growing very slowly). Corrected values for the protein and energy concentration of cattle gains are tabulated in Tables 1.23 and 1.24.

Table 1.23.   Protein concentration of empty-body weight gains in cattle (g/kg), as estimated from the mean equations given in Table 1.21 and correction factors for rate of gain, sex and breed given in Table 1.22

Empty-body weight (kg)	Negative correction (%)					(Mean)	Positive correction (%)				
	25	20	15	10	5	0	5	10	15	20	25
50	136	145	154	163	172	181	190	199	208	217	226
75	130	138	147	156	164	173	182	190	199	208	216
100	125	134	142	150	159	167	175	184	192	200	209
150	120	128	136	144	152	160	168	176	184	192	200
200	116	124	132	140	147	155	163	170	178	186	194
300	111	118	126	133	141	148	155	163	170	178	185
400	108	115	122	130	137	144	151	158	166	173	180
500	105	112	119	126	133	140	147	154	161	168	175

Table 1.24   Energy concentration of empty-body weight gains in cattle (MJ/kg), as estimated from the mean equations given in Table 1.21 and correction factors for rate of gain, sex and breed given in Table 1.22

Empty-body weight (kg)	Negative correction (%)				(Mean)	Positive correction (%)			
	40	30	20	10	0	10	20	30	40
50	4.59	5.36	6.12	6.88	7.65	8.42	9.18	9.94	10.71
75	5.23	6.10	6.98	7.85	8.72	9.59	10.46	11.34	12.21
100	5.86	6.83	7.81	8.78	9.76	10.74	11.71	12.69	13.66
150	7.08	8.26	9.44	10.62	11.80	12.98	14.16	15.34	16.52
200	8.23	9.60	10.98	12.35	13.72	15.09	16.46	17.84	19.21
300	10.42	12.15	13.89	15.62	17.36	19.10	20.83	22.57	24.30
400	12.46	14.54	16.62	18.69	20.77	22.85	24.92	27.00	29.08
500	14.41	16.81	19.21	21.61	24.01	26.41	28.81	31.21	33.61

The methods described immediately above are obviously arbitrary and involve considerable approximations. The differentiation of equations from individual sources that differ considerably in slope from the mean equations can lead to estimates of the composition of gains that bear little relation to the mean estimates. For example, several equations for protein from sources 133 to 140 are illustrated in Fig. 1.2 as crossing the line for the mean equation, and these provided estimates of the protein concentration of gains made by medium-sized bullocks of 350 kg growing at about 0.9 kg/day that were as low as 90 g/kg (against 140 g/kg from Table 1.23).

With regard to fat and energy concentrations, these sources were less at variance with the mean estimates. Animals in several of these source groups were being "finished" during a period of rapid growth following slower growth. While the estimates in Table 1.23 and 1.24 are applicable to animals making steady growth, they are unlikely to provide a good description of the composition of gains made by animals with fluctuating patterns of growth.

A further approximation arises from the use of simple percentage correction factors rather than factors varying with empty-body weight. As Fig. 1.2 shows, variations in composition are small at low empty-body weight and increase as animals grow. The correction factors for energy reproduce this pattern quite well, but those for protein lead to a greater range in the composition of gains of lighter animals than appears to be justified.

In general it was found more difficult to describe adequately the protein content of gains than it was to describe fat and energy contents. For heavier animals there may be some overestimation of proteins in gains, as discussed above for sources 133 to 140. In practice, this is unlikely to lead to the provision of excessive protein in the diet, for reasons which will be made clear in Chapter 4.

(c) Comparison with other estimates

From the representative equations given in Table 1.21, the crude protein content of fat-free tissue is calculated to vary from 213 g/kg at 50 kg empty-body weight to 203 g/kg at 300 kg and to 224 g/kg at 500 kg. These values are comparable with the figure of 216 g/kg reported for cattle by Reid et al (1968), and with the values for sheep given earlier in this chapter.

In the previous edition the liveweight gains of cattle were calculated (from data sources 107–112) to contain 2.4% N. If empty-body gain is accompanied by 90 g/kg of gut contents (see below), this is equivalent to 165 g crude protein per kg empty-body gain. The present estimate falls from 181 g/kg gain at 50 kg empty-body weight to 140 g/kg at 500 kg, the mean being approximately 155 g/kg. The present estimates are more realistic in allowing for the reduction in the protein content of gain that occurs at higher weights.

Fig. 1.3   Energy concentration of gains in empty body weight made by cattle growing at about 0.6 kg. empty body weight per day; comparison of estimates.

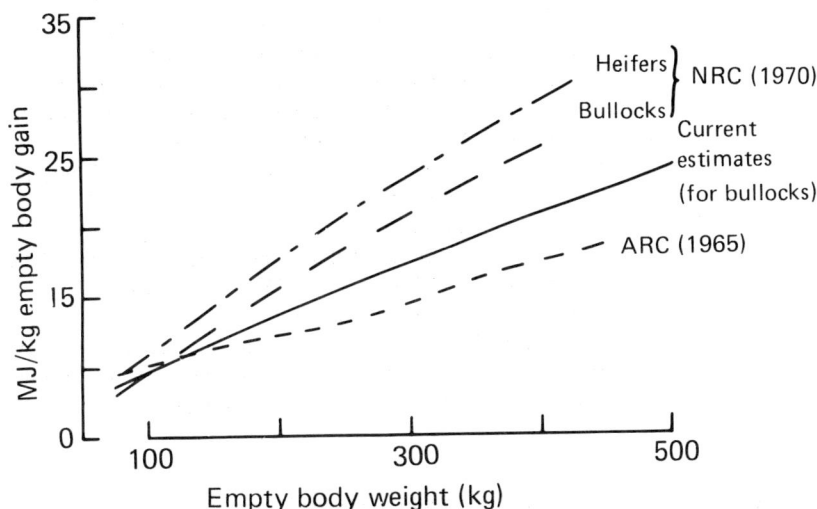

In Fig. 1.3, our current estimates of the energy concentration of gains made by bullocks growing at 0.6 kg empty-body weight per day are compared with the estimates of the previous edition and with those of the National Research Council (1970); values for both live weight and liveweight gain have been converted to an empty-body basis by means of the factors suggested later in this chapter. For young calves, all estimates are similar but at 450 kg empty-body weight the current estimates are about 20% lower than those of the NRC (1970) and 20% higher than those from the first edition. In so far as the current estimates are based both on the older Missouri and Minnesota data analysed for the previous edition and on the newer California data used by the National Research Council (1970), their intermediate position is to be expected. The difference in energy concentration between current and previous UK estimates is consistent with the difference in protein concentration noted above.

With regard to sex differences in the energy concentration of gains, the National Research Council (1970) suggest that heifers have a higher concentration than castrates; the difference varies with rate of gain but averages 13% (see Fig. 1.3). With regard to the effect of rate of gain on its energy concentration, the NRC figures indicate an interaction with sex; for each increment of 0.1 kg/day in the rate, energy concentration increases by 1.2% in bullocks and 2% in heifers (compare our current estimate of 2%). In the previous edition, the relationship between rate of gain and energy concentration was curvilinear, but the average correction for each 0.1 kg gain/day was about 3%.

Recent estimates of the composition of empty-body gain in bulls of large breeds suggest that protein concentration is uniformly high (200 g/kg) and that the energy concentration increases from the low level of 8.6 MJ/kg at 325 kg empty-body weight to 12 MJ/kg at 450 kg and 16 MJ/kg at 550 kg (Robelin & Geay 1976, Geay et al 1976).

(d) Compensatory growth

Two experiments have been reported on the effects of compensatory growth on body composition in cattle; in both, body composition was estimated from carcass density. Meyer et al (1965) grew bullocks at high, medium and low rates for various periods, the final period being always a phase of rapid growth ("realimentation"). Growth restriction reduced the fat and energy concentrations of empty-body gain, sometimes making them negative (i.e., animals gained protein and water but lost fat and energy). Realimentation increased the energy concentration of gains, although the values reached appear not to have been excessive for animals gaining so rapidly (about 1.6 kg/day). Average rates of gain over the whole experiment, for high, medium and low treatments, respectively, were 1.11, 0.75 and 0.66 kg empty-body gain per day, and the average energy value of gain was 16.5, 14.6 and 15.4 MJ/kg. From Table 1.24, the energy value of the gains of these groups, whose mean empty-body weight was 300 kg, would be estimated as 19.1, 17.9 and 17.5 MJ/kg, respectively.

The bullocks of Fox et al (1972) were restricted to a live-weight gain of 0.2 kg/day from 240 to 265 kg, but then grew at 1.2–1.6 kg/day to either 341 or 454 kg. For the first part of the period of realimentation each unit of gain contained less fat and more protein than those of control cattle, but for the second part the differences were reversed. At 454 kg, the two groups did not differ in composition.

3. Mineral content of growing cattle

(a) The data and their analysis

The sources of data are listed in Appendix Table 1.6. As for newborn calves (p. 24), the Missouri data for growing cattle (110–112) were corrected for the discrepancy in phosphorus analyses between the results of Haigh et al (1920) and those of Hogan & Nierman (1927). Allometric equations were fitted to each set of data (Table 1.25) and used to estimate the mineral composition of the empty body at a series of weights.

(b) Calcium

The calcium concentration of the body would be expected to increase in early life, owing to bone growth and mineralization, and then to decline as the deposition of fat increased. There is little evidence of this in Table 1.25, except that the isolated group of heavy, fat animals had a much lower calcium concentration. For animals in the normal range, up to 500 kg empty-body weight, it seems reasonable to assume a single representative value, of 14 g Ca/kg, which applies both to empty-body weight and to gain in empty-body weight.

Table 1.25. Estimates of the mineral concentration of the empty body of cattle

| Element | Source* | Regression of $\log_{10}$ mineral (g) on $\log_{10}$ empty-body weight (kg) | | | Concentration (g/kg) estimated from regression at empty-body weight (kg) of: | | | | | | |
		Coefficient	Intercept	Residual SD	75	100	200	300	400	500	(800)†
Calcium	110 (High plane)	0.6325	1.900	0.030	16.2	14.6	15.4	15.8	16.2	16.4	
	111 (Medium plane)	1.072	1.021	0.030	14.3	14.6	15.9	16.6	17.2		
	112 (Low plane)	1.115	0.936	0.030	14.2	14.7					9.8
	114 (Light)	0.9161	1.294	0.033	13.7	13.4	14.1				
	115 (Heavy)	1.072	0.969	0.033		13.4	13.4	13.4	13.4	13.4	
	141	0.9973	1.133	0.020		(13.7)‡					
Phosphorus	110 (High plane)	0.8482	1.217	0.031	8.6	8.2	7.4	6.9	6.6	6.4	
	111 (Medium plane)	1.022	0.869	0.031	8.1	8.2	8.3	8.4	8.4	8.5	5.3
	112 (Low plane)	1.061	0.792	0.031	8.1	8.2	8.6	8.8	8.9		
	114 (Light)	0.9378	1.006	0.030	7.8	7.6					
	115 (Heavy)	1.041	0.800	0.030		7.6	7.8	6.9	6.8	6.8	
	141	0.9727	0.906	0.023		(7.2)‡	7.0				
Magnesium	110–112	0.9821	−0.298	0.065	0.47	0.46	0.46	0.45	0.45	0.45	0.31
	141	0.9164	−0.177	0.034		(0.50)‡	0.43	0.42	0.40	0.40	
Potassium	110–112	0.9122	0.540	0.026	2.37	2.31	2.18	2.10	2.05	2.01	1.30
	141	0.8189	0.760	0.028		(2.14)‡	2.20	2.05	1.94	1.87	
Sodium	110–112	0.9064	0.478	0.022	2.01	1.95	1.83	1.76	1.72	1.68	0.98
	141	0.8242	0.535	0.015		(1.38)‡	1.35	1.26	1.20	1.15	
Chlorine	110–112	0.9017	0.338	0.019	1.42	1.38	1.29	1.24	1.21	1.18	0.75

* For sources see Appendix Table 1.6
† estimates at 800 kg are mean values for a group of four animals i.e. they were not derived by regression analyses
‡ estimates based on mean values for a separate group of six animals

(c) Phosphorus

For phosphorus, there was evidence of a fall in concentration in fast-growing animals, leading to a much reduced concentration in cattle of 800 kg empty-body weight (source 110, high plane). For this group of cattle, and also for source 141, a value of 7 g P/kg gain would be adequate, but for the remaining, slower-growing animals each kg of gain might contain 9 g P. A representative value of 8 g P/kg empty-body weight gain has been chosen.

(d) Magnesium

From sources 110–112 the magnesium concentration appears to be about 0.45 g/kg empty-body weight gain, whereas source 141 suggests 0.4 g/kg or less. However, the higher value is preferred.

(e) Potassium

The estimates for German bulls (141) are about 10% lower than the Missouri values, but are supported by the data of Salem (1970), who found the potassium content of 43 bulls with a mean empty-body weight of 485 kg to be 1.79 g/kg. Both sources in Table 1.25 indicate a fall in potassium concentration with increasing body weight. A value of 2.0 g K/kg gain seems appropriate, but would probably overestimate the potassium retention of animals above 400 kg empty-body weight.

(f) Sodium

The two sets of data in Table 1.25 differ considerably. The lower estimates of source 146 are supported by the value of 1.18 g Na/kg found in bulls of 485 kg empty-body weight by Salem (1970). Both equations indicate a fall with increasing body weight, and gains appear to contain 1.6 g Na/kg (110–112) or 1.1 g/kg (141). Until further data become available it seems prudent to take a fairly high estimate, 1.5 g Na/kg empty-body weight gain.

(g) Chlorine

The only estimates available show the concentration of chlorine to be 70% of the concentration of sodium. A value of 1.0 g Cl/kg empty-body weight gain is assumed.

(h) Conclusions

Although there are few data on the mineral composition of cattle, different sources provide estimates that agree reasonably well with one another for all elements except sodium. In Table 1.26, the estimates of the mineral composition of gains are compared with those made previously for growing cattle, and also with current estimates for the newborn calf.

**4. Composition of
body weight changes
in adult cattle**

(a) *General*

Dairy cows normally lose weight in early lactation and gain during late lactation and
the dry period. For beef cows the pattern is similar except that a lower rate of feeding
in late pregnancy commonly leads to losses in maternal body weight, as in ewes. Both
types of cow would be expected to make an appreciable net gain in body weight (50
or 100 kg) during their first annual cycle and smaller gains (about 50 kg) for the next
2 or 3 cycles.

Although these cyclical changes in weight are well recognized, their composition
has not been precisely determined. Ellenberger et al (1950a) analysed the bodies of
49 cows of several breeds. An analysis of data for 22 Holsteins, 12 of which were
pregnant, gave the following estimates for the composition of maternal empty-body
weight change:

Empty-body weight (kg)	Protein (g/kg)	Fat (g/kg)	Energy (MJ/kg)	Ca (g/kg)	P (g/kg)
300	163	299	15.6	14.9	8.1
400	157	431	20.6	14.2	7.8
500	152	573	26.1	13.7	7.5

At 400 kg, the values for fat and energy are close to the means for growing cattle
given in Table 1.21, but the value for protein is higher by 13 g/kg. However, the
change in the composition of gain with increasing empty-body weight is considerably
greater for the cows. The concentrations of calcium and phosphorus are close to those
of 14 and 8 g/kg empty-body weight gain, respectively, that were chosen as representative values for growing cattle (Table 1.26).

Table 1.26.   Estimates of the mineral composition of gains in cattle

Component	Newborn calf (g/kg)	Growing cattle (g/kg gain)	
		Present estimate	Previous estimate*
Calcium	14	14	11–17
Phosphorus	8	8	5–10
Magnesium	0.55	0.45	0.33–0.44
Potassium	2.1	2.0	1.8
Sodium	2.4	1.5	1.5
Chlorine	1.7	1.0	1.0

* Agricultural Research Council (1965) values, increased by 10% to convert
from live weight to empty-body weight basis.

(b) *Lactating cows*

Bath et al (1965) fed lactating first-calf heifers at rates which caused them to lose
41.4 kg empty-body weight between weeks 6 and 13 of lactation and a further 28.4 kg
by week 21. Body composition was estimated from carcass specific gravity, but later
Reid & Robb (1971) recalculated the original estimates using more appropriate
equations. The recalculated values suggest that between weeks 6 and 13, empty-body
weight losses contained 188 g protein, 555 g fat and 26.2 MJ/kg. For losses between
weeks 13 and 21 the values were 175 g protein, 705 g fat and 31.8 MJ/kg. From
calorimetric data and from weight losses intended not to be complicated by changes
in gut fill, Moe et al (1971) estimated the energy content of weight loss in lactating
cows to be approximately 25 MJ/kg.

Trigg (1974) estimated the composition of body gains and losses in 18 beef cows by means of deuterium oxide dilution. On average, losses contained 130 g protein, 370 g fat and 18 MJ/kg, and gains contained 170 g protein, 225 g fat and 13 MJ/kg. Both sets of values appear low, possibly because weight changes included intestinal contents.

(c) Conclusions

No firm conclusion can be drawn for the composition of empty-body weight gain or loss in adult cattle. Estimates of energy concentration, even if the values of Trigg (1974) are excluded, range from 20 to 30 MJ/kg. Estimates for protein concentration of changes in adult cattle (150–190 g/kg) are all higher than those for adult sheep. Until more information becomes available it is suggested that the energy value of 26 MJ/kg adopted for adult sheep should be applied also to cattle. For protein, a concentration of 150 g/kg is suggested.

Gut Contents of Ruminants

To translate nutrient requirements for each unit of empty-body weight gain into requirements per unit gain in live weight, account must be taken of the contents of the alimentary tract and their contribution to liveweight gain. In ruminants, gut contents (or "fill") can contribute as much as 300 g/kg live weight and, although they consist largely of water, they also contain appreciable concentrations of other nutrients. In general, the nutrients present in gut contents are assumed to constitute temporary "stores" outside the body proper, and are not included in the factorial calculation of the animal's requirements. For example, if an animal gaining 1 kg in empty-body weight, containing 150 g protein, also gains 0.1 kg of gut contents, no allowance is made for protein "stored" in the extra gut contents, and the protein concentration of its liveweight gain is calculated as $150 \times 1/1.1 = 136$ g/kg. This accounting procedure is less satisfactory for nutrients present in gut contents in relatively high concentrations, and for which there may be considerable interchange with true body stores. Sodium, for example, may on occasions be rapidly transferred between body stores and gut contents, and its concentration in gut contents may be higher than in the empty body.

(a) Data

Total gut contents can be measured only *post mortem*. Animals slaughtered in commercial abattoirs commonly have their gut contents reduced by being deprived of feed (and water), and even without fasting a rapid transfer from farm to slaughter point may diminish gut fill by accelerating excretion. Average fill in the living animal is therefore best estimated by weighing the animal midway between meals and slaughtering it with minimal delay. Fill may be measured directly, by weighing gut contents, or indirectly as the difference between live and empty-body weights; the indirect measure will also include the contents of the bladder.

The sources of data considered were experiments in which ruminants fed on adequately defined diets were slaughtered without fasting. For growing animals many sources of data on empty-body composition included information on gut fill, and

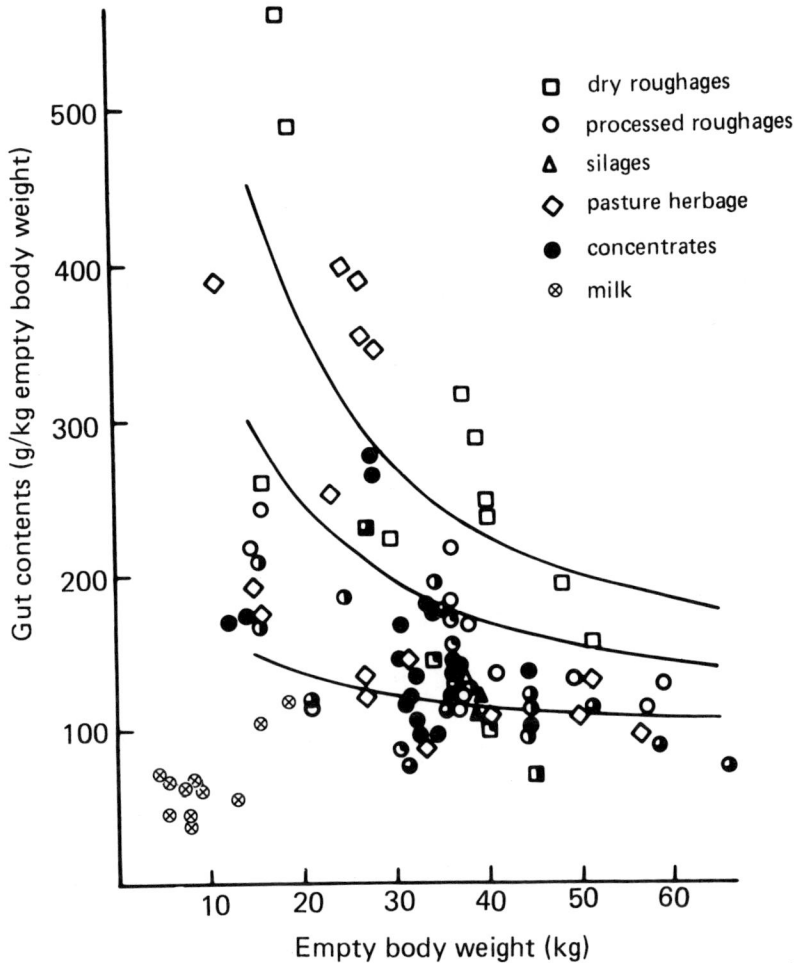

Fig. 1.4   Gut contents of sheep. Shading shows proportion of concentrates in mixed diets. Curves are for initial contents at 15 kg of 150, 300 or 450 g/kg, followed by increments of 90 g/kg increase in empty body weight.

additional sources were obtained by a selective search of the literature. Data for pregnant and lactating ruminants also were examined.

Gut contents have been expressed as g/kg empty-body weight. One hundred sets of data for growing sheep are illustrated in Fig. 1.4, and ninety sets for growing cattle in Fig. 1.5. These data have not been tabulated or identified by references, since gut fill is a topic of only peripheral interest in the assessment of nutrient requirements.

(b) Animal factors affecting gut contents

Figs. 1.4 and 1.5 suggest that gut fill per unit empty-body weight declines as animals increase in weight. This was confirmed by analysis of data from individual experiments involving a uniform diet and a range of slaughter weights; all regression coefficients relating fill per unit weight to empty-body weight were negative. From an analysis of work at Cornell University, Bensadoun et al (1968) calculated that, in sheep fed on

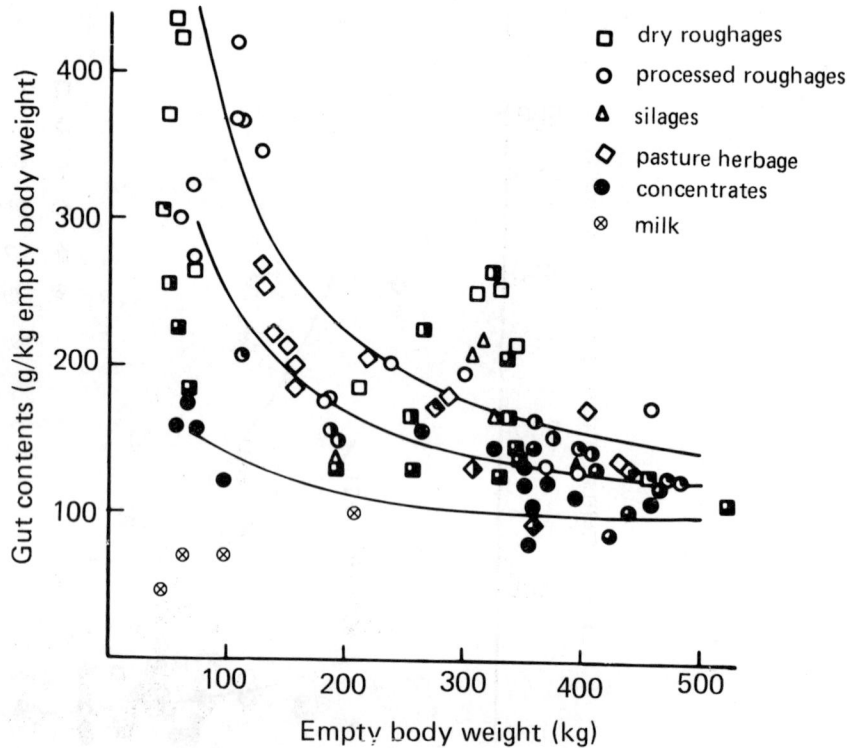

Fig. 1.5   Gut contents of cattle. Shading shows proportion of concentrates in mixed diets. Curves are for initial contents at 75 kg of 150, 300 or 450 kg, followed by increments of 90 g/kg increase in empty body weight.

chopped hay and fasted for 20 h (and hence not included in Fig. 1.4), gut fill declined from 300 g/kg empty-body weight at 15 kg to 180 g/kg at 60 kg. On pelleted hay the decline was from 240 g/kg at 15 kg to 120 g/kg at 60 kg.

It seems logical that relative fill should decline as the animal matures, since increases in body weight due to fattening are unlikely to be associated with directly proportional increases in the growth of the gut. Indeed, there is evidence for a negative relationship between fill and either internal body fat (Tayler 1959) or total body fat (Lofgreen et al 1962; Foot & Greenhalgh 1970). Furthermore, sheep grown from 50 to 70 kg live weight, partially starved back to 50 kg, and then grown again to 70 kg had 50% more gut contents during the second period of gain, when they were less fat, than during the first period (Burton et al 1974).

In pregnancy, the expanded uterus may reduce the volume of the alimentary tract (Forbes 1969a) and hence might be expected to reduce fill. But the extensive data of Rattray et al (1973a) for pregnant ewes on a pelleted diet suggest that any decline in relative fill with advancing pregnancy is no greater than would be expected from the increasing empty-body weight of the ewes. The same conclusion is drawn from a comparison of pregnant and nonpregnant ewes fed on chopped dried grass (Foot 1969). In lactation, feed intake increases and there is an associated hypertrophy of the gut in both sheep (Fell 1972) and cattle (Mowat 1963). Extensive data from Finland for cows fed mainly on hay or on hay and concentrates in equal amounts confirm the negative relationship shown in Fig. 1.5 between relative fill and empty-body weight (Paloheimo 1944, Mäkelä 1956, 1960, Paloheimo & Mäkelä 1959). For a cow of 400 kg empty-body weight, fill was on average 200–220 g/kg, which is about 30% greater than for growing cattle of the same weight (Fig. 1.5). Judging from their feed intakes these cows appear not to have been lactating. Much higher values for fill (about 400 g/kg for cows averaging 340 kg empty-body weight) may be calculated from the data of Ellenberger et al (1950a), but the feeding, preslaughter treatment and physiological state of these cows are not known. The reticulo-rumen fill of lactating cows was found by Mowat (1963) and Tulloh (1966) to be 75–100% greater than that of nonlactating cows; both sets of data suggest a total fill of 300–400 g/kg empty-body weight. But other data on rumen fill in lactating cows (Balch & Line 1957, Bath et al 1966) are more consistent with the 200–220 g total fill per kg empty-body weight found in Finnish cows.

Plane of feeding seems to have a surprisingly small effect on the gut fill of ruminants. The cattle of Moulton et al (1922) that were fed at high, medium and low rates contained at 400 kg empty-body weight 120, 120 and 122 g fill/kg, respectively. In cattle of the same weight, Crabtree (1976) found gut fill to be 126, 107 and 133 g/kg for high, medium and low rates of feeding. The fill of the cows of Mäkelä and Paloheimo was not closely related to previous dry matter intake, but appeared to increase from about 180 g/kg to 250 g/kg as intake increased from 5 to 25 g/kg empty-body weight per day. For sheep given a submaintenance diet, gut fill was only 25% less than that of animals consuming *ad libitum* four times the amount of the diet (Burton et al 1974).

(c) Dietary factors affecting gut contents

Many diets are represented in Figs 1.4 and 1.5, and some general relationships between diet composition and fill may be discerned.

Milk diets

In both sheep and cattle these give a gut fill of about 60 g/kg empty-body weight. One higher value for calves of 200 kg is probably due to slaughter immediately after feeding. Two higher figures for suckled lambs may be due to consumption of grass.

Roughages

Hay and dried grass promote a large gut fill, particularly in young animals. The highest fill recorded is that of 560 g/kg for lambs of 19 kg empty-body weight fed on dried grass (Milford & Minson 1965). According to McCarrick (1966, 1967) these dry roughages give a larger fill than silages made from the same herbages. The fill of grazing animals is generally intermediate to those for dried herbage and silage. For each type of roughage there is a negative relationship between digestibility (or metabolizable energy concentration) and fill. For example, the data of McCarrick (1966) suggest that fill with early-cut hay or silage (dry matter digestibility, 69%) was about 20% less than with late-cut roughages (62% digestible).

Ground and pelleted roughages give a smaller fill than long roughages. This difference is proportionately greater in mature animals than in young (Bensadoun et al 1968), possibly because long-continued feeding with pelleted roughage leads to a relative diminution in the weight and volume of the gut (Greenhalgh & Reid 1974).

Concentrates

The gut fill of ruminants given all-concentrate diets is generally in the range 100–180 g/kg empty-body weight. In lambs, it is about 30% greater when the diet contains whole rather than ground grains (Ørskov et al 1974a). The addition of concentrates to a roughage diet almost invariably reduces fill, the reduction being greatest for roughages that give a large fill, such as long hay and straw, and smallest for high-quality pelleted roughages (Lonsdale et al 1971; Rattray et al 1973a; Greenhalgh & Ørskov 1974, unpublished data).

(d) Prediction of gut contents and their contribution to body weight gain

Gut fill can vary widely according to the animal's weight and physiological state, and to the physical and chemical characteristics of its diet. The contribution of gut fill to the liveweight gain of a growing animal given a constant diet appears to be much less variable. As Figures 1.4 and 1.5 show, if the diet is such as to promote a relatively small fill in the young animal, then fill per unit of empty-body weight changes to only a small extent as the animal grows. Conversely, if the diet causes a large fill in the young animal, relative fill declines rapidly as the animal matures. This means that for both types of diet, fill per unit of empty-body weight *gain* is relatively constant. As shown by the curves drawn on Figures 1.4 and 1.5, a value of 90 g fill per kg empty-body weight gain fits most of the data quite well. Rates of empty-body gain can therefore be converted to rates of liveweight gain by multiplying by the factor 1.090, and concentrations of the components of empty-body gain can be related to liveweight gain by dividing by 1.090.

This simple approximation for predicting gut fill per unit of gain will not by itself provide predictions of fill per unit body weight. The latter may be required, for example, when changes of diet are made. On Figures 1.4 and 1.5, three curves consistent with the factor of 1.09 have been shown. The upper curve is based on the assumption that in younger and smaller animals (sheep of 15 kg and cattle of 75 kg empty-body weight) gut fill amounts to 450 g/kg empty-body weight; this curve would be appropriate for diets of long dried roughage. The middle curve starts from 300 g/kg and would represent gut fill for green roughages, pelleted dry roughages and many mixed diets. The lower curve, starting from 150 g/kg, would be suitable for diets consisting entirely or largely of concentrates. For suckled ruminants, the appropriate value for fill is 60 g/kg empty-body weight.

To relate empty-body weight (EBW, kg) to live weight (LW, kg) the following approximate equation may be used:

$$LW = 1.09 \ (EBW + a),$$ [1.2]

where *a* is calculated from initial gut fill at 15 kg EBW (sheep) or 75 kg (cattle) as follows:

Initial fill (g/kg EBW)	Value of *a*	
	Sheep	Cattle
150	0.83	4
300	2.9	14
450	5.0	25

It must be emphasized that the factor 1.09 should not be applied to animals subjected to changes of diet. For example, a change from milk to grass for lambs caused gut fill to change from 75 g/kg at 20 kg empty-body weight to 220 g/kg at 30 kg; an increase of 1 kg in empty-body weight was therefore accompanied by an increase of 0.51 kg in gut fill (Large 1964).

For pregnant ruminants, there seems no justification for assuming that gut fill will differ from that of growing animals of the same weight. The data for lactating cows are somewhat contradictory, but suggest that their gut fill is about 50% greater than that of nonlactating cattle of the same weight.

Composition of Animal Products

1. Composition of milk

The gross composition of the milk of both cattle and sheep varies with breed, strain, individuality, age, stage of lactation, udder disease and nutrition. Species mean values for the composition of milk are therefore of limited use for the prediction of requirements, except possibly for large groups of animals in commercial conditions. Though additional information on sheep has appeared since the first edition, there is still insufficient to make a distinction between breeds, and only species mean values for the constituents have been calculated. The more extensive information for the cow permits the calculation of mean values for several constituents of milk from the major dairy breeds. Moreover, although the fat concentration in milk is subject to wider and more varied fluctuations than the concentration of most other constituents, if fat concentration is known, as it commonly is in commercial herds, it may be used to improve the accuracy of prediction of requirements for energy, nitrogen and calcium.

Values from the literature for the composition of ewe's milk are shown in Table 1.27 and for cow's milk in Table 1.28.

(a) Energy value of milk

There is insufficient published information on the energy concentration in the milk of the cow or the sheep to permit the calculation of reliable species or breed mean values. The alternative approach of calculation of energy concentration from the concentrations in milk of the major constituents has therefore to be adopted. Two separate procedures have been proposed. Perrin (1958) presented formulae for the calculation of energy concentration from the concentrations of fat, protein and lactose on the basis of mean values for the energy contents of each of these constituents. There are inherent weaknesses in this approach; it takes no account of the minor energy-containing materials in milk (citric acid in particular), which will result in an underestimate of about 0.5% in the energy value of milk, nor does it allow for possible differences in the energy concentration of fat from milks of different fat concentration. The alternative procedure, adopted by several workers, of derivation of predictive formulae by the regression of energy concentration on the concentrations of one or more of the major constituents has therefore been preferred.

Sheep

Varela-Alvarez et al (1970), from data for 25 Columbia×(Hampshire×Suffolk) ewes, derived the following equations for predicting energy concentration (E; kJ/kg):

$$E = 36.15\ TS - 1403.7 \qquad [1.3]$$

$$\text{and } E = 2035 + 34.45\ F, \qquad [1.4]$$

where TS is total solids concentration (g/kg) and F is fat concentration (g/kg). Brett et al (1972) applied these two equations to their own values for 68 samples of milk from Merino ewes 3 to 77 days post partum and for 24 samples of milk from Border

Table 1.27. Estimates of mean values for the fat, nitrogen and mineral composition of ewe's milk

Author	Breed	Fat	Total Nitrogen	Ca	P	Mg	K	Na	Cl	Remarks
						Composition of milk (g/kg)				
Peirce (1934)	Merino	80	7.99	2.1	1.7	—	1.26	0.50	—	6 ewes; weeks 3–9 of lactation.
Godden & Puddy (1935)	Cheviot	74	9.56	—	—	—	—	—	0.78	4 ewes; total of 22 samples throughout the lactation.
El-Sokkary et al (1949)	Rohmany and Awsemy	80	9.24	—	—	—	—	—	—	21 ewes; sampled throughout lactation.
Whiting et al (1952)	Corriedale	67	6.67	—	—	—	—	—	—	40 ewes; weeks 2–7 of lactation.
Bonsma (1939)	Various	51	8.15	—	—	—	—	—	—	
Barnicoat et al (1957)	New Zealand Romney	55	8.62	1.9	1.5	—	—	—	—	13 ewes; weeks 3–9 of lactation.
Nakanishi & Tokita (1957)	Corriedale	60.3	8.70	1.68	1.45	—	—	—	—	4 ewes; days 15–90 of lactation.
Perrin (1958)	New Zealand breeds	100	10.75	2.0	1.6	0.19	1.23	0.43	1.26	12 ewes; days 7–111 of lactation.
Slen et al (1963)	Canadian Corriedale	64	8.11	—	—	—	—	—	—	
	Hampshire	74	8.42	—	—	—	—	—	—	
	Rambouillet	67	8.19	—	—	—	—	—	—	25 ewes of each breed.
	Romnelet	69	7.64	—	—	—	—	—	—	
	Suffolk	70	8.27	—	—	—	—	—	—	
Ashton et al (1964)	Clun Forest	62	8.30	1.3	1.0	—	—	—	—	41 ewes; sampled over the first 12 weeks of lactation.
Ashton & Yousef (1966)	Clun Forest	—	—	2.0	1.4	0.15	1.68	0.46	1.1	12 ewes; 6 samples during days 7–77 of lactation.
L'Estrange & Axford (1966)	Welsh Mountain	—	—	—	1.42	—	1.26	0.31	0.76	4 ewes; week 2 of lactation; 2–4 years of age.
Camalesa et al (1967)	Merino	75.3	9.61	—	—	—	—	—	—	Sampled throughout lactation.
	Spanca	78.7	8.39	—	—	—	—	—	—	
Bouchard & Brisson (1969)	Suffolk; North country Cheviot	—	—	—	—	—	—	—	—	
Poulton & Ashton (1970)		—	7.34*	—	—	—	—	—	—	7 ewes; first 12 weeks of first lactation excluding colostral period.
	Clun Forest	—	9.24†	—	—	—	—	—	—	6 ewes; first 12 weeks of lactation excluding colostral period.
Zdanovski et al. (1970)	—	71	9.42	—	—	—	—	—	—	
Kataoka & Nakae (1971)	—	59.7	8.77	—	—	—	—	—	—	
Sebela (1972)	Improved Valachian	—	9.14	—	—	—	—	—	—	
Simple mean values		70	8.60	1.8	1.4	0.17	1.4	0.4	1.0	
Weighted mean values		70	8.21	1.6	1.3	0.17	1.4	0.4	1.1	

* non-protein nitrogen, 0.57 g/kg
† non-protein nitrogen, 0.52 g/kg

Table 1.28.   Estimates of mean values for the fat, nitrogen and mineral composition of cow's milk

Author	Breed	Fat	Total nitrogen	Ca	P	Mg	K	Na	Cl	Remarks
Overman et al (1929)	Ayrshire	41.4	5.61	—	—	—	—	—	—	Fourteen Ayrshire, 16 Guernsey, 19 Holstein-Friesian and 15 Jersey cows; sampled over a 12-month period.
	Guernsey	51.9	6.03	—	—	—	—	—	—	
	Holstein-Friesian	35.5	5.56	—	—	—	—	—	—	
	Jersey	51.8	6.05	—	—	—	—	—	—	
Rowland & Rook (1949)	Ayrshire	36.9	5.27	1.16	0.93	—	1.51	0.54	0.99	Herd bulk samples monthly throughout a year; values for each breed representative of 9 to 18 herds.
	British Friesian	34.6	5.14	1.15	0.90	—	1.58	0.58	1.13	
	Guernsey	44.9	5.57	1.30	1.02	—	1.54	0.48	0.96	
	Shorthorn	35.3	5.18	1.21	0.96	—	1.52	0.59	1.02	
Reinart & Nesbitt (1956)	Ayrshire	39.7	5.09	1.20	0.93	0.15	—	—	1.01	Three herds of each breed; samples representative of one-day's production at monthly intervals throughout a year.
	Guernsey	45.8	5.48	1.37	1.06	0.16	—	—	0.88	
	Holstein-Friesian	35.6	4.78	1.17	0.92	0.15	—	—	1.10	
	Jersey	49.7	5.74	1.45	1.40	0.15	—	—	0.98	
Dawes (1965)	Ayrshire	—					1.36	0.62	—	9 cows ⎫
	Ayrshire	—					1.44	0.49	—	28 cows ⎬ Sampled at approximately monthly intervals over a 12-month period.
	Friesian	—				—	1.36	0.56	—	31 cows ⎪
	Friesian × Jersey	—				—	1.30	0.56	—	18 cows ⎭
Sasser et al (1966)	Guernsey	—	—	—	—	—	1.48	—	—	Groups of animals in the Colorado State University herd.
	Holstein-Friesian	—	—	—	—	—	1.43	—	—	
	Jersey	—	—	—	—	—	1.57	—	—	
Comberg (1967)	Jersey	61.9	6.19	1.24	1.09	0.14	—	0.40	—	
Fisher et al (1970)	Ayrshire	—	—	1.16	—	0.15	1.48	0.66	—	Eight Ayrshire cows, 19 Holstein-Friesian cows; sampled over entire lactation.
	Holstein-Friesian	—	—	1.13	—	0.13	1.50	0.64	—	
Federation of United Kingdom Milk Marketing Boards (1974)	Ayrshire	38.6	—	—	—	—	—	—	—	Annual values for recorded herds in England and Wales, 1972–73.
	British Friesian	36.8	—	—	—	—	—	—	—	
	Guernsey	45.0	—	—	—	—	—	—	—	
	Jersey	49.0	—	—	—	—	—	—	—	
	Shorthorn	35.9	—	—	—	—	—	—	—	
Milk Marketing Board (1973)	All dairy breeds excluding Channel Island	—	5.09*	—	—	—	—	—	—	Monthly samples representing one day's intake of non-Channel Island milk from M.M.B. Creameries 1942–1970.
Werner et al (1973)	Commercial herds	—	—	—	—	0.125	—	—	—	Milk derived from a wide variety of commercial sources.
	Holstein-Friesian	—	—	—	—	0.131	—	—	—	Twelve Holstein-Friesian cows sampled throughout a single lactation.

Composition of milk (g/kg)

* non-protein nitrogen content, 0.28 g/kg

Leicester ewes 30 to 60 days post partum. Equation 1.3 overestimated energy concentration, by 297 ± 113 kJ/kg for Border Leicester milks and by 498 ± 142 kJ/kg for Merino milks. Equation 1.4 similarly overestimated energy concentration, but to a slightly smaller extent.

From their own data, Brett et al (1972) concluded that energy concentration might be predicted most satisfactorily from total solids concentration or fat and solids-not-fat concentrations, but of the various equations they derived they proposed the following as the most generally useful and reliable:

$$E = 32.80F + 2.5D + 2203.3, \qquad [1.5]$$

where D is day of lactation and the other symbols are as described above.

This equation, when applied to the values of Pierce (1934 1936), underestimated energy concentration by an average 121 kJ/kg, whereas Equation 1.3 overestimated it by 339 kJ/kg. Equation 1.5 may therefore be used to predict energy concentration from direct estimates of fat concentration or, in the absence of such information, the weighted mean value of 70 g fat/kg derived from the observations summarised in Table 1.27 may be used.

Cattle

Using observations reported by Stocking & Brew (1920), Gaines & Davidson (1923) derived the formula:

$$FCM = M\ (0.4 + 0.15F) \qquad [1.6]$$

where FCM is milk energy in terms of 'fat-corrected milk', i.e., weight of natural cow's whole milk (lb) containing 4.0% fat, M milk (lb) and F fat concentration (%). Breed differences in the formula were found to be significant but small and it was concluded that for all practical purposes a single formula could be used (Overman & Gaines 1933). Later observations (Overman & Gaines 1948, Tyrrell & Reid 1965) have demonstrated that there is not a simple linear relationship between energy concentration and fat concentration. Tyrrell & Reid (1965), on the basis of observations on milks of Holstein-Friesian cows varying in fat concentration from 16 to 64 g/kg, proposed the following equation as the most practical for the prediction of energy concentration:

$$E = 38.60F + 20.56S - 236.0, \qquad [1.7]$$

where S = solids-not-fat concentration (g/kg).

When applied to other sets of literature values, this equation gave good agreement between measured and predicted values (mean difference, -1.8 ± 1.1 kJ/kg). Similar comparisons reported by Goto & Ohashi (1969) and Matěj (1970) gave mean discrepancies of -12.3 (±12.5) and 11.4 kJ/kg, respectively. An alternative equation derived by Tyrrell & Reid (1965), which may be used when only fat concentration is available, is:

$$E = 40.60F + 1509.0. \qquad [1.8]$$

The coefficient of variation was 3.06%. In the absence of information on the fat concentration of milk, the breed mean values of the Federation of United Kingdom Milk Marketing Boards (1974) may be used (Table 1.28).

(b) Nitrogen content of milk

Values normally reported are for total nitrogen, which includes a non-protein nitrogen component. This component consists largely of waste products of nitrogen metabolism, the milk in effect providing an alternative excretory outlet to the urine. In calculating nitrogen requirements for milk protein secretion, it is therefore more correct to use vales for true protein nitrogen than those for crude protein nitrogen as was done in the first edition.

Sheep

Mean values for the total nitrogen concentration in ewe's milk are presented in Table 1.27. It is proposed that the weighted mean value of 8.21 g/kg be adopted; the corresponding value in the first edition was 9.4 g/kg. After correction for an assumed value of 0.55 g/kg for non-protein nitrogen, this gives a mean value for protein nitrogen concentration of 7.66 g/kg.

Cattle

The values included in Table 1.28 have been confined to those for large groups of cows in commercial or near commercial conditions. The most representative value for total nitrogen concentration in cow's milk in British conditions is that of the Milk Marketing Board (1973) which is for commercial, non-Channel Island cattle. For individual breeds, the most representative values for Ayrshire, Friesian, Guernsey and Shorthorn cattle are those of Rowland & Rook (1949). For Jersey cattle a value for total nitrogen concentration of 6.0 g/kg has been selected, on the basis of the observations of Overman et al (1929), Reinart & Nesbitt (1956) and Comberg (1967). After correction for an assumed value of 0.30 g/kg for non-protein nitrogen concentration the recommended mean values for protein nitrogen concentration (g/kg) are:

Non-Channel Island commercial cows	4.8
Ayrshire	5.0
British Friesian	4.8
Dairy Shorthorn	4.9
Guernsey	5.3
Jersey	5.7

The figures recommended in the first edition for total nitrogen concentration were 6.1 g/kg for Channel Island cows and 5.3 g/kg for other major British dairy breeds.

There is, however, a genetic correlation between milk fat and milk protein concentration and when information on fat concentration is available, this may be used to predict protein nitrogen concentration. Gaines & Overman (1938) calculated regression equations from 305-day samples of milk from animals of several breeds, for the prediction of crude protein concentration. When adjusted to predict protein nitrogen concentration (PN, g/kg) rather than crude protein, the equations become:

Ayrshire	$PN = 3.24 + 0.047F$	[1.9]
Holstein-Friesian	$PN = 2.07 + 0.083F$	[1.10]
Guernsey	$PN = 1.06 + 0.093F$	[1.11]
Jersey	$PN = 2.19 + 0.069F$	[1.12]

where F = fat concentration (g/kg).

(c) Mineral composition of milk

Sheep

Additional information has appeared since the publication of the first edition and the mean values are presented in Table 1.27. The values recommended for the prediction of requirements are the overall weighted means (g/kg), as follows; the figures in brackets are the values proposed in the first edition.

Calcium 1.6 (1.9)   Sodium 0.4 (0.43)
Phosphorus 1.3 (1.5)   Potassium 1.4 (1.23)
Magnesium 0.17 (0.178)

Cattle

The most comprehensive set of values for mineral constituents other than magnesium representative of British conditions, are those of Rowland & Rook (1949). Since they are largely consistent with the results of more limited studies both in the United Kingdom and overseas (Table 1.28), it is recommended that the values of Rowland & Rook (1949) should be adopted. For magnesium concentration, by far the most extensive survey, that of Wernery et al (1973), gave a mean value for commercial milk of 0.125 g/kg, which is similar to the value adopted in the first edition; a slightly higher value of 0.131 was obtained in a more limited study with Holstein-Freisian cows. The recommended mean values (g/kg) for cow's milk, together with those proposed in the first edition, are therefore:

	Breed				Values recommended in the first edition
	Ayrshire	British Friesian	Guernsey	Shorthorn	
Calcium	1.16	1.13	1.30	1.21	—
Phosphorus	0.93	0.90	1.02	0.96	0.95
Magnesium		0.125			0.126
Potassium	1.51	1.58	1.54	1.52	1.43
Sodium	0.54	0.58	0.48	0.59	0.63
Chlorine	0.99	1.13	0.96	1.02	1.15

Reported correlation coefficients between fat and calcium concentrations of cow's milk are of the order of 0.6; thus, where the fat concentration of the milk is known it may be used to predict calcium concentration. There is a better correlation of calcium concentration with milk protein concentration but the values for protein are less likely to be available than those for fat.

In the first edition, the equation proposed for the prediction of calcium concentration was an amalgamation of two other equations (Ellenberger et al 1950; A. S. Foot, unpublished). When the three equations were applied to the data of Rook & Rowland (1949), Reinart & Nesbitt (1956), Vanschoubroek (1958) and Comberg (1967), the mean differences, with standard error, between the predicted and actual calcium values (g/kg) were 0.067 ± 0.023 (first edition equation), 0.023 ± 0.022 (Ellenberger et al 1950b) and 0.106 ± 0.021 (Foot, unpublished). The equation recommended is that of Ellenberger et al (1950b), which is:

$$Ca=0.79+0.011F, \hspace{3cm} [1.13]$$

where Ca = calcium concentration (g/kg milk).

Fat concentration cannot be used for the prediction of the concentration of phosphorus or of other mineral constituents as their correlations with fat concentration are poor.

2. Composition of the fleece

The fleece of the sheep contains three fractions of body origin, the actual wool fibres (hereafter referred to simply as "wool"), suint (the secretion of the sweat glands) and wax (the secretion of the sebaceous glands). There are also the two additional fractions of water and dirt, but throughout this section the word "fleece" will mean the total dry matter of the three physiological products. The relative contributions of these three fractions are extremely variable. They are affected by breed (Daly & Carter 1955, Mercik 1970), by the nutrition of the sheep (Daly & Carter 1955), and by the washing action of rain and sheep dip solutions. Much of the information available on fleece proportions relates to Merino sheep, which appear to differ considerably in this respect from the coarser-woolled breeds common in Britain. Nevertheless, reasonably accurate estimates of the relative proportions of the three fractions are essential if the chemical composition of the complete fleece is to be precisely assessed, because the three fractions differ so much in composition. For example, the nitrogen concentration of wool is 50 times that of suint, whereas potassium is present in suint in a concentration about 2000 times that in wool. From a consideration of the references given earlier, together with those of Farnworth (1956) and Onions (1962), the relative proportions of the fleece of the average British sheep are taken to be wool 80%, wax 12% and suint 8%. These figures approximate to a clean "scoured" yield of 70%, since that value excludes wax, suint and dirt, but includes moisture.

Table 1.29.   Chemical composition of the fleece and its components

Component	Relative proportions	Chemical constituent (g/kg dry matter)							Energy (MJ/kgDM)
		N	S	Ca	P	Mg	K	Na	
Wool	80	165	33	1	0.1	0.1	0.05	0.3	23.5
Wax	12	1.5	—	0.3	0.8	0.1	6.8	0.3	40.8
Suint	8	27	4	7	1.2	2	200	10	—
Fleece	100	134	27	1.4	0.3	0.3	17	1.1	23.7

Representative values for the chemical composition of the fleece components, together with values for a fleece having the proportions suggested above, are given in Table 1.29. They have been selected from the data of Freney (1934, 1940), Drummond & Baker (1929), Farnworth (1956), Brochart & Larvor (1959), Stacy et al (1963), Burns et al (1964), Paladines et al (1964), Menke et al (1969) and Kalinin & Kozmanishvili (1970).

It should be noted that although these values are intended to apply to British breeds of sheep, many are derived from exotic breeds; it has therefore to be assumed, for example, that wax comprising 12% of the fleece of a British sheep is similar in composition to wax forming 30% of a Merino fleece. For some elements, such as magnesium, the data are limited and conflicting, but the concentrations of nutrients that are quantitatively important in wool (N, S and K) are more accurately defined.

To assess the quantities of nutrients deposited daily in the fleece, an estimate of its growth rate is required. Typical fleece weights at shearing for British breeds of sheep are given by Wiener (1967), Ryder & Stephenson (1968) and Spedding (1970), and from these it appears that the typical weight for dry fleece might be 2 kg. This is equivalent to a mean growth rate of 5.5 g dry fleece per day, which is consistent with the estimates of nitrogen retention in the fleece discussed above (see p.49). It may be argued that the weight at shearing underestimates daily production, because of losses during the year. In the first edition (pp.67 and 157), an annual yield of clean scoured wool of 1.3 to 3.5 kg was assumed; these figures would be equivalent to 1.4 to 3.7 kg dry fleece.

Table 1.30.   Nutrient and energy retention in the fleece of the sheep

Weight of fleece at shearing (kg)	Clean scoured wool* (kg)	Dry fleece†		Nutrient retention (mg/day)							Energy retention (kJ/day)
		(kg/year)	(g/day)	N	S	Ca	P	Mg	K	Na	
1.4	1.0	1.0	2.7	365	75	4	1	0.5	50	3	65
2.7	1.9	2.0	5.0	730	150	8	2	1.0	100	6	130
4.0	2.8	3.0	8.2	1095	225	12	3	1.5	150	9	195
5.4	3.8	4.0	11.0	1460	300	16	4	2.0	200	12	260

* 70 % of weight at shearing.
† 74 % of weight at shearing (i.e. omitting 10 % dirt and 16 % moisture).

Table 1.30 gives estimates of the daily retention of nutrients and energy in the dry fleece growing at rates of 2.7–11.0 g/day. They are based on the values given on the last line of Table 1.29. For calcium, phosphorus and magnesium the quantities retained in wool are extremely small when considered in relation to net requirements for maintenance and other forms of production, and may safely be ignored.

Appendix Tables

Appendix Table 1.1.   Data sources for prenatal growth in sheep

Source no.	Reference	Breed	No.	Material analysed	Components determined†
1	Mitchell et al (1928)	Rambouillet and Southdown	4	Lambs dying shortly after birth	Protein, fat
2	Weniger et al (1955)	Merino and Karakul	4	Newborn lambs	Protein, fat, Ca, P, Mg
3	Searle (1970b)	Merino	4	Lambs, 3 days old	Protein, fat
4	Jagusch et al (1970)	Dorset Horn × (Border Leicester × Merino)			
5	Langlands & Sutherland (1968)	Merino	43	Lambs, 2–5 days old	Protein, fat
			49	Uteri, membranes and fluids, and foetuses, 30–145 days old; all single	Protein, fat, Ca, P, Na, K
6	Langlands & Sutherland (1969)	Merino	14	Newborn lambs	Protein, fat, Ca, P, Na, K
7	Kellaway (1973)	Merino	4	Lambs, 4–5 days old	Protein, fat, Ca, P, Mg, Na, K
8	Sykes & Field (1972b)	Scottish Blackface	24	Newborn single lambs	Protein, fat, Ca, P, Mg, Na, K
9	Sykes & Field (1972b)	Scottish Blackface	18	Newborn twin lambs	
10	Field & Suttle (1967)	Cheviot × (Border Leicester × Cheviot)	47	Foetuses, 80–144 days old; single, twin and triplet	Fat, Ca, P, Mg, Na, K
11	Foot (1969)	Scottish Blackface	16	Foetuses, 141–143 days old; single and twin	Fat
12	Foot (1970, unpublished)	Scottish Blackface	36	Foetuses	Fat
13	Rattray et al (1973b)	Suffolk × Targhee	143	Uteri, membranes, fluids and foetuses, 70–140 days old; single and twin	Protein, fat, energy
14	Lodge & Heaney (1973)	Rambouillet × Columbia	25	Uteri, membranes, fluids and foetuses 35–140 days old; (single and twin not segregated)	Protein, fat
15	Lodge & Heaney (1973), unpublished data)	—	18	Uteri, membranes, fluids and foetuses 140 days old; single and twin	Protein, fat

† in addition to water and ash

Appendix Table 1.2.   Data sources for the body composition of growing sheep

Source No.	Reference	Breed†	Range in Age (days)	Range in Empty-body weight (kg)	No. of animals	Comments‡
	Non-Merino Males					
16	Andrews & Ørskov (1970a)	Su × Ch	72–252	23–35	58	(F); Ca, P (21 sheep only)
17	Ørskov et al (1971a)	Su × Ch	56–210	19–44	10	(F); Ca, P; low protein diet
18	Ørskov et al (1971a)	Su × Ch	56–168	19–42	7	(F); Ca, P; medium protein diet
19	Ørskov et al (1971a)	Su × Ch	56–147	18–49	9	(F); Ca, P; high protein diet
20	Hovell (1972)	Su × BF	81–252	13–51	53	Ca, P
21	Hovell (1972)	Su × BF	81–252	13–51	53	(F) but otherwise as 4(a)
22	Hovell (1972)	Su × Ch	56–262	14–38	29	
23	Loery & Zelter (1948)	IF	1–133	4–30	8	(F)
24	Bull (1969)	SD	180–365	11–53	21	
25	Bull (1969)	SD	365	12–19	12	
	Non-Merino Castrates					
26	Wood (1927)	?	?	27–48	4	Breed, sex and age unknown
27	Mitchell et al (1926)	?	350–530	29–51	12	(F); breed and sex unknown
28	Mitchell et al (1928)	?	90–310	18–36	30	(F); Ca; breed and sex unknown
29	Burton & Reid (1969)	Sh	140–613	12–58	26	
30	Kellaway (1973)	DH × BL	3–41	6–11	9	⎰ Ca, P, Mg, K, Na; includes a
31	Kellaway (1973)	DH × BL	70–332	14–34	17	⎱ few females
32	Freer et al (1972)	BL	56–125	7–18	21	
33	Field & Harris (unpublished)	BF	30–650	9–45	32	Ca, Mg, K, Cl; not protein
34	Jagusch & Mitchell (1971)	DH × RM	23	3–11	11	(F)
35	Fennessy et al (1972)	DH × RM	3–90	4–26	36	(F)
36	Mitchell & Jagusch (1972)	DH × RM	60	14–21	7	(F)
37	Fennessy et al (1972)	DH × RM	8–108	7–34	60	
38	McConnell & Jagusch (1972)	DH × RM	8–71	9–26	9	
39	Jagusch (unpublished)	RM	2–107	4–23	12	
40	Jagusch & Nicol (1970)	RM	102–218	17–36	22	
41	Rattray et al (1973c)	T & T × FL	98	23–29	19	Initial slaughter
42	Rattray et al (1973c)	T & T × FL	130–172	21–29	20	Restricted feeding
43	Rattray et al (1973c)	T & T × FL	130–182	35–40	20	Feeding *ad libitum*
	Non-Merino Females					
44	Andrews & Ørskov (1970a)	Su × Ch	113–238	33–37	30	(F); Ca, P (11 sheep only)
45	Ørskov et al (1971a)	Su × Ch	56–218	13–45	8	(F); Ca, P; low protein diet
46	Ørskov et al (1971a)	Su × Ch	56–168	17–49	9	(F); Ca, P; medium protein diet
47	Ørskov et al (1971a)	Su × Ch	70–161	20–49	7	(F); Ca, P; high protein diet
48	Leroy & Zelter (1948)	IF	1–102	4–16	4	(F)
49	Searle (1970a)	CD	63–179	15–30	8	
50	Kellaway (1973)	DH × BL	66–419	14–45	27	Ca, P, Mg, K, Na
51	Freer et al. (1972)	BL	56–125	7–18	19	
52	Bull (1969)	SD	180–365	12–47	21	
53	Bull (1969)	SD	365	11–20	12	
54	Rattray et al (1973b)	T	182	16–25	11	Initial slaughter
55	Rattray et al (1973b)	T	304	12–24	14	Restricted feeding
56	Rattray et al (1973b)	T	274–304	26–39	18	Feeding *ad libitum*
57	Rattray & Garrett (unpublished)	T	>2190	46–60	13	

Merino Males						
58	Walker & Norton (1971a, b)	DH × M	24	4–9	19	(F)
59	Faruque & Walker (1970a)	DH × M	21–28	3–8	21	(F)
60	Faruque & Walker (1970a)	DH × M	21–28	5–11	21	(F)
61	Walker et al (1967)	DH × M	16–56	3–15	68	(F); Ca, P (37 sheep only)
62	Walker & Norton (1971a, b)	DH × M	16–24	3–12	19	(F)
Merino Castrates						
63	Searle (1970a)	M	30–561	9–44	39	
64	Searle (1970b; 1972)	M	111–117	15–20	4	
65	Searle (1970b; 1972)	M	292–314	26–42	12	
66	Searle (1970b; 1972)	BL × M	107–118	15–21	5	
67	Searle (1970b; 1972)	BL × M	293–301	32–54	12	
68	Allden (1970)	DH × M	48–297	12–34	14	
69	Allden (1970)	M	55–291	12–26	14	
70	Langlands & Sutherland (1969)	M	30–3300	9–34	24	Ca, P, K, Na
71	Kellaway (1973)	M	4–10	4–10	12	⎫ Ca, P, Mg, K, Na; includes a
72	Kellaway (1973)	M	84–539	12–43	28	⎭ few females
73	Freer et al (1972)	BL × M	56–116	9–23	40	
74	Freer et al (1972)	BL × M	73–175	6–19	27	
75	Freer et al (1972)	BL × M	115–188	9–18	7	
76	Freer et al (1972)	BL × M	76–233	11–31	29	
77	Freer et al (1972)	BL × M	266–385	14–32	27	
78	Freer et al (1972)	BL × M	15–30	15–30	36	
79	Kielanowski & Lassota (1960)	M	5–32	5–32	22	
80	Weniger et al (1955)		270–630	21–40	4	(F); Ca and P only
Merino Females						
81	Searle (1970b; 1972)	M	217–240	10–21	18	Ca, P, K, Na
82	Langlands & Sutherland (1969)	M	30–3300	10–34	36	Ca, P, Mg, K, Na
83	Kellaway (1973)	M	81–709	13–38	28	
84	Freer et al (1972)	BL × M	90–200	13–18	14	

† breeds abbreviated as follows:

Suffolk	Su	Dorset Horn	DH	Targhee	T
Cheviot	Ch	Border Leicester	BL	Finnish Landrace	FL
Scottish Blackface	BF	NZ Romney Marsh	RM	Merino	M
Ile de France	IF	Southdown	SD	Corriedale	CD
Shropshire	Sh				

‡ (F) indicates fleece included; symbols for elements indicate mineral analyses

Appendix Table 1.3.   Regressions for sheep of $\log_{10}$ protein or fat content (kg) on $\log_{10}$ empty-body weight (kg)

Source No.	Protein equation Regression coefficient	Intercept	RSD	Fat equation Regression coefficient	Intercept	RSD
Non-Merino Males						
16	0.8686	−0.5623	0.027	2.087	−2.409	0.086
17	0.9450	−0.7248	0.032	2.199	−2.497	0.102
18	0.8752	−0.6037	0.025	2.046	−2.389	0.115
19	1.023	−0.8289	0.025	2.222	−2.712	0.075
20	1.000	−0.7618	0.019	2.024	−2.368	0.059
21	1.020	−0.7592	0.020	2.012	−2.378	0.058
22	0.8573	−0.5686	0.026	1.999	−2.237	0.062
23	1.069	−0.8149	0.050	2.124	−2.384	0.092
24	0.8378	−0.5921	0.026	2.063	−2.247	0.076
25	0.9730	−0.7216	0.013	2.268	−2.595	0.099
Non-Merino Castrates						
26	0.9223	−0.6221	0.016	2.221	−2.626	0.032
27	0.7942	−0.4373	0.017	2.000	−2.283	−0.062
28	0.8840	−0.5736	0.023	1.906	−2.009	0.064
29	0.7828	−0.5103	0.024	2.032	−2.157	0.103
30	0.9151	−0.7024	0.012	2.902	−2.808	0.050
31	0.8666	−0.6154	0.025	2.371	−2.697	0.073
32	1.057	−0.8318	0.015	1.560	−1.695	0.110
33	—	—	—	2.258	−2.772	0.179
34	0.9287	−0.6527	0.026	2.849	−2.944	0.095
35	0.9810	−0.7196	0.018	1.629	−1.640	0.080
36	0.7238	−0.3526	0.020	2.018	−2.388	0.119
37	0.9480	−0.6574	0.015	1.973	−2.159	0.087
38	0.9165	−0.6455	0.011	1.902	−1.896	0.039
39	1.062	−0.7868	0.021	1.696	−1.684	0.098
40	0.9941	−0.6844	0.017	1.588	−1.636	0.075
41	0.8035	−0.4891	0.014	2.078	−2.323	0.097
42	0.7937	−0.5190	0.020	2.767	−3.209	0.070
43	0.9305	−0.7294	0.016	0.7943	−0.2846	0.058
Non-Merino Females						
44	0.4345	−0.9298	0.031	3.258	−4.136	0.071
45	0.8539	−0.5961	0.015	1.963	−2.079	0.076
46	0.8939	−0.6402	0.022	1.900	−2.078	0.044
47	0.8243	−0.5187	0.031	1.892	−2.109	0.082
48	1.123	−0.8464	0.074	1.821	−1.848	0.066
49	0.8192	−0.5835	0.020	1.692	−1.541	0.051
50	0.8289	−0.5658	0.028	2.242	−2.467	0.076
51	0.9258	−0.6906	0.020	2.470	−2.685	0.141
52	0.6749	−0.4046	0.026	2.086	−2.139	0.048
53	0.8430	−0.5771	0.020	2.565	−2.830	0.098
54	0.8470	−0.6086	0.013	1.670	−1.582	0.043
55	0.8178	−0.5494	0.014	1.630	−1.560	0.063
56	0.9157	−0.7148	0.023	1.231	−0.8869	0.058
57	0.7543	−0.4393	0.032	2.024	−2.316	0.068
Merino Males						
58	0.7196	−0.5579	0.029	3.292	−2.773	0.178
59	0.9337	−0.7062	0.030	2.117	−1.797	0.103
60	0.9502	−0.6948	0.024	2.330	−2.297	0.109
61	0.9589	−0.6915	0.020	2.465	−2.547	0.128
62	0.8989	−0.6070	0.019	2.674	−2.909	0.076
Merino Castrates						
63	0.8641	−0.6071	0.030	1.909	−1.978	0.086
64	0.7196	−0.4173	0.013	2.086	−2.176	0.031
65	0.6746	−0.3421	0.028	1.899	−1.857	0.049
66	1.073	−0.8655	0.008	0.7853	−0.4776	0.018
67	0.5452	−0.1384	0.019	1.871	−1.860	0.031
68	0.8713	−0.5680	0.024	2.610	−3.081	0.086
69	0.8713	−0.5549	0.028	3.298	−3.939	0.101
70	0.9549	−0.6911	0.034	1.651	−1.743	0.176
71	0.9686	−0.7412	0.036	2.080	−2.099	0.091
72	0.9265	−0.6955	0.024	2.183	−2.461	0.083
73	0.8956	−0.6467	0.020	2.452	−2.659	0.073
74	0.9276	−0.6977	0.021	2.050	−2.121	0.073
75	0.8983	−0.6680	0.011	3.073	−3.416	0.094
76	0.8347	−0.5800	0.019	2.455	−2.694	0.050
77	1.025	−0.7963	0.015	1.350	−1.384	0.036
78	0.9073	−0.6736	0.024	2.055	−2.213	0.058
79	0.9604	−0.7096	0.012	1.866	−2.125	0.116

Appendix Table 1.3
(continued)

Source No.	Protein equation			Fat equation		
	Regression coefficient	Intercept	RSD	Regression coefficient	Intercept	RSD
Merino Females						
81	0.9399	−0.7185	0.030	1.800	−1.683	0.076
82	0.8882	−0.6228	0.040	1.469	−1.476	0.130
83	0.8894	−0.6601	0.030	2.163	−2.367	0.082
84	0.9013	−0.6663	0.014	2.503	−2.698	0.075

N.B. It may be easier to appreciate the degree of scatter of data from individual animals if the RSDs of the logarithmic equations are converted into approximate residual coefficients of variation (CV) of the weights themselves. For values of RSD up to about 0.1 this may be achieved by multiplying by 230. For example, an RSD of 0.027 is equivalent to a residual CV of approximately ±6.2%.

Appendix Table 1.4.

Comparisons for sheep between the closeness of fit of regression equations estimating protein and fat from one independent variable (empty-body weight) or from two (empty-body weight and age)

	Values of residual standard deviation			
	Estimates of $\log_{10}$ protein weight		Estimates of $\log_{10}$ fat weight	
Source	Regression on weight	Regression on weight and age	Regression on weight	Regression on weight and age
Males and castrates				
68	0.024	0.022	0.086	0.075
69	0.028	0.024	0.101	0.054
70	0.050	0.048	0.169	0.133
72	0.024	0.024	0.094	0.094
31	0.025	0.018	0.073	0.060
34-37	0.022	0.021	0.114	0.114
39-40	0.019	0.018	0.088	0.089
24	0.026	0.024	0.093	0.082
All above sources†	0.045	0.044	0.125	0.122
Females				
82	0.040	0.040	0.130	0.132
83	0.030	0.029	0.082	0.083
50	0.028	0.028	0.076	0.078
52	0.027	0.026	0.085	0.069
All above sources†	0.042	0.030	0.118	0.108

* including some additional (aged) animals omitted from the equations given in Appendix Table 1.3.
† overall regressions, based on 300 males and castrates or 124 females.

Appendix Table 1.5. Data sources for prenatal growth in cattle

Source no.	Reference	Breed	No. of calves	Material analysed	Components determined†
101	Ellenberger et al (1950a)	Mainly Holstein or Holstein cross	24	Foetuses 135–267 days old	Fat, Ca, P
102	Jakobsen (1957) and Jakobsen et al (1957)	Red Danish	21	Foetuses, membranes, fluids and uterus, 70–280 days from conception	Protein, fat, energy
103	Haigh et al (1920) and Hogan & Nierman (1927)	Jersey and Hereford	18	Calves killed at birth	Protein, fat, Ca, P Mg, K, Na, Cl
104	Ellenberger et al (1950a)	Holstein	7	Calves killed at birth	Protein, fat, Ca, P
105	Greenhalgh & Kay (1968, unpublished)	British Friesian	3	Calves killed at 3 days old	Protein, fat
106	Hensler & Jentsch (1973)	German Red Pied	16	Calves, membranes and fluids collected at birth	Protein, fat, energy

† in addition to water and ash

Appendix Table 1.6. Sources of data on the body composition of growing cattle

Source No.	Reference	Breed	Sex	Age (days)	Empty-body weight (kg)	No. of animals
107, 108	Haecker (1920)	Aberdeen-Angus, Beef Shorthorn	C	21–1040	40–610	48
109	Haecker (1920)	Aberdeen-Angus, Beef Shorthorn	C	520–760	210–470	5
110–112	Moulton et al (1922) & Hogan & Nierman (1927)	Beef Shorthorn, Hereford	C	90–1450	70–490	26†
113	Greenhalgh & Kay (Unpub.)	Friesian	M	3–192	39–95	24
114, 115	Ellenberger et al (1950a)	Holstein-Friesian	F	0–400	44–240	59†
116	Holzschuh (1966)	German Black Pied	M	180/300	140–270	12
117–120	Osinska & Ziolecka (1972 and unpub.)	Polish Black & White Lowland	M	10–500	25–390	100
121	Garrett (Unpub.)	Hereford	M	90–365	54–260	6
122	Garrett (Unpub.)	Hereford	F	90–365	57–216	6
123	Garrett & Hinman (1969)	Hereford	C	365–486	230–460	57
124	Garrett et al (1971)	Aberdeen-Angus Crosses	C	430	280–440	11
125	Garrett et al (1971)	Aberdeen-Angus Crosses	F	430	240–360	15
126	Garrett et al (1971)	Charollais Crosses	C	430	350–500	12
127	Garrett et al (1971)	Charollais Crosses	F	430	340–440	12
128	Garrett (Unpub.)	?	F	?	215–370	10
129–132	Lonsdale & Tayler (1971)	Friesian	C	175	55–125	30
133–136	Kay (Unpub.)	Friesian	C	240–1090	255–580	60
137–140	Kay (Unpub.)	Hereford × Friesian	C	250–1080	230–570	63
141–142	Schulz et al (1974)	German Black Pied	M	147–630	120–520	44†
143–146	Ayala (1974)	Aberdeen Angus & Holstein	M	252–595	180–420	21
147–150	Ayala (1974)	Aberdeen Angus & Holstein	C	252–602	190–390	15
151–154	Ayala (1974)	Aberdeen Angus & Holstein	F	252–602	180–360	16

† these sources include data on mineral components

Appendix Table 1.7.   Regressions for cattle of $\log_{10}$ protein or fat content (kg) on $\log_{10}$ empty-body weight, (kg)

Source	Description	Protein equation			Fat equation		
		Regression coefficient	Intercept	RSD*	Regression coefficient	Intercept	RSD*
107	Light	0.9735	−0.6627	0.016	1.750	−2.591	0.066
108	Heavy	0.7070	−0.0023		2.136	−3.547	
109	High plane	0.8284	−0.3033		1.993	−3.192	
110	Medium plane	0.8913	−0.4980	0.016	1.828	−2.711	0.096
111	Low plane	0.9718	−0.6589		1.542	−2.139	
112		1.012	−0.7400	0.011	1.381	−1.817	0.107
113	Light	0.9427	−0.6039	0.017	1.748	−2.590	0.095
114	Heavy	1.018	−0.7525		2.289	−3.658	
115					1.556	−2.192	
116	Initial	1.005	−0.7592	0.019	1.843	−2.923	0.043
117	Veal	1.007	−0.7335	0.010	1.135	−1.552	0.081
118					2.196	−3.399	
119	Early weaned, heavy				1.589	−2.341	
120	Early weaned, light				2.047	−3.231	
121		1.050	−0.8441	0.032	1.606	−2.259	0.054
122		0.8796	−0.4788	0.020	1.903	−2.770	0.088
123		0.7139	−0.0581	0.021	2.015	−3.158	0.059
124		0.7744	−0.2574	0.017	1.554	−1.908	0.069
125					1.554	−1.875	0.034
126		1.030	−0.8763	0.022	0.9884	−0.5444	0.025
127		0.8181	−0.3239	0.018	1.330	−1.402	0.037
128		0.9967	−0.7200	0.016	—	—	—
129	Spring grass, chopped	1.010	−0.7241	0.008	2.209	−3.658	0.077
130	Spring grass, milled	1.061	−0.8163		2.144	−3.504	
131	Autumn grass, chopped	1.072	−0.8373		2.096	−3.454	
132	Autumn grass, milled	1.091	−0.8713		1.909	−3.113	
133	Fast/pelleted grass†	0.6544	0.1329	0.020	1.943	−3.122	0.040
134	Fast/concentrates	0.6827	0.0451	0.024	1.819	−2.738	0.044
135	Slow/pelleted grass	0.7725	−0.1762	0.014	2.193	−3.786	0.031
136	Slow/concentrates	0.5467	−0.4234	0.017	2.159	−3.659	0.030
137	Fast/pelleted grass	0.7101	−0.0029	0.019	1.645	−2.289	0.032
138	Fast/concentrates	0.6228	0.1916	0.014	1.585	−2.075	0.047
139	Slow/pelleted grass	0.4508	0.6805	0.022	2.466	−4.502	0.052
140	Slow/concentrates	0.5094	0.5159	0.022	2.065	−3.403	0.047
142	Above 200 kg	0.8797	−0.4262	0.018	2.238	−4.018	0.080
143	Bulls/Aberdeen Angus	0.9626	−0.6707	0.018	1.675	−2.393	0.052
144	Bulls/Aberdeen Angus	0.9626	−0.6286	0.018	1.675	−2.518	0.052
145	Bulls/Holstein	0.9626	−0.6375	0.018	1.675	−2.530	0.052
146	Bulls/Holstein	0.9626	−0.6234	0.018	1.675	−2.607	0.052
147	Bullocks/Aberdeen Angus	0.9111	−0.5725	0.025	2.127	−3.436	0.066
148	Bullocks/Aberdeen Angus	0.9111	−0.5391	0.025	2.127	−3.474	0.066
149	Bullocks/Holstein	0.9111	−0.5075	0.025	2.127	−3.557	0.066
150	Bullocks/Holstein	0.9111	−0.5096	0.025	2.127	−3.656	0.066
151	Heifers/Aberdeen Angus	0.7862	−0.2953	0.024	2.070	−3.154	0.022
152	Heifers/Aberdeen Angus	0.7862	−0.2671	0.024	2.070	−3.184	0.022
153	Heifers/Holstein	0.7862	−0.2477	0.024	2.070	−3.323	0.022
154	Heifers/Holstein	0.7862	−0.2329	0.024	2.070	−3.378	0.022

* An approximate estimate of the residual coefficient of variation (CV) of the weights of protein or fat may be obtained by multiplying the RSD of the logarithmic requation by 230. For example, an RSD of ±0.016 corresponds to a residual CV of approximately 3.7%. This may provide a better appreciation of the variability of the data from individual animals at any fixed value of empty-body weight.

† Sources 133–136 were Friesians and 137–140, Hereford × Friesian. Animals of each breed were grown rapidly or slowly to 300 kg live weight. Some were killed at this weight and the remainder were fattened on concentrates or pelleted grass to empty-body weights up to 580 kg.

Appendix Table 1.8.   Estimated rates of accretion (g/d) of components of the sheep foetus and gravid uterus, at various ages. (All relative to production of a lamb weighing 4 kg at birth.)

Component	Age (weeks before parturition)*			
	12	8	4	0
Foetus				
Total weight	4.2	26.0	65.9	96.2
Protein	0.33	3.24	11.08	19.82
Fat	0.03	0.57	2.09	3.29
Ash	0.11	0.82	2.59	4.68
Energy (kJ/day)	10.0	98.3	334	589
Calcium	—	0.28	0.91	1.31
Phosphorus	—	0.20	0.50	0.57
Magnesium	—	0.013	0.035	0.043
Potassium	—	0.088	0.109	0.064
Sodium	—	0.070	0.156	0.210
Gravid uterus				
Total weight	20.4	42.0	75.4	120.1
Protein	1.80	5.00	11.56	22.85
Fat	0.19	0.71	1.95	4.15
Ash	0.33	1.01	2.54	5.43
Energy (kJ/day)	49.3	145	347	699
Calcium	0.05	0.30	0.85	1.45
Phosphorus	0.06	0.23	0.45	0.57
Magnesium	0.002	0.013	0.036	0.058
Potassium	0.066	0.121	0.116	0.081
Sodium	0.137	0.135	0.084	0.042

* (At 147 days).

Appendix Table 1.9.   Estimated rates of accretion (g/day) of components of the cattle foetus and gravid uterus, at various ages. (All relative to production of a calf weighing 40 kg at birth)

Component	Age (weeks before parturition)*					
	20	16	12	8	4	0
Foetus						
Total weight	58.6	116	201	310	433	559
Protein	6.5	15.2	30.8	55.4	89.9	133
Fat	0.6	2.2	5.8	12.1	21.1	31.9
Ash	1.5	3.6	7.2	12.9	20.9	30.9
Energy (MJ/day)	0.169	0.387	0.819	1.61	2.94	5.07
Calcium	0.33	0.89	2.05	4.09	7.22	11.4
Phosphorus	0.22	0.55	1.21	2.34	4.05	6.41
Magnesium	0.013	0.04	0.08	0.16	0.28	0.45
Potassium	0.10	0.21	0.38	0.64	0.97	1.37
Sodium	0.14	0.29	0.50	0.76	1.04	1.29
Chlorine	0.10	0.20	0.34	0.53	0.73	0.94
Gravid uterus						
Total weight	195	278	377	488	607	728
Protein	14.0	24.6	41.5	67.2	104	157
Fat	1.8	3.3	6.2	11.5	21.4	39.9
Ash	2.7	5.0	8.7	14.4	22.5	33.7
Energy (MJ/day)	0.356	0.621	1.08	1.89	3.28	5.70
Calcium	0.75	1.39	2.43	4.02	6.35	9.59
Phosphorus	0.46	0.85	1.49	2.45	3.81	5.66
Magnesium	0.03	0.06	0.10	0.17	0.27	0.40
Potassium	0.16	0.29	0.51	0.65	1.34	2.02
Sodium	0.20	0.36	0.64	1.07	1.72	2.67
Chlorine	0.17	0.31	0.55	0.90	1.40	2.08

* (At 281 days).

Chapter 2

Feed Intake

Introduction

Nutrient requirements may be stated either in absolute terms, as quantities per animal per unit time, or in relative terms, as nutrient concentrations in the feed. Requirements expressed in absolute terms are satisfied by providing the animal with a feed or ration of known composition at the appropriate rate (i.e., quantity of feed per unit time), and it is necessary to know whether the animal is capable of consuming the feed at the rate specified. Requirements expressed as dietary concentrations are commonly calculated from absolute requirements and estimates of feed intake. Thus whichever method of expression is used, the feed consumption of the animal must be known. In the first edition, feed intake and the expression of requirements as dietary concentrations were not discussed in detail. In this edition they are given more prominence, partly because of the growing importance of unrestricted feeding.

Numerous studies have been made of the factors determining the feed intake of ruminants. These include such animal characteristics as species, weight and physiological state, such feed characteristics as energy value and method of conservation or processing, and such environmental factors as day length and temperature. The purpose of this chapter is not to review these experiments or to build their results into a complex model of appetite control. Instead, the actual measurements of intake are used to predict the feed consumption of ruminants in specific nutritional situations. The methods of prediction are somewhat simplified and approximate; they are not intended to make allowance for all factors known to influence intake. Furthermore, the results of the analyses are presented in a form which is compatible with other expressions of nutrient requirements.

Feed Intake of Growing Ruminants

1. The data and their analysis

Data on intake of feeds commonly used in UK agriculture were obtained mainly by scanning six selected journals. The experiments chosen were restricted to those in which the animals had genuinely been fed to appetite, by having continuous access to feed, and which yielded sufficient information to allow the calculation of dry matter intake, metabolic live weight and the metabolizability (q) of the energy of the diet. The latter was usually predicted from digestibility coefficients for dry matter or energy, by a factor which varied from one feed class to another but averaged 0.81. Digestibility was generally measured *in vivo* and at maximum intake, but measurements *in vitro* were occasionally used if, as in some grazing studies, they had been carefully calibrated against digestibility *in vivo*. Metabolizability values (q) may be converted to metabolizable energy concentrations (MJ/kg dry matter) by multiplying by 18.4.

The feeds or diets were first divided into two categories, coarse and fine, and were further subdivided between sheep and cattle. Coarse diets consisted of long or chopped roughages (including grazed herbage) with or without concentrates (pelleted dried grass was here classified as a concentrate). Fine diets consisted of concentrates and milled and/or pelleted or wafered roughages, alone or in combination. Some root crops were also included in fine diets since they promoted intakes similar to those for all-concentrate diets. Feeds which did not fall clearly into either the fine or coarse categories, such as roughages coarsely milled but not pelleted, were excluded from the analysis.

Each category was analysed separately. The dependent variable, daily dry matter intake, was scaled by live weight (i.e., kg DM/100 kg W) or metabolic live weight (g DM/kg $W^{0.75}$), since absolute intake had non-uniform variability and was therefore unsuited to simple regression analysis. The independent variables examined were live weight (kg), metabolizability of the dietary energy (q) and the proportion of concentrates in the diet (as a decimal). The significance of live weight as an independent variable was a reflection on the adequacy of the scaling factors for intake. If intake was scaled by live weight, live weight was a significant independent variable for all four animal/feed categories; if intake was scaled by metabolic weight, live weight remained a significant independent variable for sheep only. For this reason, and also because its coefficient of variation was lower, intake scaled by metabolic weight was preferred. Of the remaining independent variables, metabolizability was always significant, but the proportion of concentrate in the diet was significant only for cattle on coarse diets.

Table 2.1   Summary of data on feed intake of growing sheep and cattle, and of regression equations of intake on metabolizable energy concentration of the diet, proportion of concentrates and animal live weight.

Species:	Sheep			Cattle	
Diet type:	Coarse (excluding silages)	Coarse (silages)	Fine	Coarse	Fine
No. of diets	107	38	85	151	159
Mean values					
Dry matter intake (g/kg $W^{0.75}$ per day)	57.0	46.0	90.5	87.2	89.8
Metabolizable energy (MJ/kg diet DM)	9.4	10.0	9.7	10.0	10.7
(and as q)	(0.51)	(0.54)	(0.53)	(0.54)	(0.58)
Concentrate proportion	0.05	0.05	0.48	0.13	0.70
Animal live weight (kg)	60.2	51.0	46.0	283	219
Regression coefficients ($\pm$ SE)					
Metabolizability (q)	104.7 ($\pm$ 16.0)	—	−78.0 ($\pm$ 16.6)	106.5 ($\pm$ 19.3)	−46.6 ($\pm$ 13.4)
Concentrate proportion	—	—	—	37.0 ($\pm$ 4.5)	—
Animal live weight	0.307 ($\pm$ 0.069)	—	−0.408 ($\pm$ 0.059)	—	—
Constant term	−15.0	46.0	150.3	24.1	116.8
Precision					
Multiple correlation	0.67	—	0.62	0.63	0.27
Residual SD	11.1	8.3	10.5	11.8	13.2

The data in each major category and the regression equations derived from them are summarized in Table 2.1; individual data points are shown in Figs. 2.1 and 2.2. Intakes predicted from the equations are given in Tables 2.2 and 2.3.

Coarse diets

At an early stage of the analysis it became apparent that silages promoted unusually low intakes in sheep. For example, sheep of 60 kg given a diet with q = 0.54 were estimated to consume 42 g DM/kg $W^{0.75}$ if given silage alone but 60 g DM/kg $W^{0.75}$ if given other roughages. Silage diets for sheep were therefore removed from the main analysis. It should be noted that, even with silages excluded, the relative intake of coarse diets for sheep was still appreciably lower than for cattle.

Table 2.2 Dry matter intake of growing sheep, as predicted from equations given in Table 2.1.

Diet type (and units)	Metabolizability of dietary energy (q)	Live weight (kg)				
		20	30	40	50	60
Coarse* (g/kg $W^{0.75}$ per day)	0.4	33.0	36.1	39.2	42.2	45.3
	0.5	43.5	46.6	49.6	52.7	55.8
	0.6	54.0	57.0	60.1	63.2	66.2
	0.7	64.4	67.5	70.6	73.6	76.7
Fine (g/kg $W^{0.75}$ per day)	0.5	103.1	99.1	95.0	90.9	86.8
	0.6	95.3	91.3	87.2	83.1	79.0
	0.7	87.5	83.5	79.4	75.3	71.2
Coarse* (kg/day)	0.4	0.31	0.46	0.62	0.79	0.98
	0.5	0.41	0.60	0.79	0.99	1.20
	0.6	0.51	0.73	0.96	1.19	1.43
	0.7	0.61	0.87	1.12	1.38	1.65
Fine (kg/day)	0.5	0.98	1.27	1.51	1.71	1.87
	0.6	0.90	1.17	1.39	1.56	1.70
	0.7	0.83	1.07	1.26	1.42	1.54
Silages (kg/day)	—	0.44	0.59	0.73	0.86	0.99

* excluding silages, for which intake is assumed to be 46 g/kg $W^{0.75}$ per day.

Table 2.3 Dry matter intake of growing cattle, as predicted from equations given in Table 2.1.

Diet type (and units)	Metabolizability of dietary energy (q)	Live weight (kg)					
		100	200	300	400	500	600
Coarse* (g/kg $W^{0.75}$ per day)	0.4	66.70					
	0.5	77.35					
	0.6	88.00					
	0.7	98.65					
Fine (g/kg $W^{0.75}$ per day)	0.5	93.50					
	0.6	88.84					
	0.7	84.18					
Coarse† (kg/day)	0.4	2.1	3.5	4.8	6.0	7.1	8.1
	0.5	2.4	4.1	5.6	6.9	8.2	9.4
	0.6	2.8	4.7	6.3	7.9	9.3	10.7
	0.7	3.1	5.2	7.1	8.8	10.4	12.0
Fine (kg/day)	0.5	3.0	5.0	6.7	8.4	9.9	11.3
	0.6	2.8	4.7	6.4	7.9	9.4	10.8
	0.7	2.7	4.5	6.1	7.5	8.9	10.2

* increase each value by 3.7 g/kg $W^{0.75}$ for each 10% of concentrates (i.e., proportion of 0.1) in the diet.
† for each 10% of concentrates in the diet increase intake as follows:

Live weight (kg)	100	200	300	400	500	600
Increase (kg/day)	0.12	0.20	0.27	0.33	0.39	0.45

Fig. 2.1   Dry matter intake of cattle in relation to metabolizable energy-concentration of the diet.
(a) Fine diets, including ground roughages and concentrates,
(b) Coarse diets, including long roughages, with or without supplementary concentrates (O, roughages other than silages; □, silages; ●, concentrates; ◐ ◪, mixed diets, showing approximate proportion of concentrates. Continuous lines join roughages from a single experiment; broken lines join a single roughage supplemented with increasing amounts of concentrates).

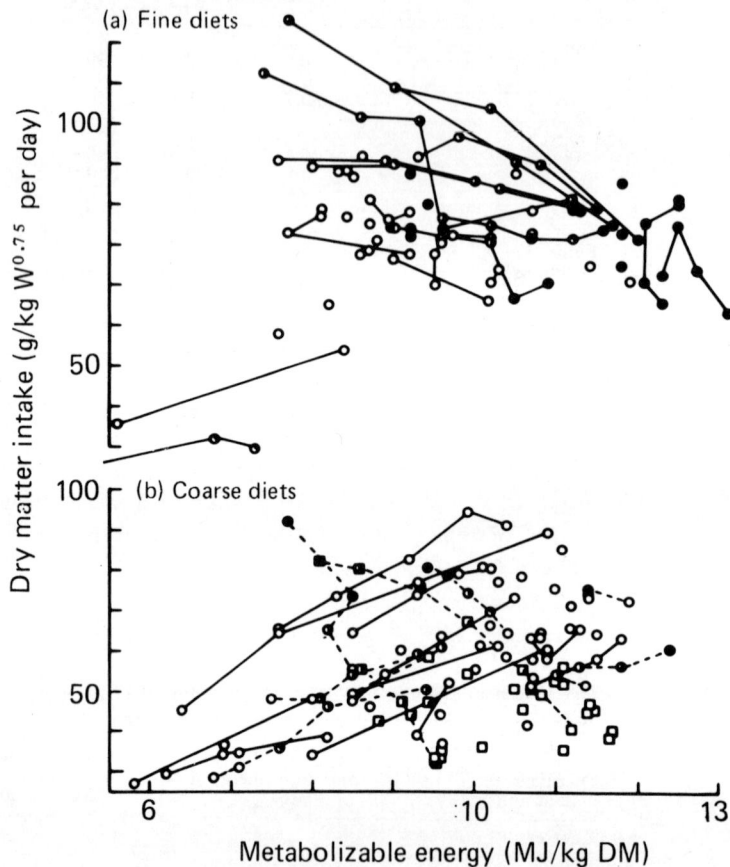

For both species, intake was positively related to metabolizability. For cattle there was an even closer relationship between intake and the proportion of concentrates in the diet (partial correlation coefficient, 0.56). This indicated that of diets equal in metabolizability, those containing a higher proportion of concentrate would be eaten in greater quantity.

The range in live weight of sheep given coarse diets was 18–87 kg, but there was a preponderance of animals weighing 60–70 kg (i.e., the mature castrates commonly used for experiments). The significant positive effect of live weight on intake per unit of metabolic weight is probably due to the difficulties experienced by young lambs in dealing with long roughages. For sheep given fine diets (see below) the effect of live weight was negative.

For the special category of silage diets for sheep, none of the regression coefficients was significant at $P < 0.05$. It is therefore suggested that the intake of silages by sheep should be estimated as the mean for the data surveyed, i.e., 46 g DM/kg $W^{0.75}$ per day.

Fig. 2.2   Dry matter intake of cattle in relation to metabolizable energy concentration of the diet.
(a) Fine diets, including ground roughages and concentrates, and also root crops.
(b) Coarse diets, including long roughages, with or without supplementary concentrates (O, roughage other than silages; □, silages; ●, concentrates; ◐ ◧, mixed diets, showing approximate proportion of concentrates; △, root crops,. Continuous lines join roughages from a single experiment; broken lines join a single roughage supplemented with increasing amounts of concentrate.

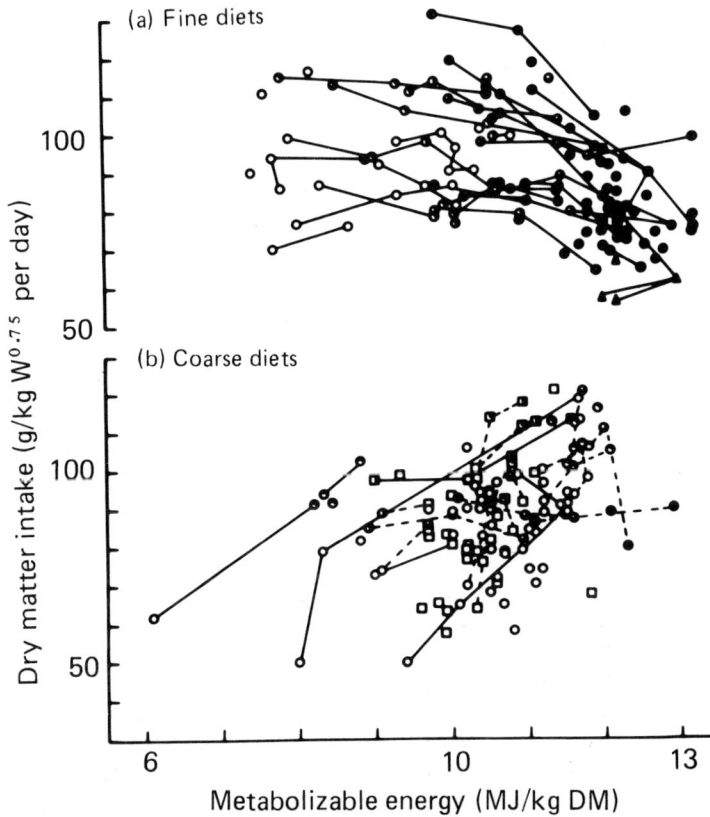

Fine diets

Data for these diets were almost as numerous as those for coarse; this may be a true reflection of recent research on the two categories, but not of their relative practical importance. In contrast to coarse diets, fine diets promoted similar intakes per unit of metabolic weight in sheep and cattle. For both species, intake was negatively related to metabolizability, but the relationships were not improved by employing the proportion of concentrates as an additional variable. These two independent variables were, in fact, closely related since all-concentrate diets provided the upper part of the range in metabolizability and pelleted roughages the lower part.

Animals given fine diets weighed less, on average, than those given coarse (Table 2.1), the disparity presumably being due to the use of young animals for intensive beef or lamb production from fine diets. For sheep there was a negative effect of live weight on intake per unit of metabolic weight.

2. Conclusions

In the first edition, estimates of intake were based on an equation derived with sheep fed on long roughages (Blaxter 1964):

$$I_T = 910\,q - 740\,q^2 - 194 \qquad (RSD \pm 10.0), \qquad [2.1]$$

where I_T = dry matter intake, g/kg $W^{0.73}$ per day.

For 50-kg sheep given diets of metabolizability 0.4 and 0.7 this predicts intakes of 0.90 and 1.40 kg DM/day, whereas the corresponding values from Table 2.2 are 0.79 and 1.38 kg/day. For cattle, equation [2.1] was increased by 15%, but even with this addition, values predicted by it are generally lower than those in Table 2.3 for cattle given coarse diets, especially diets of low metabolizability.

The estimates of Tables 2.2 and 2.3 are based on more numerous data, which were derived with cattle as well as sheep and include feeds other than roughages. They should therefore have a broad general application, but may be inaccurate in some circumstances. The residual standard deviations in Table 2.1 indicate that in general intake is predicted by the equations with a coefficient of variation of 15%. To a large degree, intake is determined by additional factors, some of which are known but could not be included in the equations without complicating them considerably. There appears also to be an element of between-experiment variability which is difficult to quantify.

In Britain, grass silages are commonly regarded as the feeds presenting most problems of intake, and silages are generally considered to promote lower intakes of dry matter than fresh or dried roughages of the same digestibility (or metabolizability). From the data reviewed here, this view appears to be justified for sheep but not for cattle. The contrast between the species in their intake of silages is sharpened by the fact that for coarse diets in general, intake per kg $W^{0.75}$ is lower for sheep than for cattle, and that for sheep — but not for cattle — intake of silages is lower even than that of the other coarse feeds. It is interesting to note that Heaney et al (1968), who made a survey of Canadian feeds, found the mean intake of sheep given fresh or dried roughages ($n=217$) to be 62 g DM/kg $W^{0.75}$ per day, whereas sheep given silages ($n=134$) ate only 34 g DM/kg $W^{0.75}$ per day; energy values were not reported. In France, Demarquilly (1973) found the daily intake of 80 grass silages by sheep to be 47 g DM/kg $W^{0.75}$ (compare our mean value of 46 g), which was 30% lower than the mean intake of the fresh herbages from which they had been prepared.

The characteristics of silages which determine their intake have been reviewed by Wilkins (1974a) and Tayler & Wilkins (1976). A strong fermentation, which gives rise to relatively high concentrations of volatile fatty acids and ammonia, leads to a low intake. Conversely, silage-making practices which limit fermentation (wilting the herbage, chopping it finely and adding acids) all lead to increases in intake. For sheep, but probably less for cattle, fine chopping has a direct beneficial effect on intake which may be demonstrated by chopping after fermentation has taken place (Dulphy & Michalet 1975). Finally, it should be borne in mind that legume and maize silages give higher intakes than grass silages having the same metabolizability.

Concentrate supplements generally reduce the intake of a roughage but increase total intake; this effect may be expressed as a "replacement rate", the change in roughage DM intake per unit change in supplement DM intake. The equations in Table 2.1 provide estimates of such replacement rates, and some examples are given in Table 2.4. The equations can be used quite simply to calculate, for a specific change in metabolizability, the quantity of concentrate required and the effect of this on total, and hence roughage, intake, but the effect on intake of a specific weight of concentrate can be calculated only by iteration. The examples show that for both sheep and cattle a given weight of concentrates reduces the intake of high-quality roughage ($q=0.6$) more than that of low-quality roughage ($q=0.4$), which agrees with many experimental observations. The examples also suggest that when concentrates form about 15% of the intake of the two species, they reduce roughage intake less in cattle; this is a consequence of including the concentrate proportion of the diet as an additional variable in the equation for cattle.

Table 2.4   The effects on dry matter intake of sheep and cattle of adding concentrates to coarse roughages of low and high metabolizability (based on the equations of Table 2.1).

	Sheep (50 kg)		Cattle (300 kg)	
Metabolizability (q)				
Roughage	0.40	0.60	0.40	0.60
Roughage + concentrate	0.45	0.61	0.45	0.61
Dry matter intake (kg/day)				
(a) Concentrate	0.15	0.15	1.00	1.00
(b) Roughage given with concentrate	0.74	1.06	4.67	5.82
(c) Roughage given alone	0.79	1.19	4.81	6.34
Replacement rate				
$(b - c)/a$	−0.33	—0.87	−0.14	−0.52

One reason why cereal concentrates reduce intake of roughages is that the rapid fermentation of the starch they contain lowers rumen pH and reduces or retards the digestion of cellulose. Processing of cereal grains can influence the rate at which their starch is digested, and hence modify their effects on intake of roughages (Ørskov & Fraser 1975). The replacement of starch in concentrates by cellulose, for example as pelleted dried grass, may also reduce the extent to which supplements depress roughage intake (Tayler & Wilkins 1976). A final point to note is that when added to low-protein roughages, concentrates commonly increase, rather than reduce, intake by virtue of their higher protein content.

General relationships between intake and feed characteristics may be influenced by features of the animal and its environment. In Chapter 1, distinctions were made between sexes and between breed types when estimating the composition of empty-body gain. With regard to intake, however, sex appears to have a relatively small influence. From data reviewed by Preston & Willis (1974) and Kay & Houseman (1975) the feed intake of bulls is calculated to be only about 2% greater than that of castrate males. Differences in intake between males and females also appear to be small and inconsistent. In three experiments made by Hedrick et al (1969), castrates had, on average, the same intake per unit of metabolic weight as heifers; bulls had intakes 5% higher. Bogart & England (1971) found bulls to eat the same amount as heifers, while Schwark & Kunert (1971) found them to eat 4% less. Ayala (1974) found no significant difference between bulls, bullocks and heifers.

With regard to breed type, Garrett (1971) reported that Holsteins ate 5% more than Herefords of the same weight (but fatter). Bogart & England (1971), however, found no difference between Hereford and Aberdeen Angus cattle. Ayala (1974) reviewed 6 experiments that showed Holstein-Friesian to eat 5–10% more than Hereford or Aberdeen Angus bullocks, but no more than Shorthorns. In his own trial, Holsteins did not eat significantly more than Aberdeen Angus cattle of the same weight. The UK Meat and Livestock Commission are currently making large-scale performance tests with cattle sired by bulls ranging in size from Aberdeen Angus to Charollais. Preliminary results (D. M. Allen, personal communication) suggest that as cattle approach their mature weight intake per unit of metabolic weight begins to decline and hence that at greater weight (> 350kg) there may be differences between breeds. When these differences are more thoroughly quantified, it may be necessary to take them into account in predictions of intake.

Feed Intake of Lactating Ruminants

1. Lactating cows

For lactating cattle, as for non-lactating, feed intake is partly determined by live weight and by diet type and metabolizability. For reasons discussed later, however, the effects of these factors are often small in relation to the special effects of lactation on intake. Lactating cows eat 35–50% more than non-lactating animals of the same weight and on the same diet. In the individual cow, intake and milk yield are not closely related; intake is low after calving but rises for the first 3–5 months of lactation, even though milk yield falls after 1–2 months. Between cows, a positive relationship exists between yield and intake; in the short term yield may determine intake, but in the longer term the roles are likely to be reversed.

Intake in lactating cows is therefore determined by the four factors; cow size and milk yield, stage of lactation and the energy content of the diet. As these factors are to some extent interdependent, multiple regression analyses may give misleading results. The approach used has therefore been to estimate first the average intake of cows from several large sources of data listed in Table 2.5; with the exception of the first, each source includes a large number of cows studied in several experiments at the same centre. Secondly, analyses applied to each source, and also the results of other experiments, have been used to estimate the separate effects on intake of the four factors listed above.

(a) Stage of lactation

The pattern of intake over the lactation is best determined from experiments with cows given a uniform diet throughout. Data from the seven experiments referred to collectively as source 1 in Table 2.5 were used to produce Table 2.6. The 14 diets included grazed pasture herbage, hay and mixtures of 30–75% silage and concentrates, or ground roughage and concentrates, but all gave a similar pattern of intake. Within experiments, there is some evidence that the increase in intake that occurs in early lactation is greater for high-yielding cows than for low yielders, and the apparent plateau of intake in months 3–6 is probably due in part to individuals reaching maximum intake at different stages of lactation. The increase is also greater for diets of higher metabolizability (Journet & Rémond 1976). Monteiro (1972) has proposed a relationship between intake and live weight, milk yield and liveweight change that includes time delay factors for the last two variables; however, it appears not to have been tested on a large set of data.

During the dry period, intake may fall as parturition approaches (Forbes 1970). From week 4 to week 1 prepartum, the daily organic matter intake of the cows of Curran et al (1970) fell by 0.6 kg (i.e., 0.2 kg/week), and similar declines were found by Johnson et al (1966) and Journet & Rémond (1976). Coppock et al (1972) found the decline to be related to the proportion of concentrates in the diet; as this rose from 25 to 70%, the decline increased from about 0.15 to 1.0 kg DM/week.

(b) Live weight

Individual cows can range in weight from perhaps 250 kg for small Jerseys to 750 kg for large Friesians. Milk yield varies with live weight and the effect of weight itself on intake needs to be considered separately. Yadava et al (1973) found intake in Jerseys (mean weight 380 kg) and Holstein-Friesians (570 kg) to be proportional to W rather than $W^{0.75}$, but differences in milk yield were not taken into account. Several

Table 2.5 Summary of data on feed intake of lactating cows

Source	Diets			Cows				
	Description	ME (MJ/kg DM)	(q)	No.	Month of lactation	Live weight (kg)	Milk (or FCM) (kg/day)	Dry matter intake (g/kg $W^{0.75}$ per day)
(1) Various whole-lactation trials†	Complete diets	10.4	(0.57)	158	1–10	515	(14.7)	139
(2) Conrad et al (1964)	Forages, alone or with concentrates	9.8	(0.53)	134	?	412	(15.4)	126
(3) Journet et al (1965)	Forages and concentrates	11.0	(0.60)	242	2	570	(16.8)	137
(4) Curran et al (1970)	Forage and concentrates	10.0	(0.54)	36–65	2–4	512	16.3	129
(5) Curran & Holmes (1970)	Grazed pasture and small amounts of concentrates	11.0	(0.60)	72	4	493	16.5	141
(6) Greenhalgh & Runcie (1962); Greenhalgh et al (1966, 1967); Greenhalgh & Reid (1969a, b) Reid et al (1972)	Grazed pasture	11.0	(0.60)	137	4–6	496	15.3	123
(7) Castle & Watson (1969–1970, 1973, 1974)	Silage and concentrates	10.6	(0.58)	72	3–6	485	16.5	132
(8) Greenhalgh & Reid (1975 and unpubl.); Amir et al (1968)	Complete diets (7–80 % forage)	10.9	(0.59)	66	3–4	519	17.2	131
(9) Bines et al (unpubl.)‡	Hay and concentrates (10–40 % hay)	10.4	(0.57)	37	1–4	576	23.5	133

†References: Graves et al (1938), Hutton (1963), Mohrenweiser & Donker (1968), Owen et al (1968), Cowan et al (1970), Coppock et al (1974) and J. N. Aitken, M. Kay & T. R. Preston (unpublished data).

‡ values are means for heifers and cows, weighted 1 : 3, to represent typical herd structure.

of the sources listed in Table 2.4 provided partial regression coefficients for intake on live weight. Journet et al (1965) found daily intake to increase by 0.74 kg dry matter for each 100-kg increase in live weight; their data appear to be generally consistent with the hypothesis that intake is proportional to $W^{0.75}$. Conrad et al (1964) reported that intake varied with $W^{0.37}$, but when their data were subdivided for class of feed, the exponent varied from 0.51 to 0.99. Curran et al (1970) found intake to be unrelated to live weight, but their cows differed relatively little in weight.

Both Yadava et al (1973) and Bines et al (1969) reported that thin cows ate more than fat ones.

In the absence of precise evidence to the contrary, it has been assumed that intake in dairy cows varies with metabolic live weight ($W^{0.75}$).

(c) Milk yield

In the long term, coefficients relating intake to milk yield are likely to be close to the theoretical energy requirement for milk production, which for a concentrate having a metabolizability of 0.7 would be 0.4 kg DM/kg fat-corrected milk. An analysis of the mean values for the whole-lactation trials grouped under source 1 of Table 2.5 gives a coefficient of 0.5 kg DM/kg FCM. McCullough (1974) refers to an analysis of 39 American experiments made with complete diets, for which the partial regression coefficient was 0.36 kg DM/kg milk.

However, when intake is to be predicted from yield, rather than *vice versa*, it seems more appropriate to use relationships derived in short trials. Journet et al (1965) recorded a partial regression coefficient of 0.28 kg dry matter consumed per kg fat-corrected milk produced. A similar value (0.25 kg organic matter) was found for grazing cows (Curran & Holmes 1970), but estimates for cows fed on roughage and concentrates varied from 0.16 to 0.42 kg organic matter/kg FCM (Curran et al 1970). For cows on good diets (i.e., dry matter digestibility > 67%) the partial regression coefficient found by Conrad et al (1964) was 0.13 kg dry matter/kg FCM

A value of 0.2 kg DM/kg FCM has been selected as representative of the results reported in the preceding paragraph.

(d) Metabolizability of dietary energy

The most comprehensive studies of dietary energy concentration and intake of dairy cows have been made with fine diets, made from milled roughages mixed with concentrates in various proportions and, in most instances, pelleted or wafered. In general, these studies show intake to be fairly constant over a wide range of metabolizability (0.45–0.65); for diets of lower or higher metabolizability, intake is less (Ronning & Laben 1966, Amir et al 1968, Nelson et al 1968, Cowan et al 1970, Bull et al 1976). McCullough (1974) concluded from his analysis of data for complete diets that maximum intake was achieved with 35–55% concentrates.

In the UK, dairy cows are usually fed on coarse diets based on long roughage and concentrates. When such diets are of low metabolizability ($q < 0.55$), intake is raised by increasing metabolizability, either by improving roughage quality (Mohrenweiser & Donker 1968) or by increasing the proportion of concentrates (Conrad et al 1964) For better diets ($q > 0.55$), Conrad et al (1964) found intake to remain constant or decline slightly as metabolizability increased. Curran et al (1970) found the intake of cows in the third and fourth months of lactation to increase by about 15% when metabolizability was raised from about 0.52 to 0.60, and a similar increase was reported by Coppock et al (1974).

Table 2.6   Relative intake of dairy cows fed on the same diet throughout lactation (daily intake % of mean intake for complete lactation)†.

Month	Relative intake	Month	Relative intake
1	81	6	108
2	98	7	101
3	107	8	99
4	108	9	97
5	109	10	93

† mean values for 14 diets (see Table 2.5, source 1).

Table 2.7   Maximum dry matter intake of lactating cows†.

Fat-corrected milk yield (kg)			Dry matter intake (kg/day) for cows of liveweight (kg)		
Per lactation	Month of lactation	Per day‡	400	500	600
3000	1	14	9‡	11	12
	2	15	11	13	15
	3	14	12	14	16
	10	5	10	12	14
5000	1	24	10	12	13
	2	25	12	14	16
	3	23	13	15	18
	10	9	11	13	15
7000	1	33	11	13	14
	2	35	13	15	17
	3	32	14	17	19
	10	13	12	15	16
9000	1	43	13	14	15
	2	45	15	17	19
	3	42	17	18	20
	10	16	15	16	18

† For diets with q = 0.55–0.65.
‡ approximate values based on the lactation curve described by Wood (1969).

(e) Conclusions

Estimates of the maximal daily dry matter intake of lactating cows are given in Table 2.7. Their starting point is the value 135 g DM/kg $W^{0.75}$, which is representative of the sources listed in Table 2.5. It is a mean value for the whole lactation of cows yielding an average of 5000 kg FCM per lactation of 305 days (16 kg/day). Adjustments for average daily FCM (or milk) yields greater or less than 16 kg have been made by the use of the factor 0.2 kg DM/kg FCM. Adjustments for specific periods of lactation have been made by using the data of Table 2.6.

No allowance is made in Table 2.7 for the metabolizable energy concentration of the diet. This simplification should introduce little error into estimates for diets in the range of metabolizability 0.55–0.65. For poorer diets (q<0.55), an approximate correction would be to decrease the expected values for dry matter intake in Table 2.7 by 15% for each reduction in metabolizability of 0.05.

At lower yields the values in Table 2.7 agree with estimates made by the Ministry of Agriculture, Fisheries & Food (1975), but at higher yields intakes in Table 2.7 are higher. The most critical test of the estimates is provided by high-yielding cows. Wagner & Loosli (1967) found that cows of 610 kg yielding 6500 kg FCM in 305 days ate about 20 kg DM/day during the first 7 months; the estimate from Table 2.7

would be 19 kg DM/day. Ekern (1972) quotes daily intakes averaging 16.1 kg DM for cows weighing 550 kg and yielding 22.4 kg FCM in months 2–6 of lactation; the estimate from Table 2.7 would be 17 kg DM. Reid & Robb (1971) give intakes ranging from 18 to 23 kg DM for cows in early lactation weighing about 600 kg and yielding 32–51 kg FCM/day; these also agree with the estimates of Table 2.7.

These estimates of feed intake are necessarily based on data for cows with unrestricted access to feeds. In practice there is a danger that intake may be restricted by time of access to feed (Bines 1976), particularly when cows are given concentrate in a milking parlour.

2. Lactating ewes

Lactating ewes eat 20 to 70% more than comparable dry ewes, the difference being larger with good diets and for ewes suckling twins (Cook et al 1961, Hadjipieris & Holmes 1966, Arnold & Dudzinski 1967, Arnold 1975). The increase in intake that occurs after parturition may be exaggerated by a depression in consumption in late pregnancy (Forbes 1970). Like cows, ewes do not achieve maximum intake until peak milk yield has passed. Average values from eight sets of data indicate that for weeks 1–10, intake as a percentage of the lactation mean is: 80, 90, 100, 106, 108, 108, 105, 104, 104 and 102 (Peart 1967 and 1968, Hadjipieris & Holmes 1966). Most of these values were obtained with pelleted diets; with long hay, Hadjipieris & Holmes (1966) found consumption to increase slowly to reach a peak in weeks 8–10.

The most comprehensive data on the intake of lactating ewes are those of Peart (1967, 1968, 1970, Peart et al 1972). All were obtained from ewes fed on a pelleted diet containing 66% dried grass ($q \simeq 0.55$). Intake was measured over lactations lasting 10–12 weeks; in one experiment it was recorded for weeks 1–4 only but the values obtained have been adjusted to a full-lactation equivalent by means of the weekly distribution factors given above. Ewes suckling twins ate more than those suckling single lambs in all comparisons except one, when a low plane of nutrition had been imposed in late pregnancy. Mean daily intake for ewes suckling singles was 139 g DM/kg $W^{0.75}$, and 152 g for those suckling twins; in one experiment, ewes with 3 or 4 lambs ate 10% more than those with twins. For pelleted dried grass, Hadjipieris & Holmes (1966) reported similar values, 131 g/kg $W^{0.75}$ for ewes with one lamb and 155 g for those with two.

Lactating ewes are not normally fed on pelleted diets. Ewes given hay, alone or with grass cubes ($q = 0.5$), ate only 84 g DM/kg $W^{0.75}$ daily, the difference between single and twin-suckling ewes being small (Hadjipieris & Holmes 1966, Forbes 1969b). For ewes at pasture, intake varies widely. Six estimates for herbage (mean $q = 0.55$) range from 83 to 134 g DM/kg $W^{0.75}$ with a mean of 106 g (Cook et al 1961, Hadjipieris & Holmes 1966, Corbett 1968, Young & Newton 1974, 1975, Arnold 1975). In addition, Coop & Drew (1963) reported the consumption of 80 and 87 g digestible organic matter per kg $W^{0.75}$ daily by ewes grazing with single lambs or twins, respectively; if the herbage contained 750 g digestible organic matter per kg

Table 2.8   Representative values for the daily dry matter intake of lactating ewes (g/kg $W^{0.75}$)[†].

	No. of lambs	
Diet	1	2
Hay and concentrates	80	85
Pasture herbage	100	110
Pelleted diets	135	155

† Mean values for a lactation of 10–12 weeks. Values for individual weeks may be calculated by using the percentage factors given above.

dry matter, these would be equivalent to 107 and 116 g DM/kg $W^{0.75}$. Owen & Ingleton (1963) recorded the approximate daily intake of ewes grazing with singles and twins to be 75–90 g DM/kg $W^{0.75}$.

Representative values for the dry matter intake of lactating ewes are given in Table 2.8. They are based on the mean values given above and the assumption that for hay diets, pasture herbage and pelleted diets, respectively, the intake of ewes suckling twins will be about 5, 10 and 15% greater than those of ewes with single lambs. No clear relationship between intake and metabolizability has been found for lactating ewes.

Intake of Milk and Milk Substitutes

Calves with continuous access to whole milk will consume 70–80 g DM/kg $W^{0.75}$ per day (e.g., Marshall & Smith 1970, 1971), and similar intakes can be achieved with milk substitutes containing 120–150 g DM/kg (Pettyjohn et al 1963, Marshall & Smith 1973). In practice, milk is more commonly offered in discrete feeds. Niedermeier et al (1959) obtained intakes of 53 g DM/kg $W^{0.75}$ with whole milk offered to appetite twice daily. Lineweaver & Hafez (1969) found daily intake to be 66 g DM/kg $W^{0.75}$ when milk substitute was continuously available, but 60 g DM/kg $W^{0.75}$ when it was offered four times daily. At the National Institute for Research in Dairying, calves have been fed twice daily to appetite in many experiments made with milk and milk substitutes (Roy 1967, Roy et al 1964, 1970a, 1971, 1973a, Stobo, I. J. F. & Roy, J. H. B., unpubl.). Mean intake for diets containing 130–140 g DM/kg was 55 g DM/kg $W^{0.75}$ per day, but appeared to decline from about 58 g to 52 g/kg $W^{0.75}$ as calves grew from 50 to 100 kg in live weight (i.e., intake is more closely related to W raised to a power of approximately 0.5 than to the conventional $W^{0.75}$.

If the dry matter content of liquid diets is less than that of whole milk, dry matter intake is lower. For example, Pettyjohn et al (1963) found that dry matter intake (g/kg $W^{0.75}$) per day fell from 84 to 71 to 57 as dry matter concentration was reduced from 150 to 100 to 50 g/kg. Conversely, an increase in dry matter content, to perhaps 200 g/kg, may improve intake (Pettyjohn et al 1963, Stobo, I.J.F. & Roy, J.H.B., unpubl.).

Sheep

Lambs with continuous access to milk or milk substitutes containing at least 200 g DM/kg consume daily about 80 g DM/kg $W^{0.75}$ in their first 4 weeks of life (Large & Penning 1967, Glimp 1972, Gibb & Penning 1972, Penning et al 1971, 1977, Penning, P. D., unpubl.). Thereafter, milk intake per kg $W^{0.75}$ declines, especially if dry feeds are available. Discrete meals give lower intakes. Thus the lambs of Large (1965) consumed daily 60 g DM/kg $W^{0.75}$ when fed four times daily with milk replacers containing 200 or 250 g DM/kg. Pinot & Teissier (1965) recorded total intakes (milk plus dry feed) of 60 g DM/kg $W^{0.75}$ per day for lambs fed four times daily, and Davies & Owen (1967) reported a similar figure for lambs fed only twice daily.

In some experiments (e.g., Brisson & Bouchard 1970), milk continuously available was generally kept cold, whereas milk given in discrete feeds was warm. Penning et al (1977) found in a single experiment that lambs 0–3 weeks old consumed most (88 g DM/kg $W^{0.75}$ per day) when offered cold milk continuously. Warm and cold milk given four times daily promoted similar intakes, of 78 and 73 g DM/kg $W^{0.75}$, respectively. If the dry matter concentration of the milk is less than 200 g DM/kg, dry matter intake is reduced (Large 1965, Penning, P. D. unpubl.).

Table 2.9   Maximum dry matter intake of calves and lambs fed on milk or milk substitutes.

Species	Live weight (kg)	Daily dry matter intake	
		g/kg $W^{0.75}$	g/animal
Sheep	5	80	270
	10		450
	15		610
Cattle	50	60	1100
	75		1500
	100		1900
	150		2600

Recommended values

Per unit of metabolic liveweight, calves appear to ingest less milk dry matter than lambs, and the recommended values in Table 2.9 are based on estimates of 80 g/kg $W^{0.75}$ for lambs and 60 g/kg $W^{0.75}$ for calves. The difference between the species may be due to the generally higher dry matter concentration of milk given to lambs, to frequency of feeding, and to the fact that calves are often taken to relatively heavier weights on milk alone than are lambs. For younger calves there is evidence that an intake of 60 g milk dry matter per kg $W^{0.75}$ per day may be exceeded by increasing dry matter concentration and feeding frequency, but for the older veal calf, weighing more than 100 kg, a higher dry matter concentration may be needed merely to maintain intake at 60 g/kg $W^{0.75}$.

Chapter 3

Requirements for Energy

Introduction

In the first edition proposals were made that the starch equivalent system for the expression of energy requirements and the energy value of feeds — a system which had been used in the United Kingdom since about 1912 — should be replaced by one based on metabolizable energy devised by Blaxter (1962a) from the results of calorimetric work done at the Hannah Research Institute. These proposals were accepted in the United Kingdom and the system has gained some acceptance also in other countries. One advantage of the new system arises from the fact that it predicts animal performance from ration attributes more precisely than did the starch equivalent system and older systems (see Alderman et al 1970, Burroughs et al 1970, Levy & Holzer 1971); it also has the advantage that it can readily be modified to accommodate new information. In essence the new system provides a set of rules which link tables of feed composition to tables listing the energy requirements of animals. Modifications to the rules or to estimates of requirement thus have no effect on the primary listing of the energy values of feeds. Furthermore, the primary listing is in terms of an attribute of the feed which is simply measured, namely metabolizable energy.

A number of new systems for feeding ruminant livestock have been proposed since 1962 when the metabolizable energy system was proposed and modifications of the ARC proposals have also been suggested. They include the Californian net energy system (Lofgreen & Garrett 1968), the East German net energy (fattening) system (Schiemann et al 1971), the net energy (lactation) system of Flatt et al (1968 1972) and the modifications made by the Ministry of Agriculture, Fisheries and Food (1972 1975) to the ARC system. These are discussed here as they relate to the scientific and numerical validity of the original metabolizable energy system.

The terminology used in studies of energy metabolism has become more standardized than it was ten years ago. Even so, it is desirable to define terms which are used subsequently and to indicate a preferred notation.

. Energy units

The unit of energy is the joule, which is the product of a force of one Newton and one metre, a force of one Newton being the product of a mass of one kilogram and an acceleration of one metre per second squared. The older unit of heat, the calorie, is related to the joule by the exact expression:

$$1 \text{ calorie} = 4.184 \text{ joules}.$$

Many of the energy requirements of animals are expressed in terms of energy per unit time; the maintenance requirement of a bullock might, for example, be expressed as 50 MJ/day. This is an expression which has the dimensions of power, and could equally be expressed in terms of watts, a watt being energy at the rate of one joule per second. The maintenance requirement of the bullock would thus be 0.6 kW. Despite the greater familiarity of many with the watt as a unit of power, energy requirements per unit time have been expressed in terms of joules per unit time rather than as watts.

2. Primary measurements and simple derived quantities

The major technique used to study energy metabolism is that of the calorimetric balance. The change in the energy content of the body in a period of time due to changes in its "fat" and "protein" content is estimated indirectly from measurements of the heat of combustion of the food ingested and of the faeces, urine and combustible gas excreted, the energy of any milk secreted together with the heat produced by the animal. These methods lead to the following definitions of rates per unit time.

Gross energy intake is the heat of combustion of the feed ingested (I_E).

Apparently digested energy intake is the gross energy of the feed less that of the faeces produced (D_E).

Metabolizable energy intake is the gross energy of the feed less that of the faeces, urine and combustible gas (M_F).

Milk energy output is the heat of combustion of milk secreted (Y_E).

Heat production is the heat produced by the animal (H).

Energy retention is the rate of energy storage in the body (which can of course be negative), and in a non-lactating animal is simply the metabolizable energy less the heat production (R_E).

By convention, when an animal has been starved for such a period that absorption of energy-yielding nutrients from its gut is negligible, a measurement of heat production is termed the fasting heat production. The energy retention of such an animal is negative owing to losses of energy as heat and in the urinary solids. The sum of these losses is the fasting catabolism. When energy retention is zero, the animal is in energy maintenance, often abbreviated to maintenance. On occasion it is convenient to designate the amount of feed given to an animal as a multiple of the amount that results in energy maintenance. These multiples define the level of feeding (given the symbol L), maintenance being designated 1.0 and fasting, zero.

The above terms are defined operationally; care has to be taken in interpreting them. Apparently digested energy is not the energy absorbed from the gut since the former includes heat and methane arising from the anaerobic fermentation of feed and excludes the heat of combustion of any organic compounds secreted into the gut which contribute either directly or after microbial activity to the heat of combustion of the faeces. Metabolizable energy is not a precise estimate of that part of the energy of feed which is the substrate for oxidation in the animal's tissues. For similar reasons, in fed animals — but not in fasted ones — the heat production includes the fermentation heat, and is therefore not a precise estimate of the heat arising from the tissue metabolism of nutrients.

There are no quantitative data relating to the heat of combustion of the metabolic products in faeces, but as a proportion of the energy of feed digested that is probably small. Fermentation heat, however, is appreciable. This component has been computed from the stoichiometry of the microbial fermentation of carbohydrate by Hungate (1966) and by Ørskov et al (1968) to be about 65 kJ/MJ of the energy of the substrate fermented and relatively invariant with diet. More detailed calculations which included biosynthesis by ruminal organisms (Baldwin et al 1970) showed fermentation heat to range from 473 to 690 kJ/kg for a variety of roughage diets, corresponding to 52 to 131 kJ/MJ of the energy apparently digested. More direct measurements by Marston (1948a) of the difference between the energy of reactants and products in fermentation *in vitro* gave a value of 59 kJ/MJ cellulose fermented, while studies by Osuji (1973) and of Webster et al (1975) of the anaerobic component of the heat produced by the viscera drained by the portal vein, a component which approximates fermentation heat, gave mean values of 75 kJ/MJ and 68 kJ/MJ of the metabolizable energy supplied. No systematic variation with the amount of diet or the physical form of the two roughage diets which were studied was noted.

These data suggest that metabolizable energy overestimates the absorbed and utilizable energy of feed, by about 7%, and, conversely that the heat production of the fed ruminant includes a component due to fermentation heat equivalent to about 7% of the metabolizable energy supplied.

From the primary measurements several terms are derived. *Metabolizability of the gross energy of feed* (q) is defined as the metabolizable energy of a diet divided by the gross energy. Metabolizability at maintenance is q_m and at any feeding level q_L. *Efficiency of utilization of metabolizable energy* (k) is defined as the increase in energy retention which occurs per unit increase in the metabolizable energy supplied, i.e.   Efficiency = Δ Retention of energy/Δ Metabolizable energy.

The measurement of efficiency presupposes that at least two amounts of metabolizable energy are given to the animal. *Proportional heat increment* is the increase in heat production which occurs per unit increase in metabolizable energy given to the animal.

Proportional heat increment = Δ Heat production/Δ Metabolizable energy.

The increment of heat and the increment of energy retention obviously sum to give the increment of metabolizable energy in a growing or fattening animal.

3. Feed intake and energy retention

The energy of feed meets energy expenditures by the animal, and promotes the synthesis of organic body constituents and secretions. Most systems of meeting energy needs commence with a consideration of the energy exchanges of the mature non-lactating stall-fed animal, and then extend the approach to include juvenile growth, pregnancy, lactation, muscular work and the energy cost of maintaining body temperature in adverse climatic circumstances. This approach is used here.

In a mature ruminant energy retention increases as the amount of a particular feed given increases, but the relationship is not linear and, as shown in Figure 3.1, varies with the type of feed supplied. Much work, commencing with that of Wiegner & Ghoneim (1930), has shown that the law of diminishing returns applies to the responses of energy retention in mature animals to successive increments of feed (Brody 1945, Blaxter & Graham 1955, Blaxter & Boyne 1970).

The reasons for the curvilinearity of response are not completely understood. They involve firstly possible changes in ruminal fermentation and passage of digesta (Ørskov et al 1969b) as level of feeding rises, which result in changes in the proportions, if not the overall amount, of different energy-yielding substrates absorbed. Secondly they include a real difference, which can be explained in biochemical terms (Blaxter 1962a 1971b, Armstrong 1969), in the efficiency of energy utilization when there is a net synthesis of body fat and protein and when body constituents are being oxidized. Thirdly there may be an acceleration of metabolism due to slightly higher body temperatures at high levels of feeding, and lastly there may be differences in the efficiency of biosynthesis of protein and fat.

The ARC metabolizable energy system states that if the energy provided by feed is measured in terms of metabolizable energy, no great error is incurred by regarding the continuous curvilinear relationship between feed intake and energy retention as consisting of two direct proportionalities, one applying from fasting to maintenance and the other above. It is thus assumed that below maintenance increments of metabolizable energy of a particular diet are used with one constant efficiency in promoting energy retention and with another, lower, efficiency above maintenance.

The efficiency of utilization of metabolizable energy for maintenance, given the symbol k_m, is measured between fasting and the point at which energy retention is zero. The efficiency of utilization of metabolizable energy for adult growth and fattening (k_f) was not so rigorously defined in the previous edition. Its lowest value was obviously the maintenance datum but its upper value could vary. Here it is stated that the upper limit is twice the feed intake for maintenance. This is a convenient level of intake for growing animals.

The ARC system states that in the mature animal the two efficiency values k_m and

Fig. 3.1   Relationship between energy retention and energy intake for contrasting foods.

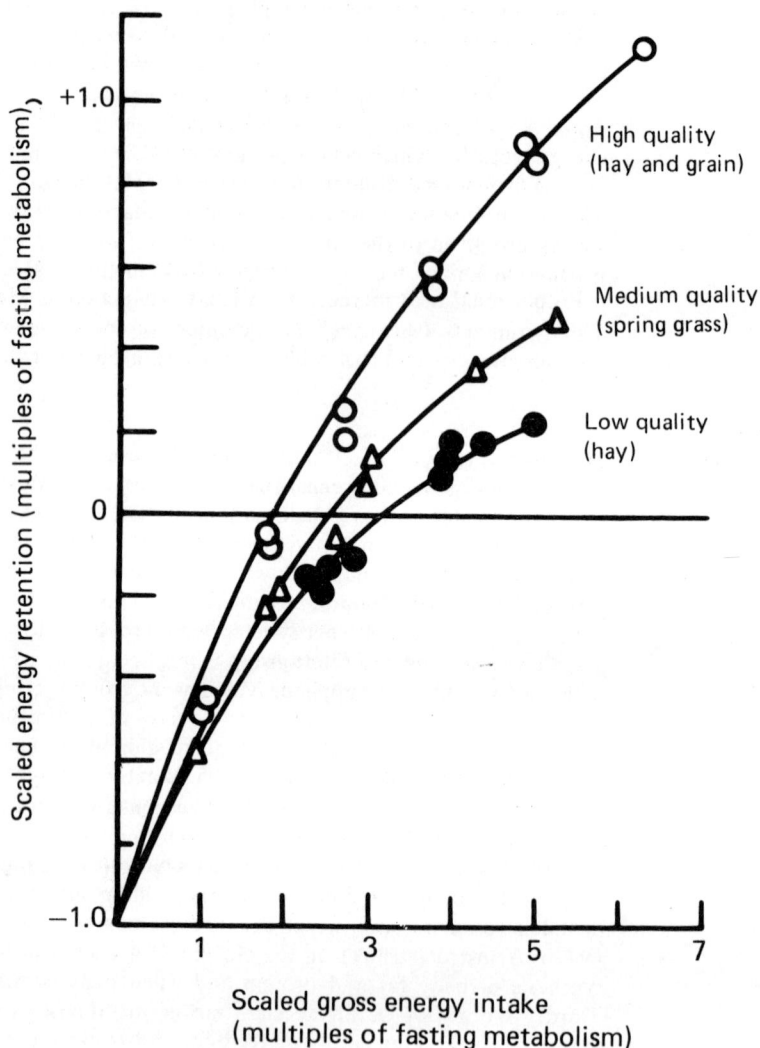

kf are functions of the nature of the total diet. In addition, the original system state: that the metabolizability of the energy of a diet is not constant but declines as feeding level is raised above the maintenance datum and that this also is related to attribute of the total diet. Decline in metabolizability with feeding level confers a curvilinearit: on the response of energy retention to feed intake at high levels of intake, that is, i provides a better approximation to the underlying curvilinear relationship. To use the system to estimate the energy retained by a mature animal receiving a given amoun of a diet also requires that the animal's fasting metabolism is known.

The factors affecting metabolizability of the gross energy of diets, the net availabilit: of metabolizable energy for maintenance (k_m) and for fattening (k_f), are discusse: first and efficiency of utilization for lactation (k_l) is discussed later.

4. Metabolizability and feeding level

There is no doubt that as the level of feeding of ruminants rises the proportional los of energy in the faeces increases and apparent digestibility declines. Older experiment with cattle and sheep have been reviewed by Brown (1966) to show that the depression in apparent digestibility is greater for finely ground roughages and for mixed diets containing grain than for long roughages. Work with dairy cows in which feeding levels of 5×maintenance have been achieved showed that when feeding level rises by one unit, that is, one multiple of maintenance, apparent digestibility averages 0.96 of that estimated at maintenance (Tyrrell et al 1966, Moe et al 1965, Wagner & Loosli 1967, Flatt et al 1969b). In the above experiments the factor by which digestibility at one level should be multiplied to estimate that at a feeding level one unit higher ranged from 0.989 to 0.938. These results have been reviewed by Ekern (1972b). There is some evidence that the depression is slightly greater in sheep than it is in cattle (Blaxter et al 1966a, Leaver et al 1969).

Analysis of the results of 53 experiments with fattening cattle and sheep (Blaxter 1969) in which the depression of apparent digestibility on increasing feeding has been measured, showed that the absolute depression for unit rise in feeding, Δd_E, depended on the apparent digestibility of the diet, d_E (in joules), determined at the maintenance level. The descriptive equation was

$$\Delta d_E = 0.107 - 0.113 d_E \qquad [3.1]$$

This implies that for two feeds with apparent digestibilities at maintenance of 0.75 and 0.55, digestibilities at 2×maintenance would be 0.728 and 0.508, respectively, and the corresponding factors 0.971 and 0.918. Schiemann et al (1971) determined the depression of digestibility in 21 experiments with dairy cows. Analysis of their data showed a mean depression of digestibility for unit rise in feeding level of 0.03 ± 0.011 units over a range of apparent digestibility from 0.65 to 0.78.

There is agreement that as the level of feeding of both cattle and sheep rises the proportional losses of energy as methane and in urine decline. In studies by Schiemann et al (1971) the losses per unit rise in feeding level declined by 11 ± 5 and 10 ± 3 kJ/MJ gross energy for methane and urine energy, respectively. In the three experiments of Flatt et al (1969a) the decline in methane loss varied from 7 ± 2 to 16 ± 2 kJ/MJ gross energy and urine losses from 5 ± 2 to 15 ± 2 kJ/MJ per unit rise in feeding level. Blaxter & Clapperton (1965) showed that for 48 diets the depression in methane production on raising feeding level (Δg) was about 15 kJ/MJ gross energy and related to the apparent digestibility of dietary energy by the equation

$$\Delta g = 50 \, d_E - 23.7 \qquad [3.2]$$

Blaxter et al (1966a) showed that the mean decrease in the loss of energy in urine on raising feeding level one unit in 37 experiments was 7.3 ± 7.7 kJ/MJ.

The increase in the proportional faecal loss of energy on raising feeding level thus tends to be balanced by decreased proportional losses of methane and of energy in the urine. The net effect is that metabolizability of the gross energy is less affected that might be inferred from a consideration of apparent digestibility. Combination of the results of the linear regression analyses and the assumption that at maintenance the ratio of metabolizable energy to digested energy is on average 0.8 (ARC 1965) leads to the equation (Blaxter 1969):

$$q_L = q_m + (L-1) \, (0.20 \, [q_m - 0.623]), \qquad [3.3]$$

where q_L is metabolizability of the gross energy at any feeding level, L, and q_m is that determined at maintenance. The equation implies that only if q_m is less than 0.62 kJ/kJ will metabolizability of energy fall on raising the feeding level. More direct analysis of the results of 72 experiments in which level of feeding was varied confirmed that

Fig. 3.2   The effect of level of feeding on the metabolizability of dietary energy.

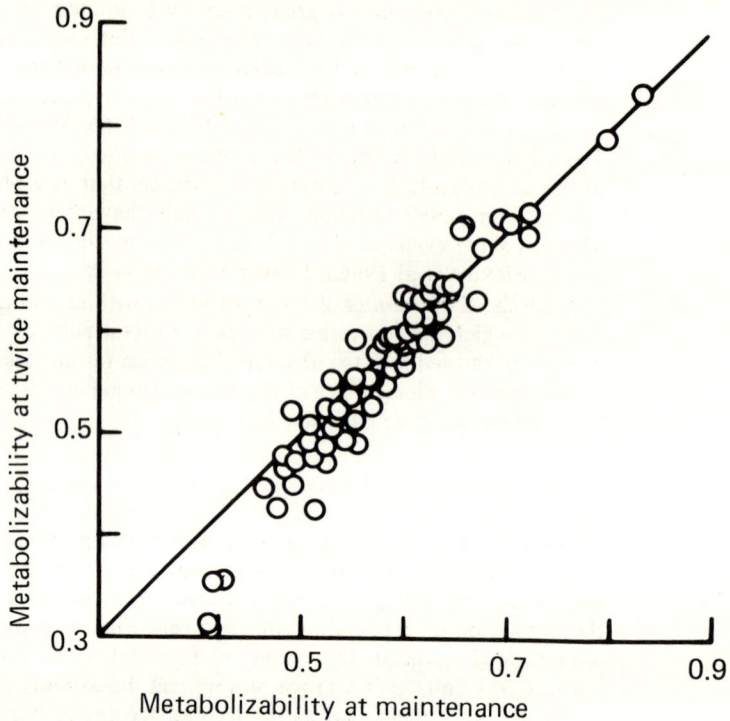

depressions of metabolizability occurred only with the poorer diets and the results are shown in Figure 3.2. Similar conclusions about the compensatory nature of faecal, methane and urinary responses to feeding level have been made by others (Flatt et al 1969a, Schiemann et al 1971, Hoffmann et al 1972).

A decline in the metabolizable energy of feed with rise in feeding level is one of several factors responsible for the curvilinearity of the response of energy retention to increased feed. The correction of estimates of energy need for effects of feeding level is given later (see page 103).

The Efficiency of Utilization of Metabolizable Energy

1. Efficiency for maintenance

Many experiments have now been made with several species of animal to determine the efficiency with which nutrients replace body fat and protein as a source of energy for maintenance. In the ruminant particular attention must be paid to the efficiency of utilization of the steam-volatile fatty acids since their heat of combustion accounts for about 65% of the energy absorbed. Experiments have been made with acetic, propionic and n-butyric acids given by direct infusion singly or in mixtures to fasted sheep (Armstrong & Blaxter 1957a, Armstrong et al 1957 1961), and to fasted cows (Holter et al 1970). The results obtained in the two species agree well and Table 3.1 summarizes those obtained with sheep. They show that the energy of those mixtures of acids likely to be absorbed from the rumen is used with an efficiency for maintenance of about 0.85. Experiments in sheep with a soluble protein, casein, showed that it was used with an efficiency of 0.81 (Martin & Blaxter 1961), a value that agrees well with those obtained when protein is given to fasted nonruminant species. Glucose given to sheep by methods which avoid its fermentation gave an efficiency of 1.00

(Armstrong & Blaxter 1961), a value that again agrees with those obtained in simple-stomached species (see review by Blaxter 1971). No data are available on the utilization of lipids by fasting ruminants; the concordance for carbohydrate and protein suggests that an efficiency of 0.97 can be assumed (Chudy & Schiemann 1969).

Table 3.1.   Utilization of mixtures of steam volatile fatty acids for maintenance of sheep (Armstrong et al 1957).

Mixture in molar proportions			Efficiency of utilization for maintenance
Acetic	Propionic	n-Butyric	
0.25	0.45	0.30	0.872
0.50	0.30	0.20	0.831
0.75	0.15	0.10	0.856
0.90	0.06	0.04	0.847

These findings show that absorbed nutrients do not replace one another in proportion to their heats of combustion as proposed by Rubner (1902). Rather they replace one another in proportion to the extent that they provide free energy to the cells of the body (Krebs 1960, Blaxter 1961) and this can be computed from a knowledge of the stoichiometry of the oxidations involved.

From the above, bearing in mind that the ruminant absorbs relatively little glucose from its gut and that the lipid content of diets rarely exceeds 70 g/kg, one might expect the efficiency with which the energy of the end products of digestion is used for maintenance to be about 0.83. It is not legitimate to conclude from this that the efficiency of utilization of metabolizable energy for maintenance is likely to be about 0.83 and the proportional increment of heat 0.17 since metabolizable energy includes fermentation heat and some muscular work has to be done to prehend, masticate, ruminate and propel feed through the digestive tract. If fermentation heat is about 7% of metabolizable energy, it can be calculated that efficiency of utilization of metabolizable energy is 0.77. The energy expended in the physical work of ingesting and propelling feed through the gut is difficult to separate from other events in the digestive process. Continuous measurement of the respiratory exchange during the eating of a meal shows that heat production increases by about 50% and this continues for the duration of the meal (Blaxter & Joyce 1963, Graham 1964b, Young 1966, Webster 1967). Webster (1971), Webster & Hays (1968) and Webster et al (1969) have shown that this increase is associated with substantial shifts of body fluids into the gut resulting in endocrine adjustments; the increase is thus not entirely due to physical work. Such increases in energy expenditure, notwithstanding their cause, are related to the duration of eating and might be expected to vary with the nature of the diet. They would reduce the efficiency of utilization of the metabolizable energy of natural diets to below the 0.77 calculated above.

The early work of Forbes (1933) showed that on average every joule of metabolizable energy provided by diet prevented the loss of 0.75J from the body of a bullock given a submaintenance diet. Blaxter (1961) showed that in both sheep and cattle efficiency averaged 0.738. Armstrong (1964) and Blaxter & Wainman (1964) later found that efficiency of utilization of metabolizable energy for maintenance (k_m) was not constant; it tended to increase slightly with improving quality of the diet as exemplified by its metabolizability (q_m) according to the equation

$$k_m = 0.546 + 0.30 \; q_m. \hspace{3cm} [3.4]$$

This equation was used in the first edition. Evidence from practical and calorimetric trials was adduced to support the relative constancy of the efficiency of utilization of metabolizable energy for maintenance, which contrasts with the variability of utilization for adult growth and fattening. Graham (1966) derived from the above equation a similar one,

$$k_m = 0.55 + 0.24 \, d_E, \tag{3.5}$$

where d_E is the apparent digestibility of dietary energy, which is on average 1.25 q_m.

The results of experiments with 78 diets with mature sheep or cattle (Blaxter & Boyne 1974 — see Blaxter 1974a), in which fasting data and sufficient additional observations were available to permit a complete description of the relationship between feed intake and energy retention, allowed Blaxter & Boyne (1974) to derive efficiencies of utilization for maintenance and these are summarized in Table 3.2 where they are compared with preferred values of the first edition. In this series it was evident, firstly that efficiency increased with increase in metabolizability at maintenance, and secondly that there were real differences between broad classes of diets. The coefficient b in the equations for different classes of feed $k_m = bq_m + a$, was 0.207 and invariant with class. The intercepts, however, varied from 0.56 for regrowths of grass to 0.62 for mixed diets. The regression was highly significant ($P<0.001$). For the pooled data the combined regression coefficient was higher, at 0.35, and the intercept lower than in the original equation. It will be noted that the newer analysis based on additional data resulted in estimates of efficiency very little different from those given in the first edition.

Table 3.2   Efficiency of utilization of metabolizable energy for maintenance (k_m): values predicted from the metabolizability of the gross energy of diet (q_m) and compared with estimates made in the first edition.

Class of feed	No. in class	Values predicted when q_m is:			
		0.40	0.50	0.60	0.70
Pelleted feeds	12	0.653	0.673	0.695	0.715
Forages	36	0.642	0.663	0.683	0.704
Mixed diets	30	—	0.728	0.749	0.769
All diets†	78	0.643	0.678	0.714	0.750
ARC (1965)‡ prediction	18	0.666	0.696	0.726	0.756

† $k_m = 0.35 \, q_m + 0.503 \, (\pm 0.064)$.
‡ $k_m = 0.30 \, q_m + 0.546 \, (\pm 0.026)$.

2. Metabolizable energy for maintenance

The metabolizable energy required to maintain energy equilibrium in an animal kept in a stall, where conditions are probably similar to those in a calorimeter or respiration chamber, is simply the sum of the fasting heat production and fasting urinary loss of energy divided by the efficiency of utilization of metabolizable energy. Efficiency can be estimated from attributes of the feed and, as will be discussed later, fasting heat production can be estimated from body weight.

Many calorimetric experiments have been made in which energy requirements for maintenance have been estimated from measurements of feed intake which result in energy retention slightly below or above energy equilibrium. Two methods of calculation have been used (see van Es 1972). In the first, "the correction method", single observations of metabolizable energy intake have been "corrected" to the quantity required for zero energy retention on the assumption that the mean efficiency of utilization of metabolizable energy below maintenance is about 0.70 and above maintenance is about 0.55. Provided departures from zero energy retention are small, the error from neglecting variation in efficiency with type of diet is also small. In the second method of estimating maintenance needs, "the regression method", a series of retentions are regressed on metabolizable energy to estimate that needed for zero retention. Breirem's (1953) analysis of the older calorimetric data with bullocks, made by the correction method, showed that maintenance requirements of metabolizable energy varied with weight raised to the power 0.6 and that the higher the fibre content of the diet (and hence the lower on average its metabolizability) the more metabolizable energy was needed. Van Es & Nijkamp (1969), also using the correction method, analysed 189 trials with mature bullocks made in the Netherlands. They assumed maintenance needs to vary with weight raised to the power 0.75 and found metabolizable energy required for maintenance tended to vary inversely with the apparent digestibility of energy and directly with the percentage of digested energy present as protein. From these data it can be computed that an increase in metabolizability of 0.01 unit increased net availability by about 0.004 unit, a value slightly greater than that given for the pooled data in Table 3.2.

Using the regression method to determine maintenance in bullocks, Lofgreen & Garrett (1968) analysed the results of 34 experiments and showed that maintenance dry matter requirements expressed per unit of weight raised to the power 0.75 fell with increase in the concentration of metabolizable energy in the dry matter of the diet. Their equation again implies that more metabolizable energy is required for maintenance when metabolizability of the diet is low than when it is high. Taking the whole range of data and making the assumption that the gross energy of feed is about 18 MJ/kg we may compute that the efficiency of utilization of metabolizable energy increases by about 0.003 unit per 0.01 increase in metabolizability, a result which agrees with the direct estimates made by Blaxter & Boyne (1974).

The experiments in which the amount of metabolizable energy required for maintenance has been estimated directly thus also show that efficiency of utilization of metabolizable energy for maintenance depends on diet quality. The preferred values for predicting efficiency are given in Table 3.3.

Table 3.3.   Preferred values for the efficiency of utilization of metabolizable energy of normal diets by ruminants.

Function	Efficiency when metabolizability (q_m) is:			
	0.40	0.50	0.60	0.70
Maintenance (k_m)	0.643	0.678	0.714	0.750
Growth and fattening (k_f)				
(a) All diets	0.318	0.396	0.474	0.552
(b) Pelleted diets	0.466	0.470	0.474	—
Lactation (k_l)[†]	0.560	0.595	0.630	0.665

[†] $k_l = 0.35 \, q_m + 0.420$.

3. Problems in meeting maintenance requirements

Three problems arise in using the ARC approach to estimate maintenance energy requirements. Firstly, as pointed out by van Es (1972), the weight of a fasted animal is less than that of one given feed to meet its maintenance needs, the difference varying with the ration. With sheep given pelleted feed, weight at maintenance was 1.08 times fasting weight (Blaxter et al 1966d). In cattle given roughage diets values varying from 1.06 to 1.10 have been obtained and the value of 1.08 is preferred. Estimation of the maintenance requirement from an estimate of fasting metabolism which is based on weight on feed clearly results in an error. Alderman et al (1970), in using the system of the first edition, have taken this into account by assuming that fasted weight, from which fasting heat production is estimated, is 0.90 live weight, that is, by assuming a factor of 1.11.

Secondly, it has been stated by some (see Kromann 1973) that maintenance energy expenditures increase with increased production. This may well prove to be a matter of semantics; the curvilinear relationship between energy retention and feed intake could be regarded as due to either a decrease in the efficiency of utilization of increments of feed given above a constant maintenance, or a constant efficiency and a progressive increase in a component analogous to a maintenance cost. If maintenance is rigorously defined as that amount of feed which results in zero energy retention the problem does not arise. Even so, when linear regression methods are used to estimate maintenance needs in high-producing dairy cows estimates of the metabolizable energy required to result in zero retention can be in error if the rate of change of production with metabolizable energy is not a constant. It will be appreciated that if activity or muscle tonus of an animal differs as between two feeding levels, estimates of efficiency will not reflect solely the metabolism of feed.

Thirdly, to use the system the fasting metabolism of the animal must be known or capable of prediction. Although the heat production of mature ruminants kept in thermoneutral environments expressed on a live weight basis is fairly constant when they have been starved for more than 72 hours, this is not so for very young animals. Thus, in milk-fed calves Blaxter & Wood (1951a) found a continuing decline in metabolism per unit weight on prolonged fasting, and Graham et al (1974) noted that the stability of metabolism found in mature sheep after fasting for 72 hours did not apply to milk-fed lambs. At this time, however, the metabolism of the lamb is 60 to 65% less than the initial heat production, the fall being similar to that noted in adults. There is thus some doubt about the validity of measurements of fasting metabolism in sucking ruminants. In addition, the measurement of fasting metabolism is not necessarily independent of previous nutrition. Marston (1948a) showed that in sheep the amount of feed consumed over the weeks preceding a fast affected the subsequent measure of fasting metabolism. Graham et al (1974) have demonstrated in sheep that the amount of feed given immediately before fasting affects the result. In dry cows the association of a previously high level of feeding with high fasting metabolism has not been noted (Flatt & Coppock 1963). For these reasons fasting metabolism is usually measured after a period of maintenance feeding.

The heat production of the individual animal at constant weight is subject to variation due to other factors. An adult sheep may have a fasting metabolism exceeding by up to 30% that noted when it is accustomed to the experimental procedure (Blaxter 1962a 1973, Graham 1962); in young lambs this elevation of metabolism may not occur (Webster et al 1972). A similar effect has been noted in the maintenance requirement of cows by van Es & Nijkamp (1969). Ritzman & Benedict (1938) drew attention to a lability of the fasting heat production of individual cows which could not be attributed to "training" and a similar type of variation in maintenance energy requirements has been noted by Brouwer & Nijkamp (1966). Lastly, there is considerable variation in the fasting metabolism between different animals accustomed to the conditions of measurement and differences between individuals are repeatable (Blaxter 1962a). Analysis of published data by van Es (1961) showed that between-animal variation in maintenance requirements expressed as a coefficient of variation was ±8 to 10%. The same extent of variation emerges from the analysis of published data given later in the section.

4. Efficiency for muscular work

Determination of the efficiency of utilization of metabolizable energy in muscular work is difficult because it involves the measurement of the amount of work done by an animal in moving its body. The work done in ascent can, however, be measured from the mass of the body, the vertical distance moved and the acceleration due to gravity, and the energy expended by the animal in undertaking this work can be estimated as the difference between walking on the level and walking on a gradient. The ratio of work done in ascent to energy expended represents the efficiency with which the energy derived from oxidation of the fat and carbohydrate of the body is used to provide energy for the contraction process. For man, dog, sheep, bovine and horse the values obtained are close to 0.30, and this value agrees with that which may be calculated from biochemical considerations assuming close conservation of energy in the process of mechano-chemical coupling.

The work done in other types of muscular work — standing, walking and running — is best expressed in terms of an energy expenditure, since it is impossible to measure the complex components involved. The utilization of metabolizable energy for the energy expended in work can thus be measured in a way analogous to that adopted in dealing with maintenance energy expenditures, that is, as the change in energy retention per kJ increase in metabolizable energy intake. Experiments in which the same amount of work was done by sheep fed above and below maintenance showed that muscular work resulted in a smaller reduction in energy retention in animals retaining energy than in those at or below the maintenance level (Clapperton et al 1960, Clapperton 1964a). This arises because muscular work is a primary charge on the animal's economy and because fattening is a less efficient process than is the maintenance of weight. It was also found that below maintenance the increase in energy retention with increasing intake of metabolizable energy was the same in exercised and unexercised animals. The metabolizable energy required to maintain a working animal is then the sum of the fasting energy expenditure and the additional energy lost to the body by doing work divided by the efficiency of utilization of metabolizable energy for maintenance. The efficiency of utilization of metabolizable energy for muscular work is thus somewhat of a misnomer; the efficiency is in reality one for meeting the energy expenditure incurred in doing work.

When animals are cold they shiver and at low temperatures their heat losses are independent of the amount of feed they are given (Graham et al 1959). The efficiency with which metabolizable energy is used in these conditions is 1.00. This means that the fermentation heat, the heat increment associated with the work of shivering, and the heat equivalent of the work of shivering are all available to keep the animals warm and combat the heat loss which environmental conditions demand.

5. Efficiency for growth and fattening

(a) The adult animal

As shown in Chapter 1, the composition of the body gains of ruminants, particularly of sheep (Searle et al 1972), varies relatively little after sexual maturity is reached, and it is with such animals that deposit energy as fat in a fairly constant ratio to protein that most calorimetric work has been done during the last 50 to 60 years. Accordingly, the efficiency with which metabolizable energy is used by ruminants which have reached sexual maturity and have reached, or are approaching, adult size will be dealt with first.

From the early work by Kellner in Germany and Armsby and Forbes in the United States it has long been known that the metabolizable energy derived from poor feeds promotes less energy deposition than does that from good ones. The early experiments, 145 in number, are not reviewed in detail. They were analysed by Breirem (1944), among others, to show that the efficiency of utilization of metabolizable energy for growth and fattening (k_f) declined with increasing crude fibre content of the diet according to the equation

$$k_f = 0.656 - 0.070C,$$

where C is the crude fibre content of the dry matter of the diet in g/kg. This implies that the metabolizable energy of fibre-free sources — such as starch or glucose —

would be used with an efficiency of 0.66 and direct experiments in cattle and in sheep (Nehring et al 1961, Hoffman et al 1971) give values which are close to 0.65.

The factor k_f has been related to attributes of feed other than crude fibre. An analysis of results for 17 diets (Blaxter 1964) showed that k_f could be related to the metabolizability of the energy of the diet (q_m) by the relationship

$$k_f = 0.656 - 0.070C,$$

$$k_f = 0.81\ q_m + 0.03, \hspace{3cm} [3.6]$$

and this was used as the basis of predicting energy retention in the first edition.

Work with mature sheep and cattle has now increased the number of experiments to over 70 and for all these experiments the simplest relationships found (Blaxter 1974a) are in terms of metabolizability of dietary energy measured at maintenance (q_m),

$$k_f = 0.78\ q_m + 0.006, \hspace{3cm} [3.7]$$

and in terms of crude fibre where C_o is the crude fibre content of the feed organic matter in g/kg,

$$k_f = 0.650 - 0.084\ C_o. \hspace{3cm} [3.8]$$

In fact a slightly better fit is obtained from a quadratic relationship in crude fibre.

The new equations relating k_f to q_m result in estimates of k_f slightly lower than those given in the previous edition. This arises in part from the more precise definition of k_f (see above). Table 3.4 gives values for k_f computed from the preferred general equations and from those given in the first edition. The table shows that with the proposed relationships animals require slightly more metabolizable energy for maintenance and use metabolizable energy surplus to maintenance slightly less well. The seven interspecific comparisons made to date (Blaxter & Wainman 1961, 1964) suggest that values obtained with fattening sheep are equally applicable to fattening cattle.

As with efficiency of utilization of energy for maintenance, so for fattening, there is now evidence of differences between classes of feeds with respect to the relationship between k_f and q_m. Table 3.4 shows values predicted from equations which are significantly different from class to class of feeds. The major difference between feeds relates to pelleted diets. There is additional evidence from calorimetric studies with milking cows (which are less sensitive to variation in efficiency with metabolizability) that pelleting results in more efficient utilization of metabolizable energy, though it

Table 3.4   Efficiency of utilization of metabolizable energy for growth and fattening (k_f): values predicted from the metabolizability of the gross energy of the diet (q_m) and compared with estimates made in the first edition.

Class of feed	Equation No.	No. in class	Value predicted when q_m is:			
			0.40	0.50	0.60	0.70
Pelleted feeds	(1)	12	0.475	0.477	0.477	—
Forages (first growth)	(2)	25	0.211	0.342	0.474	0.606
Forages (aftermaths)	(3)	11	0.157	0.273	0.388	—
Mixed diets	(4)	30	—	0.472	0.510	0.548
All diets	(5)	78	0.318	0.396	0.474	0.552
ARC (1965) (all diets)	(6)	32	0.354	0.435	0.516	0.597

Equations (1) $k_f = 0.024\ q_m + 0.465\ (\pm 0.030)$
(2) $k_f = 1.32\ q_m - 0.318\ (\pm 0.087)$
(3) $k_f = 1.16\ q_m - 0.308\ (\pm 0.045)$
(4) $k_f = 0.38\ q_m + 0.282\ (\pm 0.079)$
(5) $k_f = 0.78\ q_m + 0.006\ (\pm 0.097)$
(6) $k_f = 0.81\ q_m + 0.030\ (\pm 0.046)$

reduces metabolizable energy per unit weight (Van Es 1969, Van der Honing & Van Es 1974). In precise work, cognisance should be taken of these differences between types of feed and the extent of departures from the general relationship given in Table 3.4.

Another approach to the estimation of efficiency of energy utilization which derives from that of Kellner is the one adopted by workers at the Oskar Kellner Institute at Rostock in East Germany (see summary by Schiemann et al 1971), who related the increment of energy retained by an animal to the amounts of apparently digested nutrients present in the increment of feed. They found, as indeed had Kellner many years before, that such an equation, while it predicted the energy retention of animals given good diets, failed to predict retention in animals given roughages. They therefore introduced an additional correcting term which has the effect of reducing estimates of energy retention if the digested energy as a proportion of the gross energy is below 0.67. For example, when digestibility is 0.50 the true retention is calculated to be 82% of that predicted. The bases of the East German system are summarized in Appendix 3.1. Apart from the correction term, these equations imply that the efficiency of utilization of the energy of digested crude protein for body gain is about 0.45, of digested crude fat 0.95 to 0.97 and of digested carbohydrate about 0.55. It follows that k_f is unlikely to fall much below 0.50 or, since fat is rarely more than 50 g/kg of the digested nutrients, to rise much above 0.58. The correction term can reduce values to 0.40. Their interest lies in the fact that they show that the net availability of metabolizable energy varies with ration composition and with attributes of the ration associated with its "quality" as measured by its apparent digestibility.

The three methods of prediction of k_f, from fibre content, from metabolizability and from nutrients apparently digested together with digestibility, are all empirical relationships. The differences between feed classes in the constants of prediction equations based on metabolizability further emphasize this empiricism. The reasons for the association between efficiency of utilization of metabolizable energy for fattening and other attributes of the diet are important and a number of studies have been made of the utilization of the energy of the compounds actually absorbed by the ruminant animal in an attempt to elucidate these.

Steam-volatile fatty acids both singly and in mixtures (Armstrong & Blaxter 1957 a & b, Armstrong et al 1958), glucose (Armstrong et al 1960) and casein (Blaxter & Martin 1962) were added to maintenance diets of sheep in a series of 52 experiments. The results showed that efficiency of retention of metabolizable energy varied from nutrient to nutrient. The highest values of over 0.70 were obtained when glucose was given either by continuous infusion into the abomasum or by continuous intravenous infusion, methods of administration which avoid fermentation of the sugar by micro-organisms in the rumen. The lowest values (0.30) were obtained when acetic acid was given. Studies with lipids given in emulsion to sheep showed, in agreement with the work at Rostock described above, that their energy was used with high efficiency (0.797 ± 0.027) for fattening (Czerkawski et al 1966).

From the above experimental observations it appears that, since the molar proportion of acetic acid in the absorbed end products of ruminal fermentation tends to fall as the apparent digestibility of the diet increases and as the fibre content falls, the variation in efficiency of energy utilization for fattening could be due to the low efficiency with which acetic acid is used. This hypothesis receives support from the observations of Blaxter & Wainman (1964) that the efficiency of utilization of the metabolizable energy of diets falls as the molar proportion of acetate in rumen liquor increases. The data of Daccord (1970) obtained with sheep given different rations agree with the relationship given by Blaxter (1962a), namely,

$$k_f = 1.07 - 1.09A, \qquad [3.9]$$

where A is the molar proportion of acetic acid in the steam-volatile acids of rumen liquor (moles/mole).

Several experiments, however, are not in full agreement with this hypothesis, notably those of Ørskov & Allen (1966a, b, c) and Ørskov et al (1966) which led to the conclusion that acetate added to a basal ration is not always poorly utilized. Subsequent work by Hovell & Greenhalgh (1972) and Hovell et al (1976) showed that carcass weight as used by Ørskov & Allen was a poor indicator of fat deposition and that the metabolizable energy of diets containing salts of acetic acid was utilized significantly less well than that of a basal ration. Eskeland et al (1973), in experiments with lambs in which up to a quarter of the energy supply was given as intravenous infusions of fatty acids, fully confirmed the earlier results of calorimetric work which showed acetic acid to be used with low efficiency. Hovell (1972) concluded that the efficiency of utilization of acetate is not constant but decreases as the contribution of acetate to the total metabolizable energy supply increases. This view is consistent with the concept that fat synthesis from acetate depends on a concomitant supply of glucose, propionic acid or glucogenic amino acids to provide reduced coenzymes for fatty acid synthesis. In experiments with dry cows in positive energy balance, Tyrrell et al (1975) found on infusing acetic acid that the partial efficiency of its utilization was 0.26 when the ration was of roughage and 0.69 when the feed contained 70% of grain. This synergism is in agreement with the earlier work on the utilization of the steam-volatile acids (Armstrong et al 1957–1958, Daccord 1970), and suggests that efficient synthesis of tissue constituents from acetate requires the presence of adequate glucogenic precursors (Annison & Armstrong 1970).

The possibility that fermentation heat or the work of digestion might contribute to the low and variable efficiency of utilization of metabolizable energy from different diets has already been discussed and seems unlikely. The explanation of the variation in terms of the balance of nutrients absorbed relative to cellular demand is a sufficient one; it receives support from the observation that in fasted animals acetate as the sole source of energy is poorly used, leads to ketosis and hypoglycaemia, and increases urinary N excretion, and that supplementation with small amounts of propionic acid abolishes these effects to increase efficiency in a non-additive way (Armstrong et al 1957, 1961). However, although there are general relationships to predict k_f from crude fibre, metabolizability, digestibility or molar proportion of acetic acid, there is as yet no prediction procedure to provide exact estimates of k_f for every dietary combination.

(b) The growing animal

In the adult animal the energy retained as fat varies from 85% to 95% of total energy retention. In some young animals, fat retention may account for as little as 50% of the retained energy.

Experiments in several species of animal during growth have been concerned with separating the energy cost of synthesizing protein from that of synthesizing fat. These studies, summarized by Breirem & Homb (1972) and by Kielanowski (1976), show that the efficiency with which simple-stomached species or infant ruminants deposit fat is about 0.70, but that the efficiency of protein deposition is much lower, about 0.45. In rats, values for protein deposition of 0.40 (Schiemann 1970) and 0.43 (Pullar & Webster 1974) have been obtained and with pigs a value of 0.43 (Thorbek 1970). The high efficiency of fat deposition is in accord with biochemical calculation (Armstrong 1969) and the assumption that it constitutes a reserve turning over slowly. With protein the values are less than can be predicted on biochemical grounds and presuppose high rates of turnover. Few estimates of the relative energy costs of protein and fat synthesis in an adult ruminant have been published. Rattray et al (1974c) found efficiency values of 0.845 ± 0.291 for fat and 0.123 ± 0.023 for protein and in a series of slaughter experiments Rattray & Joyce (1976) found efficiency values for fat and protein deposition to range widely, some values for fat being negative! These authors comment, as do most authors, on the difficulty of separating the terms statistically when there is so much autocorrelation and the values obtained obviously have to be interpreted with caution.

Such observations suggest that there would be large differences in the efficiency of utilization of energy in growth as contrasted to fattening in ruminant animals. Studies by Blaxter et al (1966b), however, showed that there were no differences in the values of k_f obtained during the growth of Ayrshire bullocks from 15 to 81 weeks and fasted weights of 70 to 290 kg. The energy deposited as protein in the total energy gain varied but slightly from 32% to 25% as the animals grew. The protein energy contents of the gains of productive ruminants generally fall within the range 30 to 15%; it may be computed that if the value of k_f in a mature animal is 0.50 then in a very young animal it could, on the basis of the studies quoted for other species, fall only to 0.47. Graham (1969), in studies with very fat sheep, found no effect of fatness itself on efficiency, while direct comparison of energy metabolism in 3-month-old lambs and a 4-year-old wether sheep given the same pelleted diet by Bouvier & Vermorel (1975) showed no significant difference between them in the overall efficiency with which metabolizable energy was used to support gain.

In studies with veal calves by van Es et al (1967) the energy retained as protein fell from 40% to 25% of total retention as the animals grew. No evidence of low efficiency of utilization of energy for growth and fattening was found in the younger animals; the overall efficiency of utilization was 0.69, little different from the values for fat deposition — admittedly from carbohydrate and not from lactose, fat and protein — found in other species. Had protein been the substrate for growth k_f would be expected to be 0.60. In view of the high fat content of the diets a greater efficiency might have been expected.

These results suggest that no great error is incurred when values for k_f obtained in sexually mature animals are applied to younger ruminants in the growing stage and consuming fibrous diets. For calves and lambs subsisting on milk or milk replacers efficiency is higher than in ruminating animals. Table 3.5 summarizes estimates that have been obtained. The high value obtained by Gonzalez-Jimenez & Blaxter (1962) is anomalous and a mean value of 0.70 has been taken.

Table 3.5   Efficiency of utilization of metabolizable energy for growth in ruminants given milk or milk substitutes.

Authority	Species	Efficiency
Gonzalez-Jimenez & Blaxter (1962)	Calf	0.77–0.81
van Es et al (1969)	Calf	0.69
Vermorel et al (1974)	Calf	0.69
Webster (1976)	Calf	0.72
Holmes et al (1975, 1976)	Calf	0.67
Walker & Jagusch (1969)	Lamb	0.71
Walker & Norton (1971b)	Lamb	0.69

6. Efficiency during pregnancy

The growth of the foetus and the uterus during pregnancy together with the growth of the mammary gland represents an obligatory component of growth which can be modified to a limited extent by diet, sheep being more susceptible to modification than cattle. The later stages of pregnancy are associated with an increase in heat production of the pregnant animal even though diet is kept constant (Brody 1945, van Es 1961, Ritzman & Benedict 1938, Flatt et al 1969a, Hashizume et al 1965 a & b). In the previous edition, these indications of low efficiency of utilization of metabolizable energy for gain during pregnancy were noted but for simplicity it was assumed that the efficiency of the gain in pregnancy was the same as that in normal growth and fattening and that the maternal maintenance requirement increased. Here data relating to utilization of metabolizable energy for deposition are considered.

Moe et al (1970) in the USA related metabolizable energy intake when cows were neither gaining nor losing maternal tissues (estimated by assuming the efficiency of utilization of metabolizable energy for a gain in weight to be 0.60 and for a loss in weight, 0.85) to estimates of the energy content of the foetus and foetal membranes as determined by Jakobsen ct al (1957). The latter is an exponential function of time from conception. They found that the efficiency of utilization of metabolizable energy (k_p) was 0.11 to 0.12. Henseler et al (1973) made similar trials in East Germany, also using Jakobsen's exponential equation to describe the energy content of conception products in the cow. In 11 animals involving 68 measurements, the mean efficiency of utilization was 0.149 ± 0.075, a value close to that of Moe et al (1970).

Graham (1964a) measured the total heat produced during pregnancy above that noted in the nonpregnant state in sheep given constant feed, together with the heat of combustion of the conceptus at term. Efficiency of utilization of the metabolizable energy for pregnancy was on average 0.20. When ewes were given an increment of feed, only 1% of the metabolizable energy of the increment was stored in the conceptus and 46% in the maternal body, a joint efficiency of 0.47 compared with 0.53 in the nonpregnant animal given the same diet in similar amounts. The difference in the increments of heat associated with pregnancy for ewes given 600 g and 900 g feed per day gave an estimate of efficiency of utilization of metabolizable energy for growth of the conceptus of 0.13. In this respect, Lodge & Heaney (1970) showed that a similar low efficiency of utilization for anabolism during pregnancy applies when maternal body tissue rather than feed is the energy source. Graham's (1964a) data were recomputed so as to apply to the course of pregnancy by Langlands & Sutherland (1968); they used their own data on the rate of deposition of energy in the conceptus. These workers suggest that efficiency might be lower at term than earlier in pregnancy, but the data are not a sufficient basis on which to estimate such a relationship.

Using comparative slaughter methods, Sykes & Field (1972c) calculated the energy required to form the conceptus during the last 112 days of pregnancy in underfed ewes, which lost body fat and protein during this time. They found values of 0.142 and 0.124 for the efficiency of ewes given diets containing 118 and 60 g protein per kg. Also by comparative slaughter methods, Rattray et al (1974a) estimated efficiency of utilization of metabolizable energy for conceptus development in the last 85 to 100 days of gestation to be 0.120 to 0.135, depending on the statistical model used.

None of the estimates (if Graham's differential quantity is accepted rather than his absolute one) of the efficiency of utilization of metabolizable energy for the production and maintenance of the conceptus (k_p) and for meeting any augmentation of metabolism of the mother departs greatly from a mean of 0.133. There is no evidence to show whether this efficiency varies with diet, other than the suggestion from the work of Sykes & Field (1972c), nor do the data allow estimation of the efficiency of conversion of maternal tissue to foetal tissue. In the absence of such data, pregnancy energy requirements have been assessed to be $1/0.133 = 7.5$ joules per joule conceptus deposited per unit time.

Such an estimate ignores energy required during pregnancy for growth of the mammary gland, and for any obligatory or desirable increase in maternal energy or protein stores. It also ignores possible changes in the metabolizability of feed during pregnancy. Experiments relating to changes in apparent digestibility of diets during pregnancy have been reviewed by Henseler et al (1973) who themselves had found no effect; generally any effects found have been small though the passage of feed through the gut of pregnant sheep is faster than in nonpregnant ones (Graham & Williams 1962).

7. Efficiency in lactation

The efficiency of utilization of metabolizable energy was defined earlier as the increase in energy retention which occurs for unit increase in the metabolizable energy supplied. This can be measured in a fattening or growing animal but in a lactating animal the comparable measure of increase in milk energy secretion for unit increase in metabolizable energy is not an unique attribute of a diet, for increments of feed result in changes in both milk secretion and the deposition of energy in the body. Lactation is not determined solely by feed intake, and the relation between energy intake and milk energy yield consists not of a single curve but of a series which reflects the potential of the animal to produce milk, a potential which varies with the individual, her age and stage of lactation, and her nutritional history. To predict lactational performance entails assembling information about the factors which affect the incremental response of milk secretion and the corresponding incremental response of deposition of energy in the body, that is the incremental partition of metabolizable energy. Little calorimetric work has been done to establish the determinants of this partition, but measurements of milk energy yield and live weight change (Blaxter 1956, 1966, Broster 1972, 1974) provide useful information of practical value. The implications of these studies will be dealt with later.

Most of the calorimetric work with milking animals has been descriptive, and as such can be manipulated to estimate the efficiency with which either metabolizable energy or the energy reserves of the body are used to promote milk secretion and with which feed energy is used to increase energy reserves in the lactating animal. The problem of partition is not considered at all in such studies. Accordingly, while the results of these many trials permit good estimates of the energy required by a lactating animal which neither loses nor gains body energy, they give limited guidance to the practical feeding of the cow. The extent of these limitations and the relevance of the descriptive data compared with the more dynamic approach to the problem of partition are discussed later. The three efficiency terms in the lactating animal are:

(a) the efficiency of utilization of metabolizable energy for milk secretion in the absence of change in the energy content of the body;

(b) the efficiency of utilization of metabolizable energy in promoting gain of energy by the body when lactation occurs simultaneously;

(c) the efficiency with which the energy of body fat and protein is used to promote milk secretion when the metabolizable energy supplied is less than that needed to achieve zero energy retention.

They are discussed below, but in reverse order. Since the resolution of most of these efficiency terms has been achieved by simple or multiple regression methods, some comments on these methods are made.

If metabolizable energy intake, M_E, and milk energy yield, Y_E, in animals not gaining or losing body energy are measured, the relation between them can be expressed as:

$$Y_E/W^{0.75} = k_l (M_E/W^{0.75}) - a \qquad [3.10]$$

$$M_E/W^{0.75} = (1/k_l) (Y_E/W^{0.75}) + \alpha \qquad [3.11]$$

where k_l is the efficiency of utilization of metabolizable energy for lactation, and a and α are intercepts. Maintenance ($Y_E = 0$) is given by α or a/k_l. The estimates of k_l

and of maintenance needs of metabolizable energy obtained by these two approaches will agree only if Y_E and M_E are free of error. Generally k_l and maintenance energy estimated from equation 3.10 will be lower than values from equation 3.11. It can be argued that if the objective is to estimate metabolizable energy requirements from a knowledge of milk energy yield, then equation 3.11, which minimizes variation in metabolizable energy, should be used. If, however, the objective is to estimate the milk energy yield which would occur (in the absence of change of body energy) for a given amount of metabolizable energy, then equation 3.10 should be chosen. Mathematical techniques are available to determine the true underlying relationship and this results in values of the constants between the two estimates of k_l and of maintenance. Unfortunately, without access to original data the computation cannot be made and few experiments have provided estimates of parameters of this nature (Brouwer et al 1965). The problem has been discussed by Moe et al (1972) who have published the results of their trials conducted up to 1972 in both the ways illustrated by equations 3.10 and 3.11 above. The East German School (see Hoffmann et al 1974) have for the most part used constrained regressions of the form:

$$M_E = (1/k_l) Y_E + b\, W^{0.75} \qquad [3.12]$$

and estimated the constants k_l and b. It can be expected that the value of k_l so obtained will lie between the values obtained by the other methods.

(a) Utilization of body energy for milk secretion

By regression analysis of data on metabolizable energy intake and milk energy yield of cows which were losing energy from their bodies, Moe & Flatt (1969) obtained two alternative estimates of the efficiency of utilization of body energy for milk secretion, 0.84 ± 0.02 (milk energy the dependent variable) and 0.86 ± 0.08 (metabolizable energy the dependent variable). The same workers in a subsequent analysis with 126 records found a value of 0.840 for which an error of ± 0.036 may be computed. Using a different approach involving calculation of the presumed lactational performance, Schiemann et al (1974) found a value of 0.808 ± 0.039, while van Es et al (1970) obtained a value of 0.90. A value of 0.84 appears to be highly representative of the data and has been taken.

(b) Concomitant energy deposition in lactation

A number of older calorimetric experiments, reviewed in the first edition, showed that when cows were given constant rations and their milk yield was allowed or encouraged to decline, a reduction in milk energy yield by 1J resulted in an increase in energy deposition in the body of 0.78J (Nehring & Schiemann 1956) or 0.84J (Møllgaard 1929). Such results indicate that the utilization of energy for milk secretion is a more efficient process than is fattening in the adult. In experiments with goats, however, in which besides a secular trend in yield of milk, its composition was changed by infusion of steam-volatile fatty acids (Armstrong & Blaxter 1965), a fall in milk energy yield of 1J was associated with an increase in body energy retention of 0.96J, a much higher figure. The higher value obtained was interpreted to mean that during lactation the deposition of body fat was energetically as efficient as the secretion of milk fat. In addition, these results showed that the efficiency of body fat deposition during lactation ($k_{f(l)}$) was 0.66, considerably higher than that noted when the same diets were given while the animals were dry (0.44 to 0.52) and of the same order as that found for lactation itself ($k_{l(o)}$=0.69). Regression analysis of the results of 350 experiments with lactating cows by Moe et al (1970) showed that the efficiency with which metabolizable energy was used for lactation in the absence of gain in body energy, $k_{l(o)}$, was 0.644, and for body gain during lactation $k_{f(l)}$, 0.727. When the cows were not lactating the efficiency of utilization of metabolizable energy for body gain (k_f) was significantly lower at 0.596. These observations agreed with those of Armstrong & Blaxter (1965) in showing a higher efficiency of fat deposition during lactation than in the non-lactating animal, although in the American experiments the range of milk energy secretion involved was due not so much to changes in fat secretion as to changes in yield with time.

In a statistical analysis with metabolizable energy intake as the independent variable and a complex variable (milk energy plus either the positive concomitant energy retention or 0.8 or 0.9 times the negative concomitant energy retention), van Es et al (1970) came to the conclusion that positive energy retentions should be weighted by a factor of up to 1.1, inferring that the efficiency of the concomitant retention of energy in the body of the cow is not equal to the efficiency of her utilization of energy for lactation but 90% of that. The authors realized the difficulty of estimating this quantity. In the East German studies of lactation (see Hoffmann et al 1974, for references), it has been usual to ascribe to concomitant energy deposition during lactation the same efficiency as given to energy deposition in fattening oxen. The fact that this group found a significant positive correlation between $k_{l(o)}$ and energy retention in the body (Hoffmann et al 1972) suggests that the assumption may not be correct though the correlation was attributed to a higher metabolism of the cow in early lactation. The effect of assumptions about maintenance needs and the correction factors applied is well illustrated by results of Flatt et al (1965), later recalculated (Moe et al 1972). In the first calculation the efficiency of utilization of energy for body gain was assessed in dry cows to be 0.50 and from body loss to be 0.79. The intake of metabolizable energy so corrected when regressed on milk energy yield gave values for the efficiency of milk secretion of 0.45 to 0.56. In the second calculation, the actual amount of metabolizable energy consumed was related to the sum of milk energy secreted, body energy gained, and body energy loss multiplied by 0.84. The values for efficiency of milk secretion were increased to 0.57 to 0.66.

It seems reasonable from the above to accord to concomitant energy storage during lactation an efficiency which is 0.95 times that for milk secretion.

(c) Efficiency in the absence of change in body reserves

From the above it is clear that assumptions about the methods to be used to adjust for changes in body reserves of energy in the lactating animal could affect the estimates of the efficiency of utilization of energy when there is no change in the body energy content. Provided energy retentions are small relative to the amount of energy secreted in milk, errors from injudicious choice of adjustment factors are not likely to be great. Generally, in most recent work milk energy secretion has exceeded energy retention in the body by a factor of 3 to 4, thus reducing the effects of errors of assumption. Nevertheless, as between the American and Dutch workers on the one hand and the East German workers on the other (see van Es 1975, for references) there is a difference in approach. The American and Dutch workers estimate the quantity

$$\left\{ \begin{array}{ccc} \text{milk} & \text{energy} & \text{energy} \\ \text{energy} + \text{retention} + 0.8\,\text{retention} \\ \text{yield} & \text{(positive)} & \text{(negative)} \end{array} \right\} = Z,$$

and regard this as the quantity which is related to metabolizable energy above a maintenance datum by an efficiency term, k_l, according to equations 3.10 and 3.11. The East German workers (Hoffmann et al 1972) reduce the measured metabolizable energy intake by an amount calculated to be required to meet a positive energy retention (i.e., energy retention/k_f) and reduce the milk energy secretion by either 0.75 (Hoffmann et al 1972) or 0.81 (Schiemann et al 1974) times any negative energy retention. The metabolizable energy and milk secretion thus apply specifically to a computed zero body energy retention. The metabolizable energy above maintenance in many of the separate trials has been estimated by the East Germans by assuming that metabolizable energy is used for maintenance with an efficiency measured for fattening (k_f) rather than for maintenance (k_m) (see p. 78). In the general treatment of the data, however, equation 3.12 above has been used by the East German workers. When energy retention is negative the two approaches are similar. The American and Dutch approach assumes that concomitant energy deposition in lactation is as efficient as the secretion of milk; the East German approach assumes that it is considerably less.

Regression analyses of the results of the various groups, with either milk secretion at zero energy balance (East Germany) or the complex measure of lactational energy yield (USA, The Netherlands) are summarized in Table 3.6. These estimates apply to all diets given irrespective of their quality. A major question relates to the extent to which the values of k_l vary with the diet composition. Analysis by the East German workers (Hoffmann et al 1972) of their own results gave a mean value of k_l in 134 trials of 0.619 with a range from 0.510 to 0.814, but no correlation between the metabolizability of gross energy and the efficiency. Even so, by regression analysis they calculated that the energy of digested nutrients was apparently used with different efficiencies as given in Table 3.7 from which it might be expected that starchy feeds would result in higher efficiency than fibrous ones. An analysis of 198 determinations of energy retention made at Wageningen by Van Es (1972), in which metabolizability of the gross energy (q) was used as an additional independent variable in equation 3.10 above, showed lactational efficiency to be 0.597 when q was 0.564, and to increase with increasing q. An increase of q from 0.56 to 0.66 would increase

Table 3.6.   Equations for estimating the efficiency of utilization of metabolizable energy for milk secretion.

Symbols
$Y_{E(0)}$,   milk energy yield corrected to zero balance
$M_{E(0)}$,   metabolizable energy corrected to zero balance
M_E,   observed metabolizable energy
$Y_{E(c)}$,   corrected milk energy yield, including energy retention in the body duly weighted
All data are divided by weight raised to the power 0.75 and expressed as MJ/day

Authority	Equation	Derived efficiency
Moe et al (1972)	$M_E = 1.467\ Y_{E(c)} + 0.561$	0.681
	$Y_{(E)} = 0.629\ M_E - 0.316$	0.629 ± 0.008
Hoffmann et al (1972) Mid lactation	$M_{E(0)} = 1.62\ Y_{E(0)} + 0.429$	0.617 ± 0.051
Schiemann et al (1974) Early lactation	not stated	0.653 ± 0.056
van Es (1975) Long forages	$Y_{E(c)} = 0.61\ M_E - 0.301$	0.61
Long forages and pellets	$Y_{E(c)} = 0.52\ M_E - 0.180$	0.52
Fresh and frozen grass	$Y_{E(c)} = 0.72\ M_E - 0.485$	0.72
Holter et al (1972)*	$Y_{E(c)} = 0.685\ M_E - 0.362$	0.685

* mean of 4 equations.

Table 3.7.   Efficiency of utilization of the metabolizable energy of different nutrients for lactation and for fattening (Hoffman et al 1972).

	Efficiency of utilization of metabolizable energy for:	
	Lactation	Fattening
Digested crude protein	0.51	0.40
Digested crude fat	0.72	0.93
Digested crude fibre	0.56	0.53
Digested N-free extractives	0.66	0.55

k_l from 0.597 to 0.635, implying that:

$$k_1 \simeq 0.38q + 0.385. \qquad [3.13]$$

Van Es has also analysed data from Beltsville experiments (Van Es, private communication) using the same basic equation. This can be interpreted to imply that:

$$k_1 \simeq 0.28q + 0.466. \qquad [3.14]$$

Values computed from these two equations show that for a value of q of 0.40, k_l is about 0.56, when q is 0.60, k_l is about 0.62 and when q is 0.7, k_l is about 0.66, the data from Beltsville being systematically higher than those from Wageningen. It thus appears that the decreases in metabolizable energy needs for lactation with increase in the metabolizability of the diet as estimated from q are very similar to those which occur in maintenance. From the form of the relationship used by van Es, separation of these two effects is not feasible. It is therefore proposed that the slope of the relationship between k_l and q is taken to be 0.35, the same as for the maintenance function, but that the intercept term is lower than that for maintenance at 0.42, rather than 0.50. Preferred values were given earlier, in Table 3.3.

There is further evidence that efficiency of utilization of metabolizable energy for lactation is subject to variation which is not simply related to the metabolizability of the diet. Thus, van der Honing & Van Es (1974) showed that pelleting diets increased the efficiency with which they were used, while Moe et al (1972) showed that diets rich in protein depressed utilization. Furthermore, there are real and statistically significant differences between diets in both the American and the East German series of experiments.

(d) Partition of metabolizable energy in lactation

The efficiencies with which the energy of body reserves or the metabolizable energy of feed is used to support milk secretion, and the efficiency with which metabolizable energy is used to promote energy retention in the body during lactation, can be used constructively to estimate the requirements of energy of a lactating animal only by stating either that there is no change in daily body energy retention or that energy retention has some fixed value. In the latter instance the value might be negative for animals in early lactation and positive later, or might vary with the absolute rate of yield. Such devices imply that ideal patterns of concomitant change in body energy content can be stated for animals of different lactational propensity. This is debatable.

There is practical evidence (see Burt 1957, Broster 1972, 1974) that the response of the milk yield of the individual animal to an increase in feed is not that to be expected on the assumption that the metabolizable energy of the feed is converted to milk with a constant efficiency. Ekern (1972b) summarized 35 experiments made in different countries in which the response of milk production to additional feed was measured. These showed that the increase in milk yield ranges from about one tenth to about two thirds that to be expected were the additional feed used solely to support lactation. They also showed that the response in milk yield is greater when the cows are initially underfed, implying, as has long been known and was well demonstrated by Jensen et al (1942), that the law of diminishing returns applies to the relationship between milk secretion and feed intake.

In feeding trials, the concomitant response of milk secretion and of weight change to unit change in feed intake was measured by Blaxter (1966) and by Broster et al (1969b, 1975) to show that in the absence of change in live weight, the response of milk yield to feed was that to be expected from estimates of feed requirement derived from calorimetric trials, that in the absence of a response of yield the increase in weight was also within the range of expectation from energy metabolism studies, and that the observed responses in yield and weight change represented an inverse proportionality.

Both series of studies also showed that for the particular feeding systems employed the responses to additional feed increased with the milk yield of the cow. In Blaxter's experiments, the response of milk energy yield in kg/day per MJ increment in the metabolizable energy supply when the Woodman standards were employed was related to the yield of the cow (Y, kg/day):

$$dY/dQ_s = 0.0074Y_s - 0.0131, \qquad\qquad [3.15]$$

where $dY/dQ = $ kg milk/MJ metabolizable energy and

$\qquad$ $Y_s = $ milk yield when the cow was fed according to Woodman's standards.

If the heat of combustion of milk is taken to be 3.1 MJ/kg, the marginal energetic efficiency of lactation increases from 0.31 when daily yield is 15 kg to 0.66 when it is 30 kg. It was further shown that this relationship applied not only to cows of different producing ability but to the individual cow when yield varied with lactation stage. Broster et al (1969b) found precisely the same regression in their studies. In later experiments Broster et al (1975) considered only effects of downward changes in ration. The slope was 0.0108 ± 0.0011 kg milk/MJ metabolizable energy per kg of initial yield, a value slightly greater than the earlier estimates.

These results indicate that information is accruing from practical feeding trials which permits a more dynamic approach to the feeding of cows. Such an approach was given in the previous edition in which an econometric analysis was used to formulate a feeding system (K. L. Baxter & H. Ruben, unpublished).

(e) Conclusions on energy utilization in lactation

There is general agreement that body reserves are used to sustain lactation with an energetic efficiency of 0.84, and that the concomitant deposition of energy in the tissues is more efficient than that which occurs in normal fattening of the adult nonpregnant animal. The efficiency may be less than that for lactation, probably about 0.95 lactational efficiency. Lactational efficiency varies around a mean value of 0.62. At present precise estimates for different types of diet cannot be given, but evidence suggests that it varies with the metabolizability of the diet to about the same extent as does efficiency for maintenance. Preferred values are given in Table 3.3 (p. 81). These values are lower than those given in the previous edition.

Even so, such data allow an approach to the problem of meeting the energy requirements of lactation only *post hoc ergo propter hoc*, in that they allow energetic description of what has occurred. Their use to estimate optimal needs predicates stating ideal relations between body energy retention and milk secretion. An alternative approach is to examine the response on milk yield to giving feed above the maintenance level to animals of different potential for lactation. In such an approach the calorimetric data provide one means of assessing the resultant changes in body reserves of energy.

Estimates of Requirements

The components of the total energy requirement of an animal are the loss of energy from the body during fasting, comprising the fasting heat production and the fasting urinary energy loss, the energy expended in muscular work, including that of standing and lying, the accretion of energy by the body resulting from growth, fattening and pregnancy, and the energy lost to the body in secretions such as milk. Data relating to the energy gain of the body, that which occurs during growth of the foetus and the energy value of milk, have been summarized in Chapter 1; this part deals with the remaining components.

1. Fasting metabolism *(a) Cattle*

Table 3.8 summarizes 11 sets of observations on the fasting metabolism of cattle. The estimated values at different weights were based on linear regressions of fasting metabolism on weight. There were significant differences between laboratories in these regressions, but none of the attributes of the animals — breed type or sex — could be used as a basis for pooling the data. The linear slopes, however, grouped around a mean value of 55 kJ/kg per day and in most instances the power of weight with which metabolism varied was less than 0.75, the interspecies mean. Despite the significant differences the data were pooled for subsequent analysis, with the exception of those of Ritzman & Colovos (1943) and those from Webster's laboratory. In the former case corrections had been made to the data since the fasts were of short duration while in the latter the animals had been fasted immediately after having

Table 3.8   Estimates of the fasting metabolism of cattle: results of linear regression analysis of observations made on a single laboratory basis.

Source	No. of observa- tions	Fasting metabolism (MJ/day) at body weight (kg):					
		100	200	300	400	500	600
(1) Ritzman & Benedict (1938) (Holstein)	8	—	—	—	29.0	32.2	35.5
(2) Forbes et al (1927 1928 1931) (Shorthorn)	16	—	—	24.5	30.8	37.1	—
(3) Hashizume et al (1962) (Holstein)	6	—	—	—	31.2	33.1	—
(4) Blaxter & Wainman (1966) (Ayrshire)	14	12.4	19.2	25.9	32.6	—	—
(5) Blaxter & Wainman (1966) (Aberdeen Angus)	8	—	—	—	27.0	32.3	—
(6) Mitchell & Hamilton (1940 1941) (Shorthorn)	19	10.9	16.5	22.1	27.8	33.4	39.1
(7) Flatt & Coppock (1963) (Holstein)	6	—	—	—	—	34.3	39.0
(8) Vercoe & Frisch (1974) (Hereford × Shorthorn)	11	—	21.0	29.7	—	—	—
(9) Ritzman & Colovos (1943) (Holstein)	8	17.1	22.3	27.5	32.7	37.8	—
(10) Webster et al (1974) (Friesian)	8	17.0	24.0	31.0	38.0	—	—
(11) Webster et al (1974) (Aberdeen Angus × Friesian)	8	17.0	23.5	30.0	36.5	—	—
Values computed from 0.53 $W^{0.67}$		11.6	18.4	24.2	29.4	34.1	38.5

Fig. 3.3   Fasting metabolism of cattle (see Table 3.8 for key to sources).

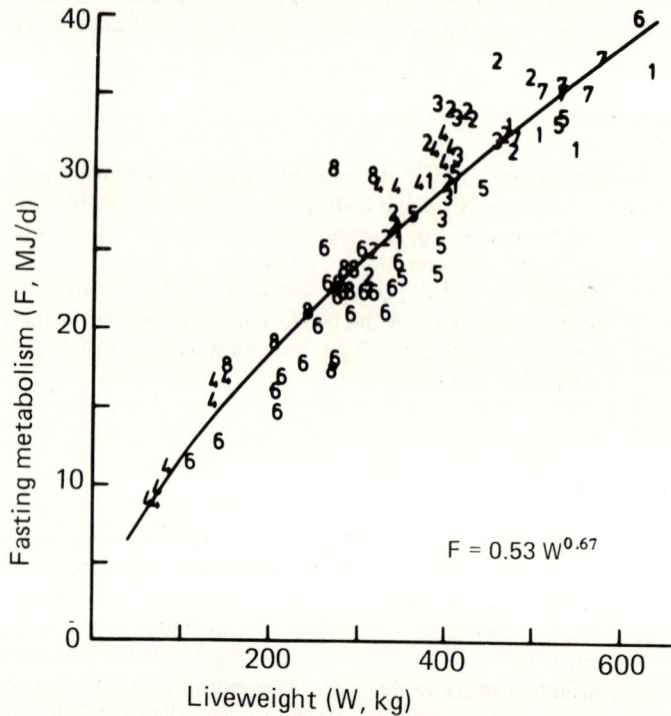

$$F = 0.53 \, W^{0.67}$$

been fed to gain at very high rates. The best fitting relationship for values obtained from 88 animals (Fig. 3.3) was $F = 0.53 W^{0.67}$ (residual standard deviation ± 2.5), where F is fasting metabolism (MJ/day) and W is fasted weight (kg). A variety of other descriptive equations was examined and it was found that the power equation gave the smallest standard deviation. Values predicted from the equation are given in Table 3.9; they are not divergent from those given in the previous edition at live weights of 400–600 kg, but are much less at weights of 200 kg and below. This largely stems from the heavy reliance placed in the earlier edition on the results of Ritzman & Colovos (1943), excluded in the present analysis.

The number of direct observations of fasting metabolism made with animals weighing less than 200 kg is in fact small, and with very young animals during the first weeks of life it is difficult to obtain fasting values which are stable. Judgement

Table 3.9.   Values of fasting metabolism of heifers and castrates (F, MJ/day) predicted from the equation $F = 0.53W^{0.67}$. (Values for bulls are greater by 15%).

Fasted body weight (kg)	Fasting metabolism (MJ/day)	Values from ARC (1965)* (MJ/day)
50	7.3	10.2
100	11.6	16.3
200	18.4	25.0
300	24.2	29.5
400	29.4	31.5
500	34.1	33.2
600	38.5	35.7

* An estimation of weight for age was necessary since the ARC (1965) scheme related fasting metabolism to age rather than fasted weight.

about the reliability of the estimates of fasting metabolism in young animals can, however, be made from published estimates of energy requirements for maintenance. Most of these have been obtained by regressing energy retention on metabolizable energy intake, or alternatively heat production on energy retention, all variables being scaled by dividing by metabolic body size ($W^{0.75}$). From such data the metabolizable

Table 3.10 Estimates of the metabolizable energy required to maintain weight in calves given liquid diets of milk or milk substitute.

Source	Breed and diet	Maintenance requirement (MJ/day per kg metabolic size, $W^{0.75}$)
Blaxter & Wood (1952)	Ayrshire, whole milk	0.455
Gonzalez-Jimenez & Blaxter (1962)	Ayrshire, whole milk	0.469
van Es et al (1969)	milk replacer	0.436
Vermorel et al 1974)	milk replacer	0.402
Holmes et al (1975)	Jersey, whole milk	0.409
Holmes & Davey 1976)	Friesian, whole milk	0.393
Webster et al (1976)	Friesian, milk replacer	0.675
Kirchgessner et al (1976)	German Brown, milk replacer	0.431

energy required or heat produced when energy retention is zero can be estimated and this is the maintenance requirement. Table 3.10 summarizes eight sets of observations relating maintenance energy requirements of calves given milk or milk substitutes to live weight.

If the anomalous data of Webster et al (1976) are excluded, the unweighted mean of the remainder is 0.428 MJ/kg $W^{0.75}$ per d. The net availability of the metabolizable energy of liquid diets in calves for maintenance is likely to be high, comparable to that in species with simple digestive tracts, and from the general equation in Table 3.2 is probably in the range 0.80–0.90. This suggests the fasting metabolism is 0.342–0.385 MJ/kg $W^{0.75}$ per d. For an animal weighing 50 kg this corresponds to 6.4–7.2 MJ/d. Although the lower limit is somewhat less than the independently estimated fasting metabolism of 7.3 MJ/d given in Table 3.9, these data provide support for the present estimates of fasting metabolism at the lower body weights as opposed to those given in the previous edition.

In a similar way, the reliability of the estimates of fasting metabolism for mature and ruminating cattle may be checked from statistical estimates of maintenance requirements. These cannot be precise since the calculations necessarily assume that the net availability of metabolizable energy for maintenance can be regarded as constant. Table 3.11 summarizes the results of analysis of compilations of maintenance data in chronological order. The results are complicated since some data have been included in both earlier and later analyses. Table 3.11 shows that the mean estimate

Table 3.11 Estimates of the maintenance energy requirements of cattle made by statistical methods.

Authority	Source of data	Mean metabolizability (q)	Maintenance (kJ/kg $W^{0.75}$ per day)	Derived fasting metabolism
Breirem (1953	All US data on fattening cattle	0.50	464	314
van Es (1961)	All US and European data on fattening cattle	0.50	469	318
van Es & Nijkamp (1967)	Own data on dry cows	0.54	443	306
van Es (1970)	All published data on milking cows	0.59	489	346
Moe et al (1970)	Lactating cows, own trials	0.50	510	346
	Dry cows, own trials	0.50	418	283
Mean				319

of fasting metabolism of all animals derived from the maintenance data is 319 kJ/W$^{0.75}$ per day. For a 500-kg cow this corresponds to 33.7 MJ/day which agrees we with the value of 34.1 given in Table 3.9.

It is possible that the metabolic rate of milking cows is higher than that of d animals; work in India (Patle & Mudgal 1976, 1977) suggests a 20% great requirement. It will be apparent from the discussion on the efficiency of utilizatic of metabolizable energy in lactation that a large element of autocorrelation is implic in the analyses which resulted in these higher estimates. It is considered that tl estimates of fasting metabolism in Table 3.9 should be taken as the best available f fasting metabolism in female cattle and castrates. There is evidence from calorimetr trials (Webster et al 1976) that the maintenance requirements of bulls are about 20 greater than those of castrates, while some of the fasting data of Vercoe (1970) al suggest a value higher by about 16%. Rams have a higher fasting metabolism tha castrated male sheep and ewes (see p.99), and until further information is availab the 15% higher fasting metabolism adopted for intact male sheep has been taken apply to intact male cattle.

When all the data were pooled there was no clearcut difference in fasting metabolis which might be ascribed to breed. Nevertheless, it is probable that such bree differences exist. Dairy animals were claimed by Ritzman & Benedict (1938) to ha a 9% higher metabolism than beef animals; Blaxter & Wainman (1966) four differences in the metabolism of dairy (Ayrshire) and beef bullocks, the beef anima having a metabolism 19% lower than dairy animals, and crosses between the bree being intermediate. In three comparisons using comparative slaughter methods mac between Holstein and Hereford bullocks, Garrett (1971) found that the Holstei had the higher maintenance cost, individual differences being 5%, 13% and 11% Undoubtedly, as more data accrue, these differences will be established with precisio and so explain the variation associated with the present mean estimate of fastir metabolism.

(b) Sheep

Data presented in the first edition showed good agreement between measuremen from seven laboratories of the fasting metabolism of weaned sheep. There was le information about the sucking lamb. A number of additional compilations of fastir data on sheep in post-weaning life are now available together with estimates of fastir metabolism arrived at by regression analysis of information obtained during feedin,

Graham et al (1974) measured fasting metabolism in 28 sheep on 212 occasior and noted that almost 90% of the variation in metabolism was associated wit variation in weight raised to the power 0.75. Subsequent analysis showed significar effects of age and of the amount of feed given immediately before the fast, but n of intake over the preceding 1 to 2 months. The final equation for weaned sheep kep at maintenance before the fast was

$$H_F/W^{0.75} = 257 \exp(-0.08t) + 18.4, \qquad [3.16$$

where H_F is fasting heat production (kJ/day), t=age in years and W is weigh Estimates of fasting metabolism expressed per kg $W^{0.75}$ from this equation ar compared with those of Blaxter (1962a 1966) in Table 3.12. The latter were the bas on which an age decline in fasting metabolism was included in the previous editio In the age range covered by Graham et al (1974), the agreement between the two se is excellent. In support are estimates of fasting heat production derived from regressio methods. Rattray et al (1973b) estimated that for 40 yearling wethers fastin metabolism was 263 kJ/kg $W^{0.75}$ per day, and direct determinations by Ursescu et al (1972) gave a value of 211 kJ/kg $W^{0.75}$ per day. It can be assumed that the data for weaned wether sheep accumulated since 1965 agree with those given in the previous edition.

Table 3.12.   Fasting metabolism of weaned wether sheep or ewes of different ages estimated from the equation of Graham et al (1974) and mean data obtained in Scotland (kJ/kg $W^{0.75}$ per day).

Age of sheep (years)	Graham et al predicted from Australian experiments‡	Values found in Scotland
0.75	260	251†
1.5	246	244*
3	220	226*
5	191§	214*
>6–7	165§	202*

* Blaxter (1966).
† From ARC (1965) preferred values.
‡ $H_F/W^{0.75} = 257 \exp(-0.08t) + 18.4$, where t = age in years.
§ These values involve considerable extrapolation beyond the data of Graham et al (1974).

The data of Graham (1968) and those of Rattray et al (1973b) showed no significant difference in metabolism as between ewes and wether sheep. There is, however, evidence that rams have a higher fasting metabolism than ewes. Joshi (1973) found the mean metabolism of 14 rams aged 2 to 5 years to be 254 kJ/kg $W^{0.75}$ per day which is higher by about 12% than the expectation from Table 3.12. Graham (1968) found that rams had a metabolic rate of 268 kJ/kg $W^{0.75}$ per day compared with 226 kJ/kg $W^{0.75}$ for ewes and wethers of the same breed, which suggests a metabolism some 18% higher. It is proposed that until more information is available the fasting metabolism of the intact male is taken to be 15% higher than that of the castrate. Breed differences have been established by Blaxter et al (1966c), but not with sufficient numbers to establish breed norms.

With young lambs given milk the only observation available on fasting metabolism when the first edition was prepared was that of Walker & Faichney (1964b) which suggested fasting metabolism to be 284 kJ/kg per day or 425 kJ/kg $W^{0.75}$ per day for a 5-kg lamb. Graham et al (1974) measured metabolism in lambs after fasts of 72 to 120 hours and showed that it varied from 279 to 322 kJ/kg per day according to the amount of liquid diet given before the fast began. Generally the fasting metabolism of lambs given liquid diets was some 23% higher than that of lambs of the same weight given dry diets. Irrespective of feeding level Graham et al (1974) found fasting

Table 3.13   Estimates by regression methods of the maintenance requirements of lambs.

Authority	Heat production at maintenance		Fasting heat production ($k_m = 0.8$) (kJ/kg $W^{0.75}$ per day)
	Actual (kJ/day)	Calculated (kJ/kg $W^{0.75}$ per day)	
Walker & Norton (1971b)	420 $W^{0.73}$	406	324
Chiou & Jordan (1973)	—	382	305
Alexander (1962a)	489 $W^{0.73}$	473	378
Jagusch & Mitchell (1971)	145 $W^{0.57}$	454	363
Estimated values			
ARC (1965): preferred value for fasting metabolism			469
Graham et al (1974): predicted value from			
Fasting metabolism = 706 $W^{0.58}$	5 kg		536
	10 kg		477

metabolism of liquid-fed lambs to be 706 kJ/kg $W^{0.75}$ per day, which for lambs weighing 5 kg and 10 kg corresponds to 536 and 477 kJ/kg $W^{0.75}$ per day respectively; the latter value is in general agreement with the preferred value of 466 kJ/kg $W^{0.75}$ given in the previous edition. From the results of the regression analyses shown in Table 3.13, however, a much lower value of about 350 kJ/kg $W^{0.75}$ can be inferred. This difference is difficult to resolve; the value from the regression data is higher than that predicted from the equation of Graham et al (1974) for weaned sheep, and indeed is intermediate between their estimates for weaned and liquid-fed animals. In view of the difficulties surrounding estimates of fasting metabolism in young animals, it is suggested that the preferred value for lambs receiving liquid diets is 350 kJ/kg $W^{0.75}$ per day.

The preferred values for fasting metabolism of sheep are thus about the same as those given in the previous edition when the latter have been corrected from a basis of weight raised to the power 0.73 to the power 0.75. For lambs given liquid diets, the value is reduced. Preferred values are given in Table 3.14.

Table 3.14   Preferred values for fasting metabolism in sheep.

Age and state of sheep	Ewes and wethers (kJ/kg $W^{0.75}$ per day)	Rams[†]
Unweaned liquid diet	350	400
Weaned 6 mos.	260	300
12 mos.	245	280
24 mos.	230	265
48 mos.	215	245
>48 mos.	210	240

† 15% greater than values for ewes and wethers.

2. Muscular work   (a) Standing and lying

Estimates of the increase in energy expenditure due to standing above that of lying in both cattle and sheep are summarized in Table 3.15.

Table 3.15   Estimates of the energy cost of standing over lying.

Author	Energy cost (kJ/kg per day)
Cattle Forbes et al (1927b)	11.7
Hall & Brody (1933)	8.8
Blaxter (1966)	5.9
J. McLean (unpublished)	10.0
Clark et al (1972)	11.1
Sheep Hall & Brody (1933)	11.7
Joyce & Blaxter (1964)	7.1
Webster & Valks (1966)	11.8

The variation in Table 3.15 together with the observations of Pullar (1963) and of Brockway et al (1969), who noted that the heat emission of a sheep can be increased on standing by up to 70%, may in part be explained by states of vigilance of the animal. Thus, in studies with sheep undergoing training Blaxter (1974a) showed that the difference between standing and lying metabolism decreased with the progress of training. Observations by Toutain & Webster (1975) indicate that in recumbent calves, metabolism during paradoxical sleep and sleep associated with short encephalographic waves is reduced to 90% of that noted when the animal is awake. In sheep the effect may be slightly greater. A mean value for the additional cost of standing of 10 kJ/kg per day has been taken to apply to average conditions for both species.

(b) Changing body position

Three estimates have been made of the energy cost of the double movement of lying and standing up again (Hall & Brody 1933, Colovos et al 1979, Clark et al 1972). The mean values obtained were 0.11, 0.32 and 0.26 kJ/kg, respectively. The value of Clark et al (1972) of 0.26 kJ/kg is preferred.

(c) Walking on the level and on gradients

The energy cost of walking in sheep has been measured by Clapperton (1964b) and by Farrell et al (1972b). The values obtained, the single observation on the ox by Hall & Brody (1934) and those of Ribeiro et al (1977) together with observations on the horse are given in Table 3.16.

Table 3.16   Energy expenditure in moving 1 kg body weight 1 metre forward or upwards.

Species	Authority	Work of walking ($J kg^{-1} m^{-1}$)	Work of ascent ($J kg^{-1} m^{-1}$)
Sheep	Clapperton (1964a, & b)	2.5	27.0
Sheep	Farrell et al (1972b)	2.9	32.0
Cattle	Hall & Brody (1934)	2.0	—
Cattle	Ribeiro (1976)	2.1	27
Horse	Zuntz & Hagemann (1898)	1.5–1.6	28
	Hoffmann et al (1967)		

The preferred values for the sheep are 2.6 J/kgm for horizontal movement and 28 J/kgm for vertical movement. The value for cattle for horizontal movement is certainly less since there is evidence that the cost of locomotion per kg weight at preferred speeds and gaits is smaller in larger species (Blaxter 1972) and values of 2.0 J/kgm for horizontal movement and 28 for vertical movement have been taken.

A 400-kg bullock travelling 3 km and ascending 200 m in the course of 24 h would thus expend $400 \times 3000 \times 2.0 + 400 \times 200 \times 28 J = 4.64$ MJ. This is an appreciable addition to a fasting metabolism of 32 MJ. A 50-kg sheep moving the same distance would expend $50 \times 3000 \times 2.6 + 50 \times 200 \times 28 J = 0.67$ MJ, which is a slightly greater relative increase over its fasting metabolism of about 4.2 MJ. No information is available for the cost of descent in ruminant animals.

(d) The energy cost of eating

The increases in heat production consequent upon eating have been discussed earlier (see p. 79). While these can be measured they are usually regarded as a tax on food consumed and included in the heat increment of food. The only occasion when their inclusion might be warranted is in the grazing situation when the ingestion of feed takes longer than it does in stall feeding. Even so, the additional expenditure is then confounded with the effects of body movement. Direct comparison of grazing with eating from a manger by Graham (1965) revealed no measurable difference in energy expenditure per minute, but grazing took twice as long. This agrees with the view that the increase in heat production is simply proportional to the time spent eating as Osuji et al (1975) have suggested.

(e) Effects of environment

In cold conditions, the amount of heat produced by metabolism may not be sufficient to keep the body warm. The animal then augments its metabolism by shivering. Since the efficiency of utilization of metabolizable energy to meet the expenditure of shivering is 1.0, the increase in heat production is equal to the increase in metabolizable energy intake required to prevent the animal from shivering. The environmental temperature at which shivering occurs, the critical temperature, is determined by the animal's resistance to cold, that is, by its total insulation, and by the amount of heat which it produces in conditions which do not elicit these regulatory responses. The increase in metabolism which occurs when environmental temperature falls below the critical temperature is determined entirely by its total insulation. The subject of tolerance to cold and the magnitude of the insulation terms have recently been extensively reviewed (Monteith 1973, Monteith & Mount 1974, Webster 1975).

In cold the loss of heat by vaporization of moisture from the skin and respiratory passages is reduced to a minimum. This minimal loss (E) remains very constant at temperatures below the critical temperature. At and below the critical temperature blood supply to the skin is at a minimum, and the insulation interposed between the deep tissues of the body and the skin surface, the tissue insulation (I_T), is also relatively constant. The interface at the skin and hair coat and the hair or wool coat itself provide a constant insulation in conditions of constant air velocity and radiation intensity and this insulation is termed the external insulation, I_E. The sum of I_T and I_E is the total insulation.

Computation of the critical temperature of an animal (T_C) and of the additional heat an animal must produce at temperatures below the critical temperature to maintain deep body temperature (T_R) constant at 39°C can be made using the following equations (Blaxter 1977):

$$T_C = T_R + E \, I_E - H^*(I_T + I_E) \qquad [3.17]$$

Additional heat produced for every degree fall in environmental temperature below the critical temperature (MJ m^{-2} d^{-1}°C^{-1}) $= 1/(I_T + I_E)$ where T_C = critical temperature, T_R = rectal temperature (39.0°), E = minimal loss of heat as water vapour (MJ m^{-2} d^{-1}), I_T = tissue insulation (°C m^2 d MJ^{-1}) and H* = heat production (MJ m^{-2} d^{-1}). It will be noted that the insulation and heat terms necessary to solve these equations are expressed in terms of surface area. The formula:

$$A = 0.09 \, W^{0.66},$$

where A = body surface area (m^2) and W = mass in kg, is that usually adopted. The heat produced from a given diet expressed per m^2 body surface by a fattening animal is

$$H^* = \frac{M_{E(m)} + (M_E - M_{E\,(m)})\,(1 - k_f)}{A}, \qquad [3.18]$$

where $M_{E(m)}$ is the maintenance energy expenditure, M_E is the daily intake of metabolizable energy, and k_f is the efficiency of utilization of metabolizable energy for growth and fattening.

External insulation (I_E) depends on the depth of the coat and its nature, wool being a more effective insulator than hair. Wind and rain reduce external insulation and so does a decrease in incoming radiation. Numerical values for the insulation terms for cattle and sheep are given in Table 3.17. A more detailed description is given in the publications already cited.

Table 3.17   Numerical values for the factors required to compute critical temperatures and energy requirements at environmental temperatures below the critical temperature.

Factor	Cattle	Reference	Sheep	Reference
E, MJ m^{-2} d^{-1}	1.5	1	1.3	2
I_T, °C m^2 d MJ^{-1}	0.7 (calf)	3	1.3	2, 5
	1.6 (adult)	4		
I_E, °C m^2 d MJ^{-1} in still air where F = coat depth in cm†	$I_E = 1.64 + 0.88$ F	4	$I_E = 1.35 + 0.141$ F	7
	$I_E = 2.33 + 0.57$ F	6		

† Insulations of cattle and sheep in winds of different velocities are summarised in these (SI) units in Blaxter (1977).
(1) Blaxter & Wainman (1961)
(2) Joyce & Blaxter (1964)
(3) Gonzalez-Jimenez & Blaxter (1962)
(4) Blaxter & Wainman (1964)
(5) Webster & Blaxter (1966)
(6) Webster et al (1970)
(7) Joyce et al (1966)

3. Correction for feeding level

Earlier it was shown that the energy retention of animals was not linearly related to their intake, and that not the whole of this curvilinearity was explained by changes of the metabolizability of feed with feeding level. Schiemann et al (1971) and van Es (1976) have estimated that the percentage decline in the metabolizability of feed with feeding level lies between 0.3 and 1.4% per unit increase in feeding level; to Schiemann et al (1971) the decline appeared to be greater for the less fibrous diets, that is those with a higher value of q, the metabolizability of the diet. Van der Honing, (1977) adopted a decline of 1.8% per unit increase in feeding level above maintenance.

From an analysis of the relationship between energy retention and feed intake (Blaxter 1974a) in fattening and growing animals, it was apparent that curvilinearity increased as the metabolizability of the diet decreased. Since the value of k_f was determined between maintenance and 2×maintenance, the use of k_f as a proportional factor to estimate energy retention at levels of feeding above 2×maintenance results in overestimation of energy retention.

The extent of the overestimation can be computed from the equations used to arrive at the estimates of k_m and k_f given in Tables 3.2, 3.3 and 3.4. These were exponential equations of the form

$$R = B(1 - \exp(-kl)) - 1,$$

where R is the retention of energy and I the intake of energy both scaled by dividing throughout by the determined fasting metabolism (see Blaxter & Boyne 1970, Blaxter 1974a). The appendix to the latter paper gives the derivation of the equation and certain aspects of its manipulation. The equation shows a falling off in rate of retention as energy intake increases.

To provide continuity with the previous edition and with the Ministry of Agriculture, Fisheries and Food's (1975) application of the work decribed there, the efficiency values for maintenance (k_m) and for fattening (k_f) were derived from the above equation, k_f being measured at precisely twice the maintenance datum. This could be done algebraically (see Blaxter 1974a).

Maintenance $= I^* = (1/K) \operatorname{Log} (B/B-1)$
Efficiency for maintenance $= k_m = 1/I$
Efficiency for fattening between maintenance and twice maintenance $=$ $(B-1)/BI^* \simeq k_f$

The linear relations between the efficiency values and the metabolizable energy of the diet were estimated statistically from the 78 individual estimates of k_m and k_f.

The estimation of retention of energy on this linear model is correct only when the level of feeding is precisely 2. Above this, gain is progressively overestimated. It may be shown that the true retention is a proportion of that estimated from the linear relationship (i.e. from k_m and K_f), the correction factor, C_1, being calculated as:

$$C_1 = (B/L-1). (1-D^{L-1}),$$

where L is the feeding level in terms of multiples of maintenance, and $D = k_f/k_m$. This expression corrects apparent retentions to true retentions; it does not permit the reverse computation of the true amount of metabolizable energy required to promote a particular energy retention at feeding levels other than precisely 2.

It may be shown that this correction factor, C_2, is calculated as:

$$C_2 = D/(D+R). (\log_e(D-R/B)/\log_e D.$$

Both these correction factors are readily tabulated, the first in terms of level of feeding (L) and the constant of the original equation (B) and the second in terms of the desired retention (R) and the constant of the original equation (B).

This approach, while correct, is cumbersome: to estimate requirements it seems more sensible to use the original equation directly. This entails transforming the statistically estimated values of k_m and k_f to provide best estimates of B and k in the primary equation. It is easily shown that

$$B = k_m/(k_m - k_f).$$

and that

$$k = k_m \log (k_m/k_f).$$

These two equations enable values of B and k to be tabulated from functions which describe the relationship between metabolizability (q) and the efficiency of utilization of metabolizable energy for maintenance and fattening, as follows:

q	B	k
0.35	1.81	0.505
0.40	1.98	0.453
0.45	2.18	0.406
0.50	2.40	0.365
0.55	2.67	0.326
0.60	2.98	0.291
0.65	3.36	0.258
0.70	3.82	0.227
0.75	4.39	0.198
0.80	5.12	0.170

With these constants, which are the transformed values of k_m and k_f, the feed intake necessary to promote a particular rate of energy retention can be estimated from the relationship:

$$I = 1/k \times \operatorname{Log} B/(B-R-1),$$

and the retention obtained from a given intake from the primary equation.

It can be argued that there would be advantage in rejecting the linear approximations embodied in the concept of constant net efficiencies for maintenance and production for estimation of requirements and of energy allowances, in favour of the more descriptive exponential model, and no doubt in many applications such an approach is desirable. The linear model has, however, been retained since no data on a similar basis are available for lactating or pregnant animals. The step of computing requirements from an exponential model may, be regarded as a first approach to a more refined approach.

4. The energy value of gains in weight

To compute energy requirements the energy retained or secreted by the animal in different circumstances must be known. These factors have been reviewed in Chapter 1 and here are summarized, largely in tabular form to facilitate computation.

(a) Growth in cattle

The composition of the empty bodies of cattle and its variation with sex, age, ultimate body size and the quantity and quality of the feed consumed were summarized in Table 1.21 in terms of "mean" equations in which the energy concentrations (MJ/kg) of gains in weight of the empty body at different empty-body weights were described. These applied to animals gaining approximately 0.6 kg of empty body daily. To allow for growth at other rates and for differences due to sex, a set of correction factors was included and was given in Table 1.22. Together these tables allow for effects of sex, breed size and rate of gain of empty-body weight.

For greater convenience in calculation, the estimates for a castrate of average breed size in Table 1.21 and the influence of rate of gain (Table 1.22) can be used to derive the equation:

$$E_{\Delta x} = 5.0 + 0.0389x = 0.0000114x^2 + 0.1608R_E, \qquad [3.19]$$

where $E_{\Delta x}$ is the heat of combustion of a gain in empty-body weight (MJ/kg), x is empty-body weight and R_E is energy retention (MJ/day).

This can be converted into an equivalent liveweight version,

$$E_{\Delta w} = 4.1 + 0.0332W - 0.000009W^2 + 0.1475R_E, \qquad [3.20]$$

by using equation [1.2] (p. 42), taking $a = 14$, the middle value appropriate for green roughages and many mixed diets. Replacing R_E by $E_{\Delta w} \cdot \Delta w$, the equation can be rewritten as

$$E_{\Delta w} = (4.1 + 0.0332W - 0.000009W^2)/(1 - 0.1475\Delta W). \qquad [3.21]$$

The equation may be used for other types of cattle by applying the percentage corrections given in Table 1.22 (+15% for small breeds and heifers, −15% for large breeds and bulls).

Substitution of a continuous function with its implied precision for the discrete "best estimates" of Table 1.24 is convenient, but the comments on variation in the composition of the body and its gains in Chapter 1 still apply. The estimates of energy value of gains derived from the discrete relationship were compared with those used in the previous edition and with those adopted by the National Research Council (1970) in Chapter 1, Figure 1.3; the estimates made from the equations above which are summarized in full in Table 3.18 obviously accord with those comparisons. Not included in those comparisons was the relationship adopted by the Ministry of

Table 3.18   The energy concentration in liveweight gain in cattle values estimated from the equation in the text (p. 105) for medium castrates (MJ/kg).

Daily gain (kg)	Weight (kg)						
	50	100	200	300	400	500	600
0.25	6.0	7.6	10.8	13.8	16.6	19.2	21.6
0.50	6.2	7.9	11.2	14.3	17.2	19.9	22.4
0.75	6.5	8.2	11.7	14.9	17.9	20.7	23.4
1.00	6.7	8.6	12.2	15.5	18.7	21.6	24.3
1.25	7.0	9.0	12.7	16.2	19.5	22.6	25·5
1.50	7.4	9.4	13.3	17.0	20.5	23.7	26.7

Agriculture, Fisheries and Food (1975). There the comparable equation based on the ARC (1965) data with a removal of gut fill corrections was

$$E_{\Delta W} = (6.28 + 0.0188W)/(1 - 0.3\Delta W).$$ [3.22]

This equation predicts that the effect of body weight on the energy concentration in gains is less and the effect of rate of gain somewhat more than that now proposed. In 100-kg animals gaining at high rates (1 kg/day) the MAFF (1975) estimate is higher (11.7 against 8.6 MJ/kg) and with 600-kg animals at low rates (0.5 kg/day) the MAFF (1975) estimate is lower (20.7 against 22.4 MJ/kg).

(b) Growth in sheep

The composition of gains in the empty bodies of sheep was summarized in Table 1.8 which was based on equations given in Table 1.7. The protein concentration declined with weight and that of fat increased. The energy concentration in empty-body gains derived from these two components, increased with body weight and the rate of increase was virtually constant. The energy content of gains of empty-body weight can thus be taken to be linearly related to body weight, and these may be converted to a live weight basis using the corrections described in Chapter 1, p. 42, assuming that initial fill at 15 kg empty-body weight is 300 g/kg empty-body weight. The derived equations are

$$\text{non-Merino males}\quad E_{\Delta W} = 2.5 + 0.35W,$$ [3.23]

$$\text{castrates}\quad E_{\Delta W} = 4.4 + 0.32\dot{W},$$ [3.24]

$$\text{females}\quad E_{\Delta W} = 2.1 + 0.45W.$$ [3.25]

(c) Pregnancy in cattle

Table 1.20 gives estimates of the parameters of Gompertz equations which describe the time course of deposition of energy-yielding constituents in the gravid uterus. The data apply to production of a calf weighing 40 kg at 281 days of gestation. Chapter 1 shows that the course of growth is such that for calves of weights other than 40 kg at birth, rates of energy deposition can be obtained by direct proportion. Differentiation of the energy equation in Table 1.20 with respect to time describes the energy deposition per unit time,

$$dE/dt = E(t)\, 0.0201 \exp(-0.0000576t)$$ [3.26]

where E(t) is the energy stored in the gravid uterus up to time t obtained from the original equation (MJ) and t is time from conception in days. Values computed from this equation are given in Appendix Table 1.9 and are extended to include values for Ayrshire and Jersey calves in computation of Table 3.30.

(d) Pregnancy in sheep

Methods similar to those used for cattle were used with the equation in Table 1.6, when

$$dE/dt = E(t) \, 0.0737 \exp(-0.099643t) \qquad [3.27]$$

Values apply to a standard lamb weighing 4 kg at 147 days of gestation and are given in Appendix Table 1.8.

Energy Requirements for Production

. Cattle

(a) Growth

The estimates of requirements for growth of cattle necessitate a multiple classification because firstly bulls have a higher fasting metabolism than heifers and castrates, and secondly there are differences in the composition of gains which arise from differences in mature size and sex. Seven tables are required for the various classes of growing cattle (Tables 3.19 to 3.25). These tables show that energy requirements for growth in animals of the same weight can differ by up to 30% according to sex and mature body size. The intact male has a higher maintenance requirement than the castrate and the heifer, but the gains it makes have a lower energy content.

Table 3.19   Metabolizable energy requirements (MJ/day) of cattle for maintenance and growth†. (a) Bulls of breeds of large mature size.

M_E/G_E (q)	Live weight (kg)	\multicolumn{7}{c}{Live weight gain (kg/day)}
		0	0.25	0.50	0.75	1.00	1.25	1.50
0.4	100	20	24	27	32	—	—	—
	200	33	37	42	49	—	—	—
	300	43	49	56	64	—	—	—
	400	53	59	68	78	—	—	—
	500	61	69	79	90	—	—	—
	600	70	78	89	102	—	—	—
0.5	100	19	22	25	29	34	—	—
	200	31	35	39	44	51	—	—
	300	41	46	51	58	66	—	—
	400	50	56	62	70	80	—	—
	500	58	65	73	82	93	—	—
	600	66	74	82	93	105	—	—
0.6	100	18	21	24	27	30	35	40
	200	30	33	37	41	46	52	59
	300	39	43	48	54	60	67	76
	400	47	52	58	65	73	82	92
	500	55	61	68	76	84	95	107
	600	63	69	77	86	96	107	121
0.7	100	18	20	22	25	28	31	36
	200	28	31	34	38	42	47	53
	300	37	41	45	50	55	62	69
	400	45	50	55	61	67	75	83
	500	53	58	64	71	78	87	97
	600	60	66	72	80	88	98	109

† In Tables 3.19–3.25 all values include an allowance for activity, of 4.3 kJ/kg liveweight per day.

Table 3.20   Metabolizable energy requirements (MJ/day) of cattle for maintenance growth. (b) Bulls breeds of medium mature size.

M_E/G_E (q)	Live weight (kg)	Live weight gain (kg/day)						
		0	0.25	0.50	0.75	1.00	1.25	1.50
0.4	100	20	24	29	36	—	—	—
	200	33	38	45	54	—	—	—
	300	43	50	59	70	—	—	—
	400	53	61	71	84	—	—	—
	500	61	71	83	98	—	—	—
	600	70	80	94	111	—	—	—
0.5	100	19	23	27	31	38	—	—
	200	31	36	41	48	56	—	—
	300	41	47	54	62	72	—	—
	400	50	57	65	75	88	—	—
	500	58	66	76	88	102	—	—
	600	66	75	86	99	115	—	—
0.6	100	18	21	25	28	33	39	46
	200	30	34	38	44	50	58	67
	300	39	44	50	57	65	75	87
	400	47	54	61	69	79	90	104
	500	55	62	71	80	92	105	121
	600	63	71	80	91	104	119	137
0.7	100	18	20	23	26	30	35	40
	200	28	32	36	40	46	52	59
	300	37	42	47	53	60	68	77
	400	45	51	57	64	72	82	93
	500	53	59	66	75	84	95	108
	600	60	67	75	84	95	107	122

Table 3.21   Metabolizable energy requirements (MJ/day) of cattle for maintenance and growth. (c) Bulls breeds of small mature size.

M_E/G_E (q)	Live weight (kg)	Live weight gain (kg/day)						
		0	0.25	0.50	0.75	1.00	1.25	1.50
0.4	100	20	25	31	39	—	—	—
	200	33	39	48	58	—	—	—
	300	43	52	62	76	—	—	—
	400	53	63	75	92	—	—	—
	500	61	73	87	106	—	—	—
0.5	100	19	23	28	34	42	—	—
	200	31	36	43	51	62	—	—
	300	41	48	56	67	80	—	—
	400	50	58	68	81	96	—	—
	500	58	68	80	94	112	—	—
0.6	100	18	22	26	31	36	44	54
	200	30	34	40	46	54	64	76
	300	39	45	52	61	71	83	98
	400	47	55	63	73	85	100	118
	500	55	64	74	85	99	116	137
0.7	100	18	21	24	28	33	38	45
	200	28	32	37	43	49	57	66
	300	37	43	49	56	64	74	85
	400	45	52	59	68	78	89	103
	500	53	60	69	79	90	104	120

Table 3.22   Metabolizable energy requirements (MJ/day) of cattle for maintenance and growth. (d) Bullocks of breeds of large mature size.

M_E/G_E (q)	Live weight (kg)	Live weight gain (kg/day)						
		0	0.25	0.50	0.75	1.00	1.25	1.50
0.4	100	18	22	27	34	—	—	—
	200	29	34	41	50	—	—	—
	300	38	45	54	65	—	—	—
	400	46	54	65	79	—	—	—
	500	54	64	76	92	—	—	—
	600	61	72	86	104	—	—	—
0.5	100	17	20	24	29	36	—	—
	200	27	32	37	44	53	—	—
	300	36	42	49	58	68	—	—
	400	44	51	62	70	83	—	—
	500	51	59	69	81	96	—	—
	600	58	67	78	92	108	—	—
0.6	100	16	19	22	26	31	38	46
	200	26	30	35	40	47	55	65
	300	34	39	45	52	61	71	84
	400	42	48	55	64	74	86	101
	500	48	56	64	74	86	100	117
	600	55	63	72	84	97	113	132
0.7	100	15	18	21	24	28	33	39
	200	25	28	32	37	43	49	57
	300	32	37	42	49	55	64	73
	400	40	45	52	59	67	77	88
	500	46	53	60	68	78	89	103
	600	52	60	68	77	88	101	116

Table 3.23   Metabolizable energy requirements (MJ/day) of cattle for maintenance and growth. (e) Bullocks of breeds of medium mature size and heifers of breeds of large mature size.

M_E/G_E (q)	Live weight (kg)	Live weight gain (kg/day)						
		0	0.25	0.50	0.75	1.00	1.25	1.50
0.4	100	18	22	29	38	—	—	—
	200	29	35	44	56	—	—	—
	300	38	46	57	72	—	—	—
	400	46	56	69	87	—	—	—
	500	54	65	80	101	—	—	—
	600	61	74	91	114	—	—	—
0.5	100	17	21	26	32	41	—	—
	200	27	33	39	48	59	—	—
	300	36	43	52	62	76	—	—
	400	44	52	62	75	92	—	—
	500	51	61	73	88	107	—	—
	600	58	69	82	99	121	—	—
0.6	100	16	20	24	28	35	43	54
	200	26	31	36	43	51	62	76
	300	34	40	48	56	67	80	97
	400	42	49	58	68	80	96	117
	500	48	57	67	79	94	112	135
	600	55	65	76	89	106	126	152
0.7	100	15	18	22	26	31	37	44
	200	25	29	34	40	46	54	64
	300	32	38	44	52	60	70	82
	400	40	46	54	62	73	85	99
	500	46	54	63	73	84	98	115
	600	52	61	71	82	95	111	130

Table 3.24   Metabolizable energy requirements (MJ/day) of cattle for maintenance and growth. (f) Bulloc of breeds of small mature size and heifers of breeds of medium mature size.

M_E/G_E (q)	Live weight (kg)	Live weight gain (kg/day)						
		0	0.25	0.50	0.75	1.00	1.25	1.50
0.4	100	18	23	31	43	—	—	—
	200	29	36	47	62	—	—	—
	300	38	48	61	80	—	—	—
	400	46	58	73	96	—	—	—
	500	54	67	85	111	—	—	—
	600	61	76	96	126	—	—	—
0.5	100	17	22	27	35	47	—	—
	200	27	34	42	52	66	—	—
	300	36	44	54	67	85	—	—
	400	44	54	65	81	103	—	—
	500	51	62	76	94	119	—	—
	600	58	71	86	107	134	—	—
0.6	100	16	20	25	31	38	49	66
	200	26	31	38	46	56	70	89
	300	34	41	50	60	73	89	112
	400	42	50	60	73	88	108	135
	500	48	58	70	84	102	125	156
	600	55	66	79	95	115	141	176
0.7	100	15	19	23	28	33	41	51
	200	25	30	35	42	50	60	72
	300	32	39	46	55	65	77	92
	400	40	47	56	66	78	93	111
	500	46	55	65	77	91	108	129
	600	52	62	74	87	103	122	145

Table 3.25   Metabolizable energy requirements (MJ/day) of cattle for maintenance and growth. (g) Heifers breeds of small mature size.

M_E/G_E (q)	Live weight (kg)	Live weight gain (kg/day)						
		0	.25	0.25	0.75	1.00	1.25	1.50
0.4	100	18	24	33	50	—	—	—
	200	29	37	50	69	—	—	—
	300	38	49	64	88	—	—	—
	400	46	60	78	106	—	—	—
	500	54	69	90	123	—	—	—
0.5	100	17	22	29	39	54	—	—
	200	27	34	44	56	75	—	—
	300	36	45	57	73	96	—	—
	400	44	55	69	88	115	—	—
	500	51	64	80	102	133	—	—
0.6	100	16	21	26	33	43	57	81
	200	26	32	40	49	62	79	105
	300	34	42	52	64	79	100	132
	400	42	51	63	77	96	121	157
	500	48	60	73	90	111	140	182
0.7	100	15	19	24	30	36	46	58
	200	25	30	37	45	54	66	81
	300	32	40	48	58	70	85	103
	400	40	48	58	70	84	102	124
	500	46	56	68	82	98	118	144

Table 3.26 compares these estimates of energy requirements for growth in catt with those in the previous edition and with those published by MAFF (1975), NR (1970) and Deutsche Demokratische Republik (1970). The NRC (1970) data we computed from the listed values of requirements of net energy for maintenance ar for body gain, using efficiency values (k_m and k_f) of 0.63 and 0.40, which accordin to the NRC are those applicable to diets containing 50% concentrates. The DD

(1970) values were computed from the declared equivalence of 1 energy feed unit for ruminants to 1 g starch equivalent or 10.5 kJ net energy for fattening. The net availability of metabolizable energy when a diet has a metabolizability (q) of 0.6 was taken to be 0.5, and values given by the DDR (1970) in terms of energy feed units (kEF$_r$) were multiplied by 21 to convert to metabolizable energy (MJ).

Table 3.26   Comparison of present estimates of energy requirements for growth and fattening of cattle with those estimated in the first edition (ARC, 1965) and those published by MAFF (1975), NRC (1970; 1971) and DDR (1970). The comparison is for diets with a metabolizability (q) of 0.6 which has been equated to a value of metabolizable energy/kg dry matter of 2.6 kcal/kg (ARC, 1965) and to 11 MJ/kg (MAFF, 1975).

Source	Description Live weight gain (kg/day):	Estimated requirement (MJ/day)							
		Live weight 200 kg				Live weight 400 kg			
		0	0.25	0.75	1.00	0	0.25	0.75	1.00
Present estimates	Bullock, medium-sized breed..	26	31	43	51	42	49	68	80
	Heifers, large ,, ,, ..	26	31	46	56	42	50	73	88
	Heifers, medium ,, ,,	30	33	41	46	47	52	65	73
	Bulls, large ,, ,,								
ARC (1965)		33	38	51	60	48	55	73	85
MAFF (1975)		27	33	49	56	45	53	74	889
NRC (1970)	Growing & finishing bullocks	27	39	52	60	46	66	94	101
(beef)	Growing & finishing heifers	27	40	56	65	46	66	101	110
NRC (1971)	Large bulls	—	—	—	57	47	—	—	100
(dairy)	Large heifers	—	—	51	—	47	—	79	—
DDR (1970)		—	—	49	53	48	—	76	88

The comparison shows that for the smaller animals, the new estimates are below those of the first edition. They are similar to MAFF estimates (which include a safety margin of 5%), but are lower than the DDR estimates and considerably lower than those for beef cattle made by the NRC. The discrepancy is not so great when comparisons are made with the NRC estimates for young dairy heifers.

When comparison is made at a weight of 400 kg, present estimates for bullocks of medium-sized breeds are somewhat less than those made in the first edition, with which MAFF estimates agree. The estimates made in DDR for growing bulls are higher than the present ones. The major discrepancy relates to the allowances made for growing beef bullocks and heifers by the NRC, which exceed the present ones by 20 to 30%. In this respect, an analysis of the results of feeding trials in which actual gains made by cattle were compared with those predicted by the system of the first edition showed a slight tendency to underestimate gain in animals weighing up to 250 kg and a distinct tendency to overestimate it in heavier animals gaining at high rates (Alderman et al 1970). The present estimates correct the former discrepancy but appear to accentuate the latter. The same authors' modifications of the earlier system (ARC 1965) which increased the energy value of liveweight gain and which were incorporated in the MAFF (1975) estimates of requirement, reduced the discrepancy but did not eliminate it. More work is required to provide better prediction of the energy value of gains as affected by age and previous and current rates of gain.

(b) Milk production

Table 3.27 summarizes the estimates made of the metabolizable energy required for secretion of milk by a 500-kg cow producing milk containing 40 g fat/kg and undertaking exercise by walking (1.3 km/day) and standing for 4 h more than a

Table 3.27   Metabolizable energy requirements of a cow weighing 500 kg, producing milk with 40 g/kg fa
(MJ/day). The cow is assumed to have an additional expenditure of energy in body movement o
4.3 kJ/kg per day. No change in body weight is allowed.

Metabolizable energy/gross energy (q)	Milk yield (kg/day)						
	0	5	10	15	20	30	40
0.5	51	78	106	134	162	—	—
0.6	48	74	100	126	153	209	—
0.7	46	70	95	120	146	198	252

Table 3.28   Metabolizable energy requirements of British Friesian, Ayrshire and Jersey cows (MJ/day)
Allowances for body movement as in Table 3.27.

Breed	Live weight change (kg/day)	M_E/G_E (q)	Milk yield (kg/day)					
			5	10	15	20	30	40
British Friesian, weighing 600 kg, producing milk containing 36.8 g fat/kg	Zero	0.5	84	110	137	164	—	—
		0.6	79	104	129	155	207	—
		0.7	76	99	123	147	197	248
	−0.5	0.5	65	91	117	144	—	—
		0.6	62	86	111	136	188	241
		0.7	59	82	106	130	178	229
	+0.5	0.5	108	134	161	189	—	—
		0.6	102	127	153	179	232	—
		0.7	97	121	145	170	220	—
Ayrshire, weighing 500 kg, producing milk containing 38.6 g fat/kg	Zero	0.5	78	105	132	160	—	—
		0.6	74	99	125	151	206	—
		0.7	70	94	119	144	195	248
	−0.5	0.5	59	85	112	140	—	—
		0.6	56	81	106	133	186	241
		0.7	53	77	101	126	176	229
	+0.5	0.5	102	130	157	185	—	—
		0.6	96	122	148	175	230	—
		0.7	91	116	141	166	218	—
Jersey, weighing 400 kg, producing milk containing 49.0 g fat/kg	Zero	0.5	74	105	137	169	—	—
		0.6	70	99	129	160	—	—
		0.7	67	94	123	152	212	—
	−0.5	0.5	55	86	117	149	—	—
		0.6	52	81	110	141	203	—
		0.7	50	77	105	134	193	—
	+0.5	0.5	98	130	162	—	—	—
		0.6	93	123	153	184	—	—
		0.7	88	116	145	175	236	—

tethered animal. Table 3.28 provides estimates of requirements for Friesian, Ayrshir
and Jersey cows weighing 600, 500 and 400 kg and producing milk containing 36.8
38.6 and 49.0 g fat/kg, respectively. This table includes calculations of the require
ments of cows which gain or lose 0.5 kg while sustaining particular rates of production

In the first edition the estimated requirements were compared with other estimate
published before 1965. In Table 3.29 comparisons are made with present feedin
standards in the UK and in the USA. The comparisons apply to a ration with
metabolizability of 0.6 (M/D = 2.65 kcal/g or 11.0 MJ/kg dry matter) and to stabili
of weight.

Table 3.29   Comparison of estimated requirements for metabolizable energy of a 500 kg lactating cow neither gaining nor losing weight made by different authorities (MJ/day); milk produced contains 40 g fat/kg; q = 0.6.

Authority	Milk yield (kg/day)				
	5	10	20	25	30
Present estimates	74	100	153	181	209
ARC (1965)	70	96	154	189	—
MAFF (1975)	81	107	160	187	213
NRC (1971)	75	99	146	169	193
DDR (1971)*	87	117	177	206	236

* Computed from EFʀ, on the assumption that q = 0.6.

The present estimates provide about 5% more metabolizable energy at the lower yields and 4% less at yields of 25 kg than those in the previous edition. They provide up to 8% less at low milk yields than the MAFF (1975) standards which include a 5% safety margin for maintenance and are mostly within 4% of the values given by NRC. The DDR values are considerably higher than those now estimated; this may reflect the assumption made regarding the mean metabolizability of energy used to convert the DDR estimates to metabolizable energy. Neither MAFF nor NRC provides for variation of requirement with diet quality and MAFF does not include any correction for feeding level. This is included in the NRC system which advises for certain types of feeds a 3% increase in intake for yields over 20 kg. On this basis the final figure in the penultimate line of Table 3.29 for a cow yielding 30 kg would be 199 MJ/day. Because of appetite limitations the range of quality in diets given to cows yielding 25 kg milk or more a day is likely to be small. With lower yielders and cows rationed to lose weight, however, differences in the amount of metabolizable energy required with ration quality can approach 10%.

(c) Pregnancy

There are no data available to estimate the effect of source of metabolizable energy on its utilization to support pregnancy. Hence the additional amount has to be taken to be invariant with type of diet. Table 3.30 summarizes the increments of metabolizable energy to be added to a maintenance diet to support pregnancy. They may equally be added to requirements calculated for growing heifers or lactating cows to cover needs of the pregnant heifer or the pregnant lactating cow.

Table 3.30   Additional metabolizable energy required in pregnancy by cattle (MJ/day).

Breed	Calf birthweight (kg)	Weeks before parturition			
		12	8	4	Term
Standard	40	8.2	14.2	24.7	42.9
Friesian	42	8.6	14.9	25.9	45.0
Ayrshire	32	6.5	11.4	19.7	34.3
Jersey	24	4.9	8.5	14.8	25.7

Generally the requirement at term is equivalent to about 80% of the maintenance requirement of the cow. The requirements so calculated differ very little from those in the previous edition which were given as mean amounts required during periods of 2 or 4 weeks rather than as amounts at particular times. The requirements in the last month (weeks 4–2 and weeks 2–0) were 30 rising to 37 MJ/day for a "standard cow. MAFF (1975) estimates mean requirement in the last month of pregnancy be 20 MJ/day which is lower than those inferred from Table 3.30.

2. Sheep

(a) Growth

Estimates of the metabolizable energy requirements of growing and fattening shee kept under outdoor conditions (but in thermoneutral environments) and given sol feed are summarized in Table 3.31. They are compared with previous estimates (AR 1965), with estimates made by MAFF (1975), with those of the NRC (1975), whic were stated to be based on the equations computed by Garrett et al (1959), and wi those published in the DDR (1970) in Table 3.32. The last were converted fro German Energy Feed Units to metabolizable energy (1 kEF$_r$=21 MJ, net availabili for fattening=0.5).

The present values are about 10% less than those estimated previously (ARC 1965 and the MAFF estimates which were derived from the latter are likewise higher tha the present ones. At high rates of daily gain estimates by the NRC agree well wi those now presented but those of the DDR are much higher. For 60-kg rams gainir 200 g/day estimates of metabolizable energy requirement in the DDR are 20 MJ/da by the NRC 19.8 and present estimates for a metabolizability of 0.6 are 19.3 and fo one of 0.5, 22.5 MJ/day, indicating better agreement for animals closer to adu weight than for lambs.

Table 3.31　Metabolizable energy requirements (MJ/day) of lambs for maintenance and growth, when kep out-of-doors (activity allowances, 10.6 kJ/kg liveweight per day), and given solid feed.

M_E/G_E (q)	Sex	Live weight (kg)	Live weight gain (kg/day)				
			0	0.05	0.10	0.20	0.30
0.5	Male	20	4.1	5.1	6.2	—	—
		30	5.6	7.0	8.5	12.6	—
		40	7.0	8.7	10.7	15.9	—
	Castrate	20	3.6	4.8	6.2	—	—
		30	5.0	6.4	8.2	—	—
		40	6.2	8.0	10.2	16.3	—
	Female	20	3.6	4.8	6.2	—	—
		30	5.0	6.6	8.7	—	—
		40	6.2	8.3	11.0	19.4	—
0.6	Male	20	3.9	4.8	5.7	7.9	—
		30	5.3	6.5	7.8	10.9	14.9
		40	6.7	8.2	9.8	13.7	18.8
	Castrate	20	3.4	4.4	5.6	8.4	—
		30	4.7	6.0	7.4	11.0	—
		40	5.9	7.5	9.2	13.6	—
	Female	20	3.4	4.5	5.6	8.6	—
		30	4.7	6.1	7.8	12.0	—
		40	5.9	7.7	9.9	15.4	—
0.7	Male	20	3.7	4.5	5.3	7.1	—
		30	5.1	6.1	7.3	9.8	12.7
		40	6.4	7.7	9.1	12.3	16.0
	Castrate	20	3.3	4.2	5.1	7.3	—
		30	4.5	5.6	6.9	9.7	13.2
		40	5.6	7.0	8.5	12.0	16.2
	Female	20	3.3	4.2	5.2	5.2	—
		30	4.5	5.8	7.2	10.4	14.6
		40	5.6	7.3	9.1	13.3	18.8

Table 3.32.   Comparison of estimates of the metabolizable energy requirements of a growing sheep weighing 30 kg given diets with a metabolizability of 0.6.

Source and description		0	0.05	0.10	0.20	0.30
		Daily gain in weight (kg)				
Present estimate:	Males	5.3	6.5	7.8	10.9	14.9
	Castrates	4.7	6.0	7.4	11.0	
	Females	4.7	6.1	7.8	12.0	
ARC (1965)		5.1	6.7	8.6	13.2	(19.4)
MAFF (1975)		5.9	7.6	9.5	13.9	(18.2)
NRC (1975)		5.3	7.5	9.1	12.1	15.2
DDR (1970)		5.5			13.0	15.9

Table 3.33   Metabolizable energy requirements of milk-fed lambs kept out of doors (MJ/day).

Sex	Live weight (kg)	0	0.1	0.2	0.3
		Live weight gain (kg/day)			
Male	5	1.6	2.4	3.2	4.2
	10	2.8	3.8	4.9	6.0
	15	3.8	5.0	6.4	7.8
	20	4.7	6.2	7.8	9.5
Female	5	1.4	2.3	3.2	4.3
	10	2.4	3.6	4.8	6.2
	15	3.3	4.8	6.4	8.1
	20	4.1	5.9	7.8	10.0

Estimates of energy requirements of lambs given liquid diets are summarized in Table 3.33 for animals kept outside. Diet quality does not vary: the estimates apply specifically to milk diets.

Feeding standards for these classes of animal have not been published by the NRC, MAFF or DDR.

(b) Pregnancy

Estimates of the energy requirements of pregnancy have been made for large ewes (75 kg) and small ewes (40 kg) producing single lambs (6 kg or 4 kg) or twins (9.6 kg or 6.4 kg). The results are summarized in Table 3.34.

Table 3.34   Metabolizable energy requirements of pregnant ewes (MJ/day).

Ewe and lamb weights	M_E/G_E (g)	63 (12)	91 (8)	105 (6)	119 (4)	133 (2)	147 Term
		Days pregnant (and weeks before term)					
40 kg ewe	0.4	6.3	7.0	7.7	8.6	9.7	11.2
4 kg lamb	0.5	6.0	6.7	7.4	8.2	9.4	10.9
	0.6	5.7	6.4	7.1	8.0	9.1	10.6
	0.7	5.5	6.2	6.8	7.7	8.9	10.4
40 kg ewe	0.4	6.5	7.7	8.7	10.1	12.0	14.3
5.4 kg twin	0.5	6.2	7.4	8.4	9.8	11.7	14.0
lambs	0.6	6.0	7.1	8.1	9.5	11.4	13.8
	0.7	5.7	6.8	7.9	9.3	11.2	13.5
75 kg ewe	0.4	10.2	11.3	12.3	13.6	15.4	17.6
6 kg lamb	0.5	9.8	10.8	11.8	13.1	14.9	17.1
	0.6	9.3	10.4	11.3	12.7	14.4	16.6
	0.7	8.9	10.0	10.9	12.2	14.0	16.2
75 kg ewe	0.4	10.6	12.3	13.8	16.0	18.8	22.3
9.6 kg twin	0.5	10.1	11.8	13.3	15.5	18.3	21.8
lambs	0.6	9.6	11.4	12.9	15.0	17.8	21.4
	0.7	9.2	11.0	12.5	14.6	17.4	20.9

Table 3.35   Comparison of estimates of the energy requirements of 75 kg ewes during pregnancy. Metabolizability of the gross energy (q) = 0.6.

| | Weeks before term | | | | |
Source	8	6	4	2	Term
Present estimates					
Singles	10.4	11.3	12.7	14.4	16.6
Twins	11.4	12.9	15.0	17.8	21.4
MAFF (1975)					
Singles	9.4	10.4	11.5	12.7	14.1
Twins	9.7	11.2	13.0	15.0	17.4
DDR (1970)					
foetal burden not stated					17.4
(requirement stated by months)		14.8			
NRC (1975)					
foetal burden not stated (requirement stated for last 6 weeks)				18.9	

In the first edition, requirements for pregnancy in the ewe were not stated. Table 3.35 compares for 75 kg ewes the present estimates with those of other authorities. Estimates made by MAFF (1975) are clearly lower than those made here. Both the NRC and DDR requirements are of the same order as those now calculated, particularly if it is assumed that twinning is commonplace.

(c) Lactation

Table 3.36 gives the requirements of lactating ewes weighing 40 kg and 75 kg for different milk yields. Comparisons with other feeding standards are made difficult because milk yield is not usually stated. Nevertheless, in the MAFF tabulations, which are in terms of recommended allowances for each month, mean milk yields are given. Calculations shows that MAFF estimates are marginally below those given in Table 3.36, possibly by 5%.

Table 3.36   Metabolizable energy requirements of lactating ewes kept out of doors (MJ/day)†.

| Ewe live weight (kg) | Ewe live weight change (kg/day) | M_E/G_E (q) | Milk yield (kg/day) | | | | |
			1.0	1.5	2.0	2.5	3.0
40	Zero	0.5	13.7	17.9	22.2	—	—
		0.6	13.0	16.9	21.0	25.1	—
		0.7	12.3	16.1	19.9	23.8	—
	−0.10	0.4	10.4	14.7	—	—	—
		0.5	9.8	13.9	18.1	22.4	—
		0.6	9.3	13.2	17.1	21.2	—
		0.7	8.8	12.5	16.2	20.1	—
75	Zero	0.5	17.2	21.3	25.4	29.6	33.9
		0.6	16.3	20.2	24.1	28.0	32.0
		0.7	15.5	19.1	22.8	26.6	30.4
	−0.10	0.4	14.1	18.4	22.8	27.2	—
		0.5	13.4	17.4	21.5	25.6	29.8
		0.6	12.7	16.5	20.4	24.3	28.2
		0.7	12.1	15.7	19.3	23.0	26.8

† mean values for months 1–3 of lactation; values for month 1 will be approximately 1% lower, and values for month 3 about 1% higher, than the means, because of changes in milk fat content.

The NRC standards do not state milk yield, appear to have been calculated as multiples of maintenance needs, include changes in weight and cannot be compared directly. The DDR estimates, when converted to metabolizable energy, are about 15% higher than those given in Table 3.36; largely because the East Germans add a 25% safety margin to the maintenance requirement of the ewe.

Appendix 3.I
The East German System of Estimating the Energy Values of Feeds for Cattle and Sheep.

Energy retention is predicted from the amounts of digested nutrients (Weende analysis), the body weight of the animal raised to the power 0.75 and the apparent digestibility of the energy of the ration. Metabolizable energy of the ration is estimated from the digested nutrients alone. The equations used are in the original terms.

Cattle: Energy retention
$$y = (1.72x_1 + 7.35x_2 + 1.90x_3 + 2.01x_4)(-0.513 + 0.03962v - 0.0002596v^2) - 57.4x_5$$

Sheep: Energy retention
$$y = (2.06x_1 + 8.83x_2 + 1.12x_3 + 2.05x_4)(-0.107 + 0.03355v - 0.0002540v^2) - 39.56x_5$$

Cattle: Metabolizable energy
$$y_1 = 4.32x_1 + 7.73x_2 + 3.59x_3 + 3.63x_4 - 6.26x_5$$

Sheep: Metabolizable energy
$$y_1 = 4.49x_1 + 9.05x_2 + 3.61x_3 + 3.66x_4 - 6.24x_5$$

These equations allow estimation of the efficiency of utilization of energy below and above maintenance provided that digested energy can be predicted from apparently digested nutrients. The Rostock workers provide appropriate equations:

Cattle: Digested energy
$$y_2 = 5.79x_1 + 8.15x_2 + 4.42x_3 + 4.06x_4$$

Sheep: Digested energy
$$y_2 = 5.72x_1 + 9.05x_2 + 4.38x_3 + 4.06x_4$$

In the above equations:
y = energy retained (kcal)
x_1 = digested crude protein (g)
x_2 = digested crude fat (g)
x_3 = digested crude fibre (g)
x_4 = digested N-free extractives (g)
x_5 = body weight (kg) raised to the power 0.75
v = digestibility of energy (%) = (digested energy/gross energy)×100
y_1 = metabolizable energy (kcal)
y_2 = apparently digested energy (kcal)

Appendix 3.II
Calculation of Estimates of Metabolizable Energy Requirements (MJ/day except where otherwise stated)

(1) Cattle, maintenance and growth (Tables 3.19 to 3.25)

Fasting heat production $F=0.53 \, (W/1.08)^{0.67}$ for castrates and heifers (15% more for bulls).

Expenditure on standing and walking $0.0043W$.

Minimal metabolism Z is total of these two amounts
$$k_m = 0.35q + 0.503$$

Requirement for maintenance $= Z/k_m$

Heat of combustion of live-weight gain (MJ/kg) is calculated from $(4.1+0.0332W-0.000009W^2)/(1-0.1475 \, \Delta W)$ where ΔW is rate of gain (kg/day) for castrates of medium breeds. (Estimate to be reduced by 15% for bulls and for large breeds or increased by 15% for heifers and for small breeds.) The relationship assumes the animals to be adjusted to a diet for which the gut contents would be 300 g/kg empty-body weight at 75 kg EBW. Hence energy retention, scaled by minimal metabolism, is

$$R = (\text{Heat of combustion}) \times \Delta W/Z$$
$$k_f = 0.78q + 0.006 \text{ (At } L = 2)$$
$$B = k_m/(k_m - k_f)$$
$$p = k_m \log_e (k_m/k_f)$$

$$\text{Requirement} = \frac{Z}{p} \log_e \{B/(B - R - 1)\}$$

(2) Cattle, lactation (Tables 3.27, 3.28)

Z calculated as in (1)

Energy retention R calculated from product of milk yield and energy concentration (MJ/kg) taken as $1.509+0.0406$ (g fat/kg). Energy retention reduced by 10.92 MJ/day to allow for weight loss of 0.5 kg/day or increased by 13.68 MJ/day to allow for weight gain of 0.5 kg/day.

Efficiency of utilization for lactation $k_l = 0.35q + 0.420$.

Approximate level of feeding $L = 1 + (R/k_l)/(Z/k_m)$.

Correction factor for level of feeding $= 1 + 0.018 \, (L-1)$.

Requirement $= (1+0.018 \, (L-1))(R/k_l + Z/k_m)$.

(3) Additional requirement for pregnancy (cattle) (Table 3.30)

For a 40 kg calf, the daily energy retention is

$$E(t) \times 0.0201 \exp (-0.0000576t)$$
where $\log_{10} \{E(t)\} = 151.665 - 151.64 \exp (-0.0000576t)$

and t is no. of days from conception.

For other calf weights retentions are in proportion.

Efficiency of utilization $= 0.133$.

(4) Maintenance and growth of sheep on ruminant diets (Table 3.31)

Fasting heat production $F = 0.251 \, (W/1.08)^{0.75}$
(15% more for intact males).

Expenditure on walking $0.0106W$.

Heat of combustion of liveweight gain (MJ/kg)

Males	$2.5+0.35W$	
Castrates	$4.4+0.32W$	Diet such that at EBW of 15 kg, gut fill would be 300 g/kg EBW.
Females	$2.1+0.45W$	

Other constants and method of calculation as for cattle.

**5) Growth of milk-
ed lambs (Table 3.33)**

Fasting heat production $F = 0.35 \, (W/1.05)^{0.75}$
Expenditure on walking $0.0106W$.
Heat of combustion of liveweight gain (MJ/kg)

Males	$3.79 + 0.365W$
Castrates	$5.60 + 0.338W$
Females	$3.67 + 0.472W$

Assuming gut fill of 60 g/kg
EBW at EBW of 15 kg.

Efficiencies of utilization $\begin{cases} k_m = 0.85 \\ k_f = 0.70 \end{cases}$

Other constants and calculations as for (1) and (4).

**6) Ewes, requirement
or pregnancy (Table
.34)**

Fasting heat production $0.226 \, (W/1.08)^{0.75}$.
Expenditure on walking $0.0106W$.
Efficiency for maintenance $k_m = 0.35q + 0.503$.
For a 4kg lamb, the daily energy retention is

$$E(t) \times 0.07372 \exp(-0.00643t)$$
$$\text{where } \log_{10}\{E(t)\} = 3.322 - 4.979 \exp(-0.00643t)$$

and t is no. of days from conception.
For other lamb weights retentions are in proportion.
Efficiency of utilization 0.133.

**7) Ewes, lactation
able 3.36)**

$Z = 0.226 \, (W/1.08)^{0.75} + 0.0106W$.
Energy retention R calculated from product of milk yield and energy concentration
(MJ/kg) taken as
   $0.0328 \text{ (g fat/kg)} + 0.0025 \text{ (day of lactation)} + 2.203$ with a fat content of 70 g/kg
   $k_l = 0.35q + 0.420$.
Approximate level of feeding $L = 1 + (R/k_l)/Z/k_m)$.
Level of feeding correction factor $= 1 + 0.018 \, (L-1)$.
Requirement $= (1 + 0.018 \, (L-1)(R/k_l + Z/k_m)$.

Chapter 4

Requirements for Protein

Methods of Expressing Requirements and the Value of Feeds

The method most widely used for expressing the ruminant's requirement for protein and the extent to which feeds could meet these requirements is one based on the measurement of digestible crude protein. Since this method was considered to give too much weight to non-protein nitrogen, the Report of the Departmental Committee on Rationing of Dairy Cows (Ministry of Agriculture & Fisheries 1925) proposed the adoption of the protein equivalent, i.e., 0.5 (digestible crude protein) + digestible true protein, in place of digestible crude protein; thus, non-protein N was assumed to be fully digestible but to have only half the value of digestible true protein. Later, Evans (1960) considered protein equivalent to underestimate the value of the non-protein nitrogen of silages and the use of digestible crude protein values was again proposed for this class of feeds. The responses obtained from varying the digestible crude protein intake are influenced by the source of the protein and also by other factors, in particular the energy supply. Thus, determination of digestible crude protein requirements is time-consuming and necessitates practical trials for each rate of productivity and for different classes of livestock.

In the first edition of this publication (ARC 1965), in an attempt to overcome some of the disadvantages associated with the digestible crude protein system, protein requirements for ruminants were expressed as available protein, the amount of crude protein of a defined biological value that would have to be absorbed from the digestive tract to meet the calculated requirements of the tissues for maintenance and production. Available protein values were calculated by the classical factorial system from inevitable losses in the urine (endogenous urine N), the retention of N in the body and foetus, output of N in the milk and an average biological value of dietary N. Compared with digestible crude protein requirements, those for available protein had the advantage of being independent of the value of any particular feed. Requirements for various rates of productivity and for different breeds of livestock could be readily computed without recourse to numerous trials.

To meet the available protein requirements with a particular ration, the values had to be converted into amounts of digestible crude protein and this conversion required the use of metabolic faecal N values. Moreover, the available protein concept has a number of limitations. It assumes that the faecal loss of N can be divided into a component of indigestible feed N and one which represents unabsorbed secretions of N-containing compounds into the tract (metabolic faecal N). In fact, faecal N consists mainly of microbial N and no such division is possible. For this reason, the validity of the measurement of true digestibility and, to some extent, of biological value is in doubt. Although it was realised that biological value varied with the nature of the feed, there was insufficient evidence in 1965 to permit an allowance for such variation. It is now apparent that much of this variation relates to the extent to which amino acids from microbial protein are supplemented by those arising from undegraded protein. It was realised in 1965 that the extent of microbial protein synthesis could be limited by the supply of N, indeed a lower limit of 90 g crude protein/kg dry matter of the diet was set to prevent such limitation. Nevertheless, the framework of ideas did not permit a rational basis for stating the conditions in which non-protein N might be used. In addition, although protein requirements were given (ARC 1965) for different rates of production and, by inference, for specific energy inputs, no attempt was made to relate protein requirements directly to energy input, either as total energy intake or as dietary energy concentration. For these reasons a new conceptual approach has been made to the problems of meeting the protein requirements of ruminants.

The New Approach

In ruminants, as in other animals, the needs of the tissues are met by amino acids absorbed from the small intestine. The ideal system for calculating the nitrogen requirements of ruminants must provide, therefore, estimates of the total and individual amino acids absorbed from the small intestine. These amino acids are supplied partly by microbial protein synthesized in the rumen and partly by dietary protein which has escaped fermentation in the rumen. The value of dietary urea or similar NPN sources depends entirely on degradation to ammonia in the rumen by bacteria and the subsequent use of this ammonia for microbial protein synthesis. The extent of the synthesis depends on the energy available to the microorganisms. Dietary protein also is degraded in the rumen by microbial attack; the pathways of this degradation are poorly understood but the nitrogenous products include peptides amino acids and ultimately ammonia. These products are used for the synthesis of microbial protein and there is evidence that mixed bacteria growing in the rumen incorporate considerable amounts of preformed amino acids as well as ammonia when the diet contains protein (Nolan & Leng, 1972). It is possible that this results in better microbial growth than the use of ammonia alone but, as a net effect for the host animal, good dietary protein which is degraded in the rumen is used inefficiently.

Degradability of dietary proteins in the rumen varies among different natural protein sources and with different processing treatments. Degradability is ideally defined as:

1 − (dietary protein entering duodenum/dietary protein consumed).

That part which escapes degradation supplements microbial protein in providing a source of amino acids for digestion and absorption in the small intestine of the host animal. Although amounts of microbial protein are limited by the dietary energy intake and are often insufficient for high rates of production, amounts of undegraded dietary protein are limited by the dietary protein consumed. It is especially important that any new system should take into account contributions from both microbial protein and undegraded food protein.

The case for consideration of the N requirements of the microorganisms as well as the N requirements of the tissues was probably first suggested by Johnson et al (1942), who showed that the addition of urea to a basal ration in amounts to produce the equivalent of 12% crude protein in the dry matter induced a retention of N that could not be improved by further urea addition, but could be raised by increasing the true protein content of the ration.

In 1971 a group of workers in the United States suggested a system for estimating the dietary requirements of ruminants for nitrogenous compounds which was based on assessing the animals' requirement for "metabolizable protein". In this scheme which has been described in detail (Burroughs et al 1975 a,b), "metabolizable protein is defined as protein digested (or amino acid absorbed) in the post-ruminal portion of the alimentary tract. Metabolizable protein supplied by different diets is estimated by taking into account the extent of degradation of feed protein in the rumen, the amount of microbial protein synthesized in the rumen and the digestibility of these components in the small intestine. Potential microbial protein synthesis in the rumen is estimated by assessing, from the TDN content of the diet, the energy available for this purpose. Workers in Germany (Kaufmann & Hagemeister 1973, Hagemeister et al 1976) have adopted a similar approach in attempting to relate protein availability for digestion in the small intestine of the dairy cow to the degradability of dietary protein in the rumen and the microbial protein synthesis which could be supported by a particular starch equivalent intake. Both systems therefore relate energy available for microbial synthesis in the rumen to the nitrogen available for this purpose. The extent of energy excess determines the possible response in microbial protein synthesis to non-protein nitrogen supplementation. This is referred to by Burroughs et al (1975 a,b) as Urea Fermentation Potential (UFP). If UFP is negative, the diet already provides enough rumen-degradable protein and added urea is valueless. In the United Kingdom, similar schemes have been developed but with a greater emphasis on the need to consider separately the nitrogen requirements of the microbial population and the host animal in relation to energy intake (Ørskov 1970 1976, Miller 1973). This approach has been adopted in developing the present system. It is recognised that the data available are inadequate in many respects; there are, for example, insufficient data to permit statements of requirements to be expressed in terms of individual amino acids, the calculations being limited to total amino acid N absorbed from the intestine. It is an important argument in favour of this system that data for essential amino acids can be incorporated into the system as they become available. The new approach should be regarded, therefore, as a framework for future research efforts and as a means of focussing attention on factors for which additional data are required. General reviews of the N metabolism of the ruminant are provided by Visek (1973), Smith (1975), Mercer & Annison (1976) and Buttery (1976).

The main argument for adopting this new approach is that digestion in the rumen is an essential component of feed utilization in ruminants and for its execution, the rumen microflora need N, mainly in the form of amino or ammonia N. If insufficient degradable N is available, the rate of digestion in the rumen of both fibrous feeds (Glover & Dougall 1960, Moir & Harris 1962) and high-concentrate diets (Ørskov et al 1972) will be reduced, especially if the diets are of small particle size (Ørskov et al 1972 1974b). This, in turn, leads to a reduction in voluntary intake of fibrous feeds (Campling et al 1962) and of high-concentrate diets (Ørskov et al 1972). The combined effects of the reduction in digestibility and in voluntary intake will lead to a decreased energy supply to the animal and to inefficient feed utilization.

The rumen microorganisms, therefore, need N in a form that can readily be used and is available at the same time that energy for microbial growth is being supplied from fermentation of organic matter. The amount of N required will depend largely on the amount of this energy, in conditions where the requirements for sulphur, phosphorus and micronutrients have been fulfilled. Thus, if the amino acids arising from the microbial proteins are in excess of N requirements of the tissues, the rumen need for N becomes the determinant of the N requirement for a particular energy intake. This need is met in part from dietary N and in part from recycled endogenous N.

. Factors involved in he new approach

The general scheme of the new approach is to calculate the amount of amino acid N of microbial origin that could be retained in the body for tissue synthesis when the maximal rate of fermentation of a particular energy input is achieved. This amount of amino acid N is compared with the total tissue needs for the particular energy input. Two alternatives present themselves:

(i) if the amount of microbial amino acid N available to the tissues is greater than the tissue needs, then the N requirement is the amount of degraded N needed by the rumen microorganisms;

(ii) if the microbial amino acid N is less than the tissue needs, then the difference must be supplied by amino acids from undegraded dietary protein.

For practical calculation of the N requirements, simple summary equations can be derived as follows:

Rumen degradable N (RDN) requirement $(g/day) = 1.25 \, M_E$ [4.1]

Amino acid N supplied to the tissues (TMN) by microbial
synthesis from RDN $(g/day) = 0.53 \, M_E$ [4.2]

If total tissue N requirement, calculated by the factorial
method (TN) is $>$ TMN, then

Undegraded dietary N (UDN) requirement $(g/day) = 1.91 \, TN - 1.00 \, M_E,$ [4.3]
where M_E is metabolizable energy requirement (MJ/day).

Total dietary N requirement is thus the sum of the RDN and UDN requirements. For the derivation of these equations, the following information is required.

(a) Degraded nitrogen (RDN) required by the microorganisms

(i) The metabolizable energy requirement for a particular dietary metabolizable energy concentration to give the rate of production required (see Chapter 3). From this value, the apparently digested organic matter intake is calculated.

(ii) The proportion of apparently digested organic matter that is apparently digested in the rumen.

(iii) The microbial N yield (g/kg organic matter apparently digested in the rumen).

(iv) The efficiency of conversion of degraded dietary N into microbial N.

(b) Net amino acid nitrogen supplied to the tissues by microbial N (TMN)

(i) The proportion of amino acid N in microbial N.

(ii) The apparent absorbability of microbial amino acid N in the small intestine.

(iii) The efficiency of utilization of absorbed microbial amino acid N.

(c) Total tissue nitrogen requirement of the animal (TN)

(i) The amount of N expected to be retained in the tissues or excreted in milk (see Chapter 1) (R_N or L_N, g/day).

(ii) The maintenance requirement, assumed to be equal to the endogenous urinary nitrogen excretion ($U_{N(E)}$, g/day) together with losses of N in hair and scurf.

If total tissue needs of the animal (TN) are less than the amount of amino acid N supplied to the tissues from microbial protein (TMN), no further calculation is required and RDN is the minimum N requirement of the animal.

If total tissue needs of the animal are greater than the amount of amino acid N supplied to the tissues from microbial protein, the difference will have to be supplied by amino acids from undegraded dietary protein as follows.

(d) Undegraded dietary nitrogen (UDN) requirement

(i) The efficiency of utilization of absorbed dietary amino acid N.

(ii) The apparent absorbability of dietary amino acid N in the small intestine.

The method of formulating rations to meet these requirements is shown later in this chapter, where it is suggested that crude protein (N×6.25) rather than nitrogen should be used and that feed tables should contain values for rumen-degradable protein and undegraded protein. But in explaining the principles of the system, nitrogen rather than crude protein has been used.

In a proposed ration, the weights of degradable N (RDN) and undegraded N (UDN) are calculated, and these are compared with the RDN and UDN requirements of the animal.

1. If the UDN requirement of the animal is greater than the UDN content of the ration, then additional protein must be given to correct the deficiency of UDN and the weight of RDN supplied by the new ration must be calculated.

2. If the RDN requirement of the animal is greater than the RDN content of the ration, then the deficiency could be made up by urea. Alternatively, if the deficiency arises after correcting for a deficiency of UDN by supplementary protein, a source of protein of higher degradability could be used in example 1 above.

Each factor needed in the system other than those discussed under Body Composition (Chapter 1) and Energy (Chapter 3) is considered separately in the subsections that follow. From a review of the literature and unpublished results a single value for each factor (except degradability) has been chosen for use in the calculations, although for some factors the decision is based on rather slender evidence.

2. Proportion of apparently digested organic matter that is apparently digested in the rumen

The apparently digested organic matter that is apparently digested in the rumen has been chosen as an estimate of the energy available to support microbial synthesis. This factor is the most widely published measure of that part of the organic matter that is fermented to volatile fatty acids, methane and carbon dioxide to yield high-energy phosphate bonds, usually referred to as ATP (adenosine triphosphate), needed for the maintenance and growth of the microorganisms. To obtain this value from the digestible organic matter intake, it is necessary to know the proportion that is apparently digested in the rumen. It is defined as $(I_O - Du_O)/(I_O - F_O)$, where, in unit time, I_O = intake of organic matter, Du_O = duodenal organic matter and F_O = faecal organic matter.

The proportion of apparently digested energy that is apparently digested in the rumen would be a preferable measure because it would permit a more direct relationship with metabolizable energy intake, but few values have so far been reported.

An alternative and possibly more appropriate measure would be the proportion of apparently digested organic matter that is truly digested in the rumen,

$$(I_O - (Du_O - Mi_O))/(I_O - F_O),$$

where Mi_O is the outflow of microbial organic matter. But few estimates of truly digested organic matter have been published because of the difficulty of accurate identification of microbial organic matter. Microbial organic matter has been estimated from the flow of microbial protein at the duodenum but there may be a wide variation in the protein content of microbial organic matter; Smith (1975) gives values of 320 to 570 g crude protein/kg dry matter.

Similarly, difficulties occur at present in the measurement of true digestibility of carbohydrates, such as starch and cellulose, the main source of ATP in the anaerobic conditions of the rumen; lipids and protein in such conditions are poor sources of energy.

Appendix Table 4.1 shows mean values for the proportion of apparently digested organic matter that was apparently digested in the rumen, classified according to type of animal and type of diet. In a few instances where results were published in terms of digested energy, the proportion of digested energy was converted to the proportion of organic matter by multiplication by 1.05 (based on Nicholson & Sutton 1969). A wide variation exists among individual published values for the same type of diet within a particular class of animal and the number of values available from cattle is small. Thus, no difference is apparent between sheep and cattle or between castrates, dry cows and milking cows. Evidence for differences among the types of diet is insufficient to justify the use of more than one factor at present. The only consistent difference is the lower value for complete diets of ground and pelleted artificially dried herbages and for uncooked maize diets (Waldo 1973). The poorer rumen digestion for the ground and pelleted forages is probably due to more rapid passage out of the rumen (Thomson 1972). It is not possible to state definitely whether artificial drying itself has any effect in the absence of grinding and pelleting, although Beever et al (1971a, b) obtained evidence of a reduction in rumen digestion when chopped grass was dried.

Table 4.1 summarizes the values for cattle and sheep. With the present evidence, *a value of 0.65 has been adopted* in the calculation of nitrogen requirements. The probability must be borne in mind that this value is an overestimate when the exceptional rations cited above are used.

Table 4.1.   Mean proportion of apparently digested organic matter apparently digested in the rumen (for details see Appendix Table 4.1)

	Cattle	Sheep
Pasture		
Grass	0.70	0.62
Grass/clover	0.62	–
Clover	–	0.66
Dried grass		
Chopped	–	0.58
Ground and pelleted	0.43	0.51
Dried legumes		
Chopped	–	0.55
Ground and pelleted	–	0.42
Hay		
Long or chopped	0.73	0.67
Ground and pelleted	0.82	0.65
Silage	–	0.62
Straw (alkali-treated) ground and pelleted	–	0.73
Mixed rations		
⩽500 g roughage/kg	0.65	0.69
<500 g roughage/kg	0.65	0.67
All-concentrate diets	0.68	0.60
Semipurified diets	–	0.66
Mean of above values	0.66	0.62
Mean of above values, excluding ground and pelleted artificially dried herbage	0.69	0.64

3. Microbial nitrogen yield in the rumen

Ideally, an estimate of the amount of energy made available for microbial growth as a result of anaerobic fermentation in the rumen should be expressed in terms of the number of high-energy phosphate bonds synthesized, usually referred to as moles of ATP formed. In practice it must be assumed that the amount of organic matter digested in the rumen is a measure of this energy, and, for reasons mentioned earlier, organic matter apparently digested in the rumen has been adopted. For exceptional feeds, this may not be true but for common ruminant feeds, low in fat and of average protein content, the energy value of digested organic matter is about 19.0 MJ/kg. Microbial N yield has been measured by the use of a number of different markers, commonly diaminopimelic acid (DAPA), ribonucleic acid (RNA) and ^{35}S. The microbial need for sulphur is considered in Appendix 4.1.

Appendix Table 4.2 gives the values from the literature and from unpublished results for microbial N yields in terms of organic matter apparently digested in the rumen; where necessary the results have been recalculated to give apparently digested organic matter. A wide variation in values is apparent. This may be due in part to the technique used in determination of the value, though there are good theoretical grounds for differences between diets. Efficiency of ATP production may vary with the type of substrate and with the type of fermentation (Henderickx et al 1972). The yield of bacterial protein per unit of ATP may vary with the rate of cell growth (Hobson & Summers 1972, Isaacson et al 1975), with the extent to which the products of partial protein degradation, such as peptides, are used by bacteria (Bryant 1970), with variations in the chemical composition of the bacteria (Smith 1975) and with the extent of recycling of bacterial components within the rumen (Nolan & Leng 1972).

It has not proved possible to show any clear difference in the microbial N yield that can be associated with class of ruminant or type of diet. The results are summarized in Table 4.2 and *a value of 30 g microbial N/kg organic matter apparently digested in the rumen has been adopted* for all diets, whether given to sheep or to cattle. It must be stressed that this value is not a biological constant but a mean based on widely varying individual results. It offers a guide to what can be expected in normal conditions, but special diets or other circumstances may need separate consideration.

Table 4.2. Mean microbial nitrogen yield/kg organic matter apparently digested in the rumen (for details see Appendix Table 4.2)

	Cattle	Sheep
Pasture		
Grass	27	21
Grass/clover	26	–
Clover	–	32
Dried grass		
Chopped	–	30
Ground and pelleted	–	25
Dried legumes		
Chopped	–	36
Hay (legume)	–	44
Silage	–	45
HCHO treated	–	15
Straw (alkali-treated, ground)	–	44
Mixed rations containing		
⩾500 g roughage/kg	33	32
<500 g roughage/kg	33	27
All-concentrate diets	–	33
Semipurified diets	–	34
Mean of above values	30	32

4. Efficiency of conversion of degraded dietary nitrogen into microbial nitrogen

After a feed, ammonia is produced in the rumen and, depending on its rate of production and utilization, an appreciable amount may enter the blood. It is well established that, with high intakes of a source of readily degradable N and where energy is limiting in the diet, there is considerable apparent absorption of dietary N anterior to the duodenum (e.g., Clarke et al 1966). When intakes of N and energy are better balanced there still may be appreciable absorption if synchronisation between N release and energy availability is inefficient. Even when dietary N is given largely as protein and is limiting, there is evidence that incorporation of degraded N into rumen microorganisms is not complete. For instance, the proportions of ^{15}N incorporated from the rumen ammonia pool were 0.86 for a barley diet given to sheep (Mathison & Milligan 1971) and 0.70 for a straw, tapioca and decorticated groundnut meal diet given to calves (Salter & Smith 1977).

Concurrently with the absorption of ammonia, there is addition of urea into the rumen from the blood by way of saliva and by passage across the rumen wall, and this addition tends to compensate for the loss of ammonia from the rumen. The quantitative importance of recycling is uncertain (Nolan et al 1973) but it is probably of greater importance at low N intakes.

Measurements of the difference between dietary N intake and flow of N at the duodenum are also affected by N, largely protein, entering as digestive secretions into the abomasum (Harrop 1974).

Although there is considerable variation, it is generally true that for diets containing mainly protein N, and which are well balanced in energy and protein, the amounts of non-ammonia N entering the duodenum are roughly equal to or somewhat greater than the N intake (e.g., Weston & Hogan 1973, Roffler & Satter 1975). In the calculation of the N requirements it has therefore been assumed that, as an approximation, any inefficiency in the incorporation of degraded N into microbial N, when N intake is just limiting, will be compensated by the recycling of N into the rumen; thus *an apparent efficiency value of 1.0 has been adopted for the conversion of degraded dietary protein N into microbial N.*

Evidence is accumulating (Nolan & Leng 1972, Gawthorne & Nader 1976, McMeniman et al 1976, Daneshvar et al 1976) that preformed amino acids are used to a large extent in microbial protein biosynthesis. Thus, amino acids and peptides, which may accumulate in considerable amounts in silages and are strictly NPN sources, could provide rumen bacteria with amino acids as well as ammonia for their growth. In the absence of other information, *an efficiency value of 1.0 has been adopted for the conversion of amino acid and peptide N into microbial N.* Despite this tentative conclusion, it must be stressed that information on the relative values of these NPN sources is meagre.

In comparison with diets containing protein, diets containing urea given in practical conditions, even when they are well balanced in energy and N content, generally lead to a less efficient utilization of dietary N. This may be due in part to the rapidity with which urea N is broken down to ammonia N in the rumen relative to the rate of fermentation of carbohydrates, and in part to a shortage of certain amino acids or the carbon skeletons from which they can be synthesized.

There is little information on the efficiency of utilization of the nitrogen in other NPN sources. In so far as these consist of either preformed ammonia or compounds yielding ammonia in the rumen, it can reasonably be assumed that their utilization would be similar to that of urea. Compounds which release ammonia slowly in the rumen are less liable than urea to lead to accidental toxicity (see Appendix 4.II), but are not generally used more efficiently than urea (Smith 1975). Indeed, if they are very resistant to degradation, as biuret is in some conditions, they may be used less efficiently (Oltjen et al 1968). Preformed nitrate in forages, unless present in toxic amounts (see Appendix 4.II), rapidly yields ammonia and can be assumed to be used with an efficiency similar to that of urea.

For the calculation of the requirements of urea or similar NPN sources when there is a deficiency in degradable protein intake, an *apparent efficiency of incorporation into microbial N of 0.8 has been adopted.* This is an arbitrary value. In ideal conditions of urea supplementation, such as with continuous feeding and with suitable energy sources, it may be possible for the apparent efficiency of utilization of released ammonia to approach that for degraded dietary protein. On the other hand the value could be much lower than 0.8, for example if energy was given mainly in the form of cellulose rather than as starch. It must be stressed that it is essential for effective utilization of urea that an adequate amount of a suitable available carbohydrate is present. It appears that the most satisfactory carbohydrate is starch, preferably cooked (Borgida et al 1976). Sugars, such as those present in molasses, are also of value but appear to be less effective than starch. It is possible that suitably modified non-protein N supplements may be developed which will be used as efficiently as protein but as yet this has not been achieved consistently.

5. Proportion of total microbial nitrogen present as amino acid nitrogen

Rumen bacteria contain appreciable amounts of nucleic acids, and also of other non-amino acid N compounds, such as muramic acid and N-acetylglucosamine in the cell wall. Thus, even when an efficiency value of 1.0 for the incorporation of degraded N to microbial N in the rumen is adopted (see above), the synthesis of microbial protein leads to a decrease in the amount of total amino acids entering the duodenum in comparison with the amount which would have been obtained if the dietary protein had not been degraded. It must also be emphasized that much of the non-amino acid N will be absorbed from the small intestine to supply the animal with unspecific N; it has, for example, been shown that degradation products from rumen bacterial nucleic acids may be incorporated into tissue nucleic acids of the animal (Smith et al 1974).

The true significance of the change in nutritional value involved in conversion of dietary protein to microbial N compounds depends on the relative amounts of essential amino acids in the dietary and microbial proteins, and on the amino acid requirements for the particular productive process. With present knowledge, however, the relative importance of the contributions of essential amino acids from the two protein sources cannot generally be satisfactorily determined. It is therefore considered, at present, that total amino acid N flow should be the criterion for determining the nutritional value of the N compounds entering the duodenum.

Estimates of proportions of microbial N present as amino acid N, made either by direct measurement or by subtracting nucleic acid N from total N, are given in Appendix Table 4.3. Values by the two methods differ slightly. The mean value for the former method, which takes account of non-amino acid N other than nucleic acid N, is 0.79. *A value of 0.80 has been adopted* for calculating N requirements. In using this factor, it must be borne in mind that the value used subsequently for efficiency of utilization of absorbed N derived from bacteria should refer to amino acid N and not to total N.

6. Apparent absorbability in the small intestine of amino acids derived from microbial and dietary protein

Apparent absorbability of amino acid N in the small intestine is the net disappearance of amino acid N between the duodenum and the terminal ileum as a fraction of the amino acid N entering the duodenum. It has been measured by the use of simple or reentrant cannulas in both (a) the abomasum or proximal duodenum and (b) the terminal ileum; such measurements include endogenous secretions into the small intestine.

Values from the literature and unpublished results are shown in Appendix Table 4.4; some are for apparent absorbability of non-ammonia N, which includes nucleic acid, and some are for total amino acid N absorbability which is the required value for the proposed system of estimating N requirement. In experiments in which absorbability of both non-ammonia N and total amino acid N have been measured, the values do not differ greatly.

Some of the values for cattle given in Appendix Table 4.4 are for total N absorbability, but they have been included because of the paucity of values for cattle.

Although the values given in Appendix Table 4.4 cover a wide variety of diets of differing protein content given to different classes of animal, it is not possible at present to associate the differences between individual values with any particular factor. There appears to be no observable difference between values for sheep and for cattle. There are undoubtedly differences in apparent absorbability between microbial protein and dietary protein and between different sources of dietary protein, but it is not possible to quantify them. There is no evidence that the grinding and pelleting of artificially dried herbage, which reduced organic matter digestibility in the rumen, had any effect on the absorption of protein N in the small intestine. However, excessive heat or chemical treatment with the intention of protecting protein sources from degradation in the rumen may result in reduced absorbability of protein N in the small intestine.

Table 4.3 summarizes the results from Appendix Table 4.4. In the computation of N requirements, *a value of 0.70 for apparent absorbability of protein N in the small intestine has been adopted.*

Table 4.3   Mean values for apparent absorbability of protein in nitrogen from the small intestine (for details, see Appendix Table 4.4)

	Cattle	Sheep
Pasture		
Grass	0.67	0.73
Clover	—	0.69
Dried grass		
Chopped	—	0.70
Ground and pelleted	—	0.72
Dried legumes		
Chopped	—	0.63
Ground and pelleted	—	0.63
Hay	—	0.45*
Mixed rations containing		
$\geqslant$500 g roughage/kg	0.65	0.68
<500 g roughage/kg	0.64	0.68
All-concentrate diets	—	0.64
Semipurified diets	0.67	0.71
Mean of above values	0.66	0.66

* Duodenal and ileal collections were made from different animals.

7. Efficiency of utilization of absorbed amino acid nitrogen

In the factorial system used in the first edition, the efficiency of utilization of absorbed amino acid N was defined as the biological value of the dietary protein, namely, the proportion of truly digested N that was retained in the tissues and was necessary to meet the obligatory loss in faeces and urine and was represented by the value

$$(R_N + F_{N(M)} + U_{N(E)})/(D_N + F_{N(M)})$$

where $R_N = $ N retention, $F_{N(M)} = $ metabolic faecal N, $U_{N(E)} = $ endogenous urinary N and $D_N = $ apparently digested N.

A value of 0.70 was used for cattle and 0.65 for sheep irrespective of diet, but was realised that in ruminants biological value had grave shortcomings as a measure of efficiency of utilization of absorbed amino acid N because of the extensive conversion of dietary N into microbial N in the rumen and also because metabolic faecal N, as conventionally measured, was partly of microbial origin.

The measure of efficiency of utilization that is required in the present method of calculating N requirements is the proportion of amino acid N apparently absorbed from the small intestine which can be used to meet tissue needs and endogenous N excreted in the urine. It is represented by the value

$$(R_N + U_{N(E)})/D_{N(SI)},$$

where $R_N = $ N retention, $U_{N(E)} = $ endogenous urinary N and $D_{N(SI)} = $ apparently absorbed amino acid N in the small intestine.

Based on the results of J. L. Black et al (unpublished), who measured amino acid N apparently absorbed from the small intestine and N retention in calves of the same weight, a value for efficiency of utilization of 0.75 was obtained with a diet limiting in N, in which apparent absorption of amino acid N in the small intestine was similar to apparent digestion of N throughout the whole alimentary tract.

At present it is not possible to quantify differences in the efficiency of utilization of microbial protein and of undegraded dietary protein. Several experiments have shown that microbial protein appears to be deficient in the sulphur-containing amino acids, methionine and cysteine (Nimrick et al 1970a, b, Armstrong & Annison 1973, Williams & Smith 1974), so that undegraded dietary protein with a relatively high content of sulphur-containing amino acids would be used most efficiently. Ørskov & Fraser (1969) and Ørskov et al (1970) have shown in lambs a higher efficiency of retention of supplementary N from white-fish meal than from soya bean meal, sunflower meal or single-cell protein. Similarly, J. L. Black et al (unpublished) have shown in calves a higher efficiency of retention of supplementary N from white-fish meal than from decorticated groundnut meal. Thus, although undoubtedly there are differences in efficiency of utilization of different protein sources, which cannot be accounted for by differences in the extent of their degradation in the rumen (Ørskov et al 1970), and may therefore be attributed to their amino acid composition, it is not possible at present to quantify these differences. Similarly, there is insufficient knowledge to allow for differences in efficiency of utilization in different productive processes.

A value of 0.75 for efficiency of utilization of apparently absorbed amino acid N from the small intestine has therefore been adopted for both cattle and sheep in the calculation of N requirement. However, this is clearly an area where much more research is needed to determine the efficiency values of undegraded protein from different dietary sources. In the previous edition, some variations between diets in biological value, as conventionally determined, were indicated but it was not possible to estimate separate values for specific diets. The variation in biological value for ruminant animals is likely to be much less than for the nonruminant, because of microbial synthesis of essential amino acids in the rumen.

Endogenous urinary nitrogen excretion

5. Endogenous urinary nitrogen excretion; dermal losses of nitrogen by cattle.

The concept of the division of the total urinary excretion into a relatively constant endogenous component and an exogenous component varying with protein intake has been generally accepted since its formulation by Folin (1905), and was examined by Mitchell (1962). The constant endogenous component is assumed to arise from the degradation and replacement of protein structures and of simple nitrogenous components of the tissues in irreversible reactions typified by the dehydration of creatine to creatinine. The exogenous component, on the other hand, is considered to represent wastage of N from the metabolic pool of absorbed N and tissue N involved in reversible biochemical processes of synthesis and degradation.

The endogenous component is assumed to be equal to the minimum rate of urinary N excretion of an animal maintained for some time on a diet containing little or no protein but adequate in all other respects. This method of measurement may underestimate the endogenous excretion of animals on diets containing larger amounts of protein. Endogenous urinary excretion has also been estimated from the value of the

intercept of the regression of urinary N on apparently digested N, or of $\log_{10}$ urinar
N on N intake with animals given diets in which N is limiting or nearly so. Mea
values for endogenous urinary N are given in Appendix Tables 4.5 and 4.6 for catt
and sheep and individual values for cattle and sheep in Figures 4.1 and 4.2. Son
early German results for adult cattle included in the first edition have been exclude

Fig. 4.1   Endogenous urinary nitrogen excretion of European cattle (O, pre-ruminant; ●, ruminant).

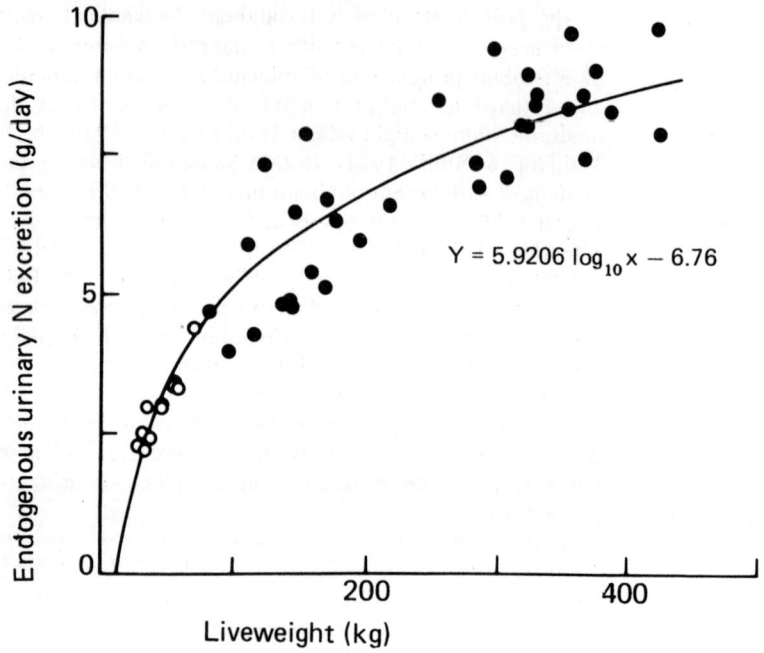

$$Y = 5.9206 \log_{10} x - 6.76$$

Fig. 4.2   Endogenous urinary nitrogen excretion of sheep (O, pre-ruminant; ●, ruminant).

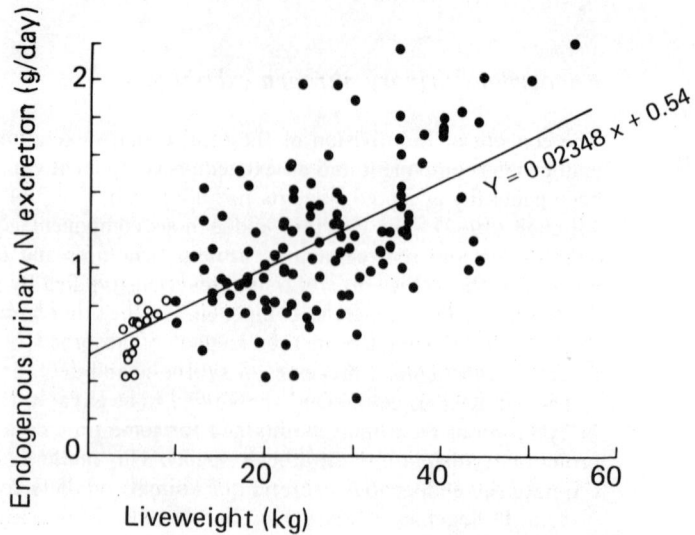

$$Y = 0.02348 x + 0.54$$

from the Table, since the N intakes were considered to be too high (>16 g N/day for a 400-kg animal) to give a valid estimate of endogenous urinary N. As a result there are no values for cattle above 450 kg live weight. The values for sheep are similar to those given in the first edition but the proposed values for cattle above 400 kg live weight are somewhat lower.

No obvious difference is apparent between preruminant and young ruminant animals in sheep or cattle. However, it is clear that zebu cattle (*Bos indicus*) have lower endogenous urinary N excretion than European breeds, and there is some indication in sheep that non-European breeds also may have lower values.

The most appropriate relationship of endogenous urinary N and live weight was for cattle

$$U_{N(E)} = 5.9206 \ (\pm 0.3442) \ \log_{10} W - 6.76, \quad SD = \pm 0.835$$

and for sheep

$$U_{N(E)} = 0.02348 \ (\pm 0.00253) \ W + 0.54, \quad SD = \pm 0.323,$$

where $U_{N(E)}$ = endogenous urinary N excretion (g/day) and W = live weight (kg).

Whereas for cattle the proportion of the variation (r^2) that could be accounted for by live weight was 0.87, for sheep it was only 0.37.

The accepted endogenous values for cattle and sheep are given in Table 4.4.

Table 4.4   Proposed values for endogenous urinary nitrogen excretion and comparison with those given in ARC (1965).

	Cattle				Sheep		
		Endogenous urinary N				Endogenous urinary N	
Live weight (kg)	Proposed value (g/day)	ARC (1965) (g/day)		Live weight (kg/)	Proposed value (g/day)	ARC (1965) (g/day)	
50	3.3	3.5		5	0.7	0.6	
100	5.1	4.9		10	0.8	0.8	
200	6.9	5.7		20	1.0	1.1	
300	7.9	7.7		30	1.2	1.1	
400	8.7	9.5		40	1.5	1.3	
500	9.2	11.2		50	1.7	1.6	
600	9.7	12.8		75	2.3	2.1	

Dermal losses of nitrogen by cattle

The values adopted for losses of nitrogen in hair and scurf by cattle are those given in the first edition; they are reproduced in Table 4.5 and are equivalent to a loss of nitrogen (g) of $0.018 \ W^{0.75}$ kg.

Table 4.5   Loss of nitrogen in hair and scurf by cattle.

Live weight (kg)	N loss* (g/day)
50	0.3
100	0.6
200	1.0
300	1.3
400	1.6
500	1.9
600	2.2
700	2.4

* Calculated from the expression $0.02 \ W^{0.73}$ kg (ARC, 1965) $\equiv 0.018 \ W^{0.75}$ kg.

9. Extent of degradation of dietary protein in the rumen

Dietary proteins entering the rumen are partly degraded, at first to peptides and amino acids, which play an uncertain but probably minor role in the nutrition of the animal, and finally to ammonia. Some dietary protein escapes degradation and the extent of this survival influences both the amount of degraded N available as a nutrient to the rumen microbial flora and the amount of amino acids available to the animal.

Some food proteins, e.g., casein and zein, have chemical properties which allow direct estimation of the amounts in digesta. More usually, the only means of estimating the survival of dietary protein is by deducting microbial N from total N or from total non-ammonia N passing the duodenum. On this basis, the proportion of intake that survives or is undegraded is:

$$(\text{duodenal N}-\text{microbial N})/I_N,$$

and that degraded is:

$$1-(\text{duodenal N}-\text{microbial N})/I_N.$$

Values for the proportion of N degraded, usually calculated in this way, are shown in Appendix Table 4.7. As mentioned earlier, the accuracy of measurements of microbial N is uncertain and may vary with the method used. Any error in this determination is usually exaggerated in the calculation of feed protein survival by difference, and a further error is introduced by the presence of an unknown endogenous N contribution from abomasal digestive secretions. Since no correction for this contribution has been made to the values in Appendix Table 4.7, values for the proportion of dietary N that is degraded are slightly underestimated.

There are wide variations in the estimates of dietary protein degradability, given in Appendix Table 4.7. These may be due to the many differences that exist between diets, classes of animal and methods of assessing microbial N, and also to special conditions such as the fact that Hagemeister & Pfeffer (1973) and Hagemeister & Kaufmann (1974) measured duodenal flow posterior to the entry of the bile and pancreatic ducts, which probably increased the error associated with endogenous N secretion. In spite of these variations, some fairly consistent differences between diets are apparent.

Because of the variability in the values within feeds and because values are available for only a small number of feeds, attempts have been made to relate degradability in the rumen to values for protein solubility *in vitro*. Some support for this approach is provided by reports that protein solubility appears to be reasonably well correlated with rumen ammonia concentration and N retention *in vivo* (Chalmers et al 1954, Chalmers & Synge 1954, Chalmers & Marshall 1964, Preston & Pfander 1961 1963, Preston R. L. et al 1965, Hudson 1969, Hudson et al 1970). But protein solubility varies with pH so that solubility measured *in vitro* at a fixed pH may not reflect solubility *in vivo* where pH may differ markedly with type of diet and time after feeding (Preston et al 1963). Values for solubility of a number of proteins in conditions simulating those in the rumen have been published by Wohlt et al (1973) and further values have been obtained by McMeniman (1976) and Smith et al (1977). For four protein supplements ranging in solubility from 15% (fish meal) to 54% (groundnut meal), McMeniman found a significant positive relationship between protein solubility and the extent of degradation in the rumen of sheep. On the other hand, Smith et al (1978), who examined four supplements with a similar range of solubilities, observed only a small and not significant relationship in young cattle. It is clear that factors other than solubility are involved in determining the degradability, such as the molecular structure of the protein (Mangan 1972) and the other components of the diet; Schoeman et al (1972) found a greater degree of protein degradation in the rumen of sheep given a diet containing high rather than medium proportions of roughage. A high plane of nutrition may increase the proportion of undegraded protein reaching the small intestine (Ørskov & Fraser 1973), especially for relatively insoluble proteins such as fish meal (Miller 1973).

Thus, it appears that many more data are needed before the possibility of using determinations of solubility *in vitro* to predict degradability in the rumen can be assessed. Another possible, simple way of assessing the rate at which a protein source is degraded in the rumen is to measure its rate of disappearance from a synthetic fibre bag suspended in the rumen (Schoeman et al 1972, Mehrez 1976, Mehrez & Ørskov 1977, Ørskov & Mehrez 1977, Mathers et al 1977). The few direct comparisons of this measurement of degradability with that obtained from cannulated animals are conflicting. Mathers et al (1977) stated that at short incubation times (4 to 6 h), the synthetic fibre bag technique gave estimates for protein degradability similar to those obtained from estimates of residual dietary protein flow to the duodenum. On the other hand, although Smith & Mohamed (1977) found rates of disappearance from synthetic fibre bags of different protein sources similar to those found by Ørskov & Mehrez (1977) and Mathers et al (1977), they did not find a relationship between these rates of disappearance and estimates of the amount of the sources recovered at the duodenum (Smith et al 1978). A number of comparative values for different feeds obtained by this method are given in Appendix Table 4.8 but it should be stressed that before it can be accepted the technique must be validated by comparison of the values with those obtained for the same feeds in cannulated animals. Even for the dietary components examined *in vivo*, for which results are given in Appendix Table 4.7, it is not at present possible to predict the extent of protein degradation for a particular feed with any degree of precision. Nevertheless, it is possible to come to

Table 4.6   Suggested grouping of different dietary protein sources into four classes, based on the extent of protein degradation in the rumen

Class	Range of degradability (dg)	Forages	Cereals	Protein supplements
A	0.71–0.90	Grass hay Legume hay Grass silage (*wilted* and unwilted) Artificially dried grass (chopped) (*Swede roots*)	Barley (*Wheat*) (*Barley*)	Casein Wheat gluten Groundnut meal Sunflower meal Soya bean meal (unheated) Rapeseed meal Field bean meal Yeast protein
B	0.51–0.70	Grass (fresh or frozen) Legumes (fresh or frozen) Artificially dried grass (ground & pelleted) Artificially dried legume (except sainfoin) (chopped) (*Maize silage*) (*Clover silage*)	Maize (*Maize*)	Soya bean meal (cooked) Lupin meal Coconut meal Fish meal* (*Cottonseed meal*) (*Sunflower seed*) (*Ground peas*) (*Linseed cake*)
C	0.31–0.50	Artificially dried legume (ground and pelleted)	Milo	Zein Casein (formaldehyde treated) Fish meal* (*Ground winter beans*) (*Meat and bone meal*) (*Guar meal*) (*White-fish meal*)
D	<0.31	Grass silage (formaldehyde treated) Artificially dried sainfoin (chopped)		(*Peruvian fish meal*)

* Results for fish meal are divergent, probably because of different treatments in processing and/or differences between types of fish. It may be reasonable to allocate white-fish meal to Class B, and herring meal and Peruvian fish meal to Class C.
In italics, results from synthetic fibre bag technique (see text).

some general conclusions and to allocate some feed ingredients into broad groups. Observed differences within feeds include those due to natural variation, to variations in the severity of heat treatment in the artificial drying of fodders or in the processing of protein supplements, such as fish meal, and to chemical treatments, such as formaldehyde or tannin (Ferguson 1975; Leroy & Zelter, 1970).

A suggested classification is given in Table 4.6. It is proposed that *degradability factors (dg) of 0.8, 0.6 and 0.4 be used for components in Groups A, B and C, respectively*. For practical purposes, it is suggested that, in the absence of evidence to the contrary, untested protein supplements should be assigned to Group A.

It is clear that in the present state of knowledge degradability values for particular ingredients of a ration may have to be assumed. If the assumed value for degradability were too low, rumen microbes would receive more degradable N than they needed, and the amount of dietary protein reaching the small intestine would be underestimated. This might lead to reduced performance in high-producing animals. An overestimation of degradability could lead to a shortage of degradable protein for microbial needs. This could have serious repercussions on voluntary feed intake if the error was large, for instance if overheating of dried grass or cereals had caused a reduction in degradability. The overestimation would also mean that more than the required amount of protein would enter the small intestine so that dietary protein intake would be excessive. If the voluntary intake of feed was being reduced because of inadequate N for maximum rate of fermentation in the rumen, the protein would possibly not be used as a source of amino acids.

The application of the proposed system for formulation of diets to meet the N requirements of the animal should be treated with caution if the diet contains a component, whether of natural occurrence of after processing, which has a degradability value of 0.50 or less, until it has been shown that its digestibility in the small intestine is not abnormally low. It has, for example, been shown that zein (group C) is poorly digested in the small intestine of cattle (Little & Mitchell 1967). Similarly, too severe treatment with formaldehyde can reduce digestibility in the small intestine, although by careful control of conditions, it is possible to prepare a dietary ingredient resistant to rumen degradation but with digestibility in the small intestine largely unimpaired (Ferguson 1975).

10. Summary of the calculation of nitrogen requirement

The three summary equations given on p. 124 were calculated from the values adopted for the various factors, as follows:

(a) Degraded nitrogen (RDN) required by rumen microorganisms

RDN required (g/day) = $M_E \times F \times$ proportion of D_O apparently digested in the rumen $\times$ microbial N yield/kg D_O apparently digested in the rumen $\times$ efficiency of conversion of degraded N to microbial N,

where (1) M_E = Metabolizable energy intake (MJ);

(2) F = factor for conversion of M_E to D_O on the assumption that 18% of apparently digested energy is lost in CH_4 and urine, and that 1 kg D_O = 19.0 MJ D_E, therefore F = $1/(0.82 \times 19)$;

(3) proportion of D_O that is apparently digested in the rumen = 0.65;

(4) microbial N yield (g/kg D_O apparently digested in the rumen) = 30;

(5) efficiency of conversion of degraded N to microbial N = 1.0.

Thus RDN required (g/day) = $M_E \times 1/(0.82 \times 19) \times 0.65 \times 30$
$$= 1.252 \ M_E \ (1.25 \ M_E \ \text{adopted}).$$

(b) Net amino acid nitrogen supplied to tissues by rumen microorganisms (TMN)

TMN (g/day)=RDN× proportion of total microbial N as amino acid N× apparent absorbability in the small intestine of amino acid N from microbial protein×efficiency of utilization of absorbed amino acid N,

where (1) proportion of total microbial N as amino acid N=0.88;

(2) apparent absorbability in the small intestine of amino acid N from microbial protein=0.70;

(3) efficiency of utilization of absorbed amino acid N=0.75.

Thus TMN (g/day) = RDN requirement×0.80×0.70×0.75

= RDN requirement×0.42

= 0.526 M_E (0.53 M_E adopted).

(c) Undegraded dietary nitrogen (UDN) requirement

(i) If tissue nitrogen required (TN) < net amino acid N supplied to tissues by microbial protein (TMN), i.e. <0.526 M_E,

where TN is the sum of $\begin{cases} R_N = \text{N retention} \\ L_N = \text{Lactation N} \\ N_m = \text{N required for maintenance etc.,} \end{cases}$

no undegraded dietary N is required and the value for the UDN requirement is therefore 0. Total N requirement will then equal RDN requirement=1.252 M_E.

(ii) If tissue nitrogen required (TN) >net amino acid N supplied to tissues by microbial protein (TMN), i.e. >0.526 M_E,

UDN requirement=(tissue nitrogen−net amino acid N supplied to tissues by microbial protein)/(apparent absorbability of amino acid N in small intestine×efficiency of utilization of absorbed amino acid N)

= (TN−0.526 M_E)/(0.70×0.75)

= 1.91 TN−1.00 M_E.

(d) Total nitrogen requirement

= RDN requirement+UDN requirement.

Provided that the UDN requirement is not zero,

total N requirement = 1.25 M_E+1.91 TN−1.00 M_E

= 0.25 M_E+1.91 TN.

(e) Crude protein concentration in dry matter

The minimum concentration of crude protein in the dry matter required by the animal is thus ((RDN+UDN) requirement×6.25)/I_T, where I_T=total dry matter intake (kg/day), and this would also be the minimum concentration of crude protein in the dry matter of a diet if the degradability of the protein allowed it to supply exactly the RDN and UDN requirement.

11. Formulation of a ration to meet the protein requirements of the animal

A detailed example of the calculation of the N requirements of an animal is shown in Table 4.7. However, for ease of formulation of rations, it is proposed that protein (N×6.25) should be used rather than nitrogen. Then from the Summary of the Calculation of the N requirement (above),

Rumen-degradable protein (RDP) required (g/day) $= 7.8\ M_E$ [4.4]

   (RDN $= 1.25\ M_E$)   [4.1]

Tissue protein supplied by microbial protein (TMP) (g/day) $= 3.3\ M_E$ [4.5]

   (TMN $= 0.53\ M_E$)   [4.2]

Undegraded protein (UDP) required (g/day) $= 1.91\ TP - 6.25\ M_E$,   [4.6]

   where $TP = 6.25 \times TN$

   (UDN $= 1.91\ TN - 1.00\ M_E$)   [4.3]

Crude protein (CP) required (g/day) $= RDP + UDP$ [4.7]

Degradability of protein (dg) $= RDP/CP$.   [4.8]

If the value of UDP is negative and sufficient RDP is provided for maximum utilization of ME, it means that RDP supplies sufficient protein (TMP) to cover tissue needs and the value of UDP should be ignored.

TN tissue N. requirement

TMN tissue microb. N provided

Do digested organic matter intake

Table 4.7

Example of the detailed calculation of the N requirement of a 600 kg Friesian cow giving daily 30 kg of milk with 36.8 g fat/kg, assuming no change in body weight.

Assuming:	
Metabolizable energy concentration of ration (MJ/kg I_T)	11.0
(equivalent to q = 0.60, where q = M_E/G_E	
Metabolizable energy required (MJ/day)	207
Then:	
Dry matter intake (kg/day) (I_T)	18.82
Apparently digested organic matter intake (kg/day)	13.29
($D_O = M_E/(19.0 \times 0.82) = M_E/15.58$)	
N requirement of rumen microorganisms (RDN)	
D_O apparently digested in rumen (kg/day)	8.64
(13.29 × 0.65)	
Microbial N yield (g/day) (RDN)	259
(8.64 × 30)	
Tissue N requirement supplied by microorganisms (TMN)	
Amino acid N from microbial N (g/day)	207
(259 × 0.80)	
Apparently absorbed amino acid N in small intestine (g/day)	145
(207 × 0.70)	
N retention (g/day) (TMN)	109
(145 × 0.75)	
Total tissue N requirement of animal (TN)	
N in milk (g/day)	144
(4.8 g protein N/kg) (p. 149)	
Endogenous urinary N ($U_{N(E)}$) (g/day)	10
(Table 4.4)	
N in hair and scurf (g/day)	2
(Table 4.5)	
Total (TN)	156
Tissue N requirement (TN) not supplied by microbial N (TMN)	47
(156–109)	
Undegraded dietary N (UDN) requirement	
Apparently absorbed amino acid N required (g/day)	63
(47/0.75)	
Undegraded dietary N intake required (g/day) (UDN)	90
(63/0.70)	
Total dietary N requirement (g/day)	
RDN requirement + UDN requirement	349
(259 + 90)	
Minimum crude protein content of diet (g/kg)	
((RDN requirement + UDN requirement) × 6.25)/I_T	116
((349 × 6.25)/18.82)	

In the example in Table 4.7 for a nonpregnant 600 kg Friesian cow giving daily 30 kg of milk containing 36.8 g fat/kg, the following values for daily requirement are given:

Fig. 4.3   Crude protein required for rumen microbes with diets differing in
digestibility and protein degradability.

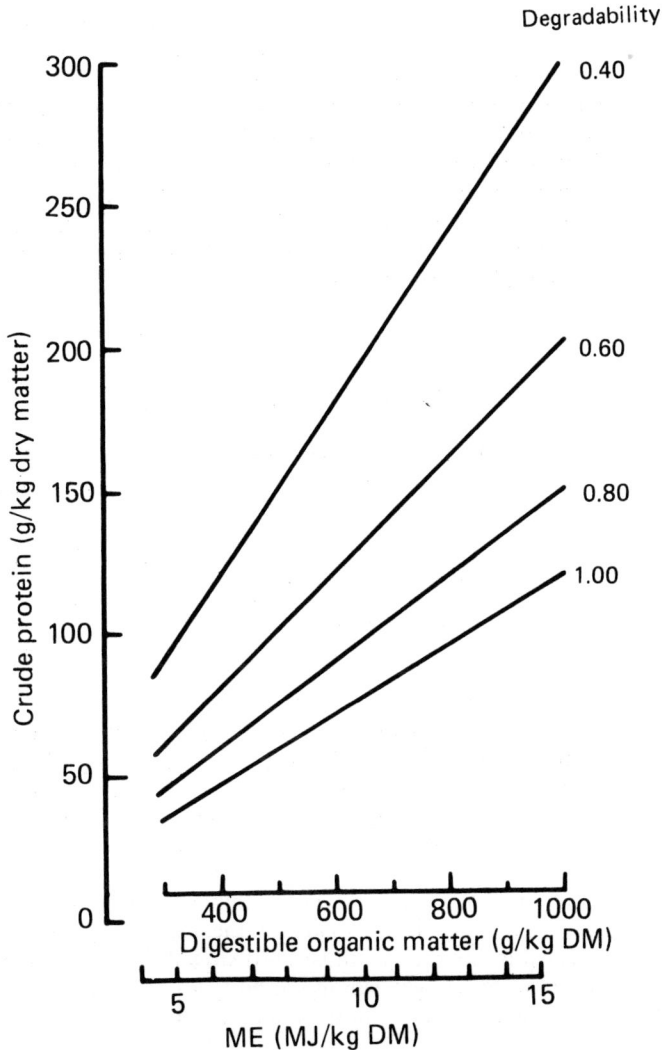

M_E intake (MJ/day) = 207
(for ration of q = 0.60, i.e. 11 MJ/kg dry matter)
Dry matter intake (kg/day) = 18.82
Tissue protein required (TP, g/day) = 975

Then, RDP = 1615 g/day from [4.4] above; UDP = 569 g/day from [4.6] above; CP = 2184 g/day from [4.7] above; dg = 0.74 from [4.8] above. Minimum CP concentration = 116 g/kg DM.

For the formulation of rations to meet the protein requirements, feed tables should list the following data for each feedstuff:

Dry matter (DM, g/kg)
Metabolizable energy (M_E, MJ/kg DM)
Rumen-degradable protein (RDP, g/kg DM)
Undegraded protein (UDP, g/kg DM).

In addition it might be useful to tabulate:

Crude protein (g/kg DM)

Degradability (dg, as a coefficient)

Degradable protein energy index (RDP/M_E, g/MJ M_E).

The degradable protein energy index would be a useful description of a feedstuff with regard to rumen fermentation. A value of $\geqslant 7.8$ g/MJ M_E would indicate sufficient degradable N in the feedstuff for the rumen microflora to make full use of the readily fermentable carbohydrate that was present. It is possible to represent such relationships graphically. An example is given in Fig. 4.3, which shows the total protein required in the dry matter of a feedstuff to supply the degradable N necessary to meet rumen microbial need, in relation to the digestible organic matter content of the dry matter of the feed and the degradability of its protein.

Ration formulation procedure

The DM, M_E, RDP and UDP of the individual ingredients of the proposed ration for a particular productive process are calculated and summated. Thus,

	DM (kg)	M_E (MJ)	RDP (g)	UDP (g)	dg
Feed 1 data x W_1 (kg DM) →	x	x	x	x	
Feed 2 data x W_2 (kg DM) →	x	x	x	x	
Feed 3 data x W_3 (kg DM) →	x	x	x	x	
Total	x	x	x	x	x

The totals together with the overall degradability of the protein in the ration are then compared with the requirement. Whilst M_E should equate fairly closely to the requirement and DM should be within the maximum appetite of the animal given such a ration (see Chapter 2), RDP and UDP should not be less than the requirement.

The amount of crude protein in the ration may exceed the minimum requirement if the degradability of the total ration, available on the farm, is greater or less than that needed by the animal. For example, if RDP alone satisfies the tissue needs of the animal and the degradability of the ration is 0.80, the amount of crude protein received in the ration will be $(1/0.80) \times RDP = 1.25$ RDP, and UDP will be wasted. However, for the efficient use of protein it is desirable that the RDP and UDP in the ration should be close to the requirements of the animal. Thus,

(a) if there is an excess of RDP or UDP in the ration, a new ration can be formulated to reduce the excess;

(b) if there is a deficit of RDP only, this may be overcome by inclusion of a source of non-protein nitrogen. If urea is the compound used, then the amount required is:

Weight of urea (g/day) = Deficit of RDP/$(6.25 \times 0.80 \times 0.46)$

= Deficit of RDP/2.30,

where 0.46 = N content of urea (g/g) and 0.80 = efficiency of conversion of urea to microbial N;

(c) if there is a deficit of both RDP and UDP in the ration, the deficit of UDP must be corrected first by selection of a protein supplement of appropriate degradability. If there is still a deficit of RDP, this may be supplied either by recalculating the ration using a protein supplement of higher degradability or by calculating the amount of urea to overcome the deficit as in (b) above.

Thus for the requirements of the 600 kg cow given in detail in Table 4.7 and summarized on p. 138 a ration can be formulated as follows:

Composition of ingredients of ration

	DM (g/kg)	M$_E$ (MJ/kg DM)	RDP (g/kg DM)	UDP (g/kg DM)	dg
Grass hay	850	8.4	68	17	0.80
Grass silage	200	10.2	136	34	0.80
Rolled barley	860	13.7	86	22	0.80
Flaked maize	900	15.0	66	44	0.60
Soya bean meal	900	12.3	302	201	0.60

Requirements per day†

DM (kg) (at M$_E$ concentration of 11 MJ/kg)	M$_E$ (MJ)	RDP (g)	UDP (g)	dg
18.8	207	1615	569	0.74

Ration

	DM(kg)	M$_E$ (MJ)	RDP (g)	UDP (g)	dg
Grass hay	7.5	63	510	128	
Grass silage	3.2	33	435	109	
Rolled barley	3.2	44	275	70	
Flaked maize	4.1	62	271	180	
Soya bean meal	0.41	5	124	82	
Total	18.41	207	1615	569	0.74

†600 kg Friesian cow giving 30 kg milk containing 36.8 g fat/kg in mid-lactation (q=0.60≡11 MJ/kg DM). Maximum dry matter intake (kg/day)=19.0 (Table 2.7).

In practice in formulating this ration, the composition of the basal ration (grass hay, grass silage, rolled barley and flaked maize) would be calculated first and then the amount of protein supplement (soya bean meal) needed to bring the total UDP in the ration up to the animal's requirements. If the total ration then failed to meet the M$_E$ requirement or was much in excess of appetite, the basal ration would be modified and the amount of protein supplement recalculated. For the cow giving 30 kg milk, it is difficult to formulate a ration that does not contain an exessive amount of RDP, so the possibility of using non-protein N to overcome a deficit of RDP is unlikely, except if ration ingredients of very low degradability are used.

Thus, to obtain an optimum degradability of 0.74, as required in the example, feeds of low degradability such as flaked maize and soya bean meal would need to be included in the ration. However if, as shown in Fig. 4.4, the degradability of a ration was 0.80, an overall crude protein content of 150 g/kg dry matter intake would be required to supply the UDP requirements for a cow yielding 30 kg milk rather than the value of 116 g/kg dry matter for the ration of 0.74 degradability given in the example.

Similarly, for the young growing animal, below 300 kg live weight for cattle and 30 kg live weight for sheep, failure to achieve the optimum degradability in the ration will result in an increase in the protein requirement. This is illustrated in Fig. 4.5 for the 100 kg bullock of a medium-sized breed at a metabolizability of energy of 0.70, where the minimum protein requirement for a weight gain of 0.75 kg/day is 165 g protein/kg dry matter at an optimum degradability of 0.61. If the degradability of the ration was 0.70, the protein concentration in the diet would have to be increased to 215 g protein/kg dry matter to supply the UDP requirement of the animal.

Fig. 4.4   Relationship for Friesian cows (600 kg live weight), given a diet with q = 0.6, between the required dietary concentration of crude protein, milk yield, and rumen degradability of protein (optimum degradability for a given milk yield shown by broken line).

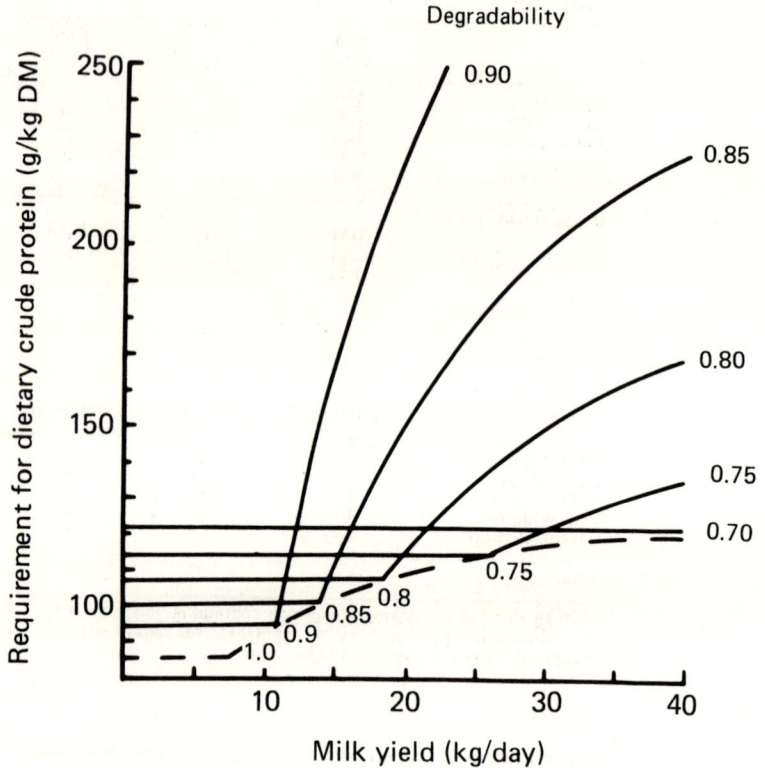

On the other hand, for the older growing animal, the amount of RDP required wi usually supply enough microbial protein to the tissue to satisfy the tissue protei requirement. In these conditions, a deficit of RDP can be overcome by the use c urea.

This is illustrated in the following example.

Bullock of medium-sized breed (300 kg live weight gaining weight at 1.25 kg/day)

M_E requirement (MJ) = 79.7 (from Chapter 3) (q = 0.60 ≡ 11 MJ/kg DM).
Maximum dry matter intake (kg/day) = 7.9 (Table 2.3)
Tissue protein (TP) required

Protein retention (g) (128.2 g protein/kg weight gain) (see p. 147)	160.3
Endogenous urinary protein (g) (see p. 133)	49.4
Protein in hair and scurf (g) (see Table 4.5)	8.1
Total	217.8
Tissue protein supplied by microbial protein (g) (TMP = 3.3 M_E)	263.0

Feed composition	DM (g/kg)	M_E (MJ/kg DM)	RDP (g/kg DM)	UDP (g/kg DM)	dg
Barley straw	860	7.3	30	8	0.80
Rolled barley	860	13.7	86	22	0.80
Requirements	(kg)	(MJ)	(g)	(g)	
	7.25†	79.7	622	0	1.00
Ration					
Barley straw	3.0	21.9	90	24	
Rolled barley	4.2	57.5	361	92	
Total	7.2	79.4	451	116	0.80
Deficit	0.05	0.3	171		
Excess				116	
Urea required (kg)‡	0.074		171		
Total ration	7.27	79.4	622	116	0.84

† q = 0.6, M_E = 11 MJ/kg DM.   ‡ (RDP deficit)/(2.3 × 1000).

The urea addition required is thus 10 g/kg DM, (0.074×1000)/7.2, and the final crude protein concentration is 102 g/kg dry matter.

Fig. 4.5   Protein requirements of a 100 kg bullock of a medium-sized breed gaining at various rates when given a diet of q=0.7 to meet energy requirements, at optimum and fixed rumen degradability protein. (Optimum degradability for a given growth rate shown by broken line).

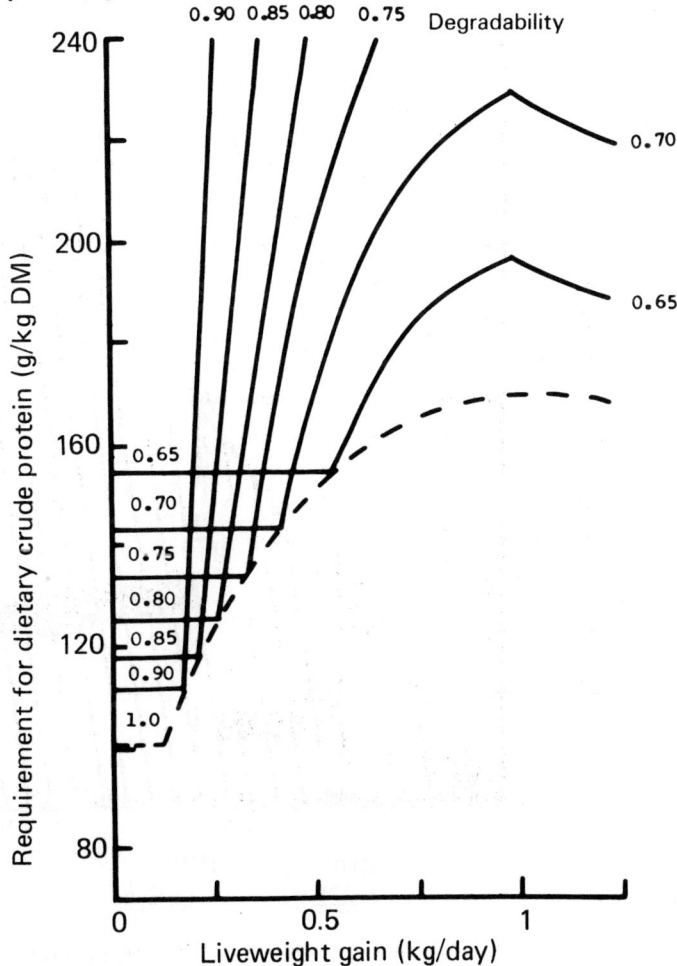

Use of the graphical method

Since the protein requirements of an animal can be calculated from the three equations, namely: RDP (g) $= 7.8$ M_E [4.4]

$$TMP (g) = 3.3 M_E \quad [4.5]$$

and if tissue protein requirement (TP) is greater than TMP

$$\text{from UDP (g)} = 1.91 TP - 6.25 M_E \quad [4.6]$$

they may be presented in the graphical form shown in Fig. 4.6. The figure shows the example given in Table 4.7 where RDP requirement of the animal is insufficient to meet tissue needs at the required energy input, and also further examples where RDP is sufficient or more than sufficient to meet tissue needs at a given energy input.

Fig. 4.6   Relationship between dietary protein requirements of ruminants and metabolizable energy intake (M_E) and tissue requirements (TP).
Requirement for rumen degradable protein (RDP, g/d) = 7.8 M_E and for undegradable protein (UDP, g/d) = 1.91 TP – 6.25 M_E. Tissue protein supplied by microbial protein (TMP) = 3.3 M_E. (Numbers between lines refer to lines immediately below them).

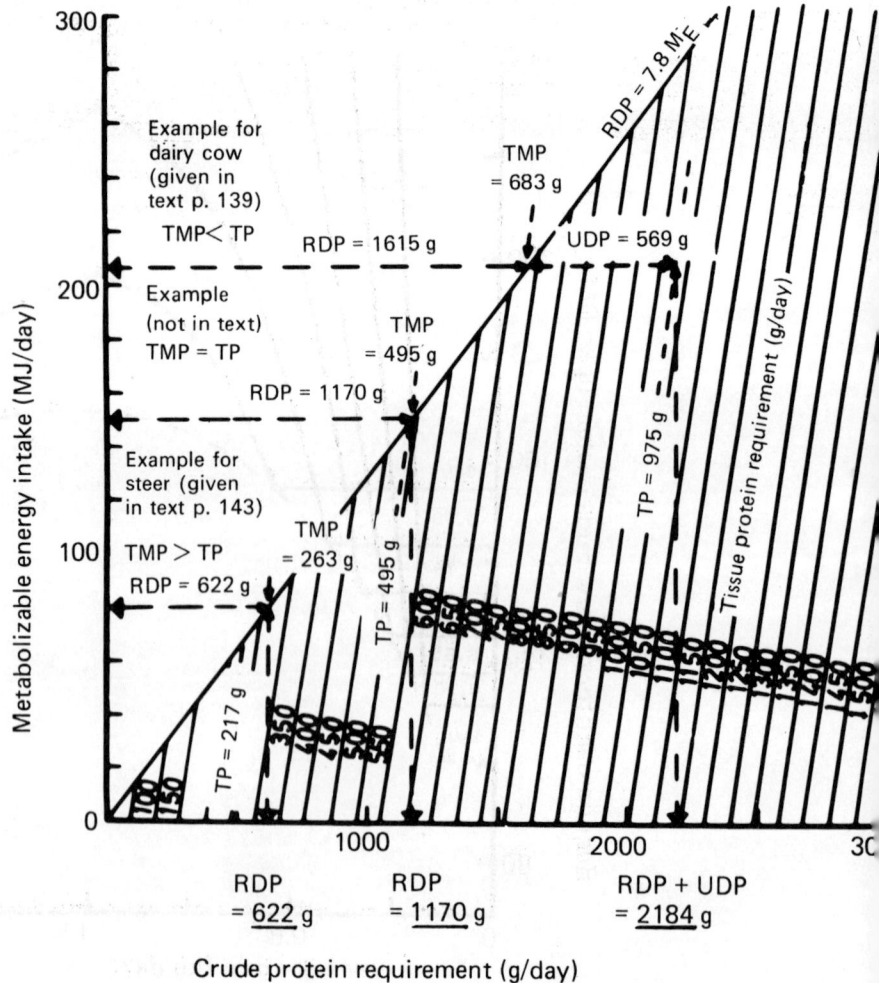

From the metabolizable energy intake for a particular rate of production, the amount of protein supplied to the tissue from microbial protein (TMP) may be read from the sloping tissue protein requirement (TP) lines at the point of intersection of metabolizable energy intake with the RDP regression line. If this value is

(a) greater than the tissue protein requirement (TP), RDP is the protein requirement;

(b) less than the tissue protein requirement, the amount of undegraded protein (UDP) is read off the graph from the difference between (1) the point of intersection of metabolizable energy and total tissue protein (TP) required and (2) the point of intersection of metabolizable energy with the RDP regression line.

12. The protein requirements of the preruminant animal

For estimating protein requirements of the preruminant animal, the factorial system, used in the first edition, does not have the same disadvantages that it has for the ruminant animal.

Since the microbial N component of the metabolic faecal N is much smaller in the preruminant than in the ruminant animal, values for metabolic faecal N are similar to that of the nonruminant. Estimates have been as low as 1.65g N/kg dry matter intake for lambs (Walker & Cook, 1967) and 1.9 g N/kg dry matter for calves (Roy et al 1964 1970b).

The protein of fresh or 'mildly' heat-treated milk has a true digestibility for both species of 1.0 (Roy et al 1964, Black et al 1973, Black & Griffiths 1975), although with processed milks the true digestibility, especially at a young age, may decline as the severity of the heat treatment used is increased (Shillam & Roy 1963). 'Severe' heat treatment in processing of milk causes denaturation of the whey proteins and a reduction in the concentration of ionizable calcium. This results in poor coagulation in the abomasum, increased escape of undigested protein into the duodenum (Tagari & Roy 1969), reduced acid and protease production in the abomasum (Williams, V. J. et al 1976) and reduced protease production by the pancreas (Ternouth et al 1974). The milder the heat treatment in processing, the more satisfactory is the diet for the newborn calf. A value of less than 170 mg non-casein N/g total N in a milk powder is likely to increase the risk of digestive disorders (Roy 1969). This value is approximately equivalent to 6 mg whey protein N/g powder (Roy 1975); whey protein N values greater than 6 are designated 'low heat powders' under the ADMI grading system (American Dry Milk Institute Inc. 1971).

The protein requirements of the preruminant animal, given in the first edition as available protein, $((R_N + U_{N(E)} + F_{N(M)}) \times 6.25)/bv$ were based on the formula:

$$N \text{ requirement (g)} = (R_N + U_{N(E)} + F_{N(M)})/(td_N \times bv),$$

where $R_N = N$ retention (g/day), $U_{N(E)} =$ endogenous urinary N (g/day), $F_{N(M)} =$ metabolic faecal N (g/day), $td_N =$ true digestibility of N (as a coefficient) and $bv =$ biological value (as a coefficient).

It is recognised that for the preruminant, the method used in the first edition, which involved the use of true digestibility of N and making an allowance for metabolic faecal N, has greater theoretical validity than the use of apparent digestibility of N and the ignoring of metabolic faecal N. However, for consistency with the approach taken for the ruminant animal and because apparent digestibility of different protein sources may be measured directly, it has been decided to adopt the latter procedure. There is little difference between the two methods in the estimate of N requirements; the proposed system results in total N requirements being about 0.5% less for a calf of 100 kg live weight and about 6% less for a calf of 200 kg live weight at maximal intakes.

With protein limiting, it has been shown that the apparent digestibility of milk protein is 0.92 for both calves (Roy et al 1970b) and lambs (based on Black et al

1973). Similarly it may be calculated, from the results given in these two sources, that the efficiency of utilization of apparently digested N, when protein is limiting, for N retention and for meeting the obligatory loss in endogenous urinary N is 0.80. But ingredients not of milk origin are now quite frequently included in milk substitute diets. Digestibility of these may be impaired owing to poor coagulation in the abomasum and reduced enzyme activity throughout the tract (Ternouth et al 1975, Williams, V. J. et al 1976, Roy et al 1977), or to other factors in the protein source leading to changes in digestive function (Smith & Sissons 1975, Sissons & Smith 1976).

The formula for calculating protein requirements of the preruminant is thus:

$$\text{Protein requirement (g/day)} = 6.25 \ (R_N + \text{Wool}_N$$
$$\text{(or dermal N loss)} + U_{N(E)})/(d_N \times k_{N(U)})$$

where $R_N = $ N retention in liveweight gain (fleece-free for lambs) (g/day), $U_{N(E)} = $ endogenous urinary N (g/day) (see this Chapter, p. 131), $d_N = $ apparent digestibility of N (0.92 for milk protein, lower value for non-milk protein sources) and, $k_{N(U)} = $ efficiency of utilization of absorbed N (0.80 for milk protein, possibly lower value for some non-milk protein sources).

In comparison with the first edition, in which the requirement of an unspecified preruminant calf of 55 kg liveweight, gaining 0.77 kg/day and having a dry matter intake of 1.0 kg/day, was 175 g available protein/day ($\equiv$ 204 g crude protein/day), the system now proposed gives a value of 237 g crude protein/day, for a bull calf of a large breed.

Similarly, Roy et al (1970b) found that 192 g crude protein/day was limiting to growth in bull calves, mostly Ayrshire, of 53 kg liveweight gaining 0.66 kg/day, which agrees well with the value of 196 g crude protein/day for bull calves of a medium-sized breed in the proposed system.

In preruminant lambs, Black & Griffiths (1975) found a relationship of N balance to N intake, when N intake was limiting, as follows:

$$\text{N balance} = 0.723 \text{ N intake} - 0.148 \ W^{0.75}.$$

A comparison of the N intakes corresponding to the N retentions used in the proposed system with those calculated by the above equation is shown below:

N intakes (g/day)

	Black & Griffiths (1975) Liveweight gain (kg/day)		Proposed system Liveweight gain (kg/day)	
Liveweight (kg)	0.1	0.4	0.1	0.4
5	4.9	16.3	5.3	16.3
10	5.1	16.1	5.3	16.0
20	5.5	15.8	5.4	15.5
30	5.8	15.7	5.6	15.2

Table 4.8   The protein requirements of the preruminant calf (bull calf of a large breed (see Chapter 1) except for 25 kg liveweight, when bull calf of a small breed)

		Proposed system*										ARC (1965)
Live weight (kg)	Maximum dry matter intake (kg/day)†	Apparently digested protein (g/day) Live weight gain (kg/day)					Crude protein (g/day) (for animal receiving milk protein only) Live weight gain (kg/day)					Apparently digested protein (g/day)‡ Liveweight gain (kg/day)
		0	0.25	0.50	1.0	1.5	0	0.25	0.50	1.0	1.5	1.0
25	0.7	13	70	123	220	—	15	76	134	239	—	
50	1.1	28	95	157	269	367	31	103	171	292	399	225
100	1.9	44	107	167	275	368	48	117	182	299	400	235
200	3.2	61	119	174	274	360	66	130	190	298	391	

* For method of calculation, see above.
† Maximum dry matter intake = 60 g/$W^{0.75}$ kg from Table 2.9.
‡ Converted from available protein by addition of (3.9 × I_T) g protein (ARC, 1965).

Table 4.9   The protein requirements of the preruminant lamb (male).

Live weight (kg)	Maximum dry matter intake (kg/day)[†]	Proposed system*										ARC (1965)			
		Apparently digested protein (g/day)					Crude protein (g/day) (for animal receiving milk protein only)					Apparently digested protein[‡] (g/day)			
		Live weight gain (kg/day)					Live weight gain (kg/day)					Live weight gain (kg/day)			
		0	0.1	0.2	0.3	0.4	0	0.1	0.2	0.3	0.4	0.1	0.2	0.3	0.4
5	0.27	9	30	52	73	94	10	33	56	79	102	30	55	75	100
10	0.45	10	30	51	72	92	11	33	56	78	100	30	55	80	105
20	0.76	12	31	51	70	89	13	34	55	76	97	35	60	80	110
30	1.03	13	32	51	69	88	15	35	55	75	95	35	60	85	110

* For method of calculation, see text.
† Maximum dry matter intake = 80 g/$W^{0.75}$ kg from Table 2.9.
‡ Converted from available protein by addition of $(3.9 \times I_T)$ g protein (ARC, 1965).

The suggested apparently digested protein requirements, and the crude protein requirements, for use where the N source is entirely milk protein, are presented in Tables 4.8 and 4.9 for the preruminant calf and lamb respectively.

For their calculation, the protein retention is based on a quadratic approximation to the estimates from the allometric equations given in Tables 1.21 and 1.8, relating protein content to empty-body weight, the estimates for cattle being increased by 10% for bulls and a further 10% for large-sized breeds and reduced by 10% for heifers and a further 10% for small-sized breeds. In conversion from empty-body weight to liveweight, a gut content of 60g/kg for the preruminant animal has been assumed (p. 41).

The equations are as follows:

Calves

Protein in liveweight gain (g/day) =
$\Delta W(170.22 - 0.1731W + 0.000178W^2) \times (1.12 - 0.1258\Delta W)$,
where ΔW is rate of liveweight gain (kg/day).

Lambs

Males and castrates:
Protein in fleece-free liveweight gain (g/day) = $\Delta W (161.0 - 1.22W + 0.0114W^2)$
Females:
Protein in fleece-free liveweight gain (g/day) = $\Delta W (154.3 - 1.94W + 0.0188W^2)$,
where ΔW is rate of liveweight gain (kg/day).

In addition to these protein retentions, dermal N loss as protein in cattle (g/day) = 6.25 $(0.018W^{0.75}$kg) (Table 4.5) and protein retained in wool (g/day) = 3 + (0.1 × protein retained in fleece-free body) (p. 17).

The endogenous urinary N as protein (g/day) = 6.25 $(5.9206 \log_{10} W — 6.76)$ for cattle or 6.25 $(0.02348W + 0.54)$ for sheep (p. 133).

Thus the apparently digested protein requirements are the sum of the protein retention, dermal protein loss or protein retained in wool and the endogenous urinary N as protein divided by 0.80, and the crude protein requirements are the apparently digested protein requirements divided by 0.92.

The apparently digested protein requirements for 1 kg gain per day of the preruminant calf are somewhat greater than those given in the first edition. Separate available protein requirements for the preruminant lamb were not presented in the first edition, but those values converted to digestible crude protein are also shown in Table 4.9 for comparative purposes. The proposed values for the apparently digested crude protein requirements of the preruminant lamb are slightly lower than those given for the unspecified lamb in the earlier edition.

It must be recognised that when milk substitutes are given at a restricted intake, the protein supplied may be in excess of the needs of the animal, whose growth under such a regimen will be limited by the energy supply. This situation occurs because replacement with other ingredients to reduce the protein content may result in amounts of fat and carbohydrate beyond the digestive capacity of the young animal.

13. Summary of Rumen Degradable (RDP) and Undegraded (UDP) protein requirements for various productive processes and classes of ruminant

The protein requirements of the preruminant calf and lamb are given in Tables 4.8 and 4.9, respectively.

For the ruminant animal, the rumen degradable protein requirements have been calculated from the equation: RDP (g/day) $= 7.8\ M_E$. The undegraded protein requirements, in conditions where microbial protein available to the tissues is less than the total tissue protein needs, have been calculated from the equation:

$$UDP\ (g/day) = 1.91 TP - 6.25\ M_E$$

where TP = total tissue protein requirement of the animal (g/day). The tissue protein (TP) requirements, for which representative values for bullocks, lactating cows and lambs are given in Tables 4.10, 4.11 and 4.12 respectively, have been calculated as follows:

Cattle: maintenance and growth

The tissue protein (TP) requirement is the sum of:
1. Endogenous urinary N, as protein (g/day) $= 6.25\ (5.9206\ \log_{10} W - 6.76)$ (p. 133)
2. Dermal N loss, as protein (g/day) $= 6.25\ (0.018 W^{0.75})$ (Table 4.5)
3. Protein in weight gain (g/day) $= \Delta W\ (168.07 - 0.16869 W + 0.0001633\ W^2)$
   $(1.12 - 0.1223\ \Delta W)$ for castrates of medium-sized breeds, where ΔW is rate of live weight gain (kg/day).

The equation is based on a quadratic approximation to the estimates from the allometric equation given in Table 1.21, relating protein content to empty-body weight together with the adjustment for empty-body gain from Table 1.22; the estimates being increased by 10% for bulls and a further 10% for large-sized breeds or reduced by 10% for heifers and a further 10% for small-sized breeds. Conversion to a live weight basis was made by the equation: $LW = 1.09\ (EBW + 14)$ given on p. 42, which assumes a diet such that at 75 kg empty-body weight the content of the gut would be 300 g/kg empty-body weight.

Table 4.10   Tissue protein requirements (g/day) for bullocks of a breed of medium body size gaining weight at different rates. The values apply equally to heifers of large breeds.

Live weight (kg)	Liveweight gain (kg/day)					
	0	0.25	0.50	0.75	1.0	1.25
200	49	87	123	158	189	219
300	57	93	127	159	189	217
400	64	98	131	162	190	217
500	69	103	135	165	194	220
600	74	108	140	171	199	226

Cattle: lactation.

For no change in body weight, the tissue protein (TP) requirement is the sum of:
1. Endogenous urinary N, as protein, given above;
2. Dermal N loss, as protein, given above;
3. The protein secreted in milk (g/day)=milk yield (kg/day)×protein concentration in milk (g/kg),

      where protein concentration in milk for a
      standard cow=6.25×4.8 g/kg†
      Friesian cow=6.25×4.8 g/kg
      Ayrshire cow=6.25×5.0 g/kg
      Jersey cow=6.25×5.7 g/kg.

Tissue protein is assumed to be reduced by 56 g/day for a weight loss of 0.5 kg/day or increased by 75 g/day for a weight gain of 0.5 kg/day

Table 4.11   Tissue protein requirements (g/day) for cows weighing 500 kg producing milk with fat 40 g/kg.

Change in live weight	Milk yield (kg/day)					
	5	10	15	20	30	40
None	219	369	519	669	971	1270
Loss of 0.5 kg/day	163	313	463	613	913	1214
Gain of 0.5 kg/day	294	444	594	744	1044	1345

Cattle: additional requirement for pregnancy

For a 40 kg calf, the daily tissue protein retention (g/day) is

$$TP_{(t)} \times 0.03437 \exp^{-0.00262t}$$

where $\log_{10} (TP_{(t)}) = 3.707 - 5.698 \exp^{-0.00262t}$ (Table 1.20), and t=no. of days from conception.

For calves of other birth weights, the tissue protein retention is in proportion.

Sheep: maintenance and growth

The tissue protein (TP) requirement is the sum of:
1. Endogenous urinary N, as protein (g/day)=6.25 (0.02348W+0.54) (p. 133)
2. Protein retained in wool (g/day)=3+(0.1× protein retained in fleece-free body) (p. 17)
3. Protein retained in fleece-free live weight gain (g/day)
    Males and castrates ΔW (160.4−1.22W+0.0105W²)
    Females   ΔW (156.1−1.94W+0.0173W²)
    where ΔW is rate of liveweight gain (kg/day).

The equations are based on quadratic approximations to the estimates given in Table 1.8, converted to a liveweight basis by the equation: LW=1.09 (EBW+2.9) given on p. 42, which assumes a diet such that the gut contents would be 300 g/kg empty-body weight for a lamb of 15 kg empty-body weight.

---

†to simplify the equations the factor 6.25 has been used throughout, rather than the accepted factor for milk protein, of 6.38

Table 4.12   Tissue protein requirements (g/day) for lambs gaining weight at different rates.

Live weight (kg)	Sex	Liveweight gain (kg/day)				
		0	0.05	0.10	0.20	0.30
20	Male	9	17	25	40	56
	Castrate	9	17	25	40	56
	Female	9	16	23	37	50
30	Male	11	18	25	40	55
	Castrate	11	18	25	40	55
	Female	11	17	23	36	48
40	Male	12	19	26	40	55
	Castrate	12	19	26	40	55
	Female	12	18	24	36	47

Ewes: pregnancy

The tissue protein (TP) requirement is the sum of:
1. Endogenous urinary N as protein, given above;
2. Protein retained in wool (g/day) = 6.25×0.85 (Table 1.30);
3. The protein retained in the lamb (g/day)

For a 4 kg lamb, the tissue protein retention (g/day) is

$$TP_{(t)} \times 0.06744 \exp^{-0.00601t}$$

where $\log_{10} (TP_{(t)}) = 4.928 - 4.873 \exp^{-0.00601t}$ (Table 1.6), and t = no. of days from conception.

For lambs of other birth weights, the tissue protein retention is in proportion.

Ewes: lactation

For no change in body weight, the tissue protein (TP) requirement is the sum of:
1. Endogenous urinary N, as protein, given above;
2. Protein retained in wool (g/day) = 6.25×0.85 (Table 1.30)
3. The protein secreted in milk (g/day) = milk yield (kg/day) × protein concentration in milk (g/kg),

where protein concentration = 6.25×7.66 g/kg (p. 47)

Protein retention is reduced by 9.8 g/day for a weight loss of 0.1 kg/day. The separate tabulation for month of lactation arises from differences in estimated energy concentration in milk from month to month.

The tables of protein requirements, with the corresponding tables of energy requirements in parentheses, are as follows:

Cattle, maintenance and growth
  (a) Bulls of breeds of large mature size   Table 4.13 (Table 3.19)
  (b) Bulls of breeds of medium mature size   Table 4.14 (Table 3.20)
  (c) Bulls of breeds of small mature size   Table 4.15 (Table 3.21)
  (d) Bullocks of breeds of large mature size   Table 4.16 (Table 3.22)
  (e) Bullocks of breeds of medium mature size and
     heifers of breeds of large mature size   Table 4.17 (Table 3.23)
  (f) Bullocks of breeds of small mature size and heifers
     of breeds of medium mature size   Table 4.18 (Table 3.24)
  (g) Heifers of breeds of small mature size   Table 4.19 (Table 3.25)
Cattle, lactation.
  Standard cow   Table 4.20 (Table 3.27)
  British Friesian   Table 4.21
  Ayrshire   Table 4.22 } (Table 3.28)
  Jersey   Table 4.23
Cattle, additional requirement for pregnancy   Table 4.24 (Table 3.30)
Sheep, maintenance and growth   Table 4.25 (Table 3.31)
Sheep, pregnancy   Table 4.26 (Table 3.34)
Sheep, lactation   Table 4.27 (Table 3.36)

Table 4.13 Rumen-degradable (RDP) and undegraded (UDP) protein requirements (g/day) of cattle for maintenance and growth: (a) Bulls of breeds of large mature size. (For energy requirements, see Table 3.19).

M_E/G_E (q)	Live weight (kg)	Form of protein	Weight gain (kg/day)						
			0	0.25	0.50	0.75	1.0	1.25	1.50
0.4	100	RDP	160	185	215	250			
		UDP		15	80	135			
	200	RDP	255	290	330	385			
		UDP				35			
	300	RDP	335	380	435	500			
	400	RDP	410	465	525	605			
	500	RDP	480	540	615	705			
	600	RDP	540	610	695	795			
0.5	100	RDP	150	170	195	225	260		
		UDP		25	95	155	205		
	200	RDP	240	270	305	345	395		
		UDP			20	65	100		
	300	RDP	320	355	400	450	515		
	400	RDP	390	435	485	550	620		
	500	RDP	455	505	565	640	725		
	600	RDP	515	575	640	720	815		
0.6	100	RDP	145	160	185	205	235	270	310
		UDP		35	105	170	230	275	310
	200	RDP	230	255	285	320	360	405	460
		UDP			35	85	130	160	180
	300	RDP	305	335	375	420	465	525	595
		UDP				10	35	55	60
	400	RDP	370	410	455	505	565	635	720
	500	RDP	430	475	530	590	660	740	835
	600	RDP	490	540	600	665	745	835	945
0.7	100	RDP	135	155	170	195	215	245	275
		UDP		40	115	185	245	295	340
	200	RDP	220	245	270	300	330	370	415
		UDP			50	105	150	190	215
	300	RDP	290	320	350	390	430	480	535
		UDP				30	65	90	105
	400	RDP	355	390	430	475	525	580	650
		UDP						5	10
	500	RDP	410	450	500	550	610	675	755
	600	RDP	465	515	565	625	690	765	850

Table 4.14 Rumen-degradable (RDP) and undegraded (UDP) protein requirements (g/day) of cattle for maintenance and growth: (b) Bulls of breeds of medium mature size. (For energy requirements, see Table 3.20).

M_E/G_E (q)	Live weight (kg)	Form of protein	Weight gain (kg/day)						
			0	0.25	0.50	0.75	1.0	1.25	1.50
0.4	100	RDP	160	190	230	280			
		UDP		5	55	90			
	200	RDP	255	300	350	420			
	300	RDP	335	390	460	540			
0.4	400	RDP	410	475	555	655			
	500	RDP	480	555	645	765			
	600	RDP	540	625	730	865			
0.5	100	RDP	150	175	205	245	295		
		UDP		15	70	120	155		
	200	RDP	240	280	320	375	435		
		UDP				25	40		
	300	RDP	320	365	420	485	565		
	400	RDP	390	445	510	590	685		
	500	RDP	455	520	595	685	795		
	600	RDP	515	585	670	775	895		
0.6	100	RDP	145	165	190	225	260	305	360
		UDP		20	85	135	180	210	230
	200	RDP	230	260	300	340	390	450	525
		UDP			10	50	75	90	90
	300	RDP	305	345	390	445	510	585	675

(continued)

Table 4.14
(continued)

M_E/G_E (q)	Live weight (kg)	Form of protein	Weight gain (kg/day)						
			0	0.25	0.50	0.75	1.0	1.25	1.50
0.6	400	RDP	370	420	475	540	615	705	815
	500	RDP	430	490	550	625	715	820	945
	600	RDP	490	550	625	710	810	925	1070
0.7	100	RDP	135	155	180	205	235	270	315
		UDP		30	95	150	200	240	265
	200	RDP	220	250	280	315	360	405	465
		UDP			25	70	100	125	140
	300	RDP	290	325	365	415	465	525	600
		UDP					15	25	20
	400	RDP	355	395	445	500	565	640	725
	500	RDP	410	460	520	585	655	740	840
	600	RDP	465	525	585	660	740	835	950

Table 4.15   Rumen-degradable (RDP) and undegraded (UDP) protein requirements (g/day) of cattle for maintenance and growth: (c) Bulls of breeds of small mature size. For energy requirements see Table 3.21).

M_E/G_E (q)	Live weight (kg)	Form of protein	Weight gain (kg/day)						
			0	0.25	0.50	0.75	1.0	1.25	1.50
0.4	100	RDP	160	195	240	310			
		UDP			30	45			
	200	RDP	255	305	370	455			
	300	RDP	335	400	485	590			
	400	RDP	410	490	585	715			
	500	RDP	480	570	680	830			
	600	RDP	540	645	770	935			
0.5	100	RDP	150	180	220	265	325		
		UDP			45	80	95		
	200	RDP	240	285	335	400	480		
	300	RDP	320	375	440	520	620		
	400	RDP	390	455	535	630	750		
	500	RDP	455	530	620	730	870		
	600	RDP	515	600	700	825	985		
0.6	100	RDP	145	170	200	240	285	340	420
		UDP		10	60	100	130	145	140
	200	RDP	230	270	310	365	425	500	600
		UDP				10	20	15	
	300	RDP	305	350	405	470	550	645	765
	400	RDP	370	430	495	575	665	780	925
	500	RDP	430	500	575	665	775	905	1070
	600	RDP	490	565	650	755	875	1020	1210
0.7	100	RDP	135	160	190	220	255	300	355
		UDP		20	70	115	155	180	195
	200	RDP	220	255	290	335	385	445	515
		UDP			5	35	55	60	55
	300	RDP	290	330	380	435	500	575	665
	400	RDP	355	405	460	530	605	695	805
	500	RDP	410	470	540	615	705	810	935
	600	RDP	465	535	610	695	795	915	1055

Table 4.16   Rumen–degradable (RDP) and undegraded (UDP) protein requirements (g/day) of cattle for maintenance and growth: (d) Bullocks of breeds of large mature size. (For energy requirements see Table 3.22).

M_E/G_E (q)	Live weight (kg)	Form of protein	Weight gain (kg/day)						
			0	0.25	0.50	0.75	1.0	1.25	1.50
0.4	100	RDP	140	170	210	265			
		UDP		20	70	105			
	200	RDP	225	265	320	390			
		UDP				5			
	300	RDP	295	350	420	510			
	400	RDP	360	425	505	615			

(continued)

Table 4.16 (continued)

M_E/G_E (q)	Live weight (kg)	Form of protein	Weight gain (kg/day)						
			0	0.25	0.50	0.75	1.0	1.25	1.50
–	500	RDP	420	495	590	715			
	600	RDP	475	560	670	805			
0.5	100	RDP	130	160	190	230	280		
		UDP		30	85	130	165		
	200	RDP	210	250	290	345	415		
		UDP			15	45	60		
	300	RDP	280	325	380	450	535		
	400	RDP	340	395	465	545	645		
	500	RDP	395	460	540	630	750		
	600	RDP	450	525	610	715	845		
0.6	100	RDP	125	150	175	205	245	295	355
		UDP		35	100	150	190	220	235
	200	RDP	200	235	270	315	365	430	510
		UDP			35	70	95	105	100
	300	RDP	265	305	355	410	475	555	655
		UDP					5		
	400	RDP	325	375	430	495	575	670	790
	500	RDP	380	435	500	575	665	780	915
	600	RDP	430	490	565	650	755	880	1035
0.7	100	RDP	120	140	165	190	220	255	300
		UDP		45	105	165	210	250	275
	200	RDP	190	220	255	290	330	385	445
		UDP			50	90	120	145	155
	300	RDP	255	290	330	380	430	495	570
		UDP				20	40	50	40
	400	RDP	310	350	400	460	525	600	690
	500	RDP	360	410	470	535	610	695	800
	600	RDP	410	465	530	605	690	785	905

Table 4.17   Rumen-degradable (RDP) and undegraded (UDP) protein requirements (g/day) of cattle for maintenance and growth: (e) Bullocks of breeds of medium mature size and heifers of breeds of large mature size. (For energy requirements, see Table 3.23).

M_E/G_E (q)	Live weight (kg)	Form of protein	Weight gain (kg/day)						
			0	0.25	0.50	0.75	1.0	1.25	1.50
0.4	100	RDP	140	175	225	295			
		UDP		5	40	55			
	200	RDP	225	275	340	435			
	300	RDP	295	360	445	560			
	400	RDP	360	440	540	675			
	500	RDP	420	510	625	785			
	600	RDP	475	575	710	885			
0.5	100	RDP	130	160	200	250	320		
		UDP		15	60	90	105		
	200	RDP	210	255	310	375	460		
	300	RDP	280	335	400	485	595		
	400	RDP	340	405	485	590	720		
	500	RDP	395	475	565	685	835		
	600	RDP	450	535	640	770	940		
0.6	100	RDP	125	150	185	225	270	335	425
		UDP		25	75	115	140	150	140
	200	RDP	200	240	285	335	400	485	595
		UDP			10	30	40	30	
	300	RDP	265	315	370	435	520	620	755
	400	RDP	325	380	450	530	630	750	910
	500	RDP	380	445	525	615	730	870	1055
	600	RDP	430	505	590	695	825	980	1190
0.7	100	RDP	120	145	170	205	240	290	345
		UDP		30	85	130	165	190	200
	200	RDP	190	225	265	310	360	425	500
		UDP			25	55	75	80	70
	300	RDP	255	295	345	400	470	545	640
	400	RDP	310	360	420	485	565	660	775
	500	RDP	360	420	490	565	660	765	900
	600	RDP	410	475	555	640	745	865	1015

Table 4.18   Rumen-degradable (RDP) and undegraded (UDP) protein requirements (g/day) of cattle for maintenance and growth: (f) Bullocks of breeds of small mature size and heifers of breeds of medium mature size. (For energy requirements, see Table 3.24).

M_E/G_E (q)	Live weight (kg)	Form of protein	Weight gain (kg/day)						
			0	0.25	0.50	0.75	1.0	1.25	1.50
0.4	100	RDP	140	180	240	335			
		UDP				15			
	200	RDP	225	285	365	485			
	300	RDP	295	370	470	620			
	400	RDP	360	450	570	750			
	500	RDP	420	525	665	865			
	600	RDP	475	595	750	980			
0.5	100	RDP	130	165	215	275	365		
		UDP		5	35	50	35		
	200	RDP	210	260	325	405	520		
	300	RDP	280	345	425	525	665		
	400	RDP	340	415	515	635	800		
	500	RDP	395	485	595	735	930		
	600	RDP	450	550	675	835	1050		
0.6	100	RDP	125	155	195	240	300	385	510
		UDP		15	50	80	90	80	30
	200	RDP	200	245	295	360	440	545	690
	300	RDP	265	320	385	470	565	695	875
	400	RDP	325	390	470	565	685	840	1055
	500	RDP	380	455	545	660	795	975	1220
	600	RDP	430	515	620	745	900	1100	1370
0.7	100	RDP	120	145	180	215	265	320	395
		UDP		20	65	95	120	130	120
	200	RDP	190	230	275	330	390	465	560
		UDP					15	20	10
	300	RDP	255	305	360	425	505	600	720
	400	RDP	310	370	435	515	610	725	870
	500	RDP	360	430	510	600	710	840	1005
	600	RDP	410	485	575	680	800	950	1135

Table 4.19   Rumen-degradable (RDP) and undegraded (UDP) protein requirements (g/day) of cattle for maintenance and growth: (g) Heifers of breeds of small mature size. (For energy requirements, see Table 3.25).

M_E/G_E (q)	Live weight (kg)	Form of protein	Weight gain (kg/day)						
			0	0.25	0.50	0.75	1.0	1.25	1.50
0.4	100	RDP	140	190	260	390			
	200	RDP	225	290	385	540			
	300	RDP	295	380	500	690			
	400	RDP	360	465	610	830			
	500	RDP	420	540	705	960			
	600	RDP	475	610	800	1085			
0.5	100	RDP	130	175	225	300	425		
		UDP			10	5			
	200	RDP	210	270	340	440	585		
	300	RDP	280	350	445	570	745		
	400	RDP	340	430	540	685	895		
	500	RDP	395	500	625	795	1040		
	600	RDP	450	565	710	900	1170		
0.6	100	RDP	125	160	205	260	335	440	630
		UDP			30	40	35		
	200	RDP	200	250	310	385	480	615	815
	300	RDP	265	330	405	500	620	785	1025
	400	RDP	325	400	490	605	750	940	1225
	500	RDP	380	465	570	700	865	1090	1415
	600	RDP	430	530	645	795	980	1230	1590
0.7	100	RDP	120	150	190	230	285	355	450
		UDP		10	40	60	70	65	35
	200	RDP	190	235	285	350	420	515	630
	300	RDP	255	310	375	450	545	660	805
	400	RDP	310	375	455	550	660	795	970
	500	RDP	360	440	530	635	765	925	1125
	600	RDP	410	495	600	720	865	1040	1265

Table 4.20   Rumen–degradable (RDP) and undegraded (UDP) protein requirements (g/day) of standard cow weighing 500 kg, producing milk with fat 40 g/kg. No allowance is made for change in body weight. (For energy requirements see Table 3.27).

M_E/G_E (q)	Form of protein	Milk yield (kg/day)						
		0	5	10	15	20	30	40
0.5	RDP	395	610	825	1040	1265		
	UDP			45	155	265		
0.6	RDP	380	575	780	985	1195	1625	
	UDP			80	200	320	550	
0.7	RDP	360	550	740	935	1135	1545	1965
	UDP			110	240	370	615	850

Table 4.21   Rumen-degradable (RDP) and undegraded (UDP) protein requirements (g/day) of British Friesian cows when losing or gaining 0.5 kg/day or static in body weight. (For energy requirements, see Table 3.28).

Breed	Change in weight	M_E/G_E (q)	Form of protein	Milk yield (kg/day)					
				5	10	15	20	30	40
British Friesian weighing 600 kg producing milk containing 36.8g fat/kg	None	0.5	RDP	650	855	1065	1275		
			UDP		25	145	265		
		0.6	RDP	620	815	1010	1210	1615	
			UDP		65	190	320	565	
		0.7	RDP	590	770	960	1145	1535	1930
			UDP		95	235	370	630	885
	Loss of 0.5 kg/day	0.5	RDP	505	710	915	1125		
			UDP		40	160	280		
		0.6	RDP	480	670	865	1065	1465	1880
			UDP		70	200	330	575	820
		0.7	RDP	455	640	825	1010	1390	1785
			UDP		95	235	370	640	895
	Gain of 0.5 kg/day	0.5	RDP	840	1045	1260	1470		
			UDP		20	135	250		
		0.6	RDP	795	990	1190	1395	1805	
			UDP		65	190	315	555	
		0.7	RDP	755	940	1130	1320	1715	
			UDP		105	240	370	630	

Table 4.22   Rumen-degradable (RDP) and undegraded (UDP) protein requirements (g/day) of Ayrshire cows when losing or gaining 0.5 kg/day or static in body weight. (For energy requirements, see Table 3.28).

Breed	Change in weight	M_E/G_E (q)	Form of protein	Milk yield (kg/day)					
				5	10	15	20	30	40
Ayrshire weighing 500 kg producing milk containing 38.6 g fat/kg	None	0.5	RDP	605	815	1030	1250		
			UDP		75	205	325		
		0.6	RDP	575	770	975	1180	1605	
			UDP		110	245	380	640	
		0.7	RDP	545	735	925	1120	1520	1935
			UDP		140	285	430	705	970
	Loss of 0.5 kg/day	0.5	RDP	455	665	875	1090		
			UDP		90	220	345		
		0.6	RDP	435	630	830	1035	1450	1885
			UDP		115	255	390	650	905
		0.7	RDP	415	600	790	980	1375	1785
			UDP		140	290	430	710	980

(continued)

Table 4.22 (continued)

Breed	Change in weight	M_E/G_E (q)	Form of protein	Milk yield (kg/day)					
				5	10	15	20	30	40
		0.5	RDP	790	1005	1225	1445		
			UDP		65	190	310		
	Gain of 0.5 kg/day	0.6	RDP	750	950	1160	1365	1795	
			UDP		110	245	375	630	
		0.7	RDP	715	905	1100	1295	1705	
			UDP	5	150	290	430	700	

Table 4.23   Rumen-degradable (RDP) and undegraded (UDP) protein requirements (g/day) of Jersey cows when losing or gaining 0.5 kg/day or static in body weight. (For energy requirements, see Table 3.28).

Breed	Change in weight	M_E/G_E (q)	Form of protein	Milk yield (kg/day)				
				5	10	15	20	30
Jersey weighing 400 kg producing milk containing 49.0 g fat/kg	None	0.5	RDP	575	820	1065	1320	
			UDP		145	290	425	
		0.6	RDP	545	775	1010	1245	
			UDP	25	180	335	485	
		0.7	RDP	520	735	955	1185	1650
			UDP	45	215	375	535	840
	Loss of 0.5 kg/day	0.5	RDP	430	665	910	1160	
			UDP	10	160	305	445	
		0.6	RDP	405	630	860	1100	1585
			UDP	30	190	345	495	785
		0.7	RDP	385	600	820	1040	1500
			UDP	45	215	380	540	850
	Gain of 0.5 kg/day	0.5	RDP	765	1010	1265		
			UDP		135	275		
		0.6	RDP	725	955	1195	1440	
			UDP	25	180	330	475	
		0.7	RDP	690	910	1135	1365	1840
			UDP	55	220	380	535	835

Table 4.24   Additional rumen-degradable protein (RDP) (g/day) required by cattle during pregnancy. Metabolizability of gross energy (q) is assumed to be 0.5. (For energy requirements, see Table 3.30).

Birth weight of calf (kg) and reference animal	Form of protein	Weeks before parturition			
		12	8	4	Term
40 kg (standard)	RDP	65	110	190	335
42 kg (Friesian)	RDP	65	115	200	350
32 kg (Ayrshire)	RDP	50	90	155	270
24 kg (Jersey)	RDP	40	65	115	200

Table 4.25   Rumen-degradable (RDP) and undegraded (UDP) protein requirements (g/day) of lambs for maintenance and growth, when kept out-of-doors, and given solid food. (For energy requirements, see Table 3.31).

M_E/G_E (q)	Sex	Live weight (kg)	Form of protein	Weight gain (kg/day)				
				0	0.05	0.10	0.20	0.30
0.5	Male	20	RDP	30	40	50		
			UDP			10		
		30	RDP	45	55	65	100	
		40	RDP	55	70	85	125	
	Castrate	20	RDP	30	35	50		
			UDP		5	10		
		30	RDP	40	50	65		
		40	RDP	50	60	80	125	
	Female	20	RDP	30	35	50		
			UDP			5		
		30	RDP	40	50	65		
		40	RDP	50	65	85	150	
0.6	Male	20	RDP	30	35	45	60	
			UDP		5	10	25	
		30	RDP	40	50	60	85	115
			UDP				10	10
		40	RDP	50	65	75	105	145
	Castrate	20	RDP	25	35	45	65	
			UDP		5	15	25	
		30	RDP	35	45	60	85	
			UDP				10	
		40	RDP	45	60	70	105	
	Female	20	RDP	25	35	45	65	
			UDP		5	10	15	
		30	RDP	35	50	60	95	
		40	RDP	45	60	75	120	
0.7	Male	20	RDP	30	35	40	55	
			UDP		5	15	30	
		30	RDP	40	50	55	75	100
			UDP			5	15	25
		40	RDP	50	60	70	95	125
			UDP					5
	Castrate	20	RDP	25	30	40	55	
			UDP		5	15	30	
		30	RDP	35	45	55	75	105
			UDP			5	15	20
		40	RDP	45	55	65	95	125
			UDP				5	5
	Female	20	RDP	25	35	40	60	
			UDP		5	10	25	
		30	RDP	35	45	55	80	115
			UDP				5	
		40	RDP	45	55	70	105	145

Table 4.26   Rumen-degradable (RDP) and undegraded (UDP) protein requirements (g/day) of pregnant ewes. (For energy requirements, see Table 3.34).

Ewe and lamb weights	M_E/G_E (q)	Form of protein	Days pregnant: 63 / Weeks before term: 12	91 / 8	105 / 6	119 / 4	133 / 2	147 / Term
40 kg ewe	0.4	RDP	50	55	60	65	75	85
4 kg lamb	0.5	RDP	45	50	55	65	75	85
		UDP						5
	0.6	RDP	45	50	55	60	70	85
		UDP						5
	0.7	RDP	45	50	55	60	70	80
		UDP					5	5

(continued)

Table 4.26
(continued)

| | | | Days pregnant: | 63 | 91 | 105 | 119 | 133 | 147 |
			Weeks before term:	12	8	6	4	2	Term
Ewe and lamb weights	M_E/G_E (q)	Form of protein							
40 kg ewe 6.4 kg twin lambs	0.4	RDP		50	60	70	80	95	110
		UDP						5	10
	0.5	RDP		50	60	65	75	90	110
		UDP						5	10
	0.6	RDP		45	55	65	75	90	105
		UDP					5	5	10
	0.7	RDP		45	55	60	70	85	105
		UDP					5	10	15
75 kg ewe 6 kg lamb	0.4	RDP		80	90	95	105	120	135
	0.5	RDP		75	85	90	100	115	135
	0.6	RDP		75	80	90	100	110	130
	0.7	RDP		70	80	85	95	110	125
75 kg ewe 9.6 kg twin lambs	0.4	RDP		85	95	110	125	145	175
		UDP							5
	0.5	RDP		80	90	105	120	140	170
		UDP							5
	0.6	RDP		75	90	100	115	140	165
		UDP							10
	0.7	RDP		70	85	95	115	135	165
		UDP						5	10

Table 4.27　Rumen-degradable (RDP) and undegraded (UDP) protein requirements (g/day) of lactating ewes kept out-of-doors† (For energy requirements, see Table 3.36).

Ewe liveweight (kg)	Ewe liveweight change (kg/day)	M_E/G_E (q)	Form of protein	Milk yield (kg/day)				
				1.0	1.5	2.0	2.5	3.0
40	None	0.5	RDP	105	140	175		
			UDP	35	55	70		
		0.6	RDP	100	130	165	195	
			UDP	40	60	80	100	
		0.7	RDP	95	125	155	185	
			UDP	40	65	85	110	
	−0.10	0.4	RDP	80	115			
			UDP	35	55			
		0.5	RDP	75	110	140	175	
			UDP	40	60	80	100	
		0.6	RDP	75	105	135	165	
			UDP	40	65	85	105	
		0.7	RDP	70	100	125	155	
			UDP	45	70	90	110	
75	None	0.5	RDP	135	165	200	230	265
			UDP	20	40	60	80	100
		0.6	RDP	125	155	190	200	250
			UDP	25	50	70	90	110
		0.7	RDP	120	150	180	205	235
			UDP	30	55	80	100	120
	−0.10	0.4	RDP	110	145	180		
			UDP	20	40	60		
		0.5	RDP	105	135	170	200	235
			UDP	25	45	70	85	105
		0.6	RDP	100	130	160	190	220
			UDP	30	55	75	95	115
		0.7	RDP	95	120	150	180	210
			UDP	35	60	80	105	125

† mean values for months 1–3 of lactation.

14. Comparison of the protein requirements calculated by the new system with the results of practical trials and with earlier estimates of protein requirements

The response of growing or lactating ruminants to the combined effects of energy and protein intakes can be represented in the form given in Fig. 4.7. The linear slope indicates that response to increasing protein intake when output is being limited by protein rather than by energy intake. At any given energy intake, an increase in protein intake above the minimum requirement will result in a curvilinear response or, according to some workers, in no further response. At very high protein relative to energy intake, there may even be a decline in response.

Fig. 4.7   The response of growing or lactating ruminants to the combined effects of energy and protein intake.

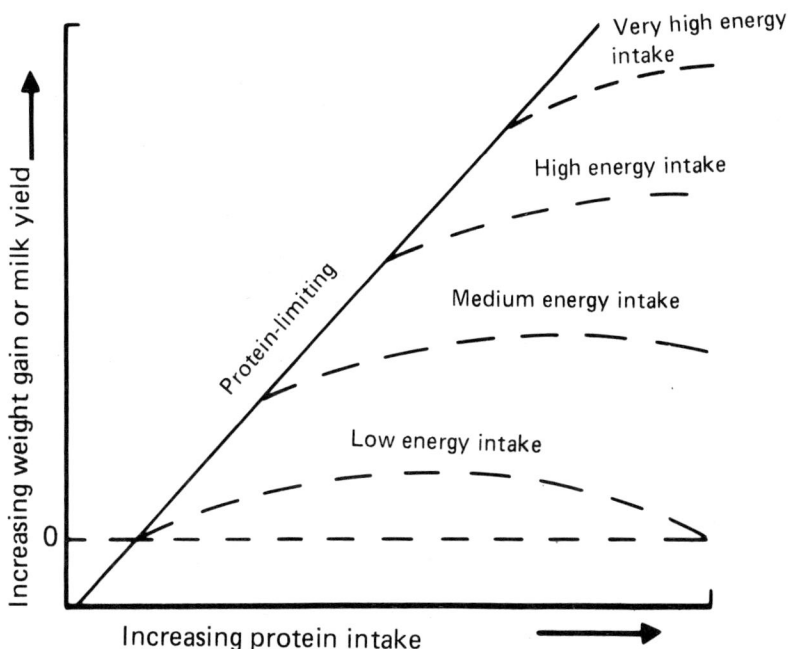

Protein intakes considered to have been limiting growth or lactation have been obtained from a survey of feeding and balance trials, made since 1963 mainly in British conditions, in which energy and protein inputs for growth at particular live weights or for lactation have been measured or estimated. These protein intakes, which are treatment mean values and which for growth rates have been scaled by dividing by live weight, have been plotted separately for cattle of 50–100 kg, 100–200 kg, 200–300 kg, 300–400 kg and over 400 kg live weight gaining weight from 0 to 1.4 kg/day; for milk yield of 4% FCM or SCM, corrected where possible to zero live weight change; for sheep of less than 20 kg, 20–30 kg, 30–40 kg, 40–50 kg and over 50 kg live weight gaining at 0 to 0.5 kg/day, for pregnant sheep and for milk yield of sheep. The published results included in this survey are prefixed in the list of references by the letter (S). However, only the following representative figures are presented here. Fig. 4.8a and b for cattle of 50–100 kg and 200–300 kg live weight, respectively; Fig. 4.9 for milk yield of cattle; Fig. 4.10 for sheep of 20–30 kg live weight; and Fig. 4.11 for milk yield of sheep.

Fig. 4.8   Comparison of proposed protein requirements with results from practical trials in which Protein appeared to be limiting performance. (Each value is a treatment mean.)

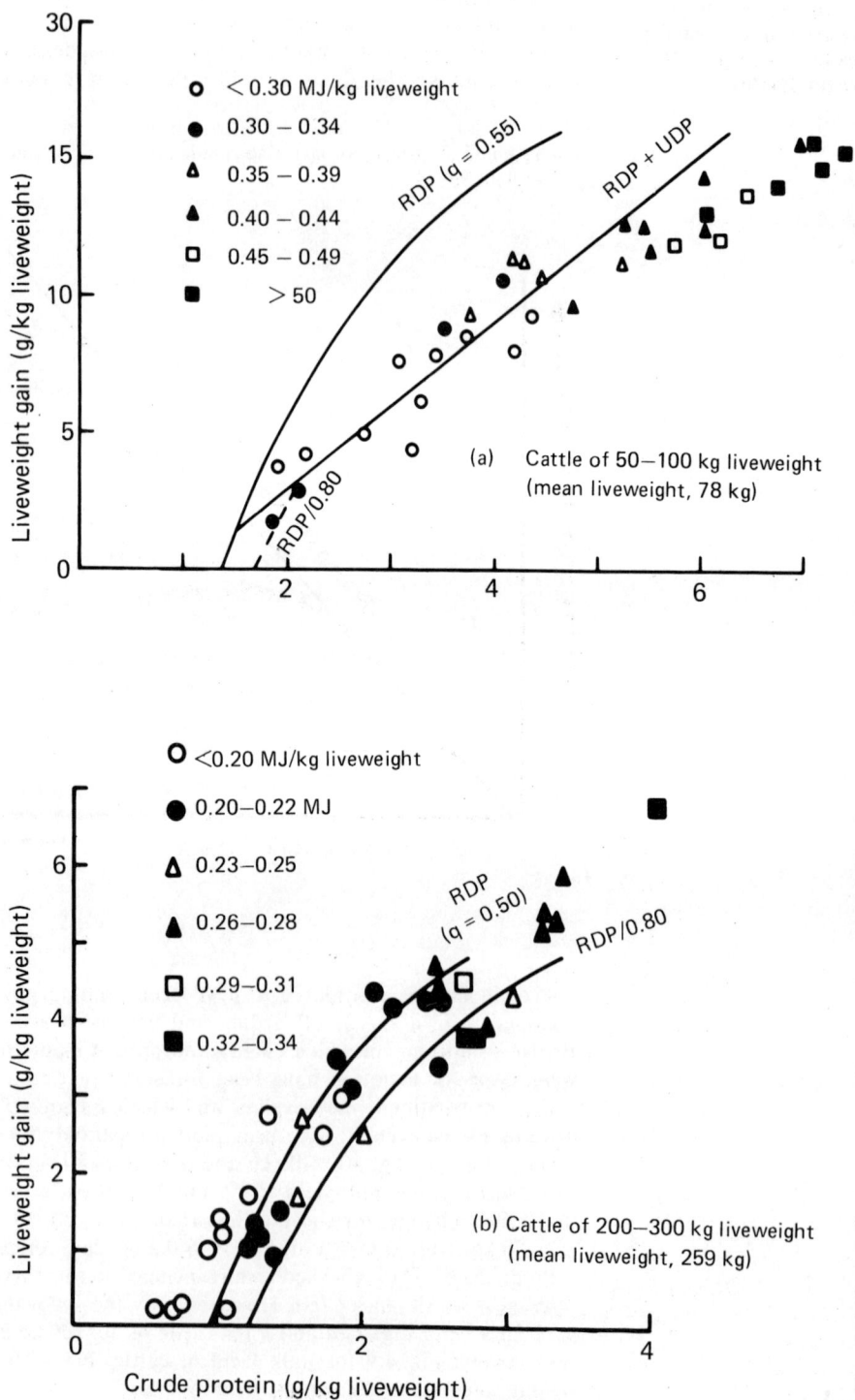

Fig. 4.9 Comparison of proposed protein requirements with results from practical trials in which protein intake appeared to be limiting performance. (Each value is a treatment mean.) Lactating cows (mean liveweight, 544 kg).

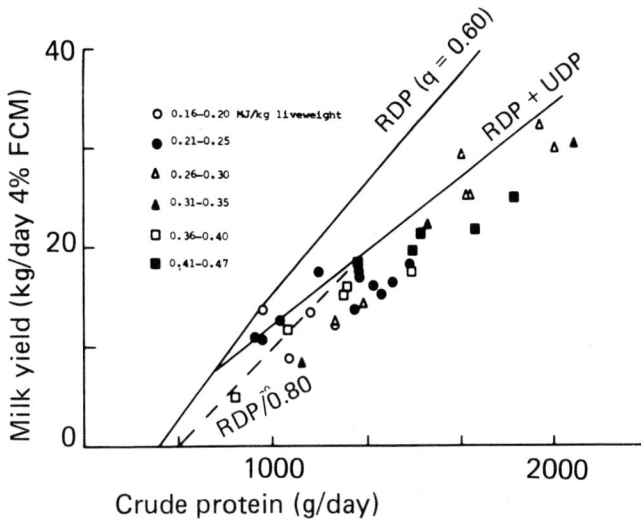

It must be borne in mind that for the milk yield data, in most cases the inputs of energy and protein were based on the yields of milk achieved rather than being independent of them, whereas with the growth data, the energy and protein inputs were independent of the resultant weight gains.

In each of the figures, the rumen-degradable protein (RDP) and, where necessary, the undegraded protein (UDP) requirement are shown at the mean liveweight of the animals and at the mean metabolizability of the gross energy of the diets used in the feeding and balance trials. Where rumen-degradable protein is sufficient by itself to cover the needs of the animal for a particular growth rate, the minimum crude protein required in a diet with a protein degradability of 0.80 is shown also.

From an examination of the figures for cattle, it can be seen that the proposed system gives results which agree with those of practical trials. For high milk yields the proposed system tends to give slightly lower values than would be indicated by the results from the practical trials. However, few results are available for high-yielding cows and it is difficult to assess whether the diets given were really protein-limiting. It must be appreciated that the degradability (dg) of a ration given to a dairy cow, in relation to the ideal dg required for a specified milk yield, is the overriding factor determining whether a response in milk yield will be obtained with a particular protein or non-protein supplement.

Fig. 4.4 (p. 142) shows that at a metabolizability (q) of 0.6, a diet of 1.0 dg, i.e. consisting only of rumen-degradable protein (RDP) and having a crude protein (CP) concentration of 86 g/kg dry matter, would support a milk yield of up to 7.5 kg/day. Undegradable dietary protein (UDP) would be required for milk yields greater than this. The CP intake needed would vary with the relative proportions of RDP and UDP provided and would be at a minimum only when each exactly matched requirements. For diets imbalanced in this respect, responses to either RDP or UDP, depending upon which was in short supply, could therefore sometimes be expected in diets supplying more than the mimimum CP requirements. Thus, although in trials in early lactation by Schwab et al (1971) and Treacher et al (1976) there was no difference in milk yields, averaging about 27 kg FCM/day, between cows fed on diets

Fig. 4.10   Comparison of proposed protein requirements with results from practical
             trials in which protein intake appeared to be limiting performance. (Each
             value is a treatment mean): Sheep of 21–30 kg liveweight (mean
             liveweight, 27.5 kg).

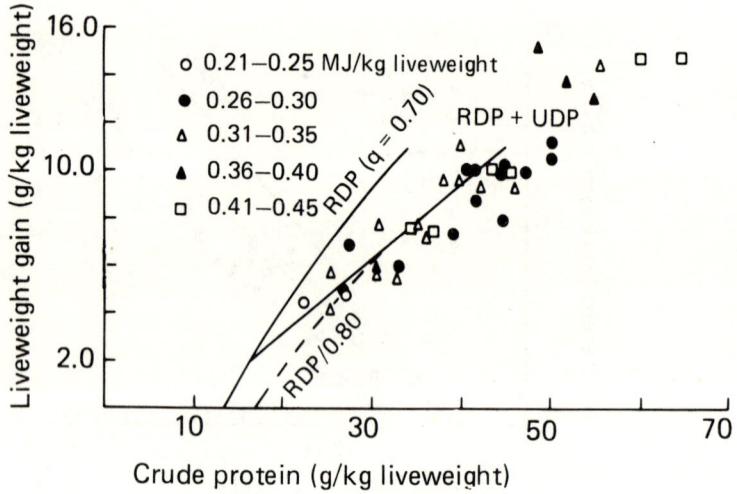

Fig. 4.11   Comparison of proposed protein requirements with results from practical trials
             in which protein intake appeared to be limiting performance. (Each value is a
             treatment mean): Lactating sheep (mean liveweight, 69 kg).

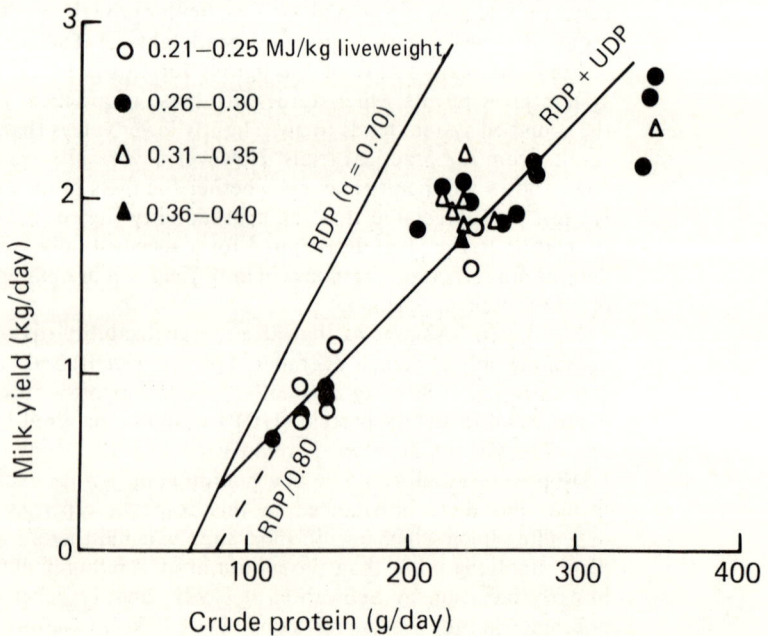

containing 127–128 g CP/kg DM and those given diets supplemented with protein to contain 165–177 g CP/kg DM, other workers have shown responses to protein supplementation of rations containing about 140 g CP/kg DM (Gardner & Park 1973, Gordon 1977). Similarly in mid lactation, Holmes et al (1956) and J. W. Thomas (1971) obtained no response in milk yield on protein supplementation of basal diets containing about 120–140 g CP/kg DM, whereas other workers obtained such responses (Frens & Dijkstra 1959, Rook & Line 1962, Broster et al 1969).

Responses to diets containing more than 125 g CP/kg DM could be accounted for by changes in the dg of the ration, either as a response to additional RDP (Rook & Line 1962, Broster et al 1969a) or as a response to UDP (Frens & Dijkstra 1959, Rook & Line 1962, Gordon 1977). High milk yields were achieved with diets of low CP concentration when the ingredients were maize silage, ground maize and soya, all of which would be expected to have a dg in the range 0.6–0.75 (Thomas, J. W., 1971, Aitchison et al 1976, Polan et al 1976). Responses to higher CP concentrations were obtained when grass silage was the main basal feed (Frens & Dijkstra 1959, Castle & Watson 1976, Gordon 1977), but responses were smaller or absent when artificially dried forage was used (Holmes et al 1956, Rook & Line 1962, Treacher et al 1976). The high yields obtained by Treacher et al (1976) with a diet of low CP concentration were with concentrate ingredients expected to have a high degradability, but which had been processed into compound nuts. It is possible that processing had reduced dg but no direct evidence of this is available.

Responses to urea are to be expected only when RDP is limiting, which is most likely in rations given to the low-yielding cow, but may also exist when the milk yield is above 30 kg/day, if other components have a dg of 0.7 or less. This may occur in typical North American diets based on maize silage, ground maize and soya. Roffler & Satter (1975) concluded that an adequate ruminal ammonia concentration was achieved (indicating adequate RDP in the proposed system) on such rations with an intake of about 130 g CP/kg DM and that urea would be useful only below this level. With diets based on maize silage, Huber (1975) found that supplementation of a basal diet containing 117 g CP/kg DM with either soya or urea to give a final CP concentration of 140 g CP/kg DM gave similar increases in milk yields. This finding would be predicted from the proposed system only if the dietary protein is assumed to have a dg of 0.6. In the experiments of Polan et al (1976), addition of urea to a basal diet with 95.1 g CP/kg DM increased dry matter intake and milk yield, whilst addition of soya gave further increases in intake and yield, which suggests that the basal diet was deficient in both RDP and UDP. Treacher (1977) compared the addition of urea with that of groundnut and white-fish meal to a basal diet containing about 140 g CP/kg DM. Cows given the diet supplemented with urea reached a peak yield of 28 kg at 4–6 weeks of lactation and yield then declined. For the diet supplemented with protein, milk yield was similar at 4 weeks, peaked at 30 kg by week 9 and then started to decline. Milk yields were significantly lower for the cows given the urea-supplemented diet between weeks 10 and 20. These findings would be predicted from the proposed scheme, on the basis of a deficit of UDP in the urea-supplemented diet.

In Table 4.28 the proposed minimum requirements for maintenance and 1 kg/day liveweight gain for cattle of 100, 200, 300 and 400 kg live weight and for milk yield are compared with those given in the earlier edition. For a q value of 0.6, the new values for maintenance are similar to the minimum value of 90 g crude protein/kg dry matter intake (ARC 1965), this value being higher than the factorial values. For 1.0 kg/day weight gain, the new values are lower at 100 kg live weight and higher at 300–400 kg. The new values are also higher for the maintenance of the lactating cow and somewhat lower for maintenance + 30 kg milk.

The National Research Council requirements for dairy cattle (NRC 1971) and

Table 4.28.   Comparisons of new ARC standards for cattle with ARC (1965), NRC (1970 1971) and with results from Wageningen.

Live weight (kg)	ARC (1965) Available protein (g)	ARC (1965) Dry matter intake (kg)	ARC (1965) Digestible crude protein† (g)	ARC (1965) Crude protein‡ (g)	ARC (1980) Dry matter intake (kg)	ARC (1980) Rumen degradable protein (g)	ARC (1980) Undegraded protein (g)	ARC (1980) Crude protein (g) (g/kg dry matter)	NRC (1971) Crude protein (g)	NRC (1970) Crude protein (g)	NRC (1970) Crude protein (g)	Wageningen* Crude protein (g)
	M_E concentration 10.9 MJ/kg				Bullocks (medium) and heifers (large) q = 0.60 ≡ M_E concentration 11 MJ/kg				Growing dairy heifers (0.75 kg/d)	Finishing heifer calves (0.9–1.0 kg/d)	Finishing steer calves (1.0–1.1 kg/d)	
Maintenance												
100	50	2.4	77	110(216)§	1.5	125	0	156¶ } 107				
200	80	3.1	101	144(279)	2.4	201	0	251¶				
300	80	3.8	131	187(342)	3.1	264	0	332¶				
400	100	4.4	159	227(396)	3.8	324	0	405¶				
1 kg/d liveweight gain												
100	260	4.3	318	454	3.2	271	141	412 } 125	370			
200	270	5.5	344	491	4.7	401	40	501¶ } 107	500	610	610	
300	290	6.6	378	540	6.1	519	0	649¶	640	890	870	
400	310	7.8	415	593	7.3	628	0	785¶	800	970	980	
Milk production (600 kg live weight)						Friesian giving 36.8 g fat/kg milk						
Maintenance (M)	135	5.0	202	289(450)	5.6	428	0	535¶ } 107	734			409
M+10 kg milk	605	9.3	730	1043	9.5	813	63	1016¶	1514			987
M+30 kg milk	1555	19.9	1822	2602	18.8	1616	565	2181 116	3074			2143

*Boekholt (1972).
†Available protein + (13.4 × dry matter intake (kg)).
‡Apparent digestibility of crude protein assumed to be 0.70.
§Values in parenthesis are a minimum of 90 g crude protein/kg dry matter intake, as suggested in ARC (1965)
¶Assuming the maximum degradability of a diet is 0.80.

Table 4.29. Comparisons of new ARC standards for sheep with ARC (1965) and NRC (1968).

	ARC (1965)				ARC (1980)					NRC (1968)
Liveweight (kg)	Available protein (g)	Dry matter intake (g)	Digestible crude protein* (g)	Crude protein† (g)	Dry matter intake (g)	Rumen degradable protein (g)	Undegraded protein (g)	Crude protein (g)	Crude protein (g/kg dry matter)	Crude protein (g)
		M_E concentration 10.9 MJ/kg			$q = 0.60 \equiv M_E$ concentration 11 MJ/kg					
					Lambs (male)					
Maintenance										
20	10.3	0.36	16	23(32)§	0.35	31	0	39‡	} 107	
30	10.4	0.47	18	26(42)	0.48	42	0	39‡		
40	12.5	0.57	22	32(51)	0.61	52	0	65‡	107	
0.20 kg/day liveweight gain										Fattening lambs
20	60	1.0	77	111	0.72	62	27	89	124	—
30	60	1.2	80	115	0.99	85	9	106‡	} 107	150
40	80	1.4	83	118	1.3	107	0	134‡	107	163
Milk production										
		70 kg liveweight (2nd month)			75 kg liveweight (2nd month)					First 8–10 weeks of lactation (liveweight loss 36 g/day)
Maintenance (M)	29	0.72	41	59(65)	0.87	75	0	94‡	} 107	
M + 1.0 kg milk	119	1.5¶	144	206	1.5	127	27	159‡		} 209
M + 2.0 kg milk	210	2.0¶	244	349	2.2	188	70	258	117	
M + 3.0 kg milk	300	2.5¶	342	489	2.9	250	112	362	125	

* Available protein + (16.8 × dry matter intake (kg));
† Assuming an apparent digestibility of 0.70.
‡ Assuming the maximum degradability of a diet is 0.80.
§ ( ) Values in parenthesis are a minimum of 90 g crude protein/kg dry matter intake, as suggested in ARC (1965).
¶ Intakes estimated from average value of 2.0 kg dry matter at 2nd month of lactation at a M_E concentration of 10.9 MJ/kg (ARC 1965).

beef cattle (NRC, 1970) are much higher than the proposed requirements for growth of beef cattle or milk yield.

The proposed minimum protein requirements for high levels of milk production agree well with the data from Wageningen (Boekholt 1972) which showed a relationship between the N deposition in milk and tissue and the intake of apparently digested N of

$$Y = 0.74x - 22.26,$$

where Y = N in milk + tissue and x = apparently digested N.

In calculating the requirements for milk yield from this equation, the tissue N requirements used were those proposed in the new ARC system.

From an examination of the figures for sheep, it can be seen that the proposed system, over the range of liveweights presented and for lactation, gives a quite consistent fit to the results for the minimum protein requirements obtained from feeding trials.

In Table 4.29 the proposed minimum requirements for maintenance and 0.2 kg/day liveweight gain for sheep of 20, 30 and 40 kg and for milk yield are compared with those given in the earlier edition. The differences are similar to those found with cattle, namely a slightly higher requirement for maintenance and liveweight gain for the heavier sheep in the proposed system. Similarly the requirement for maintenance of the lactating sheep is higher but the requirement for milk yield is considerably lower.

The National Research Council requirements for growth of sheep (NRC, 1968) are much higher than those in the proposed system, but the NRC values for lactation appear to be somewhat lower.

Appendix 4.I. The Sulphur Requirements of Ruminants

1. The relationship between sulphur and nitrogen requirements

Sulphur, like nitrogen, is an essential element for microbial synthesis because it contributes to microbial S-containing amino acids. Failure to meet the microbial need for S may depress the rate of digestion in the rumen (Bray & Hemsley 1969, Kennedy & Siebert 1973) with consequent depression of feed intake (see review by Moir 1974). The problems involved in calculating the ruminant's requirements for S are therefore similar to those outlined for N. Both organic and inorganic S in the diet can be degraded to sulphide in the rumen and subsequently incorporated into the S-amino acids of microbial protein, although S may enter microbial protein by other routes (Gawthorne & Nader 1976, McMeniman et al 1976). The factorial approach to the calculation of S requirements (e.g. Langlands & Sutherland 1973) has the same shortcomings as it has for calculating N requirements, because it makes no allowance for factors which limit the rate of microbial protein synthesis and, therefore, control the utilization of inorganic S.

An alternative approach, first used by Loosli (1952), is to state S requirements relative to those for N. Authors have expressed this ratio in different ways but it is proposed in this Appendix to express them as S/N(g/g); reported values have been converted appropriately. Loosli (1952) recommends a S/N ratio of 0.067 based on the relatively constant S/N ratio found in animal tissues and products (See Appendix Table 4.9). Moir (1970) has suggested that a ratio of 0.10 is more appropriate for the ruminant because it corresponds to the S/N ration found in microbial protein by Walker & Nader (1968). If, however, ratios of S/N required were generally as high as 0.10, as suggested by Moir (1970), few mixed diets would appear to meet that requirement (cf. Appendix Table 4.9) and it would be expected that S deficiency would be more prevalent than it is in practice. If estimates of S in microbial protein from several authors are used (Hungate 1966, Walker & Nader 1968, Bird 1973), the mean S/N ratio is close to that proposed by Loosli (1952), i.e. 0.067. Although many experiments with sheep (e.g. Bray & Hemsley 1969, Hume & Bird 1970), and some with cattle (Kennedy 1974), have shown that N retention can be improved by increasing the S/N ratio from 0.03–0.05 to 0.10, only Kennedy & Siebert (1973) have tested intermediate S/N ratios; their results did not show a consistent increase in N retention in Merino sheep above a ratio of 0.067. Leibholz (1972b) found that when a basal diet for cattle was supplemented with urea, there was a significant increase in growth rate even though the S/N ratio was only 0.03. Increasing the S/N ratio to 0.07 had no effect on growth rate or N retention. Bird (1974) obtained an apparent response from increasing the S/N ratio from 0.067 to 0.143 but this may have been attributable to the low degradability of S in the basal diet of wheaten straw. Most of the S in a diet is contained in the S-containing amino acids of dietary protein. Since dietary protein is only partly degraded it follows that the S in undegraded protein also escapes degradation and cannot serve as a source of S for rumen microbes.

2. Recycling of sulphur

Some reports suggest that sheep require proportionately more S than cattle because they recycle S less efficiently (Kennedy & Siebert 1973, Kennedy et al 1975). It is more likely that sheep recycle less S than cattle because they have a higher S requirement. The S/N ratio in wool is much higher than that in microbial protein (0.20; Langlands & Sutherland 1973). If Merino sheep and cattle are given diets with the same S and N concentrations, the former are likely to have less S available for recycling because microbial protein is a relatively poor source of S amino acids for wool protein synthesis. If the extent of S recycling in ruminants is variable, largely reflecting the excess of S supply over requirement, it is difficult and probably unnecessary to allow for recycling in the formulation of S requirements.

3. Formulation of sulphur requirement

In view of the interdependence of S and N requirements of the rumen microorganisms and the serious effects on rates of digestion when S is limiting, the use of a degradable S/degradable N ratio to determine S requirement is advocated. A value of 0.07 is proposed which is close to the S/N ratio in microbial, tissue and milk proteins. The S need is, therefore, calculated by simply multiplying the degradable N (RDN) requirement by 0.07. Calculated as a N/S ratio, as often reported in the literature, this would be equivalent to 14.1:1.

It is important to realise that S/N ratio alone is of little value unless it is related to a quantitative need for N. A diet providing an excess of degradable N can have a low S/N ratio and yet meet the requirement, whereas a diet quantitatively deficient in N will fail to meet the S requirements even if it has the recommended S/N ratio.

The practical assessment of the adequacy of a given diet is simplified by the fact that S deficiencies will be found primarily where the diet is also deficient in protein or contains a significant proportion of NPN, because dietary proteins generally contain a ratio of S/N similar to that in microbial protein (cf. Appendix Table 4.9). If a deficiency of degradable N is corrected by using a protein supplement, then S deficiency will generally not occur. If the deficiency is corrected by the use of NPN or the diet is naturally rich in NPN, S should be added. For readily utilized sources of S, such as sodium sulphate or molasses (cf. Bouchard & Conrad 1973a), the efficiency of incorporation of S into microbial protein appears to be similar to that used for the incorporation of degraded N from urea, about 0.80 (Bouchard & Conrad 1973a, Johnson et al 1970). When used in conjunction with urea, these S sources should therefore provide 0.07 g S/g N in urea (e.g., 0.13 g anhydrous sodium sulphate/g urea). For the example of the N requirement of a bullock (given on

p. 138), the need for sodium sulphate would be $0.13 \times 74 = 9.6$ g/day. Elemental S appears to be used less efficiently than sulphate (Albert et al 1956, Johnson et al 1970) and it should be added in approximately twice the quantity of S required from sodium sulphate.

For sheep with high wool production, S-containing amino acids may often be the limiting factor for wool growth because of the high S content of wool. This limitation can generally be alleviated by giving S-amino acids in ways which bypass the rumen or limit the extent of rumen degradation (Marston 1935, Reis 1970, Langlands 1972) but not by adding inorganic S to the diet in excess of microbial need.

4. Comparison of recommendations with results of practical trials

Application of the recommended S/N ratio of 0.07 would give minimum S concentrations in the diet of between 1.1 and 1.6 g/kg dry matter, depending on protein concentration and digestibility. A survey of the literature relating to S-supplementation trials suggests that the critical S concentration in the diets is about 1 g/kg dry matter. In experiments with lactating cows lasting 1–4 months, the addition of S to maize-silage diets, containing 1.0 to 1.3 g S/kg, has generally produced little or no increase in milk yield (Jacobson et al 1967, Bouchard & Conrad 1973a b, Grieve et al 1973). In growing lambs given lucerne hay, containing 1.3–1.5 g S/kg, during a period of 4 months S supplementation did not improve liveweight gain (Rendig & Weir 1957).

Appendix 4.II. Toxicity of Nitrogenous Compounds

Some naturally-occurring or processed materials may contain poisonous nitrogenous compounds; an example is ammoniated molasses. In addition to effects due to specific nitrogenous poisons, which are not further considered here, dietary nitrogen compounds may be more generally toxic if they lead to the absorption of excessive amounts of ammonia or of nitrite from the rumen.

Ammonia toxicity

Excessive amounts of nitrogenous compounds entering the blood are, in the main, converted to urea. This is not toxic and it is possible to have very high levels of urea in the blood, for example in animals consuming pasture which has been heavily fertilized with nitrogen, without ill effects. But if large amounts of ammonia are produced in the rumen and are absorbed rapidly, the capacity of the animal to convert the ammonia to urea may be exceeded; concentrations of ammonia in the blood which are above normal are toxic. In practice, this usually occurs only after a large amount of urea, or a similar NPN source which is rapidly degraded by the bacteria in the rumen, has been consumed. It is generally associated with high pH values in the rumen (above about 7) when appreciable amounts of ammonia are in the undissociated form. Undissociated ammonia readily diffuses across the rumen wall, either into the portal system or into the peritoneal fluid and hence into the systemic blood (Chalmers et al 1971). The latter pathway bypasses the liver and may be of particular importance in causing toxicity. Signs of the disorder, which include heavy, stertorous breathing, incoordination, tetany and death, occur when ammonia N in systemic blood has increased from the normal concentration of less than 1 mg/litre to more than about 7–12 mg/litre (Lewis et al 1957, Helmer & Bartley 1971, Bartley et al 1973, Soar et al 1973). It is not possible to define precisely the NPN intake which will cause these blood concentrations as this depends not only on the properties of the NPN source but also other factors including, in particular, the amount and type of the energy source ingested at the same time and errors in the prescribed feeding pattern. Toxicity is most likely to occur when NPN source is given without other feed or with a poor energy source such as fibre. Numerous experiments, generally made in these conditions, have shown toxic effects in sheep or cattle given single doses of urea ranging from 0.3 to 0.8 g/kg body weight (Helmer & Bartley 1971). If the urea is given at the same time as a feed containing a substantial amount of a readily available carbohydrate, for example a cooked cereal, intakes of urea up to about 0.5 g/kg body weight are unlikely to be harmful. Even larger amounts could theoretically be tolerated and possibly used to advantage with specially formulated feeds. For practical feeding at present, it will become apparent from the calculations of N requirements (p. 140) that an intake of 0.5 g urea/kg body weight would generally be considerably more than an animal could use; this amount should therefore not be exceeded.

Nitrite toxicity

This is generally caused in ruminants by the ingestion of nitrate. Nitrate occurs widely in plants as an intermediate in their nitrogen metabolism but not usually in very great concentrations. Amounts vary, for a number of reasons, and in some herbages, or in dried products made from them, nitrate N may approach or exceed 10 g/kg dry matter. High concentrations in pastures are particularly associated with heavy applications of certain nitrogenous fertilizers (Wilman 1965, Purcell et al 1971, O'Hara & Fraser 1975, Phipps 1975).

Fairly high concentrations of nitrate N may also be found in certain other crops, such as sugar beet, mangolds and kale, particularly if these have been grown in dry hot conditions with heavy N fertilization. Nitrate is usually present in only small amounts in drinking water although the possibility of contamination of ground water should be borne in mind. Although absorbed nitrate is itself relatively harmless it has long been known that ruminants given diets containing high nitrate concentrations may suffer severe disorders. Nitrate is reduced in the rumen to nitrite, hydroxylamine and finally ammonia. If conditions favour the accumulation of nitrite then enough may be absorbed to oxidize appreciable amounts of haemoglobin in the blood to methaemoglobin. This prevents the haemoglobin from being effectively used and ultimately may lead to death of the animal from oxygen lack. It is clear that many factors other than nitrate intake determine the extent of nitrite accumulation in the rumen; it is, for example, greater in the absence of a readily available energy source (Van Leeuwen 1972). It is, therefore, difficult to define a dangerous concentration of nitrate in the diet. Holmes (1968) calculated from the data of Wright & Davison (1964) that the LD_{50} level for nitrate N in the diet is about 5–7 g/kg dry matter. Some recent studies give general support to this as a toxic level in fresh and dried forages for cattle (Purcell et al 1971, O'Hara & Fraser 1975). Purcell et al (1971) also observed signs of sublethal toxicity with nitrate N concentrations of only 4.2 g/kg dry matter although Phipps (1975) found that cows grazing a pasture containing 7.6 g/kg dry matter were unaffected. According to Phipps (1975) there have been few authenticated reported cases of nitrite toxicity in Britain but these become more likely with the increasing use of high levels of nitrogenous fertilizers. It seems that forages containing more than about 3–5 g nitrate N/kg dry matter should be regarded with suspicion.

Appendix Tables

Appendix Table 4.1   Proportion of apparently digested organic matter (DOM) apparently digested in the rumen

Diet	Bullocks	Dry cows	Milking cows	Lambs	Sheep	Reference
Grasses (fresh or frozen)						
Young spring grass			0.69			Van't Klooster & Rogers (1969)
Meadow grass			0.70			Hagemeister & Kaufmann (1974)
Perennial ryegrass (S24)					0.66	Beever et al (1971a)
Ryegrass (*L. perenne*)					0.64	Ulyatt & MacRae (1974)
Ryegrass (*L. multiflorum × perenne*)					0.55	Ulyatt & MacRae (1974)
Grass/clover mixture (frozen)			0.61			Hagemeister & Kaufmann (1974)
Green legumes (fresh or frozen)						
Red clover					0.66	Beever et al (1971b)
White clover					0.66	Ulyatt & MacRae (1974)
Artificially dried grass						
Chopped						
Perennial ryegrass (S24)						
Early cut					0.56	Beever et al (1971a)
Medium cut					0.63	Beever et al (1972)
Cocksfoot (wafered)					0.60	Thomson & Beever (1972) cited by Thomson (1972)
					0.53	
Ground and pelleted						
Unspecified		0.43				Pfeffer et al (1972)
Perennial ryegrass (S24)						
Early cut					0.57	Beever et al (1972)
Medium cut					0.57	Thomson & Beever (1972) cited by Thomson (1972)
Cocksfoot					0.39	
Artificially dried legume						
Chopped						
Lucerne					0.47	Thomson et al (1972)
Lucerne (cobbed)					0.43	Beever et al (1971b)
Red clover (wafered)					0.65	Hogan (1973)
Subterranean clover						
Ground and pelleted						
Lucerne					0.34	Thomson et al (1972)
Red clover					0.50	Beever et al (1971b)
Hay (grass)						
Long	0.73					McGilliard (1961)
					0.72	Bruce et al (1966)
					0.70	MacRae & Armstrong (1969)
Chopped					0.60	Weston & Hogan (1968)
Ground and pelleted	0.82				0.83	Topps et al (1968)†
Hay (wheaten)						
Chopped					0.67	Hogan & Weston (1967a)
Ground and pelleted					0.62	Hogan & Weston (1967a)
Hay (legume)						
Lucerne					0.69	Hogan & Weston (1969)
Chopped						
Lucerne					0.65	Hogan & Weston (1967a)
Red clover (wafered)					0.63	Beever et al (1971b)
Ground and pelleted						
Lucerne					0.64	Hogan & Weston (1967a)
Red clover					0.50	Beever et al (1971b)

Appendix Table 4.1 (cont)

Diet	Bullocks	Dry cows	Milking cows	Sheep	Reference
Silage					
Perennial ryegrass S24 (wilted)				0.64	⎱ Beever et al (1971a)
(unwilted)				0.60	⎰
Straw					
Ground					
Wheat (alkali-treated)				0.66–0.80	Hogan & Weston (1971)
Mixed rations containing ⩾500 g hay or dried grass (chopped or pelleted) or silage/kg					
Barley, rolled		0.69	0.68		Watson et al (1972)†
		0.71	0.81		D. A. Corse (unpublished)
					G. P. Savage (unpublished)
					MacRae & Armstrong (1969)
					Watson et al (1972)†
high moisture			0.61		G. P. Savage (unpublished)
pelleted		0.72	0.62		⎱ McMeniman (1976)
micronised		0.72	0.59		⎰
rolled + field beans		0.62			
+ heated field beans					
Oats, rolled				0.72	Pfeffer et al (1972)
					Hogan & Weston (1967b)
Maize, ground	0.65		0.56		McGilliard (1961)
			0.56–0.70		Watson et al (1972)†
					G. P. Savage (unpublished)
pelleted				0.68	Bruce et al (1966)
flaked					
Maize + groundnut meal, ground and pelleted				0.73	Hogan & Weston (1967b)
Maize, flaked + soya bean protein			0.64	0.66	Bruce et al (1966)
Soya bean protein			0.69	0.65	Bruce et al (1966)
Concentrates				0.69	Van't Klooster & Rogers (1969)
					Hagemeister & Kaufmann (1974)
					Nicholson & Sutton (1969)
					Tamminga (1975)
+ cod liver oil			0.51	0.69	⎱ Sutton et al (1975)
				0.66	⎰
Mixed rations containing <500 g hay or dried grass (chopped or pelleted) or silage/kg					
Barley, rolled		0.52		0.65	MacRae & Armstrong (1969)
		0.65			Pfeffer et al (1972)
					D. A. Corse (unpublished)
					S. Papasolomontos (unpublished)
flaked		0.77		0.74	⎱ McMeniman (1976)
micronised		0.72			⎰
				0.66	⎱ S. Papasolomontos (unpublished)
Barley, rolled + urea		0.72		0.62	⎰ S. Papasolomontos (unpublished)
Maize, rolled		0.59		0.70	McGilliard (1961)
ground	0.52			0.62	Beever (1969)
					McMeniman (1976)
micronised		0.57		0.71	S. Papasolomontos (unpublished)
					McMeniman (1976)
flaked		0.69		0.74	S. Papasolomontos (unpublished)
				0.70	Coelho da Silva (1971)

Appendix Table 4.1 (cont)

Diet	Bullocks	Dry cows	Milking cows	Lambs	Sheep	Reference
flaked + concentrates (propionate fermentation)					0.69	Nicholson & Sutton (1969)
(acetate fermentation)					0.60	
Milo, rolled					0.53	Harrison et al (1975)
micronised					0.60	
flaked					0.63	
Concentrates		0.60	0.59		0.64	S. Papasolomontos (unpublished)
+ paper containing yeast			0.70		0.74	Sutton et al (1975) / Pfeffer et al (1972) / Van't Klooster & Rogers (1969)
urea			0.70			Hagemeister & Kaufmann (1974)
soya bean meal			0.72			
coconut meal			0.67			
fish meal			0.71			
rape meal			0.70			
groundnut meal			0.69			
field bean meal			0.67			
casein			0.63			
HCHO treated protein			0.68			
tannin protected			0.58			
All-concentrate rations						
Barley, pelleted, high intake	0.70				0.75	Sutton et al (1975)
low intake	0.59				0.73	
rolled				0.47	0.65	Topps et al (1968)† / MacRae & Armstrong (1969)
+ soya concentrate				0.59		Topps et al (1968)†
+ 0.7% urea				0.53		Ørskov et al (1972)
+ 1.4% urea				0.56		Ørskov et al (1974b)
+ 2.1% urea				0.60	0.76	Topps et al (1968)
+ fish meal				0.59		Ørskov et al (1972)
+ fish meal + 1% urea				0.61		Ørskov et al (1974b)
Maize, ground	0.74					McGilliard (1961)
Semipurified diets						
Semipurified diet					0.57	McMeniman (1976)
+ 3.6% urea						
propionate fermentation					0.58	Harrison (1977)
acetate fermentation					0.60	
+ field beans					0.85	McMeniman (1976)
+ heated field beans					0.84	
+ 4.2% urea						
2 × daily feeding					0.75	N. Ellis (unpublished)
24 × daily feeding					0.67	
+ lipid					0.70	
+ VFA					0.73	
+ casein					0.51	McMeniman (1976)
+ zein					0.58	
+ fish meal					0.57	

† Values derived from the proportion of digestible energy digested in the rumen × 1.05 (based on Nicholson & Sutton 1969 and Sutton et al 1975).

Appendix Table 4.2   Microbial nitrogen yield per unit of organic matter apparently digested in the rumen (g/kg).

Diet	Bullocks	Dry cows	Milking cows	Sheep	Microbial marker	Reference
Grass (fresh or frozen)						
Meadow grass			18–35		DAPA	Hagemeister & Kaufmann (1974)
Perennial ryegrass				17–24	35S	Beever et al (1974b)
Grass/closer mixture (fresh)			25–26		DAPA	Hagemeister & Kaufmann (1974)
Green legumes (fresh or frozen)						
Subterranean clover				28–36	35S	Hume & Purser (1975)
Artificially dried grass						
Chopped						
Perennial ryegrass				26	35S	Beever et al (1974b)
H.I. ryegrass				32	DAPA	Hogan & Weston (1970)
Phalaris				31	DAPA	Hogan & Weston (1970)
Ground and pelleted						
Perennial ryegrass				25	35S	Beever et al (1974b)
Artificially dried legume						
Chopped						
Subterranean clover				30	DAPA	Hogan & Weston (1970)
Berseem clover				22	DAPA	Lindsay & Hogan (1972)
Red clover				52–61	DAPA	Lindsay & Hogan (1972)
Hay (legume)						
Chopped						
Lucerne				37–50	DAPA	Lindsay & Hogan (1972)
Silage						
Perennial ryegrass						
unwilted				45	35S	Beever et al (1974b)
formaldehyde treated				14	35S	Beever et al (1974b)
formaldehyde treated, dried				15	35S	Beever et al (1974b)
Straw						
Ground						
Wheat, alkali-treated				32–55	DAPA	Hogan & Weston (1971)
Mixed rations containing ≥ 500 g hay or dried grass (chopped or pelleted)/kg						
Barley, rolled						
high moisture			30		DAPA	G. P. Savage (unpublished)
pelleted		35	29		DAPA	G. P. Savage (unpublished)
rolled + field bean meal		30	40		DAPA	G. P. Savage (unpublished)
heated				38	DAPA	McMeniman (1976)
Maize, pelleted			20–36		DAPA	G. P. Savage (unpublished)
Concentrates				24	RNA	Sutton et al (1975)
Purified diet				33	non-NH$_3$ N	Hume (1970a) / Hume (1970b)
Mixed rations containing <500 g hay or dried grass (chopped or pelleted)/kg						
Barley, rolled		21			RNA	S. Papasolomontos (unpublished)
		22			35S	McMeniman (1976)
				23	DAPA	McMeniman (1976)

Appendix Table 4.2   (cont)

Diet	Bullocks	Dry cows	Milking cows	Sheep	Microbial marker	Reference
micronised				32	RNA	Sutton et al (1975)
flaked		33		30	RNA	S. Papasolomontos (unpublished)
Barley, rolled + groundnut, soya and rapeseed tanned				27	^{35}S	McMeniman (1976)
+ casein			33	27	RNA	Sutton et al (1975)
+ casein HCHO treated			32	32	RNA	Sutton et al (1975)
+ casein + HCHO treated			47		DAPA	Hagemeister & Pfeffer (1973)
+ soya meal			41		DAPA	Hagemeister & Pfeffer (1973)
+ soya meal, HCHO treated + urea			32		DAPA	Hagemeister & Pfeffer (1973)
+ urea		23	34		^{35}S	McMeniman (1976)
+ coconut meal pellets			20		DAPA	Hagemeister & Kaufmann (1974)
+ fish meal			36		DAPA	Hagemeister & Kaufmann (1974)
+ yeast			35		DAPA	Hagemeister & Kaufmann (1974)
+ rapeseed meal			33		DAPA	Hagemeister & Kaufmann (1974)
+ groundnut meal			31		DAPA	Hagemeister & Kaufmann (1974)
+ field bean meal			30		DAPA	Hagemeister & Kaufmann (1974)
Barley, flaked, + urea		31		22	^{35}S	McMeniman (1976)
Maize, rolled		42			RNA	S. Papasolomontos (unpublished)
micronised		40		31	DAPA	McMeniman (1976)
flaked		48		22	RNA	S. Papasolomontos (unpublished)
					DAPA	McMeniman (1976)
Maize, flaked + groundnut meal	23				RNA	R. H. Smith & A. B. McAllan (unpublished)
+ soya meal	30				RNA	R. H. Smith & A. B. McAllan (unpublished)
+ fish meal , toasted	35				RNA	R. H. Smith & A. B. McAllan (unpublished)
propionate fermentation	32				RNA	R. H. Smith & A. B. McAllan (unpublished)
acetate fermentation					RNA	R. H. Smith & A. B. McAllan (unpublished)
All-concentrate rations						
Barley, rolled				24	^{35}S	Harrison et al (1975)
+ urea				32	^{35}S	Harrison et al (1975)
Semipurified diets						
Semipurified diet				35	DAPA	Ørskov et al (1972)
+3.6 % urea				29–33	DAPA	Ørskov et al (1972)
Semipurified diet				32	non-NH_3 N	McMeniman (1976)
propionate fermentation				29	^{35}S	Harrison (1977)
acetate fermentation				34	^{35}S	Harrison (1977)
+ casein				46	non-NH_3 N	McMeniman (1976)
+ zein				32	non-NH_3 N	McMeniman (1976)
+ fish meal				32	non-NH_3 N	McMeniman (1976)

Appendix Table 4.3 Proportion of rumen microbial nitrogen present as amino acid nitrogen

Samples	Source	No. of samples	Proportion of total N as amino N	Reference
Estimates based on sum of amino acid N				
Pure strains of bacteria	in vitro culture	22	0.86*	Purser & Buechler (1966)
Mixed bacteria	calf	5	0.78	D. N. Salter & K. Daneshvar (unpublished)
Mixed bacteria	calf	5	0.82	D. N. Salter & K Daneshvar (unpublished)
Mixed bacteria	cattle	2	0.74	Mason & Palmer (1971)
Mixed bacteria	cattle	11	0.77	McMeniman (1976)
Mixed bacteria	sheep	4	0.75	Weller (1957)
Mixed protozoa	sheep	4	0.79	Weller (1957)
Mixed bacteria	sheep	2	0.85	Bird (1973)
Mixed bacteria	sheep	—	0.72	Burris et al (1974)
Mixed bacteria	sheep	8	0.76	S. Papasolomontos (unpublished)
Mixed bacteria	sheep	5	0.87	McMeniman (1976)
ICI single cell protein (Pruteen)			0.74	D. G. Armstrong (personal communication)
Mean			0.79	
Estimates based on total N — nucleic acid N				
Entodinium caudatum	*in vitro*	1	0.90	R. H. Smith & A. B. McAllan (unpublished)
Mixed bacteria	defaunated calves	3	0.82	}Smith & McAllan (1974)
Mixed bacteria	faunated claves	3	0.87	
Mixed bacteria	cows	9	0.88	McAllan & Smith (1972)
Entodinium species	cow	1	0.91	R. H. Smith & A. B. McAllan (unpublished)
Mixed bacteria	sheep	8	0.87	McAllan & Smith (1972)
Mixed bacteria and protozoa	sheep	8	0.84	Ellis & Pfander (1965)

* Excludes tryptophan but includes cystine and DAPA.

Appendix Table 4.4   Apparent absorbability of protein N from the small intestine

Diet	Calves	Bullocks	Dry cows	Milking cows	Lambs	Sheep	Reference
Grasses (fresh or frozen)							
Unspecified				0.67(c)			Van't Klooster & Rogers (1969)
Perennial ryegrass						0.71(b)	MacRae & Ulyatt (1974)
Ryegrass (*L multiflorum × perenne*)						0.74(b)	MacRae & Ulyatt (1974)
Green legumes (fresh or frozen)							
White clover						0.66(b)	MacRae & Ulyatt (1974)
Red clover						0.71(b)	Beever et al (1971b)
Artificially dried grass							
Chopped							
Grass (unspecified)						0.64(a)	MacRae et al (1972)
S24 Perennial ryegrass							
Early cut						0.77(b)	Coelho da Silva et al (1972a)
Medium cut						0.70(b)	
Ground and pelleted							
S24 Perennial ryegrass							
Early cut						0.71(b)	Coelho da Silva et al (1972a)
Medium cut						0.73(b)	
Artificially dried legume							
Chopped							
Lucerne						0.71(b)	Coelho da Silva et al (1972b)
Lucerne (cobbed)						0.66(b)	Beever et al (1971b)
Red clover						0.53(b)	Beever et al (1971b)
Subterranean clover						0.61(a)	Hogan (1973)
Ground and pelleted							
Lucerne						0.70(b)	Coelho da Silva et al (1972b)
Red clover						0.55(b)	Beever et al (1971b)
						0.45(b)	Clarke et al (1966)
Hay (grass)							
Mixed rations containing ≥ 500 g hay or dried grass (chopped or pelleted)/kg							
Barley, rolled				0.61(c)			Watson et al (1972)
				0.66(b)			
				0.77(b)			G. P. Savage (unpublished)
high moisture			0.58(b)				
pelleted			0.59(b)				
rolled + unheated field beans				0.69(b)			McMeniman (1976)
+ heated field beans							
Maize, ground				0.66(c)			Watson et al (1972)
pelleted				0.70(b)			G. P. Savage (unpublished)
flaked						0.64(b)*	Clarke et al (1966)
+ soya bean protein						0.71(b)*	
						0.76(b)*	
Soya bean protein						0.65(a)	MacRae et al (1972)
Casein						0.62(a)	
HCHO treated							
Concentrates				0.62(c)			Van't Klooster & Rogers (1969)

Appendix Table 4.4 (cont)

Diet	Calves	Bullocks	Dry cows	Milking cows	Lamb	Sheep	Reference
Mixed rations containing <500 g hay or dried grass (chopped or pelleted)/kg							
Barley, rolled			0.53(b) 0.62(b)			0.63(b)	S. Papasolomontos (unpublished) } McMeniman (1976)
flaked			0.50(b)			0.69(b)	S. Papasolomontos (unpublished) McMeniman (1976)
micronised rolled + urea			0.65(b) 0.67(b)			0.74(b)	S. Papasolomontos (unpublished) } McMeniman (1976)
flaked + urea							
Maize, rolled						0.74(b)	S. Papasolomontos (unpublished)
flaked						0.67(b)	Coelho da Silva (1971)
ground						0.76(b)	Beever (1969)
micronised						0.72(b)	S. Papasolomontos (unpublished)
flaked + barley, rolled	0.71(b)					0.68(b)	J. L. Black et al (unpublished)
+ fish meal	0.71(b)						
+ decorticated groundnut	0.69(b)						
Concentrates							
Semipurified diets (110 g CP/kg)				0.62(c)		0.48(a)	Ben-Ghedalia et al (1974)
(170 g CP/kg)				0.71(c)			Van't Klooster & Rogers (1969)
All-concentrate diets							
Barley, rolled					0.60(a)		Ørskov et al (1971b)
+urea					0.65(a)		Ørskov et al (1972)
+fish meal					0.61(a)	0.58(a)	Ørskov et al (1974b)
+urea					0.67(a)	0.64(a)	Ørskov et al (1971b, 1972)
					0.69(a)		Ørskov et al (1971b, 1974b)
					0.66(a)		Ørskov et al (1974b)
Semipurified diets							
containing rapeseed meal HCHO treated		0.66(b)					Sharma et al (1974)
casein		0.65(b)					
HCHO treated		0.65(b)					
		0.72(b)					
3.6% urea							
a) propionate fermentation						0.79(b)	Harrison (1977)
b) acetate fermentation						0.76(b)	
+field bean,						0.73(b)	McMeniman (1976)
+field bean, heated						0.79(b)	
4.2% urea						0.65(b)	N. Ellis (unpublished)
2 × daily feeding						0.70(b)	
24 × daily feeding						0.68(b)	
Lipid						0.56(b)	
+VFA							

a) Values refer to non-ammonia N.
b) Values refer to total amino-acid N or amino acids.
c) Values refer to total N.
* Duodenal and ileal collections were made from different animals.

Appendix Table 4.5   Endogenous urinary nitrogen excretion of cattle.

Mean live weight (kg)	Type of nutrition	No. of observations	Technique	Mean endogenous urinary N (g/day)	Reference
European breeds					
32	Preruminant	3	N-free diet	2.6	Blaxter & Wood (1951b)
34	,,	8	Regression	2.2	Shillam & Roy (1963)
45	,,	4	N-free diet	3.0	Cunningham & Brisson (1957)
53	,,	28	Regression	3.4	Roy et al (1964)
58	,,	36	Regression	3.7	Roy et al (1970b)
70	,,	?	N-free diet	4.4	Jahn (1967)
81	Ruminant	36	Regression	4.7	Stobo & Roy (1964)
97	,,	1	N-free diet	4.0	Harris & Loosli (1944)
123	,,	12	Regression	7.3	Stobo & Roy (1973)
126	,,	2	N-free diet	6.9	Mukherjee & Mitchell (1951)
145	,,	1	N-free diet	6.5	Steenbock et al. (1915)
160	,,	4	N-free diet	5.4	Harris et al (1943)
167	,,	2	N-free diet	5.9	Hart et al (1912)
254	,,	6	Regression	8.5	Based on Vercoe (1969)
344	,,	18	N-free diet	8.4	Swanson & Herman (1943)
Zebu breeds					
243	Ruminant	4	Regression	3.9	Elliott & Topps (1963b)
255	,,	3	N-free diet	5.0	Kehar et al (1943)
285	,,	4	Regression	4.4	
341	,,	4	Regression	5.0	
394	,,	4	Regression	5.0	Elliott & Topps (1963b)
402	,,	4	Regression	5.1	
437	,,	3	N-free diet	8.5	Kehar et al (1943)
485	,,	4	Regression	6.8	Elliott & Topps (1963b)

Appendix Table 4.6   Endogenous urinary nitrogen excretion of sheep.

Mean live weight (kg)	Type of nutrition	No. of observations	Technique	Mean endogenous urinary N (g/day)	Reference
5.2	Preruminant	6	N-free diet	0.59	Walker & Faichney (1964a)
5.6	,,	6	N-free diet	0.66	Norton & Walker (1971)
6.5	,,	4	N-free diet	0.74	Jahn (1970)
19	Ruminant	3	N-free diet	1.2	Hutchinson & Morris (1936)
22	,,	6	N-free diet	0.83	Turk et al (1935)
24	,,	9	N-free diet	1.2	Smuts & Marais (1939)
24	,,	12	N-free diet	0.81	Sotola (1930)
25	,,	6	N-free diet	0.70	Hamilton et al (1948)
27	,,	13	N-free diet	1.2	Turk et al. (1934)
31	,,	16	Regression	0.31*	Elliott & Topps (1964)
32	,,	1	N-free diet	0.99	Völtz (1920)
33	,,	8	N-free diet	1.0	Ellis et al (1956)
33	,,	24	N-free diet	2.0	Jahn (1970)
34	,,	8	N-free diet	1.1	Harris & Mitchell (1941)
35	,,	6	N-free diet	1.4†	Singh & Mahadevan (1968)
35	,,	6	Regression	1.6†	Singh & Mahadevan (1968)
35	,,	6	Regression	1.3†	Singh & Mahadevan (1970)
37	,,	8	N-free diet	1.3	Smuts & Marais (1938)
40	,,	3	N-free diet	1.1	Scheunert et al (1922)
42	,,	9	N-free diet	1.8	Smuts & Marais (1939)
43	,,	2	N-free diet	1.0‡	Deif et al (1968)

* Blackhead Persian weathers.
† Indian rams.
‡ Barki rams.

Appendix Table 4.7   Extent of degradation of dietary nitrogen in the reticulo-rumen of ruminants.

Diet	Bullocks	Milking cows	Sheep	Microbial marker	Reference
		(a) Basal feeds			
Grasses (fresh or frozen)					
Perennial ryegrass (S24)			0.50	35S	Beever et al (1974b)
(Ruanui)			0.69	DAPA	Ulyatt et al (1975)
Short rotation ryegrass (Manawa)			0.70	DAPA	
Italian ryegrass		0.33–0.44† / 0.54†		DAPA	Hagemeister & Kaufmann (1974)
Clover/grass (frozen)					
Green legumes (fresh or frozen)					
White clover			0.67	DAPA	Ulyatt et al (1975)
Artificially dried grass					
Chopped					
Perennial ryegrass (S24)					
early cut			0.85*	RNA	Coelho da Silva et al (1972a)
medium cut			0.91*	35S	Beever et al (1974b)
—			0.29		
Ground and pelleted					
Perennial ryegrass (S24)					
early cut			0.58*	RNA	Coelho da Silva et al (1972a)
medium cut			0.79*	35S	Beever et al (1974b)
—			0.27		
Artificially dried legumes					
Chopped					
Lucerne			0.30*	RNA	Coelho da Silva et al (1972b)
cobbed			0.37*	35S	Beever et al (1974a)
Sainfoin			0.19*	35S	Hume & Purser (1975)
Subterranean clover			0.73		
Ground and pelleted					
Lucerne			0.15*	RNA	Coelho da Silva et al (1972b)
including barley roughage			0.65*	35S	Leibholz (1972a, b); Leibholz & Hartmann (1972)
Hay (grass)					
Brome and lucerne			0.73	Centrifugation	Mathison & Milligan (1971)
+ Dairy concentrate			0.76	RNA	Sutton et al (1975)
Hay (legume)					
Lucerne			0.59	Centrifugation	Nolan & Leng (1972)
			0.78	15N	Pilgrim et al (1970)
			0.75*‡	DAPA	Lindsay & Hogan (1972)
			0.73*‡		Hogan et al (1972)
Red clover			0.66*‡	DAPA	Lindsay & Hogan (1972)
Subterranean clover			0.47	35S	Hume & Purser (1975)
Silage					
Perennial ryegrass (S24), unwilted			0.78	35S	Beever et al (1974b)
HCHO treated			0.07		
dried			0.15		
Predominantly cereal diets					
Barley, rolled			0.59	RNA	S. Papasolomontos (unpublished)
			0.67*	—	Ørskov et al (1974b)
			0.62	Differential centrifugation	Mathison & Milligan (1971)
micronised			0.92	RNA	Sutton et al (1975)
flaked			0.53		
			0.33	RNA	S. Papasolomontos (unpublished)

Appendix Table 4.7   (cont)

(b) Protein Supplements

Protein supplement	Proportion of N intake provided by protein supplement	Bullocks	Milking cows	Sheep	Microbial marker	Reference
Maize						
yellow dent, rolled				0.49	RNA	S. Papasolomontos (unpublished)
waxy, rolled				0.39		
high lysine, rolled				0.42		
yellow dent, micronised flaked				0.71		
+hay		0.74		0.42	RNA	A. B. McAllan & R. H. Smith (unpublished)
+straw		0.67				
Milo, rolled micronised flaked				0.14 / 0.31 / 0.41	RNA	S. Papasolomontos (unpublished)
Protein supplement						
Zein	0.82–0.94			0.42	Feed N determined by lysine concentration in duodenal outflow	McDonald (1954)
	0.87–0.93			0.90	Microbial N = total N − Zein N	McDonald & Hall (1957)
Casein	0.93			0.84*	35S	Leibholz (1972a); Leibholz & Hartmann (1972)
	0.55–0.73				Feed N in duodenal outflow determined directly by estimating phosphoprotein	Williams & Smith (1976)
HCHO treated	0.50		0.74†		DAPA	Hagemeister & Pfeffer (1973)
	0.50		0.34†		DAPA	Hagemeister & Pfeffer (1973)
	0.55–0.73				Feed N in duodenal outflow determined directly by estimating phosphoprotein	Williams & Smith (1976)
Wheat gluten	0.90–0.94			0.83*	35S	Leibholz (1972a)
Groundnut meal	0.90			0.63	35S	Hume (1974)
	1.0			0.78	35S	Miller (1973)
	0.60				DAPA and 35S	A. B. McAllan & R. H. Smith (unpublished)
	0.50				RNA / DAPA	Hagemeister & Kaufmann (1974)
Groundnut meal (0.60), rape seed (0.30) tanned	0.43			0.83	RNA	Sutton et al (1975)
	0.43			0.73	DAPA and 35S	Miller (1973)
Sunflower meal	1.0			0.77	35S	Hume (1974)
Soya bean meal	0.95			0.39	DAPA	Hagemeister & Pfeffer (1973)
	0.50					
unheated	0.60	0.86	0.54†		RNA	A. B. McAllan & R. H. Smith (unpublished)
cooked	0.60	0.71				
Lupin meal	0.90			0.65	35S	Hume (1974)
Rapeseed meal	0.50		0.60†		DAPA	Hagemeister & Kaufmann (1974)
Field bean meal	0.50		0.54†			
Yeast protein	0.50		0.58†			
Coconut pellets	0.50		0.43†			
Fish meal	0.96			0.29	35S	Hume (1974)
	0.60	0.71			RNA	A. B. McAllan & R. H. Smith (unpublished)
	0.50		0.67†		DAPA	Hagemeister & Kaufmann (1974); Ørskov et al (1974b)
Fish meal, Peruvian	1.0*			0.16*	DAPA and 35S	Miller (1973)
	1.0			0.31		

* Values calculated from original data given in publication.
† Values obtained from duodenal flow posterior to entry of bile and pancreatic ducts.
‡ Based on 30 g N formed per kg apparently fermented organic matter.

Appendix Table 4.8   Degradability of feeds measured by the synthetic fibre bag technique.*

Description of feed	N concentration (g/kg DM)	Degradability (dg)	Reference
Barley (low protein)	16.0	0.80	Mehrez (1976)
(high protein)	20.0	0.73	
Maize	18.7	0.55	Mehrez & Ørskov (1977)
Wheat	17.9	0.72	
Swede roots		0.89	
Artificially dried grass			Ørskov & Mehrez (1977)
(low protein)	20.7	0.70	
(high protein)		0.60	
Silages†			
Timothy (wilted 36 h)	19.9	0.82	
(wilted 12 h + HCOOH)	18.3	0.70	
Timothy/Ryegrass			
(wilted 6 h)	21.2	0.78	
(wilted 24 h + HCHO + H_2SO_4)	24.3	0.80	
Red clover (+HCHO + H_2SO_4)	27.9	0.70	
Silage mixture	19.9	0.82	
Maize	10.3	0.60	

Protein supplements‡	After 6 h	After 12 h	
Groundnut	0.85	0.94	Mathers et al (1977)
Rapeseed	0.84	0.90	
Cottonseed	0.70	0.83	
Grassmeal	0.66	0.77	
Sunflower seed	0.56	0.76	
Soya bean meal	0.54	0.82	
Peruvian fish meal	0.17	0.21	
Sunflower meal	0.84	0.96	E. R. Ørskov & M. J. Hancock (Unpublished)
Rapeseed meal	0.78	0.85	
Ground peas	0.70	0.75	
Linseed cake	0.63	0.77	
Soya bean meal	0.60	0.83	
Ground winter beans	0.47	0.65	
Meat and bone meal	0.40	0.48	
Guar meal	0.37	0.73	
White-fish meal	0.34	0.38	

* Although within a particular experiment, feeds can be placed in rank order, there is a large variation between the results for similar feeds in different experiments.
† Disappearance of N when 90% of the digestible dry matter has disappeared.
‡ Disappearance after 6 h in the rumen appears to be the most appropriate measure of degradability.

Appendix Table 4.9   Ratio of sulphur to nitrogen in animal products and some common feeds.

Product/Feed		S/N ratio	Reference
Animal products	Beef	0.065	McCance & Widdowson (1946)
	Mutton	0.068	
	Cow's milk	0.055	
	Ewe's milk	0.060	Langlands & Sutherland (1973)
	Fleece	0.200	
Protein concentrates	Coconut	0.082	
	Cottonseed	0.068	
	Linseed	0.078	Moir (1970)
	Groundnut	0.055	
	Wheat	0.082	
Cereals	Oats	0.105	
	Barley	0.102	Mitchell & McClure (1937)
	Maize	0.091	
	Maize silage	0.063	Loosli (1952)
	Clover hay	0.074	
Roughages	Lucerne hay	0.097	
	Wheaten hay	0.122	Moir (1970)
	Oat straw	0.167	
	Barley straw	0.238	Loosli (1952)

Chapter 5

Requirements for the Major Mineral Elements: Calcium, Phosphorus, Magnesium, Potassium, Sodium and Chlorine

Introduction

The major minerals are widely distributed throughout the animal's body and each element serves a variety of functions. In practical conditions, requirements for sodium and chlorine are readily met through the provision of a supplement of common salt, and, with the possible exception of all-concentrate diets, the dietary intake of potassium is invariably in excess of requirement. Many ruminant feeds require supplementation with calcium and phosphorus but, because of the extensive skeletal reserves, animals normally can adjust to temporary imbalances, providing they are in overall balance over longer periods. The important exceptions are females in late pregnancy and early lactation, as metabolic changes associated with the onset of lactation can cause hypocalcaemia, irrespective of the dietary or nutritional status. A dietary deficiency of magnesium, on the other hand, may quickly result in hypomagnesaemia and tetany, especially in older animals. In practical conditions, therefore, the application of standards for the major minerals is most critical in relation to magnesium, especially in individual lactating animals grazing swards in the spring when the magnesium supply of the animal is frequently marginal.

A. Estimation of requirements

The factorial approach has been used to estimate the mineral requirements of animals of different classes producing at different rates. The factorial method assesses requirements in two stages. Firstly, the net requirement is calculated from estimates of the storage and secretion of the element made during growth, pregnancy and lactation and of inevitable losses from the body (endogenous losses). Secondly, the dietary requirement is calculated by dividing the net requirement by a factor that represents the proportion of dietary mineral that is absorbed as assessed from metabolism experiments. In detail:

Net minimum endogenous requirement (E)	=	the inevitable loss of the element from the body in faeces and urine.
Net requirement for body growth (G)	=	daily retention of the element at the specified rate and stage of growth.
Net requirement for pregnancy (P)	=	daily retention of the element in the foetus and adnexa at the specified stage of pregnancy.
Net requirement for lactation (L)	=	daily secretion of the element in milk at the specified yield.
Total net requirement	=	E+G+P+L
Dietary requirement, where A is the coefficient of absorption	=	(E+G+P+L)/A

Absorption is defined as the amount of a mineral supplied in the diet that enters the body from the gut, and *apparent absorption* is this entity less the net endogenous secretion into the gut. The *coefficient of absorption* (or of *apparent absorption*) is the amount *absorbed* (or *apparently absorbed*) divided by the amount ingested. Conceptually, the factorial method provides a satisfactory basis for the assessment of dietary requirement. In its application, however, there are technical problems associated with the measurement of the individual terms, especially endogenous loss, and more serious technical and interpretational difficulties due to considerable and real biological variation in certain of the components. Variation in the composition of the body and its secretions and in the absorption of an element from the gut presents problems in terms both of the amount of basic information needed to give reliable mean values and of allowing for the variation in dietary requirement of different animals in specific productive states. Of the various components of the factorial estimate, the biological variation in coefficient of absorption is likely to be of greatest significance, as it affects the calculation of dietary requirements for each individual function.

The absorption of calcium by ruminants is affected by their physiological status and is largely independent of dietary characteristics. For phosphorus, the excretion into the gut varies substantially with dietary supply and must be measured separately for each determination of coefficient of absorption; as a result there is comparatively little information, especially for the dairy cow. For magnesium the coefficient of absorption is both variable and low, and because of the dependence of the mature animal on a continuous dietary supply, a measure has been given of individual variation in dietary requirement. Because of the metabolic interrelationships between sodium and potassium, the status with respect to one may affect the requirement for the other, but, through lack of information, it has not proved possible to take account of this in the prediction of requirements.

2. The validation of factorial estimates

The critical test of the validity of estimates of nutrient requirements made by the factorial method is that they should predict reliably the requirements of animals in practical conditions, as demonstrated by the results of trials in which different nutrient allowances have been given. There are difficulties in selecting the criteria for dietary adequacy, and criteria differ in their sensitivity. For example, a diet may not depress growth or milk production and yet be deficient in that it fails to maintain normal concentrations of the element in all body fluids and tissues. Therefore, failure to affect performance by offering a diet providing less than the requirement predicted by the factorial method is not necessarily proof that the factorial estimate is too high, but if practical trials indicate that the estimates are too low, the basis of their assessment must be re-examined.

3. The presentation of data

A similar presentation has been adopted for each of the six elements. The data on accretion during growth and pregnancy and in wool and on secretion in milk are given in Chapter 1, but losses of potassium, sodium and chlorine through the skin and in saliva are covered in this Chapter. For all elements, there is a sequential consideration of endogenous loss, coefficient of absorption, calculation of requirement, comparison with the recommendations of the first edition and of the US National Research Council, and evaluation of the estimates in the light of the results of feeding trials. For magnesium, there is also a brief comment on supplementation, and for sodium, a statement on salt tolerance. There are, however, differences from element to element in the method of assessment of endogenous loss and coefficient of absorption because of differences in the amount and accuracy of the basic information.

Calcium

1. The endogenous losses of calcium from the body

(a) Cattle

The sources of data on faecal endogenous losses are listed in Appendix Table 5.1. All estimates have been obtained by methods involving radioactive calcium and represent an increase of 30% in the number of estimates available in 1965. The data are meagre and fragmentary and are not suitable for an investigation within experiments of the factors influencing faecal endogenous excretion. Despite possible errors in interpretation arising from systematic differences between experiments, the effects on endogenous faecal excretion of the amounts of calcium ingested or absorbed, live weight and the age of the animal have been calculated from the pooled data, with or without inclusion of the few estimates for milk-fed calves. In general the conclusions reached from the pooled data agree with those from individual experiments designed for specific purposes.

It has been established that the faecal endogenous loss of calcium is independent of the amount of calcium ingested or absorbed, confirming the conclusion of Visek et al (1953) and other workers. Faecal endogenous loss was directly proportional to the live weight of the animal and, when expressed as a proportion of live weight, was remarkably constant over all experiments. The mean value was 15.7 ± 3.82 mg/kg live weight per day. Part of the residual variation could be attributed to the age of the animal, since faecal endogenous loss was positively related to age. The relationship between faecal endogenous loss (E, mg/kg live weight per day) and age (X, years) was $E = 14.0 + 0.48 \ (\pm 0.116) \ X$. Because of the small effect of age and the difficulty of applying such a correction to adult dairy cattle in practice it was decided to ignore this effect and to assume that faecal endogenous loss for a given live weight was indeed constant. To obtain total endogenous loss 0.8 mg/kg live weight per day must be added for endogenous urinary loss (ARC, 1965) giving a value which differs so little from that adopted in the first edition (16 mg/kg per day) that the latter is retained. There was no indication from the limited data available that the faecal endogenous losses of milk-fed calves were significantly less than those of weaned animals.

The strongly held view of Mitchell (1962) that estimates of faecal endogenous excretion obtained by isotopic techniques are gross overestimates prompted a search of the literature of balance trials with cattle for instances where the faecal excretion of calcium was appreciably less than the isotope-derived value of 15.7 mg/kg per day. Unfortunately few of the trials involved diets low in calcium and of these, Paquay et al (1968) alone cite 10 values less than 10 mg/kg and 6 less than 5 mg/kg per day. Thus faecal endogenous losses can be less than the preferred value, but until more evidence on this point is forthcoming the isotope-derived figure will be retained.

(b) Sheep

In the first edition a figure of 40 mg/kg live weight per day was adopted as the total minimum endogenous loss. Since that date extensive work by Braithwaite and his co-workers using isotopic methods and by Field and his co-workers using diets low in calcium has shown clearly that the endogenous loss is much less than was previously thought and approximates more to the figure adopted for cattle. It is difficult to understand why essentially the same radioisotope methods in the hands of different groups of workers give such different results for sheep but not for cattle.

The data on faecal endogenous loss obtained by isotopic methods are listed in Appendix Table 5.2. Faecal endogenous losses were independent of the amount of calcium ingested or absorbed and were positively related to live weight. A total of 59 observations gave a mean value of 16.3 ± 0.53 mg/kg live weight per day.

Field & Suttle (1969) gave low-calcium diets to growing lambs and mature wethers and found that faecal endogenous losses probably lay in the range of 11–13 with a maximum of 22 mg/kg per day. A similar study with mature wethers by Nel & Moir (1974) provides estimates of 10–12 mg/kg per day. Feeding trials and carcass analyses by Sykes & Field (1972*a*) on pregnant ewes consuming low-calcium diets gave a maximum value of 13 mg/kg per day for total endogenous loss.

Estimates of faecal endogenous loss for milk-fed lambs come from feeding experiments in which synthetic diets low in calcium were used. Walker (1972) obtained values of 16.6 and 9.0 mg/kg per day, the difference depending on the method of calculation. Hodge (1973) found a value of 17.7 after 32 days on the calcium-depleted diet and one of 8.8 mg/kg per day after 32 to 64 days on the diet.

For the present a value of 16 mg/kg live weight per day for total endogenous loss, the same as that for cattle, will be adopted.

2. Efficiency of absorption of dietary calcium

(a) Cattle

To convert net requirements into dietary requirements sufficiently accurate for practical feeding purposes one must be able to predict accurately the amount of a nutrient absorbed from the amount ingested. At present there is no convincing evidence that this basic premise holds for ruminants with respect to calcium. Nor can such a relationship be expected, as it can be argued from present knowledge of calcium metabolism that ruminants absorb calcium according to bodily need.

Information on the relationship between absorbed and dietary calcium comes from a small number of studies with radioactive calcium and a large number of conventional balance trials. The previous edition used exclusively the former as they alone gave data on true absorption and on endogenous excretion. However, these trials refer nearly exclusively to studies involving beef cattle and then to a very restricted range of diets. For information on the absorption of calcium by dairy cows we have to turn to the data from the large number of conventional balance trials in the literature. The values obtained in these trials for apparent absorption can be converted to true absorption by adding a component for endogenous faecal excretion. In the present study the figure used was 16 mg Ca/kg live weight per day (ARC 1965). Resistance to use of balance data to investigate factors influencing absorption or retention of calcium has always been based on the low precision of the estimates for apparent absorption and retention. It should be realised that comparable estimates from studies of radioactivity, especially those in which the radioisotope is given parenterally, are subject to exactly the same errors. In the absence of any other relevant information we have collected published data from balance trials to examine the relationship between apparent and true absorption of dietary calcium with feed calcium, milk calcium and calcium retention (Appendix Table 5.3). The criteria used in the selection of balance data were those adopted by Duncan (1958).

A statistical analysis of 624 balance trials with lactating cows of known weight gave the total correlation matrix shown in Table 5.1 Feed calcium as a source of variation in true absorption accounted for 10.4%. Even this small component was found to be a simple reflection of the association between milk and feed calcium After adjustment to constant milk calcium secretion only 4% of the variation was attributable to feed calcium. We cannot escape the conclusion therefore that absorption cannot be predicted from ingested calcium. Translating this finding into practical terms: at a given intake, absorption of dietary calcium depends on the net requirements of the animal and, at a given net requirement, the coefficient of absorption varies inversely with feed calcium. This conclusion can be reached from existing knowledge of calcium excretion. If calcium were to be absorbed in excess of requirements the surplus would have to be excreted. But of the two possible routes, urinary excretion is small compared with absorption and faecal endogenous losses are constant. This constancy of endogenous excretion means that absorption is regulated at the gut level and is equated to body needs. Further confirmation of this conclusion comes from the relationship between calcium intake and its coefficient of absorption which was derived

Table 5.1   Correlation coefficients relating apparent (AA) or true absorption (A) of calcium to the amounts of calcium ingested, secreted in milk and retained by or lost from the body.

AA	1.000				
A†	0.997	1.000			
Feed Ca	0.300	0.322	1.000		
Milk Ca	0.581	0.588	0.358	1.000	
Balance Ca	0.688	0.677	0.052	−0.175	1.000
	AA	A	Feed Ca	Milk Ca	Balance Ca

† A = AA + 0.016 W (where W = live weight in kg).

in the first edition from studies with radioactive calcium given to non-lactating cattle. They showed that for a given requirement (i.e., similar live weight) the amount of calcium absorbed from the diet is independent of dietary intake.

In order to meet calcium requirements the diet must contain sufficient calcium in a form that can be absorbed. At present there is no information on what proportion of dietary calcium is in this form. The reason for this unsatisfactory situation is that most published balance trials related to intakes of calcium many times greater than the experimental animals' requirements, and hence dietary calcium was rarely limiting. In these conditions observed and expected coefficients of absorption should be the same and equal to R/I, where R and I are the requirement and intake of calcium, respectively. If dietary calcium is not completely in an available form, the observed coefficient would be lower than expected in those balance trials where intake and requirements are of the same order. Provided the animal's ability to absorb is not a limiting factor, the observed value would then be a measure of the proportion of dietary calcium available for absorption.

Data from a series of 778 balance trials on dairy cows (Appendix Table 5.3) were used to compare the observed and theoretical absorbability of dietary calcium at different values of I/R (Table 5.2). Since this exercise is by necessity relatively imprecise no account was taken of the requirements of pregnancy in calculating total calcium requirements. Significant deviations of the observed from theoretical occurred at I/R values <1.5 and it would appear that for the majority of diets only about 0.68 of the calcium is in a form which can be absorbed. There was evidence that the calcium in some diets was of a low availability; for instance, two diets involving a total of 6 animals gave low values (0.3−0.4) for absorbability in the group for which I was <R. However, there is no information on how to recognise such diets. Observed was higher than theoretical for values of I/R >4, but whether this difference is a true finding or whether it reflects the difficulty of measuring apparent absorption in situations where absorption is small relative to intake is not known. It is relevant to recall that most balance trials overestimate absorption and hence retention (Duncan 1958).

Table 5.2   Comparison in dairy cows between observed and theoretical coefficients of absorption of dietary calcium in relation to the ratio of dietary intake (I) to requirements (R) of calcium.

I/R	n	Mean I/R	Theoretical coefficient (R/I)	Observed coefficient	SE
<1	16	0.82		0.56	0.052
1.0 − 1.5	67	1.30	0.77	0.68	0.020
1.5 − 2.0	75	1.74	0.57	0.60	0.017
2.0 − 2.5	78	2.25	0.44	0.41	0.023
2.5 − 3.0	139	2.75	0.36	0.37	0.010
3.0 − 3.5	92	3.25	0.31	0.34	0.014
3.5 − 4.0	83	3.73	0.27	0.31	0.017
4.0 − 4.5	48	4.28	0.23	0.34	0.025
4.5 − 5.0	31	4.78	0.21	0.28	0.025
5.0 − 6.0	49	5.48	0.18	0.28	0.025
6.0 − 13.0	91	9.67	0.10	0.23	0.019
>13.0	9	15.1	0.07	0.15	—

The dietary requirements for cattle (Tables 5.3–5.5) have been calculated by using a figure of 0.68 for the absorption of dietary calcium. It must be stressed that the requirements based on these figures are minimal ones and no safety margin has been included.

For milk-fed calves normal calcium intake is probably less than the potential rate of skeletal deposition and consequently high value for the coefficients of true absorption have been reported from studies with radioactive calcium (0.98, Hansard et al 1954; 0.82, Guéguen 1964). It is therefore proposed that the values adopted in the first edition be retained, namely 0.95 for animals receiving whole milk alone and 0.90 for those receiving a high proportion of milk in the diet.

(b) Sheep

It can be concluded from the smallness of urinary excretion and the constancy of endogenous faecal losses that sheep, like cattle, absorb calcium according to their needs. Experimental verification of this thesis comes from the work of Braithwaite et al (1969) and Sykes & Dingwall (1975), who established that absorption was related positively to requirements and negatively to calcium intake.

Unfortunately at present there is not the body of evidence necessary to establish what proportion of calcium in sheep diets can be absorbed. Studies with radioactive calcium suggest an upper limit of 0.45, but they include few experiments in which animals with high requirements ingested diets low in calcium. On the other hand, work by Sykes & Field (1972a) and by Field et al (1975) in which diets low in calcium were given to pregnant and growing sheep, respectively, gave high values. It was therefore decided to adopt for sheep the same value (0.68) used to calculate dietary requirements for cattle.

There is no doubt that calcium in milk is efficiently absorbed by unweaned lambs, but at present it is not possible to put a precise figure to the proportion absorbed. Bose (1955) claimed that milk calcium is used with an efficiency approaching 1.00, whereas Walker (1972) gives a lower figure of 0.88 for lambs fed on cow's milk. In the absence of more comprehensive data we have adopted a figure of 0.95, the same as that used for milk-fed calves.

3. Requirements for calcium

(a) Cattle

Estimates of dietary requirements of calcium for growth, pregnancy and lactation are given in Tables 5.3 to 5.5.

Two points must be made regarding these requirements: firstly, certain cows cannot absorb sufficient calcium to meet their requirements for milk production, and secondly, the calcium intake of dairy cows in the dry period has a bearing on the subsequent incidence of parturient hypocalcaemia and milk fever. As early as 1916 Forbes and his workers cast doubt on the thesis that dairy cows can meet their demands for

Table 5.3   Dietary requirements (g/day) for calcium of cattle gaining at different rates.

Live weight (kg)	0	0.52	0.50	1.0	1.5
50	1 (1)	6 (4)	11(8)	20 (14)	29 (21)
100	2 (2)	7 (5)	12 (8)	21 (15)	30 (22)
200	5	9	14	24	33
300	7	12	17	26	35
400	9	14	19	28	38
500	12	17	21	30	40

( ) dietary requirements of milk fed calves.

Table 5.4   Dietary requirements (g/day) for calcium of pregnant cows.

| | Week of pregnancy | | | | | |
	20	24	38	32	36	40
1st calf	12	13	14	26	29	34
2nd or later calf	15	16	18	20	24	28

A first-calf cow was assumed to weigh 450 kg and to grow at the rate of 0.5 kg/day over the last 2 months of pregnancy.

Table 5.5   Dietary requirements (g/day) for calcium of lactating cows.

| | Milk yield (kg/day) | | | |
Breed	10	20	30	40
Jersey	30	50	71	92
Ayrshire	29	46	63	80
Friesian	31	48	64	81

Ca content of milk was taken as 1.4 for Jersey, 1.16 for Ayrshire and 1.13 g/kg for Friesian cows.

calcium, however high, by absorption from the diet. Subsequent workers have confirmed their viewpoint and the position has been summarised by Duncan (1958) who, using data from published balance trials, has shown an association between high milk yields and negative calcium balance. Thus it must be appreciated that certain dairy cows, particularly in early lactation, will be in negative calcium balance, irrespective of the calcium content of the diet, and that the resultant losses from the skeleton will be made up in the subsequent dry period. No adjustment of the dietary requirements has been made for the replacement of those skeletal losses, partly because demineralization of the skeleton is known to stimulate calcium absorption from the gut (Sykes & Dingwall 1975) and partly because we have no precise estimate of these losses. There is some evidence that the same situation holds for lactating ewes (Braithwaite et al 1969).

There is now considerable evidence that feeding on a low-calcium diet in the latter part of the dry period will reduce the severity of parturient hypocalcaemia and the incidence of milk fever (Boda & Cole 1954, Westerhuis 1974). The requirements of the pregnant cows given in Table 5.4 are within the range of intakes which are thought to be consistent with a low incidence of milk fever.

(b) Sheep

Estimates of calcium requirements for growth, pregnancy and lactation are given in Tables 5.6 to 5.8.

Table 5.6   Dietary requirements (g/day) of calcium for sheep gaining at different rates.

| | Rate of gain (kg/day) | | | | |
Live weight (kg)	0	0.1	0.2	0.3	0.4
5	0.1 (0.1)	1.6 (1.1)	3.1 (2.2)	4.6 (3.3)	6.1 (4.3)
10	0.2 (0.2)	1.7 (1.2)	3.2 (2.3)	4.7 (3.4)	6.2 (4.4)
20	0.5 (0.3)	2.0 (1.4)	3.4 (2.5)	4.9 (3.5)	6.4 (4.6)
40	0.9	2.4	3.9	5.4	6.9
60	1.4	2.9	4.4	5.9	7.3

( ) dietary requirements of milk-fed lambs.

Table 5.7   Dietary requirements (g/day) of calcium for ewes during pregnancy.

Live weight (kg)	Week of pregnancy			
	9	13	17	21
40	1.0	1.4	2.3	3.0
75	1.9	2.3	3.0	3.9

Table 5.8   Dietary requirements (g/day) of calcium for lactating ewes

Live weight (kg)	Milk yield (kg/day)		
	1.0	2.0	3.0
40	3.3	5.6	8.0
75	4.1	6.5	8.8

4. Comparison of the estimates of requirements with those of the first edition and of the National Research Council

(a) Cattle

Table 5.9

Comparison of estimates of the calcium requirements (g/day) of growing cattle.

Live weight (kg)	Daily weight gain (kg)	ARC (1965)	NRC Dairy cattle* (1971)	ARC (this edition)
55	0.40	—	4.5	9
100	0.75	24	11	17
200	0.75	25	18	19
300	0.75	28	24	21
400	0.75	33	26	24
500	0.60	31	27	23
600	0.15	25	24	17

* Growing heifers (large breeds).

Table 5.10   Comparison of estimates of the calcium requirements (g/day) of pregnant cattle in the last 2 months of pregnancy.

Live weight (kg)	ARC (1965)	NRC (1971)	ARC (this edition)
400	30	23	19
500	33	29	22
600	37	34	24

Table 5.11   Comparison of estimates of the calcium requirements (g/day) of lactating cows.

Milk yield (kg)	Jersey			Ayrshire			Friesian		
	ARC (1965)	NRC (1971)	ARC (this edition)	ARC (1965)	NRC (1971)	ARC (this edition)	ARC (1965)	NRC (1971)	ARC (this edition)
10	46	45	30	46	47	29	49	48	30
20	78	74	50	74	74	46	77	74	48
30	110	103	71	102	101	63	105	100	64
40	142	132	92	130	128	80	133	126	81

In Tables 5.9 to 5.11 estimates for growth, pregnancy and lactation given by the ARC (1965), the NRC (1971) and the present recommendations are compared and, as expected, the last suggests that the calcium needs of cattle can be met with diets much lower in calcium than previously thought.

(b) Sheep

A comparison similar to that for cattle is given in Tables 5.12 and 5.13.

Large differences in the three sets of recommendations for a given class of sheep are obvious and can be attributed to the choice of values for endogenous faecal loss and coefficient of absorption. In general, the estimates from the first edition are much higher than the present ones, notably for dry, pregnant and lactating ewes. Those from the NRC (1975) are higher than but closer to the present recommendation for growing and pregnant sheep. No comparison with the NRC recommendations for lactating ewes is possible, because the milk yields adopted by this Committee are not given, nor are the reasons why their allowances are some 77% higher than the previous ones (NRC 1968).

Table 5.12   Comparison of estimates of the calcium requirements (g/day) of growing sheep.

Live weight (kg)	Daily weight gain (g)	ARC (1965)*	NRC (1968)	ARC (this edition)
30	200	6.0	4.8	3.7
35	220	6.7	4.8	4.1
40	250	7.7	5.0	4.6
45	250	8.1	5.0	4.8
50	220	7.9	5.0	4.4
55	200	8.0	5.0	4.3

* Assumed efficiency of absorption of 0.50 (see ARC (1965) Table 2.13 p. 31).

Table 5.13   Comparison of estimates of the calcium requirements (g/day) of pregnant and lactating ewes.

Live weight (kg)	ARC (1965)	NRC (1975)		ARC (this edition)
Non-lactating and first 15 weeks of gestation				
50	4.7	3.0		1.2
60	5.6	3.1		1.4
70	6.5	3.2		1.7
80	7.4	3.3		1.9
Last 6 weeks of gestation				
50	7.6[a]	4.1		2.8[b]
60	8.5	4.4		3.0
70	9.4	4.5		3.2
80	10.3	4.8		3.4
First 8 to 10 weeks of lactation[c]				
50	10.0	10.9[d]	6.2[e]	4.3
60	10.9	11.5	6.5	4.6
70	11.8	12.0	6.9	4.8
80	12.8	12.6	7.2	5.1

(a) Lamb weighing 5.9 kg.
(b) Lamb weighing 4.0 kg.
(c) For ewes giving 1.36 kg milk/day.
(d) No milk yield given.
(e) NRC (1968) requirements.

5. Comparison of the estimates of calcium requirements with the results of practical trials

In the earlier edition practical feeding trials were used in the validation of estimates of dietary requirements derived from the factorial approach. But existing feeding trials give only imprecise estimates of dietary requirements, because they often involved only two dietary treatments, one frankly deficient and the other more than adequate and, secondly and more importantly, they rarely include any assessment of the degree of mineralization of the skeleton, an essential criterion in judging the adequacy of a diet. The sole manifestation of a marginal deficiency of calcium is a reduction in the calcium content of bone; it requires a severe and protracted deficiency to reduce the rate of gain in weight or milk secretion. The degree of bone mineralization desirable is still not clear; is it one which is adequate for the normal functioning of the skeleton or is it the degree of mineralization found in adequately nourished animals? For instance, Hodge et al (1973) found that the femurs of milk-fed lambs receiving 250 mg Ca/kg per day were not fully mineralized but were structurally sound, whereas maximum mineralization was associated with a calcium intake of 450 mg/kg per day Despite these limitations, feeding trials are useful in the sense that they provide a range of calcium intake within which the recommended dietary allowance must fall. A comparison with the practical feeding trials on cattle and sheep listed in the first edition showed that the present recommendations fall within this range. The sole exception is the feeding trial of Converse (1954) with growing cattle (as with the earlier recommendations); in fact the theoretical net requirements of Converse's cattle consuming the calcium-deficient diet were more than their dietary intake, indicating that although skeletal growth was normal, as measured by height at withers, skeletal mineralization must have been drastically reduced. A similar phenomenon has been described for growing lambs by Field et al (1975).

Recent studies by Walker (1972) and Hodge (1973) with milk-fed lambs, in which the calcium content of the body or in individual bones was measured, provide estimates of dietary requirements in close agreement with the present standards.

Phosphorus

1. The endogenous losses of phosphorus from the body

(a) Cattle

Very few estimates of the endogenous faecal loss were available in 1965 and the position has not changed materially in the intervening years. The data are summarized in Appendix Table 5.4 and show considerable variation from experiment to experiment.

In the first edition it was assumed that net endogenous faecal loss varied with live weight but not with phosphorus intake or absorption. These assumptions mean that either dietary phosphorus is not absorbed in excess of the animal's net requirements or, if it is, the excess is excreted in the urine. We know that only trace amounts of phosphorus are normally secreted by the kidney, whereas large amounts are secreted in the saliva and are subsequently reabsorbed further down the intestinal tract (Bruce et al 1966). Thus homeostasis must be achieved in the digestive tract by controlling the secretion and reabsorption of salivary phosphorus and faecal endogenous losses must be related in part to intake or absorption (Clark et al 1973). Multivariate regression of the data from cattle, excluding those for milk-fed animals, showed that faecal endogenous losses were not constant, as would be expected if absorption was equated to requirements, but varied with phosphorus intake or absorption. This conclusion as to the dominant role of faecal endogenous excretion in homeostasis is supported by previous findings with sheep (Lueker & Lofgreen 1961, Young et al 1966). A distinction must therefore be made between net endogenous faecal loss and minimum net endogenous faecal loss, the entity which is by definition equated with net requirements for maintenance.

Although phosphorus intake or absorption was by far the most important source of variation in net endogenous faecal loss, accounting for some 80% of the variation, live weight also was significant in this respect. Indeed, in the absence of data from balance trials in which phosphorus-free diets were used, the partial regression of net endogenous faecal loss on live weight (10.3 ± 4.7 mg/kg per day) can be one measure of the minimum net endogenous loss. Another estimate of 9.2 ± 1.3 mg/kg per day was obtained by extrapolating to zero intake the regression of net endogenous faecal loss, expressed as a function of live weight, on phosphorus intake (Fig. 5.1). On the basis of these two estimates a minimum endogenous faecal loss of 10 mg/kg live weight per day has been taken for weaned animals. It must be stressed that this figure is not firmly based; we do not know, for instance, if the minimum endogenous faecal loss is a constant for a given live weight. It may vary with the rate of salivary secretion and hence with quantity and physical form of diet. Since minimum endogenous faecal loss is an important determinant of dietary requirements we need much more information on this aspect. The minimum endogenous faecal loss per unit of live weight appears to be significantly less for suckling than for weaned cattle (Lofgreen et al 1952, Guéguen 1963, 1964) and it is proposed that the value of 4 mg/kg live weight per day be be adopted for the minimum endogenous faecal loss of the veal calf.

Urinary losses of phosphorus by cattle on conventional diets are generally low, although much higher values are found with individual cattle. There is some evidence from identical twin cows that urinary losses are in part genetically determined (Field & Suttle 1970), but it is not known whether minimum endogenous urinary losses also vary with the individual. For this reason the value of 2 mg/kg live weight per day used in the first edition is retained.

Fig. 5.1   Estimation of net endogenous faecal loss of phosphorus by cattle, from the regression of faecal loss on intake.

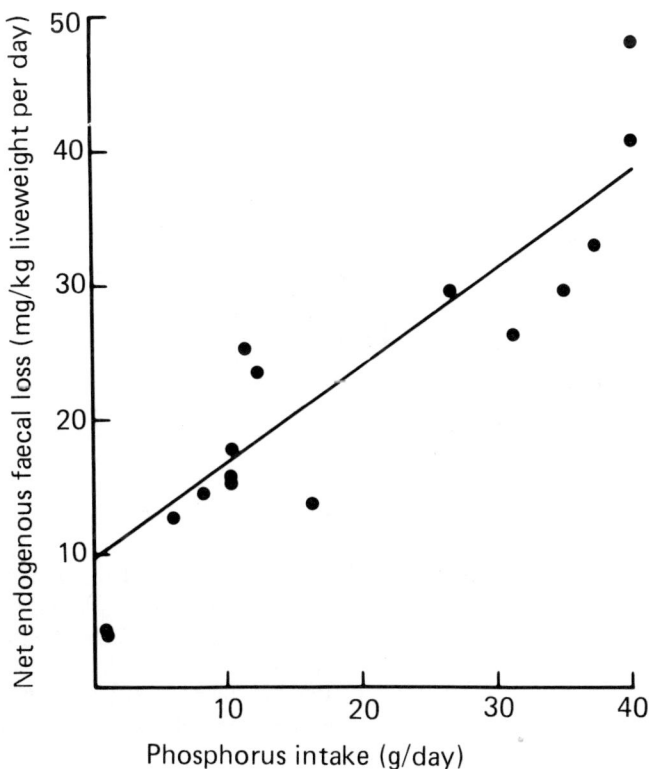

The present estimate of 12 mg/kg per day for total minimum endogenous losses i
similar to that adopted in the first edition for young ruminants but is much smalle
than those adopted for more mature or heavy animals.

(b) Sheep

In the first edition endogenous faecal excretion, the predominant component of tota
endogenous loss, was taken as a constant and equal to 42 mg/kg live weight per day
Observations on the faecal output of phosphorus by sheep consuming experimenta
or natural diets low in phosphorus demonstrated that this value for endogenous faeca
loss was a gross overestimate (Field et al 1974, Nel & Moir 1974, Sykes & Dingwa
1976). For instance, faecal excretions of 12–14 and 9–14 mg/kg per day have bee
reported by Sykes & Dingwall (1976) and Tomas & Somers (1974 and privat
communication) respectively.

That the minimum endogenous faecal loss lies in the region of 10–14 mg/kg pe
day is confirmed by the studies in which radioactive phosphorus was used. An estimat
of 11.6 ± 3.4 mg/kg per day for the minimum net endogenous faecal loss can b
obtained from the pooled data (Appendix Table 5.5) by extrapolating to zero intak
the regression of net endogenous faecal loss, expressed as a function of live weigh
on phosphorus intake (Fig. 5.2). The partial regression coefficient of net endogenou
faecal loss on live weight was not significant, mainly because of the small range i
weight present in the data.

Fig. 5.2   Estimation of net endogenous faecal loss of phosphorus by sheep,
from the regression of faecal loss on intake.

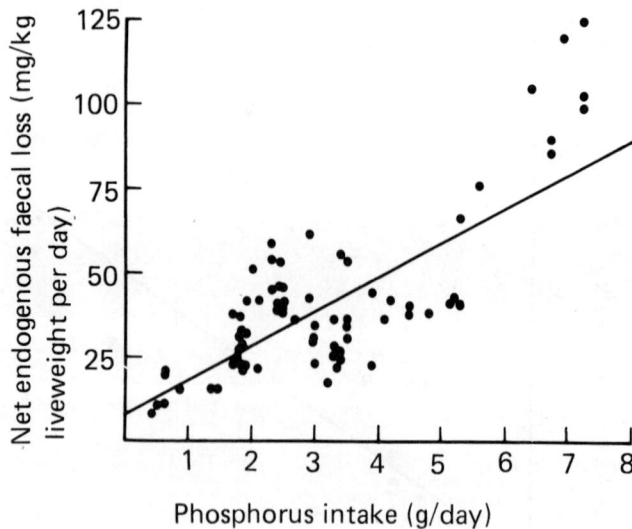

The milk-fed lamb, like the milk-fed calf, has a low net endogenous faecal los
(2–3 mg/kg per day; Compère et al 1967) and, for practical situations, this value ca
be taken as the net minimum faecal loss of milk-fed lambs.

The urinary excretion of phosphorus is generally considered to be low, but recentl
Walker (1972) has shown that the urine is an important route of excretion in th
milk-fed lamb. He has argued furthermore that his figure of 28 mg/kg live weigh
per day for urinary phosphorus loss represents the minimum net endogenous urinar

loss of milk-fed lambs, rather than that of 1.5 mg/kg per day adopted in the first edition (1965). Until investigations using milk substitute diets free of phosphorus resolve this problem the earlier estimate of 1.5 mg/kg live weight per day will be retained.

The estimate of 14 mg/kg live weight per day taken for the total endogenous loss is much lower than the one of 42.5 mg used in the first edition.

2. Efficiency of absorption of dietary phosphorus

(a) Cattle

Balance trials with radioactive phosphorus give direct measurements of the coefficient of true absorption of dietary phosphorus. Unfortunately, at the present time there are few estimates and these refer mainly to growing and not milking animals. There are indications in the data that the absorption of dietary phosphorus varies with age and that the data fall into two age classes, above and below 14 months. For both age classes, the regression coefficient in the rectilinear relationship between phosphorus absorbed and P ingested is highly significant but the constant is not. The regression coefficients are 0.78 ($r = 0.987$, $P < 0.001$) for the younger and 0.43 ($r = 0.99$, $P < 0.001$) for the older class of animal. The value of 0.78 will be adopted for growing cattle up to 1 year of age and one of 0.94, the mean of the four estimates given in Appendix Table 5.4, for the suckled calf.

The older age class contains 8 estimates, only one of which relates to a lactating cow. These are insufficient data on which to accept without further verification the figure of 0.43 for dairy cows, the predominant group in this age class. For the only other source of data on absorption we must turn to the literature on conventional balance trials with milking cows. Not all balance trials are suitable, since "absorption", i.e., apparent absorption adjusted for minimum net endogenous loss, approximates to actual absorption only in those trials where absorption itself is equal to or less than requirements. Unlike that of calcium, net endogenous loss of phosphorus is not a constant but is proportional to the amount absorbed in excess of requirements. The balance trials listed in Appendix Table 5.6 were classified according to dietary intake (I), expressed as a fraction of requirements (R) and data from trials with $I/R < 1.75$ were taken. In these trials the animals would need to absorb dietary phosphorus with an efficiency greater than 0.57 to meet their requirements. The group ($I/R < 1.5$) contained the results from 24 balance trials and these gave a value for the coefficient of absorption of 0.58 ± 0.058. For the group ($1.5 < I/R < 1.75$), the corresponding figures were 39 and 0.58 ± 0.056. Thus a more realistic value than the figure of 0.43, derived from the very limited radioactive data for this age class, is 0.58 and this value has been used to derive the phosphorus requirements for the various classes of cattle. Whether this value applies to the diets of beef cattle which are traditionally offered roughage of low digestibility, like straw, is not known.

The preferred value of 0.58 for the coefficient of absorption of dietary phosphorus is close to the one of 0.55 used in the first edition.

(b) Sheep

Data on the absorption of phosphorus in dietary sources to sheep are summarised in Appendix Table 5.5. Most studies refer to inorganic supplements, but as the absorbability of phosphorus in these sources and in natural feeds appears to be similar, the data were pooled.

To allow absorption to be predicted with precision from phosphorus intake it is necessary to divide the data into two groups according to age, above and below year. Within each age class the linear relationship between absorption and intake was highly significant ($P < 0.001$) and accounted for 85–90% of the variation in absorption. Both relationships had significant constants but little error was introduced if the constant was suppressed and the resultant regression coefficient taken as a measure of the coefficient of absorption of dietary phosphorus. The estimates were 0.73 for lambs up to 1 year of age and 0.60 for more mature sheep.

The limited studies of Compère et al (1967) with the milk-fed lamb suggest that the coefficient of absorption of phosphorus in milk is greater than 0.95 and this value has been adopted.

3. Requirements for phosphorus

Estimates of phosphorus requirements for growth, pregnancy and lactation are given for cattle in Tables 5.14 to 5.16 and for sheep in Tables 5.17 to 5.19.

In the calculation of the requirements of dairy cows no allowance was made for the replacement of the phosphorus which is frequently lost from the body of high producing dairy cows in early lactation (Duncan 1958). The reasons for not doing so are twofold: firstly we do not know when in the lactating cycle these losses are made up and secondly, and more importantly, we have little idea of the extent of these losses and have little hope of getting such data in the future. However, it is thought that the error introduced is small. For instance, replacement of a loss of 10% of body phosphorus in the first 3 months of lactation would require about an additional 3 P/per day in the diet until the next calving.

Table 5.14   Dietary requirements (g/day) of phosphorus for cattle gaining at different rates.

Live weight (kg)	Coefficient of absorption	P requirements for daily gains in weight (kg) of:				
		0	0.25	0.50	1.0	1.5
50	0.78	0.8 (0.3)	3.1 (2.2)	5.5 (4.1)	10 (8.0)	15 (11.8)
100	0.78	1.5 (0.6)	3.9 (2.5)	6.2 (4.5)	11 (8.3)	16 (12.2)
200	0.78	3.1	5.4	7.8	13	17
300	0.78	4.6	6.9	9.3	14	19
400	0.58	8.3	11	15	21	27
500	0.58	10	14	17	23	29

In parenthesis ( ) requirements of milk-fed calves.

Table 5.15   Dietary requirements (g/day) of phosphorus for pregnant cows.

	Week of pregnancy					
	20	24	28	32	36	40
1st calf	11	11	12	20	22	26
2nd or later calf	13	14	15	17	19	22

A first-calf cow was assumed to weigh 450 kg and to grow at the rate of 0.5 kg/day over the last 2 months of pregnancy.

Table 5.16   Dietary requirements (g/day) of phosphorus for lactating cows.

	Milk yield (kg/day)			
Breed	10	20	30	40
Jersey	29	51	72	94
Ayrshire	27	43	59	75
Friesian	28	44	59	75

P content of milk was taken as 1.24 for Jersey, 0.93 for
Ayrshire and 0.90 g/kg milk for Friesian cows.

Table 5.17   Dietary requirements (g/day) of phosphorus for sheep gaining at different rates.

Live weight (kg)	P requirements for daily gain in weight (kg) of:				
	0	0.1	0.2	0.3	0.4
5	0.1 (0.02)	0.8 (0.6)	1.6 (1.2)	2.4 (1.8)	3.1 (2.3)
10	0.2 (0.05)	0.9 (0.6)	1.7 (1.2)	2.5 (1.8)	3.2 (2.4)
20	0.4 (0.09)	1.1 (0.7)	1.9 (1.3)	2.6 (1.8)	3.4 (2.4)
30	0.6	1.3	2.1	2.8	3.6
40	0.8	1.5	2.3	3.0	3.8
60	1.2	1.9	2.7	3.4	4.2

In parenthesis requirements of milk-fed lambs.

Table 5.18   Dietary requirements (g/day) of phosphorus for ewes during pregnancy.

Live weight (kg)	Week of pregnancy			
	9	13	17	21
40	1.0	1.4	1.7	1.9
75	1.9	2.2	2.5	2.7

Table 5.19   Dietary requirements (g/day) of phosphorus for lactating ewes.

Live weight (kg)	Milk yield (kg/day)		
	1	2	3
40	3.1	5.3	7.4
75	3.9	6.1	8.3

4. Comparison of the estimates of requirement with those of the first edition and of the National Research Council

(a) Cattle

In Tables 5.20 to 5.22 the present estimates for growth, pregnancy and lactation are compared with those of the first edition and of the National Research Council. In general the present estimates for the different classes of cattle are appreciably lower than those in the first edition. This difference is due mainly to the choice of the value for the minimum net endogenous faecal loss; in the first edition the value was weight-dependent and ranged from 12 to 28 mg per kg live weight per day, whereas the present estimate is independent of weight and equal to 12 mg/kg per kg per day. However, for the lactating Jersey cow the phosphorus content of its milk was taken as 1.24 g/kg compared with that of 0.95 g/kg in the first edition and this difference was sufficient to cancel out the proposed decrease in maintenance requirements.

Some concern may be felt because the estimates for cattle, particularly those for dairy cows, are lower than those previously proposed by the ARC and the NRC. Although the present estimates are the best obtainable from existing knowledge a greater precision is being demanded than is warranted by the meagre data. For instance, we do not know if minimum endogenous faecal losses are proportional to the level of feeding. Furthermore, these dietary requirements refer to the average cow and by definition the standards will not be optimal for all cows. There is no evidence, however, from practical feeding trials that the proposed standards are too low.

Table 5.20   Comparison of estimates of the phosphorus requirements (g/day) of growing cattle.

Live weight (kg)	Daily weight gain (kg)	ARC (1965)	NRC (1971)*	ARC (this edition)
55	0.40	—	3.5	4.4
100	0.75	10	8.4	8.2
200	0.75	12	14	9.5
300	0.75	17	18	11
400	0.75	26	20	18
500	0.60	30	21	18
600	0.15	—	18	14

* Growing heifers (large breeds)

Table 5.21   Comparison of estimates of the phosphorus requirements (g/day) of pregnant cattle in last 2 months of gestation.

Live weight (kg)	ARC (1965)	NRC (1971)	ARC (this edition)
400	27	18	15
500	33	22	17
600	41	26	20

Table 5.22   Comparison of estimates of the phosphorus requirements (g/day) of lactating cows.

Milk yield (kg)	Jersey (400 kg)			Ayrshire (500 kg)			Friesian (600 kg)		
	ARC (1965)	NRC (1971)	ARC (this edition)	ARC (1965)	NRC (1971)	ARC (this edition)	ARC (1965)	NRC (1971)	ARC (this edition)
10	36	35	29	43	35	27	50	36	28
20	53	57	50	60	55	43	67	55	44
30	71	79	72	77	75	59	85	74	59
40	88	101	94	95	95	75	102	93	75

(b) Sheep

The origin of the difference in the two sets of ARC allowances for sheep, like cattle, lies mainly in the choice of value used to calculate minimum net endogenous faecal loss (Tables 5.23, 5.24). The present edition has adopted a value of 14 mg per kg live weight per day, much lower than the one of 42 mg used in the previous edition. Thus the magnitude of the differences between the two sets of recommendations will depend on the relative contribution of maintenance to total dietary requirements. Consequently, differences are minimal for young growing animals and maximal for pregnant and lactating sheep.

The NRC recommendations again occupy an intermediate position between the two sets of ARC recommendations for growing and pregnant sheep. No comparison with the NRC recommendations for lactating ewes is possible, because the milk yields adopted by this Committee are not given, nor the reasons why these observations are some 70% higher than the previous ones (NRC 1968).

Table 5.23   Comparison of estimates of the phosphorus requirements (g/day) of growing sheep.

Live weight (kg)	Daily weight gain (kg)	ARC (1965)	NRC (1968)	ARC (this edition)
30	0.200	2.6	3.0	2.1
35	0.220	2.9	3.0	2.3
40	0.250	3.3	3.1	2.6
45	0.250	3.6	3.1	2.7
50	0.220	3.6	3.1	2.6
55	0.200	3.8	3.1	2.6

Table 5.24   Comparison of estimates of the phosphorus requirements (g/day) of pregnant and lactating ewes.

Live weight (kg)	ARC (1965)	NRC (1975)		ARC (this edition)
Non-lactating and first 15 weeks of gestation				
50	3.6	2.8		1.4
60	4.3	2.9		1.7
70	5.0	3.0		1.9
80	5.8	3.1		2.2
Last 6 weeks of pregnancy				
50	5.0[a]	3.9		2.0[b]
60	5.7	4.1		2.2
70	6.4	4.3		2.5
80	7.2	4.5		2.8
First 8–10 weeks of lactation[c]				
50	7.0	7.8[e]	4.6[e]	4.1
60	7.7	8.2	4.8	4.2
70	8.4	8.6	5.0	4.5
80	9.2	9.0	5.2	4.8

(a) Lamb weighing 5.9 kg.
(b) Lamb weighing 4.0 kg.
(c) For ewes giving 1.36 kg milk/day.
(d) No milk yield given.
(e) NRC (1968) requirements.

5. Comparison of the estimates of phosphorus requirements with the results of practical trials

Deficiency of phosphorus in ruminants is characterized by a low concentration of inorganic phosphate in blood, poor mineralization and growth of the skeleton and anorexia or a depressed appetite, the latter leading to an intensification of any weight loss or depression in milk yield which can be attributed to phosphorus deficiency. Opinions differ on which of these signs is the most sensitive to phosphorus deficiency. Burroughs et al (1956) and Tillman et al (1959) have argued that requirements for maintenance of normal appetite and growth are greater than for bone growth or maintenance of normal plasma inorganic phosphorus concentrations. Field et al (1975) found that the appetite of growing lambs was affected within 1 week of the introduction of the low-phosphorus diet, long before any effect on plasma concentration of inorganic phosphate was apparent. On the other hand, Van Landingham et al (1935) considered that the fall of inorganic phosphorus in blood precedes any physical signs such as loss of appetite or stiffness in front and rear quarters. In face of these contradictions a diet has been deemed to be adequate if growth and concentrations of inorganic phosphorus in plasma or blood were not significantly different from those of the adequately fed group.

(a) Cattle

Most of the practical trials relate to young growing animals. The rate of weight gain of most animals on trial (ARC 1965) fell within the range 0.5–0.6 kg/day and for this rate of gain the dietary requirements adopted in the first and in the present edition are similar and deviate markedly only for live weights greater than 200–300 kg, the point of departure depending on the time taken for the animal to reach 300 kg. A reappraisal of the practical trials with growing cattle listed in the first edition, made by using a higher growth rate to calculate predicted P requirements, points to a small and possible overestimate in the present recommendations.

For more mature growing cattle the present recommendations are much lower than the previous ones and agree more closely with results of practical trials with this class of animal (Otto 1938, Archibald & Bennett 1935, Van Landingham et al 1935).

Huffman et al (1933) and Van Landingham et al (1936) have assessed the phosphorus requirements of pregnant heifers up to calving at 33 and 26 mg/kg live weight per day, respectively, with the latter figure thought to be the minimum. This contrasts with the present standards of 23 to 57 mg per kg per day according to stage of gestation.

There appear to be few practical trials designed specifically to determine the phosphorus requirements of milking cows. Those which are available provide no evidence that the present standards are too low. Huffman et al (1933) considered that an allowance of 1.65 g P/kg of milk above maintenance would provide sufficient to meet the needs of reproduction and of lactation at any rate of milk production, concentration of phosphorus in milk or Ca:P ratio of the diet. These allowances are nearly identical to those at present proposed for Ayrshire and Friesian cows. Riddell et al (1934) found that an intake of 40 g P/day was adequate for a dairy cow yielding 10 kg milk/day, whereas 13.7 and 14.2 g were not; the present estimate would be 28 g/day. Van Landingham et al (1936) suggested that an allowance less than that of Huffman et al (1933) would maintain normal concentrations of inorganic phosphorus in blood of dairy cows. More recently, Coppock et al (1976) could find no sign of deficiency when dairy cows yielding 10 to 23 kg milk/day were given for 11 weeks diets containing about 30 g P/day which is less than the proposed requirements given here for comparable cows.

Numerous balance trials with phosphorus have been done in the past but interpretation of the balance data in terms of dietary adequacy is difficult, not only because of the known bias in many of these trials (Duncan, 1958), but more importantly, because a negative phosphorus balance can be caused, not necessarily by a dietary deficiency of phosphorus, but by a concomitant negative calcium balance.

It must be concluded that there is no way at present of testing the general applicability of the estimates of phosphorus requirements of dairy cows other than by practical experience accruing from their use.

(b) Sheep

The new estimates, although less than the early estimates, particularly for the heavier sheep, are at least as good a fit with the data from practical trials with growing and nonpregnant mature sheep (ARC 1965) as the earlier recommendations. As to the requirements of pregnant sheep, a recent study by Sykes & Dingwall (1976), using constancy of body phosphorus and normality in the concentration of inorganic phosphate in plasma as the criteria of dietary adequacy, has provided estimates of dietary phosphorus requirements for the various stages of pregnancy which are in close agreement with the present standards. Unfortunately, there are no suitable practical trials on which to assess dietary requirements for lactation, as those trials which cover lactation also include pregnancy.

The Ratio of Calcium to Phosphorus in Ruminant Diets

In our assessment of the calcium and phosphorus requirements for each class of ruminants, the elements were considered independently. There is, however, a long-held view that the ratio of Ca to P in the diet is likely to be important for the proper utilization of these two elements. Certainly the Ca:P ratio is important in diets deficient in phosphorus (Theiler 1931, Lewis et al 1951, Young et al 1966, Field et al 1975), but there are few indications of adverse effects on animal performance with the ratios commonly found in practical diets. The studies of Dowe et al (1957), Wise et al (1963) and Ricketts et al (1970) have provided evidence that a high ratio of Ca:P in the diet can reduce growth rate, probably through its depression of the digestibility of the diet (Dowe et al 1957, Colovos et al 1958), but the actual ratio associated with this reduced growth rate varied among the experiments (8–14:1). Wise et al (1963) found depressed growth rates with diets having ratios less than 1:1, but this observation has not been confirmed. These findings do not eliminate the possibility that certain ratios of Ca:P in the diet may increase dietary phosphorus requirements by reducing its absorption. Evidence on this point is conflicting, but diets varying in ratio of Ca:P from 1:1 to 10:1 were without effect on efficiency of absorption of dietary phosphorus by sheep (Lueker & Lofgreen 1961, Compère et al 1965, Young et al 1966). In contrast Manston (1966) claimed that the absorption of calcium and phosphorus by pregnant heifers was better from diets with a Ca:P ratio of 2:1 than with 1:1. Some workers have suggested that a high ratio is associated with infertility in cattle (Webster 1932, Hignett 1959), but attempts to confirm this theory in controlled conditions have generally been unsuccessful (Littlejohn & Lewis 1960).

Thus it is not possible to state with certainty which is the optimal ratio of Ca:P for animal performance or whether such a ratio actually exists. However, the ratio of the elements in all recommended allowances falls within the generally accepted safe range of 1:1 to 2:1.

Magnesium

1. The endogenous losses of magnesium from the body

Estimates of the endogenous faecal loss of magnesium have been made by three separate procedures:

(i) extrapolation to zero magnesium intake of the regression of faecal loss of magnesium on magnesium intake for natural diets;
(ii) measurement of faecal excretion of magnesium on artificial diets extremely low in magnesium;
(iii) isotope dilution methods.

Values obtained by the first of these procedures are subject to gross error (see first edition) and have been rejected as unreliable. Other values are summarized in Table 5.25.

With the second procedure there are two possible sources of error. Firstly, the faecal magnesium may include small amounts of unabsorbed dietary magnesium. This would be important only in work with lactating animals for which, if the energy requirements for lactation are met, the diet unavoidably contains significant amounts'

Table 5.25.   Estimates of the endogenous faecal loss of magnesium.

Animal	No.	Age	Live weight (kg)	Diet	Intake of magnesium (mg/day)	Endogenous faecal loss of magnesium (mg/day)	Endogenous faecal loss of magnesium (mg/kg live weight per day)	Method	Reference
Cattle									
Calf	1	3–8 weeks	30–35	Low-magnesium, artificial milk diet	24, 19	76, 60	2.06	Low-magnesium diet	Blaxter & Rook (1954)
Calf	3, 1, 4	2–5 weeks; 9–11 weeks; 24–32 weeks	46, 50, 128	Low-magnesium artificial milk diet	30, 33, 55	29, 61, 315	0.63, 1.22, 2.46	Low-magnesium diet	Smith (1959)
Calf	1	—	25.5, 27.3	Whole milk	300, 375	89, 95	3.5	Isotope dilution	Simesen et al (1962)—recalculated
Dry cow	1	Mature	500†	Low-magnesium, artificial diet	500	1080, 930	2.01	Low-magnesium diet	Storry & Rook (1963)
Lactating cow	4	Mature	500†	Low-magnesium, artificial diet	3000	3000	6.00	Low-magnesium diet	Rook, et al (1964)
Lactating cow	—	Mature	—	Low-magnesium artificial diet	—	1500–2000	3.0–5.0	Low-magnesium diet	Blaxter & McGill (1956)
Lactating cow	1, 1	Mature	478, 558	Lucerne, hay and grain	—	696, 876	1.5	Isotope dilution	Simesen et al (1962)—recalculated
Sheep									
North country Cheviot wether	1, 1	1½ years	59, 60	Fresh herbage	—	100, 250	3.35	Extrapolation of rectilinear regression of urinary magnesium on dietary magnesium	Field et al (1958)
North country Cheviot wether	1	—	60†	Hay	—	205	3.4	Comparative balance with ^{28}Mg	Field (1959)
Mixed breeds	1, 1, 1	1½ years	43.0, 52.5, 41.4, 49.5	Lucerne pellets / grass diet	—	53, 59, 29, 52	1.23, 1.12, 0.70, 1.05	Isotope dilution	Hjerpe (1968a)
Kerry Hill ram	1, 1	20 months	(45), (75)	Hay	—	227, 156	5.04, 2.08	Isotope dilution	Care (1960)
Crossbred	4	1 year	40	Semipurified diet	1700	120	2.9	Isotope dilution	House & Van Campen (1971)
—	1	1½–2 years	21.4, 48.6	Low-magnesium artificial diet	—	100	2.7	Low magnesium diet	Hjerpe (1968b)
North country Cheviot wether	1, 1	4–5 years	59, 75, 66, 66	Dried grass	—	75, 33, 185, 115	1.27, 0.44, 1.29, 1.69	Values for faecal excretion at dietary magnesium intake sufficient for maintenance	Field (1962)
Suffolk × Masham wethers	8	Mature	50†)	Low-magnesium artificial diet	25	140‡	2.8	Based on low-magnesium diet with varying levels of intravenous magnesium, regression of endogenous faecal loss on serum magnesium concentration ^{28}Mg	Allsop & Rook (1972)
Lozere non-pregnant ewes	6		(47)	Dried grass; Frozen grass	—, —	—, —	3.2, 3.2	^{28}Mg	Larvor (1976)

† Assumed value.   ‡ Calculated for a serum magnesium concentration of 25 mg/litre.

of magnesium. Secondly, the net endogenous faecal loss of magnesium may be reduced in animals receiving a diet deficient in magnesium. This source of error has been examined in sheep by Allsop & Rook (1972) and has been shown to be important. Their results permit the calculation of net endogenous faecal loss in normal magnesium status. At a serum magnesium concentration of 25 mg/litre, the calculated value was 2.8 mg/kg live weight per day. This value is of the same order as those obtained for both sheep and cattle by other workers using the same procedure, with the exception of the distinctly lower values for young calves reported by Smith (1959).

Isotope dilution methods have been widely used, and though the validity of the techniques may be in doubt (Field 1961), the values are nevertheless broadly similar to those obtained by the method based on artificial diets low in magnesium.

The value of 3.0 mg/kg live weight per day recommended in the first edition for mature cattle has therefore been adopted for both cattle and sheep, with the exception that, for calves of up to 100 kg receiving a milk diet, a lower value of 2.0 mg/kg live weight has been assumed.

The urinary loss of magnesium is negligible in animals receiving artificial diets low in magnesium and the urinary endogenous loss of magnesium has therefore been ignored.

2. The coefficient of absorption of dietary magnesium

A few direct measurements of the coefficient of absorption of dietary magnesium have been made by tracer techniques and the results are presented in Table 5.26. The values for hay give a mean of 0.255.

Table 5.26

Measurements of the coefficient of absorption of the magnesium of feeds made by tracer techniques.

Food	Animal	True 'availability'	Reference
Hay	Sheep	0.258	Field (1959)
		0.263	Field (1959)
		0.233	Macdonald et al (1959)
		0.239	Care (1960)
		0.116	Care & Ross (1962)
		0.301	Hjerpe (1968a)
		0.322	Hjerpe (1968a)
Grass	Sheep	0.100	Care & Ross (1962)
		0.250	Hjerpe (1968a)
		0.290–	Hjerpe (1968a)
Hay and grass	Milking cow	0.169	Simesen et al (1962)
		0.373	
Milk	2-month-old calf	0.445	Simesen et al (1962)
		0.586	Simesen et al (1962)

Extensive information on the faecal loss of magnesium in individual animals is available from metabolic trials, and from this information and an assumed value of 3 mg/kg live weight per day for the endogenous faecal loss, the coefficient of absorption for dietary magnesium has been calculated. The results for three categories of feed, mixed (forage and concentrate) diets, hay and grass, are summarized in Table 5.27.

The values in all categories are extremely variable. This may, in part, reflect differences between animals in ability to absorb magnesium, as Field & Suttle (1970) have reported a significantly greater variation between than within monozygotic cattle twins. The observations of Rook & Campling (1962), however, suggest that differences between individuals are not consistent from diet to diet.

Mixed diets and hay

A wide range of dietary constituents and composition has been investigated, but there is little firm indication of the basis of observed differences in the coefficient of

Table 5.27 Calculated values for the coefficient of absorption of the magnesium of feeds

Feed	Animal	No. of measurements	Mean daily intake of magnesium (g)	Mean value for the calculated coefficient of absorption of dietary magnesium	Reference	
Forage/concentrate (excluding silage)	Lactating cow	21	24.2	0.308	Rook et al	(1958)
		8	32.9	0.358	Monroe & Perkins	(1925)
		8	44.8	0.284	Monroe	(1924)
		3	20.3	0.287	Forbes et al	(1916)
		12	31.6	0.199	Forbes et al	(1918)
		3	37.2	0.349	Van der Horst	(1960)
			Group mean value	0.289		
Forage/concentrate (including silage)	Lactating cow	10	21.2	0.301	Rook et al	(1958)
		4	33.1	0.257	Kemp et al	(1961)
		15	26.9	0.184	Forbes et al	(1916)
		6	34.5	0.251	Miller et al	(1925)
		6	34.1	0.318	Miller et al	(1924)
		12	32.1	0.231	Forbes et al	(1917)
			Group mean value	0.245		
Forage/concentrate	Non-lactating cow	8	12.0	0.305	Rook & Campling	(1962)
		2	5.5	0.477	Lomba et al	(1970)
	Castrated male cattle	10	11.3	0.445	Forbes et al	(1929)
	Male calf (7–13 weeks)	8	3.0	0.224	Raven & Robinson	(1964)
			Group mean value	0.344		
	Sheep (lactating ewe)	4	1.8	0.399	L'Estrange & Axford	(1966)
	Sheep (castrated male)	4	2.0	0.411	L'Estrange et al	(1967)
			Group mean value	0.405		
Hay Grass	Non-lactating cow	6	7.1	0.459	Rook & Campling	(1962)
	Lactating cow	35	16.6	0.260	Kemp et al	(1961)
		6	31.9	0.205	Hutton et al	(1965)
		4	19.6	0.302	Van der Horst	(1960)
		16	12.5	0.297	Rook & Balch	(1958)
			Group mean value	0.267		
	Non-lactating cow	18	8.4	0.330	Rook & Campling	(1962)
		14	14.0	0.251	Lomba et al	(1970)
			Group mean value	0.296		
	Sheep (lactating ewe)	7	1.4	0.224	L'Estrange & Axford	(1966)
	Sheep (castrated male)	{ 12	1.7	0.215	L'Estrange et al	(1967)
		8	1.7	0.258	Stillings et al	(1964)
			Group mean value	0.230		

Weighted mean value, with S.D.　　0.294 ± 0.135

absorption of dietary magnesium. The most extensive values are for lactating cows. Of the various dietary comparisons available, only the inclusion of silage (0.245 against 0.289) had a significant ($P<0.01$) effect. The results of Rook et al (1958) suggest an inverse relationship with dietary protein content but the other results do not support this. The overall mean value for lactating cows is 0.267.

The values of Rook & Campling (1962) for non-lactating cows, with the exception of those for hay, and those of Raven & Robinson (1964) for male calves are of the same order as the values for lactating cows, but the values of Lomba et al (1970) for non-lactating cows, of Forbes et al (1929) for castrated male cattle, of Rook & Campling (1962) for non-lactating cows receiving hay and the various values for sheep are all considerably higher. The magnesium intakes with the hay diets of Rook & Campling (1962) and in the investigations of Lomba et al (1970) and Forbes et al (1929) are comparatively low and the calculated coefficient of absorption is particularly sensitive to the assumed figure for faecal endogenous loss. The values may be anomalous. There is, however, no similar basis for the rejection of the values for sheep, though they are inconsistent with those determined by isotope dilution methods.

Grass

Four sets of results are available for lactating cows. The mean values for the results of Hutton et al (1965) and Kemp et al (1961) differ significantly ($P<0.05$) from each other and from those of Rook & Balch (1958) and Van der Horst (1960). The comparison of early- and late-cut grass within the results of Rook & Balch (1958) shows no significant effect of stage of maturity but the sward used by Hutton et al (1965) contained a high proportion of clover and this may account for the lower values.

The values for non-lactating cows are on average slightly higher than those for lactating cows. The few values for sheep are, however, significantly ($P<0.05$) lower. Two of the sets of results for sheep compare different rates of nitrogenous fertilizer application: the application of 336 as compared with 57 kg N/ha as ammonium nitrate gave a highly significant ($P<0.01$) reduction in the coefficient of absorption of magnesium (0.219 against 0.297) (Stillings et al 1964), whereas the application of 98 as compared with 18 kg N/ha as nitrochalk had no significant effect (0.235 and 0.195) (L'Estrange & Axford 1966). There is conflicting evidence from grazing trials as to the importance of a high dietary concentration of potassium in relation to the development of hypomagnesaemia, but dietary additions of potassium salts have been shown to depress magnesium absorption in sheep (Suttle & Field 1967, House & Van Campen 1971, Newton et al 1972).

The group mean values for cows and for sheep are 0.277 and 0.230, respectively; the value for cows does not differ significantly from that for mixed diets but for sheep the value is significantly ($P<0.01$) lower.

Milk diets

In calves and lambs receiving milk diets, the coefficient of absorption of dietary magnesium is high in the very young animal but falls away rapidly with age, especially when animals are unmuzzled and thus able to consume fibrous bedding (Smith 1961). Specific values calculated by Smith (1957) are 0.87 for animals 2 to 3 weeks of age, reducing in unmuzzled animals to 0.32 at 7 to 8 weeks and to 0.12 at 13 to 14 weeks, but other values (Smith 1958) suggest a less extreme variation.

Recommended values for the coefficient of absorption

The wide variation in the estimated values makes the prediction of the coefficient of absorption of magnesium an uncertain procedure. The distinctly high figure obtained with very young calves receiving milk diets justifies the use of separate values for milk-fed calves (and lambs) and it is proposed that values of 0.70 for a 50-kg calf (or a 5-kg lamb), 0.30 for a 75-kg calf (or a 7.5-kg lamb) and 0.20 for a 100-kg calf be adopted.

For other diets and classes of stock, there is insufficient information to permit reliable categorization. There is a widely held view (based largely on measurement of the coefficient of apparent absorption) that grasses and forages as a group show unusually low values for the coefficient of absorption of magnesium, but this is not supported by the calculated values presented here for which a constant endogenous loss was assumed. This does not exclude the possibility that for certain swards, particularly those grazed early in the spring and receiving heavy dressings of nitrogenous or potassic fertilizers, the coefficient of absorption of magnesium may be especially low. For the calculation of average requirements, therefore, the overall mean value for the data presented in Table 5.27, of 0.294, is proposed but for the calculation of allowances which provide a margin of safety, the lower decile value of 0.17 is recommended.

3. Requirements and recommended dietary allowances for magnesium

Estimates of average requirement and recommended dietary allowances for maintenance, growth, pregnancy and lactation, based on the above values for absorption, are given for cattle in Tables 5.28 to 5.31 and for sheep in Tables 5.32 to 5.34. Separate estimates of requirement are given for calves receiving whole milk diets. The relative requirements for the two species are similar, the major net requirements being those for lactation and maintenance. The higher magnesium concentration in the milk of the ewe as compared with that of the cow is offset by a relatively lower milk volume.

Table 5.28   Dietary requirement (g/day) for magnesium of calves receiving whole milk diets and growing at four different rates.

Live weight (kg)	Rate of gain (kg/day)			
	0	0.25	0.50	1.00
50	0.15	0.30	0.45	0.75
75	0.50	0.85	1.20	1.90
100	1.00	1.50	2.05	3.05

Table 5.29   Average dietary requirements for magnesium of cattle growing at four different rates with, in parenthesis, recommended dietary allowances (g/day).

Live weight (kg)	Rate of gain (kg/day)			
	0	0.50	1.00	1.50
100	1.0 (1.8)	1.7 (3.0)	2.4 (4.2)	3.1 (5.4)
200	2.0 (3.5)	2.8 (4.8)	3.4 (6.0)	4.1 (7.2)
300	3.1 (5.3)	3.8 (6.5)	4.5 (7.8)	5.2 (8.9)
400	4.1 (7.1)	4.8 (8.3)	5.5 (9.5)	6.2 (10.7)
500	5.1 (8.8)	5.8 (10.1)	6.5 (11.3)	7.2 (12.5)
600	6.1 (10.6)	6.8 (11.8)	7.5 (13.1)	8.2 (14.2)

Table 5.30 Average dietary requirements for magnesium of pregnant cows with, in parenthesis, recommended dietary allowances (g/day).

Pregnancy	Week of pregnancy					
	20	24	28	32	36	40
1st calf	5.9 (10.2)	6.1 (10.5)	6.4 (11.1)	6.8 (11.9)	7.6 (13.2)	8.6 (14.9)
2nd or later calf	6.2 (10.8)	6.4 (11.1)	6.7 (11.7)	7.2 (12.5)	8.0 (13.8)	8.9 (15.5)

Table 5.31 Average dietary requirements for magnesium of lactating cows with, in parenthesis, recommended dietary allowances (g/day).

Breed	Milk yield (kg/day)			
	10	20	30	40
Jersey	8.3 (14.4)	12.6 (21.8)	16.8 (29.1)	
Ayrshire	9.4 (16.2)	13.6 (23.5)	17.9 (30.9)	22.1 (38.2)
Friesian	10.4 (17.9)	14.6 (25.3)	18.9 (32.6)	23.1 (40.0)

Table 5.32 Average dietary requirements for magnesium of sheep gaining weight at different rates with, in parenthesis, recommended dietary allowances (g/day).

Live weight (kg)	Rate of gain (kg/day)				
	0	0.1	0.2	0.3	0.4
5	0.05 (0.09)	0.18 (0.32)	0.31 (0.54)	0.44 (0.76)	0.56 (0.98)
10	0.11 (0.19)	0.24 (0.42)	0.37 (0.64)	0.50 (0.86)	0.62 (1.08)
20	0.22 (0.38)	0.35 (0.61)	0.48 (0.82)	0.61 (1.05)	0.73 (1.26)
40	0.45 (0.77)	0.57 (0.99)	0.70 (1.21)	0.83 (1.44)	0.96 (1.65)
60	0.61 (1.15)	0.80 (1.38)	0.92 (1.59)	1.05 (1.82)	1.18 (2.04)

Table 5.33 Average dietary requirements for magnesium of ewes during pregnancy with, in parenthesis, recommended dietary allowances (g/day).

Live weight (kg)	Weeks of pregnancy			
	9	13	17	21
40	0.41 (0.71)	0.52 (0.89)	0.64 (1.11)	0.77 (1.33)
75	0.77 (1.32)	0.87 (1.51)	1.00 (1.72)	1.13 (1.90)

Table 5.34 Average dietary requirements for magnesium of lactating ewes with, in parenthesis, recommended dietary allowances (g/day).

Live weight (kg)	Milk yield (kg/day)		
	1.0	2.0	3.0
40	0.99 (1.71)	1.56 (2.71)	2.14 (3.71)
75	1.34 (2.32)	1.92 (3.32)	2.50 (4.32)

4. Comparison of the present estimates with those of the first edition and of the National Research Council

The primary difference between the present calculations of estimates and those of the first edition is the use of a coefficient of absorption of 0.294 instead of 0.20 for the magnesium of diets other than milk diets offered to calves or lambs. Additionally, the allowances for foetal growth in both cattle and sheep are slightly higher and calculations of body accretion of magnesium in the growing animal are based on empty body gain instead of live weight gain. The overall effect is a reduction in the estimated requirements, but in the calculation of recommended allowances, whereas in the first edition it was proposed that the estimates of requirement for mature cattle should be increased by 2 g magnesium/day, the present proposal is that a coefficient of absorption of 0.17 should be used and this results in values which are equal to or higher than those of the first edition.

The National Research Council (1970 1971) recommendations are imprecise. They quote values of 0.06% of magnesium in the diet of young calves increasing to 0.20 in the diet of lactating cows, and for beef cattle 12–30 mg/kg live weight. The present estimates fall within the range of values for beef cattle and are less than those recommended for lactating cows, but the present recommended allowances appear to cover their requirements. The values recommended for young calves are similar to the present ones.

5. Comparison of the estimates of magnesium requirements with the results of practical trials

Magnesium absorbed in excess of requirements is excreted in the urine and the renal threshold for magnesium is a serum concentration not greater than 18–20 mg/litre. Serum values in excess of this may therefore be taken as an index of adequate magnesium nutrition; though young animals subjected to magnesium deficiency tend to deplete their bones of magnesium and the development of hypomagnesaemia protracted, serum magnesium concentration is still an index of adequacy.

There are several estimates from practical trials of the dietary requirements of the milk-fed calf for the maintenance of a normal serum magnesium concentration. Huffman et al (1941) estimated dietary requirement as 12–15 mg magnesium/kg body weight for animals of 100–400 kg weight receiving a whole milk diet supplemented with maize, maize gluten or lucerne hay. The estimated requirement was higher at 30–40 mg/kg for animals receiving an unsupplemented diet, although the original data suggest that the lower figure would be adequate for young (50–100 kg live weight) calves. For calves gaining 0.4 kg per day up to 100 kg the present estimate would be 8–19 mg/kg whereas for heavier animals gaining 0.75 kg/day the estimate of requirement would be 13–21 mg/kg and the recommended allowance 25–38 mg/kg. Blaxter et al (1954) found that with calves receiving an artificial milk diet, a dietary intake of 0.75 g was adequate for calves over the weight range of 40–60 kg (a higher intake of about 1.0 g/day was necessary to maintain a high retention of magnesium in bone (Blaxter & Rook 1954)). The recorded rate of gain was 0.56 kg/day and the present estimate would therefore range from 0.4 g/day at the lower weight to 0.9 g/day at the higher weight. With suckled beef calves growing at an estimated 0.7 kg/day, an assumed intake of 9 litres/day of milk (equivalent to about 1.1 g magnesium/day) failed to maintain normal serum magnesium concentrations in young animals on some farms, whereas with a supplement of 0.5 g magnesium/day as magnesium carbonate normal serum magnesium concentrations were maintained at least until 4 weeks of age (Blaxter & Sharman 1955). For calves weighing 40 kg at birth, the present estimate of requirement at 4 weeks of age would be 1.0 g/day. When allowance is made for the assumptions used, the estimates of requirement for the young milk-fed calf and the recommended allowances for animals of more than 100 kg live weight are in reasonable agreement with the results of practical trials.

For growing cattle receiving dry diets, the only experimental assessment available is that of Ray (1942) for heifers weighing 250 kg, growing at 0.5 kg/day and receiving a diet of straw and concentrates. On the basis of maintenance of body magnesium, the dietary requirement was calculated to be 8 g/day. The present estimate is 3.3 g magnesium/day and the recommended allowance 5.7 g/day.

In balance trials with bullocks weighing 500 kg and receiving diets which maintained body weight, Forbes et al (1929) found that with lucerne hay supplying about 16 g magnesium/day and with maize meal/lucerne hay diets supplying about 8 g magnesium/day, magnesium retention varied from −0.14 to 1.0 g/day. The present estimate of requirement for maintenance of a 500-kg animal is 5.1 g/day and the recommended allowance is 8.8 g/day.

Fig. 5.3   Comparison of dietary intake of magnesium in lactating cows with the present estimates of requirement and recommended allowances in relation to the occurrence of hypomagnesaemia. (● – animals in which the serum magnesium concentration remained above 18 mg/litre; O – animals in which the serum magnesium concentration fell to <18 mg/litre: _____present estimate of requirement for a 600 kg cow; .... present recommended allowance for a 600 kg cow.)

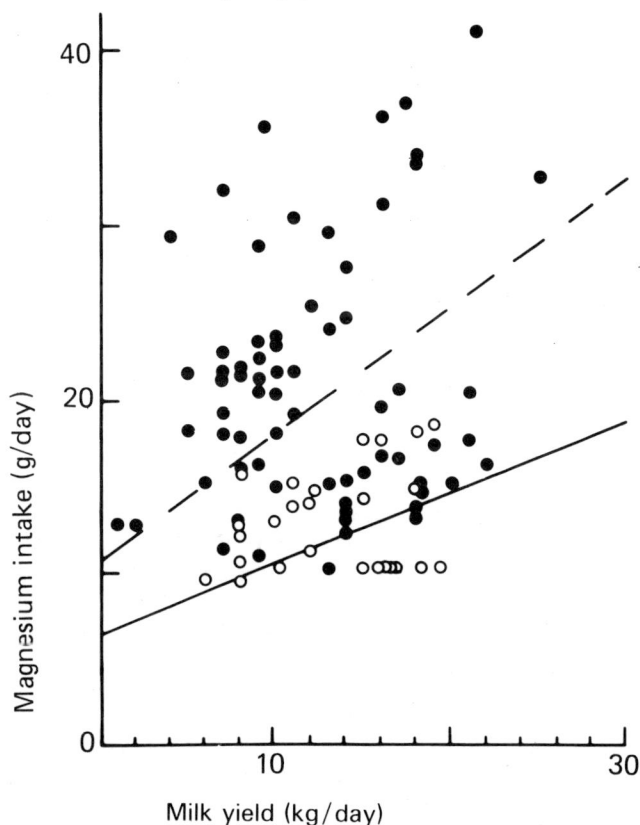

Considerably more information is available for lactating cows (Breirem et al 1949, Rook & Balch 1958, Rook et al 1958, Van der Horst 1960, Kemp et al 1961) and this has been presented in Fig. 5.3. Animals which showed hypomagnesaemia (defined as a serum magnesium concentration of less than 18 mg/litre) are identified by open circles. The lines represent the present estimates and recommended allowances for cattle of 600 kg live weight. With two exceptions, animals maintaining normal serum magnesium concentrations were receiving dietary magnesium in excess of the estimated requirements. Of those which showed hypomagnesaemia, about two-thirds were receiving a dietary intake in excess of the estimated requirement but substantially (at least 5 g/day) less than the recommended allowance. It would appear that both the estimates and the recommended allowances are of the right order.

There is little comparable information for the sheep. Evered (1961) failed to produce hypomagnesaemia in two sheep given artificial diets, one of 30 kg live weight receiving 1.4 g magnesium/day, the other of 45 kg live weight and receiving 2.0 g magnesium/day. Both of these intakes are in excess of the present recommended allowances. For sheep weighing about 40 kg and receiving a semipurified diet, Suttle & Field (1969) have reported that the intake of magnesium necessary for the maintenance of a serum magnesium concentration of 18 mg/litre was 0.5–1.0 g/day in dry ewes and 1.0–1.5 g/day in lactating ewes when the intake of potassium was low, and that these values were higher by about 0.5 g/day when additional potassium was given. The estimated requirements for dry and lactating ewes, respectively depending on the values assumed for weight gain and milk yield, are 0.5–1.0 and 1.0–1.6, and the recommended allowances 0.8–1.7 and 1.7–2.7 g magnesium/day. Of four lactating ewes receiving all-grass diets and between 1.1 and 1.6 g magnesium/day only one when receiving 1.1 g/day showed hypomagnesaemia; the estimated requirement is 0.8 g/day and the recommended allowance 1.4 g/day.

6. Supplementation

The recommended values for the coefficient of absorption of magnesium have been derived for natural diets, and do not necessarily apply to supplementary magnesium offered in mineral form. An absorption of more than 0.50 has been reported for magnesium as carbonate, oxide or sulphate added to an artificial diet offered to sheep but for commercial magnesium carbonate the value was about 0.10 (Ammerman et al 1972). For sheep receiving a diet of hay, molassed sugar-beet pulp and urea values of 0.44–0.49 were obtained for supplements of magnesium oxide of reagent grade (Fishwick & Hemingway 1973). In bullocks receiving an artificial diet, the coefficient of absorption of the magnesium of the oxide and carbonate was similar but that of dolomitic limestone much less (Moore, et al 1971). Values for two dry cows with magnesium as the sulphate, chloride, nitrate, acetate, citrate, lactate, oxide or trisilicate, varied from 0.15 to 0.50, the values for magnesium oxide, a supplement commonly used in the United Kingdom, being 0.28 and 0.36 (Storry & Rook 1963) There are, however, gross variations in the physical form of commercial magnesium carbonate (magnesite) and magnesium oxide (calcined magnesite) and both the physical form (Ammerman et al 1972), and the diet to which supplements are added may affect absorption. Fifty grammes of magnesium oxide (30 g magnesium) offered daily as a drench or dietary supplement to lactating cows grazed on spring pasture did not prevent the development of hypomagnesaemia in all animals, although serum magnesium values did not fall to concentrations at which tetany can occur (Line et al 1958). The coefficient of absorption of the magnesium of the oxide when offered to grazing cows has been estimated as on average 0.05–0.10 but it was concluded that within a group of cows absorption was poorest in those with the most marked hypomagnesaemia (Rook & Storry 1962). There is also a report that in ewes with a very low plasma magnesium concentration, drenching with 4 g magnesium per day as either oxide or nitrate did not increase the values in plasma (Ritchie & Hemingway 1963).

Whilst it is desirable that animals at risk should be given daily a mineral supplement containing magnesium in a form readily soluble in rumen liquor and in an amount sufficient to prevent the development of hypomagnesaemia and tetany in any individual animal, excessive doses should be avoided. Care (1960) has reported scouring in bullocks given daily 170 g or more calcined magnesite, and prolonged feeding with 14 g daily of calcined magnesite in yearling calves has produced stiffness and a stilted gait (Downey 1960). Excessive amounts of magnesium salts in cereal-based diets have been associated with the formation of magnesium-containing calculi in the urethra of male lambs which have caused obstruction and subsequent death from ruptured bladder (Ørskov 1975).

Potassium

1. The loss of potassium through the skin of cattle and in saliva

The loss of potassium through the skin has been estimated by Aitken (1976). For cattle in temperate conditions, a loss of 0.1 g/day has been calculated for a 200-kg male and of 0.2 g/day for a 500-kg female. The corresponding figures for European cattle in tropical conditions (assuming exposure for 7 hours each day to a temperature of 40°C at a relative humidity of 90%) are 1.1 and 1.8 g/day.

In most conditions the loss of potassium in dribbled saliva is negligible but unacclimatized cattle in tropical conditions receiving a diet adequate in sodium may lose up to 0.7 g K/100 kg per day; with diets low in sodium this loss may be increased twentyfold (Aitken 1976).

2. Inevitable losses of potassium in the faeces and urine

There are no published experiments in which sheep or cattle have been given very small amounts of potassium in an attempt to estimate minimal losses. However, the concept of an inevitable faecal loss of potassium strictly of endogenous origin is not applicable as there is substantial interchange between the potassium in the gut and that in the body. Moreover, the endocrine regulation of renal excretion and the time lag in adjustment to a change in nutritional status (St. Omer & Roberts 1967) make it difficult to interpret nutritional studies and unlikely that measurements at one level of potassium nutrition can be applied to another.

Estimates have therefore been made for the inevitable losses of potassium in faeces and urine over the normal range of dietary intake.

Faecal losses

In cattle (Paquay et al 1969b), faecal losses are highly variable and poorly correlated with potassium intake (r, 0.364; $P < 0.01$) over the range of 25–250 g/day, even though at the lower intakes retention of potassium may be negative. Faecal losses of potassium are more closely correlated with output of faecal dry matter (r, 0.635; $P < 0.01$) and with dry-matter intake (r, 0.541; $P < 0.01$); the correlation with output of faecal water has not been determined. As information on dry-matter intake is usually available, it has been selected as the basis for prediction of faecal loss of potassium. The following estimates (g potassium/kg dry-matter intake) have been derived: 2.3 ± 0.28 (Forbes et al 1916); 2.4 ± 0.22 (Forbes et al 1917); 1.6 ± 0.07 (Forbes et al 1918); 1.2 ± 0.14 (Forbes et al 1922); 3.1 ± 0.23 (Paquay et al 1969b); 3.4 ± 0.27 (Rook & Balch 1962). The overall weighted mean is 2.6 ± 0.12. The value appears to vary with diet but insufficient information is available to categorize on that basis.

Few comparable figures are available for sheep. Faecal excretion of potassium during low, medium and high potassium intakes was respectively 0.96, 1.09 and 1.93 g/kg dry-matter intake (Campbell & Roberts 1965). Warner & Stacey (1972) reported values of 0.56–0.87 g/kg dry-matter intake. In the absence of fuller information, a value of 1.0 g/kg dry-matter intake is recommended.

Urinary losses

In the first edition, minimal urinary loss of potassium was estimated from unpublished values for urinary potassium of calves on diets deficient in potassium; the derived value was 35 mg/kg body weight. St. Omer & Roberts (1967) have since reported experiments with bullocks on a standard diet receiving different potassium intakes and, from a regression of urinary potassium on potassium balance, their results give a urinary output of potassium at zero potassium balance of 37.5 mg/kg body weight. Because of the high content of potassium in sheep wool it is not possible to make a similar estimate for sheep and the same value as for cattle has been assumed.

3. Requirements for potassium and their agreement with other estimates and the results of practical trials

Practical diets for cattle and sheep, with the possible exception of all-concentrate diets, are invariably rich in potassium and provide potassium in excess of need. Detailed estimates of requirement are not therefore presented but some examples are given in Table 5.35, and where possible the values are compared with those of the first edition. The values used for body composition are those for empty-body gain but they have been assumed to apply to live weight gain, which has been adopted as the basis for the calculation of requirement. The requirements for potassium are determined mainly by the faecal loss, dependent on dry-matter intake, and the inevitable urinary loss, dependent on body size. In comparison the requirements for productive processes are small, except at high rates of milk production.

From feeding trials, the maintenance requirement for a 250-kg heifer receiving 2.57 kg dry matter/day has been estimated to be 13.0 g (St. Omer & Roberts 1967); the value calculated from the present estimates is 16.2 g. For bullocks growing at a rate of 0.9 kg/day and receiving 6.3 kg feed dry matter, a potassium concentration of 5–6 g/kg of the dietary dry matter was required (Roberts & St. Omer 1965); for a live weight of 250 kg, the present estimate would be 4.4 g/kg of dietary dry matter. Du Toit et al (1934) recommended a minimum dietary potassium concentration of 3.2 g/kg for cows producing 9.1 kg milk/day, whereas Pradhan & Hemken (1969) with cattle giving on average 12 kg milk/day, found dietary concentrations of 0.6 and 1.5 g/kg to be inadequate but one of 8 g/kg to be wholly adequate. For a dry-matter intake of 12 kg/day, the corresponding figure calculated from the present estimates is approximately 5 g/kg. The National Research Council (1971) recommends for dairy cows a dietary concentration between 5 and 8 g/kg.

In two separate trials with growing sheep (Telle et al 1964, Campbell & Roberts 1965), the dietary requirement was found to be 3–5 g/kg of the dry matter; present estimates give a value of about 3 g/kg.

Table 5.35   Examples of dietary requirements for potassium of cattle and sheep and comparison with those of the first edition

Animal	Assumed dry-matter intake (kg/day)	Faecal loss (g/day)	Inevitable loss in urine and through skin, saliva, etc. (g/day)	Production requirement (g/day)	Total dietary requirement (g/day)	Estimate from first edition (g/day)
50-kg calf growing at a rate of 0.5 kg/day	0.5	1.3	1.9	1.0	4.2	3.0–4.8
250-kg bullock growing at a rate of 0.5 kg/day	5	13.0	9.5	1.0	23.5	—
600-kg cow, weeks 36–40 of pregnancy	8	20.8	22.7	2.8	46.3	—
600-kg cow, yielding 10 kg milk/day	10	26.0	22.7	15	63.7	35–53
600-kg cow, yielding 30 kg milk/day	14	36.4	22.7	45	104.1	—
40-kg sheep, growing at a rate of 200 g/day	1	1.0	1.6	0.4	3.0	—

Sodium and Chlorine

1. Losses of sodium and chlorine through the skin of cattle and in dribbled saliva

Aitken (1976) has estimated losses of sodium through the skin of cattle in temperate and in tropical (assuming exposure for 7 hours each day to a temperature of 40°C and a relative humidity of 90%) conditions. Adopting the approach of Aitken (1976) and assuming a loss of 3.0 mg chlorine/m² per h during exposure at 40°C, 90% relative humidity (Jenkinson and Mabon 1973), we have made similar estimates for chlorine. For a 200-kg castrated male, the estimates of loss are 0.2 g sodium and 0.2 g chlorine in temperate conditions and 0.8 g sodium and 1.0 g chlorine in tropical conditions. For a 500-kg cow the corresponding values are 0.3, 0.4, 1.3 and 1.6 g/day.

Losses of sodium and chlorine in dribbled saliva are negligible, except for unacclimatized cattle in tropical conditions when the loss may be as high as 1.4 g sodium (Aitken 1975) and 0.9 g chlorine/100 kg body weight.

2. Inevitable losses of sodium and chlorine in the faeces and urine

Dietary sodium and chlorine can be freely and completely absorbed but they are present in high concentrations in the digesta of the intestine. There is, however, a substantial active absorption of sodium from the lower gut (Renkema et al 1962) and presumably a concomitant transfer of chlorine as the concentrations of both sodium and chlorine are much reduced in the faeces. As with potassium, the concept of a strict endogenous faecal loss does not apply. Lomba et al (1969) and Paquay et al (1969a), however, have reported in cattle significant correlation between dietary intake and faecal output for both sodium and chlorine, and values from the literature for cattle have been used to determine the regression of faecal output on intake. For sodium (Forbes et al 1916, Forbes et al 1917, Forbes et al 1918, Forbes et al 1922, Kemp 1964, Lomba et al 1969, Rook & Balch 1962) the pooled regression was:

$$\text{Faecal Na}_{(g/day)} = 0.086 \text{ Dietary Na}_{(g/day)} + 2.88 \qquad [5.1]$$
$$(r = 0.35; \text{ P} < 0.001; \text{ RSD} = 3.20)$$

For chlorine (Forbes et al 1916, 1917, 1918, 1922) the pooled regression was:

$$\text{Faecal Cl}_{(g/day)} = 0.154 \text{ Dietary Cl}_{(g/day)} + 3.96 \qquad [5.2]$$
$$(r = 0.67; P < 0.001; \text{ RSD} = 2.76)$$

Similar information for sheep is not available but the equation for sodium derived for cattle, after adjusting the intercept for differences in body weight (assumed value for cattle, 500 kg), gave the following prediction of faecal loss for the results of Devlin & Roberts (1963):

Intake of Na (g/day)	Faecal loss of Na (g/day)	
	Actual	Predicted
0.09	0.16	0.21
1.01	0.33	0.29
2.97	0.63	0.46

Urinary losses

Very low urinary excretions of sodium and chlorine have been reported in numerous studies with cattle, and it appears that there are highly efficient renal mechanisms for conserving sodium and chlorine. In the first edition, therefore, it was assumed that

Table 5.36　Dietary requirements (g/day) for sodium and chlorine for cattle and sheep growing at different rates.

Cattle

Growth rate (kg/day)

Live weight	0		0.25		0.50		1.00		1.50	
	Sodium	Chlorine	Sodium	Chlorine	Sodium	Chlorine	Sodium	Chlorine	Sodium	Chlorine
50	0.39	0.53	0.79	0.82	1.20	1.12	2.03	1.71	2.85	2.29
100	0.75	1.05	1.16	1.34	1.57	1.64	2.40	2.18	3.22	2.81
200	1.49	2.09	1.91	2.39	2.32	2.68	3.14	3.27	3.97	3.86
300	2.24	3.28	2.65	3.44	3.07	3.73	3.89	4.32	4.71	4.91
400	2.59	4.19	3.40	4.48	3.81	4.78	4.64	5.36	5.46	5.95
500	3.74	5.24	4.15	5.53	4.56	5.82	5.39	6.41	6.21	7.00

Sheep

Growth rate (g/day)

Live weight	0		100		200		300		400	
	Sodium	Chlorine	Sodium	Chlorine	Sodium	Chlorine	Sodium	Chlorine	Sodium	Chlorine
5	0.14	0.05	0.26	0.14	0.38	0.23	0.50	0.33	0.63	0.42
10	0.28	0.09	0.40	0.19	0.53	0.28	0.55	0.38	0.77	0.47
20	0.57	0.19	0.69	0.28	0.81	0.37	0.93	0.47	1.05	0.56
30	0.85	0.28	0.97	0.37	1.09	0.47	1.21	0.56	1.33	0.66
40	1.13	0.37	1.25	0.47	1.38	0.56	1.50	0.66	1.62	0.75
50	1.42	0.46	1.54	0.56	1.66	0.66	1.78	0.76	1.90	0.84

inevitable losses of sodium and chlorine in the urine of both sheep and cattle are negligible. This appears to be a correct assumption for sodium in cattle: Morris & Gartner (1971) have shown that bullocks receiving 0.10 g sodium/day excreted 0.02 g/day in their urine and remained in approximate sodium balance. In sheep, though there may be a negligible loss of sodium in the urine of sodium-depleted animals, there is a substantial loss in normal sodium status (Blair-West et al 1970). Devlin & Roberts (1963) reported that sheep which received only 0.09 g sodium/day continued to excrete sodium in the urine (0.2 g/day) and were in negative sodium balance. At an intake of 1.01 g/day the animals were in approximate sodium balance and the urinary loss of sodium was 0.71 g/day. In calculating requirements for maintenance account has therefore to be taken of the inevitable loss of sodium in urine associated with the maintenance of normal sodium status. A value of 0.02 g sodium/kg body weight per day has been assumed. Comparable information for chlorine is not available; because of the need to maintain ionic balance, it is possible that there may be a similar inevitable urinary loss.

3. Estimates of dietary requirements for sodium and chlorine

For the calculation of dietary requirements, the intercepts in the equations relating faecal output to dietary intake have been taken as a maintenance requirement (for a 500-kg animal) and one minus the regression coefficient as the coefficient of absorption. In addition, in calculating the sodium requirement for maintenance in sheep, account has been taken of the inevitable loss of sodium in urine. The particular values used for net maintenance requirements for sodium are as follows (mg/kg live weight per day): beef cattle, 5.8 mg (faecal loss) + 1.0 mg (dermal loss); dairy cattle, 5.8 mg (faecal loss)) + 0.6 mg (dermal loss); sheep, 5.8 mg (faecal loss) + 20 mg (urinary loss). For chlorine, the corresponding values are: beef cattle, 7.9 mg + 1.0 mg; dairy cattle, 7.9 mg + 0.8 mg; sheep, 7.9 mg (no dermal or urinary loss). For both species the coefficients of absorption used are 0.91 for Na and 0.85 for Cl. Values for empty-body gain have been assumed to apply to live weight gain (since there is accretion of Na and Cl in gut contents). The calculated requirements for growing cattle and sheep are presented in Table 5.36 and for pregnant lactating cows (British Friesian) in Table 5.37. For a 40-kg ewe producing 2.0 kg milk, the estimated daily requirements would be 2.01 g sodium and 2.96 g chlorine.

The estimated requirements for sodium relative to chlorine are higher in growing sheep than in cattle (because of the allowance for an inevitable loss of sodium in the urine) and in young than in older cattle. The narrowest ratio of sodium to chlorine is in high-yielding cows where the value is about 1:2. The ratio of sodium to chlorine in common salt is 1:1.56.

Table 5.37   Dietary requirements (g/day) for sodium and chlorine of pregnant and lactating British Friesian cows†.

Status	Requirement	
	Sodium	Chlorine
Non-pregnant, non-lactating	4.22	6.14
Pregnant, 36–40 weeks	7.87	9.25
Lactating 10 kg milk/day	10.59	19.43
Lactating 20 kg milk/day	16.96	32.72
Lactating 30 kg milk/day	23.33	46.01
Lactating 40 kg milk/day	29.70	59.30

† assumed live weight, 600 kg.

4. Comparison of the estimates of requirement with these of the first edition and other standards

The procedure adopted for the calculation of maintenance requirements differs from that of the first edition and, with the exception of values for sodium in sheep which make allowance for an inevitable urinary loss, gives substantially lower figures. But whereas in the first edition a coefficient of absorption of 1.0 was assumed for both sodium and chlorine, slightly lower values have now been adopted, and values for the composition of the animal body and its products have been modified in the light of additional information. The overall trend in the estimates of requirements is for the values for growing cattle to be lower and those for growing sheep to be higher, changes which are consistent with the evidence of feeding trials (Joyce & Rattray 1970, Morris & Gartner 1971). The values for lactating cows are of the same order.

The NRC (1970 1971) recommendations do not include specific figures for sodium and chlorine.

5. Comparison of estimates with results of practical trials

Theiler et al (1927) obtained normal growth but poor fertility in heifers which received only 1.5 g sodium/day. On the basis of growth and a range of physiological measurements, Morris & Gartner (1971) considered that an intake of 3.1 g/day is adequate for bullocks of 300 kg weight growing at 0.9 kg/day. The present estimate of 3.7 g/day is consistent with this latter figure.

There are several estimates from feeding and metabolic trials of the sodium requirements of sheep and these are compared, in brackets, with the present estimates. For animals of 30 kg, Devlin & Roberts (1963) found a maintenance requirement of 1.01 g/day (0.85 g/day), for growing sheep weighing 20–30 kg, Joyce & Rattray (1970) reported a value of 1.6 g/day (0.8–1.1 g/day, for weight gain of 200 g/day) and for sheep growing at 200 g/day and weighing 40 kg, McClymont et al (1957) found 0.88 g sodium/day inadequate and 1.2 to 2.6 g/day adequate (1.4 g/day).

Aines & Smith (1957) found that in milking cows producing 18 kg milk/day an intake of 5.9 g sodium/day depressed yield, live weight gain and forage intake whereas 23.6 g/day (Smith & Aines 1959) maintained normal production: the present estimated requirement is 15.7 g/day (for a 600-kg cow). Kemp (1964) has estimated the maintenance requirement of the lactating cow as 6.0 g/day (present estimate 4.2 g/day) and the requirement of a cow giving 25 kg milk as 18.2 g/day (present estimate, 20.2 g/day). The higher estimate by Kemp (1964) for the maintenance requirement is due to an allowance of 2.5 g sodium/day for losses through the skin and in dribbled saliva: the present allowance is 0.4 g sodium/day.

No feeding trials have been reported in which the requirement for chlorine can be assessed separately from that for sodium.

6. Salt tolerance

Sheep and cattle are able to tolerate dietary intakes of sodium which are in excess of requirement, especially when there is free access to water, but high dietary concentrations may depress feed intake and the efficiency of feed utilization. The dietary concentrations at which effects are observed appear to vary with diet and the associated anion (Kromann & Ray, 1967). Meyer & Weir (1954), Meyer et al (1955) and A D. Wilson (1966) did not observe adverse effects on feed utilization and performance at dietary concentrations of 5% or higher, whereas Jackson et al (1971) reported progressive effects on efficiency of feed conversion of increases in sodium concentration above 0.7% of the dry matter.

Appendix Tables

Appendix Table 5.1   Estimates of the net endogenous faecal calcium and coefficient of absorption of dietary calcium for cattle.

Diet	No. of animals	No. of estimates	Sex	Age (m)	Live weight (kg)	Intake (mg/kg live weight per day)	Endogenous	Coefficient of absorption	Reference
Hay, concentrates	9	11	M, F	24	258–425	32–123	9–17	0.24–0.59	Comar et al (1953)
Hay, concentrates, bone	3	8	F	48–96	288–376	48–417	16–31	0.07–0.56	Visek et al (1953)
meal or calcium carbonate	2	4	M	12	248–255	67–192	15–20	0.19–0.50	Hansard et al (1954)
Milk	6	2	—	0.3, 1	35, 36	251, 112	12, 15	0.98, 0.98	
Hay, calf meal	5		—	6	179	116	18	0.41	
Hay, concentrates	23	4	M	15–190	300–454	44–74	14–22	0.22–0.36	Hansard et al (1957)
Hay, concentrates, variety of calcium salts	15	15	M	5–7	146–253	39–92	13–15	0.41–0.68	
Milk, calcium chloride	15	15	M	36–73	359–589	32–54	13–17	0.31–0.56	
Hay concentrates	3	1	—	2	234	80	16	0.82	Guéguen (1964)
Hay, concentrates/bone flour	3	1	—	12	212	47	12	0.35	Lengemann (1965)
Oat straw, concentrates	1	1	F	12	187	163	13	0.33	Symonds et al (1966)
Commercial diet	3	2	M, F	54	500	53, 62	22, 24	0.43, 0.53	Ramberg et al (1970a)
Hay, grain mixture	7	7	F	48–120	202–436	234–332	8–10	0.25–0.30	Ramberg et al (1970b)
	7	7	F	48–120	395–550	100, 198	12–17	0.25–0.41	
			F	48–120	393–550	155–332	12–27	0.25–0.33	

Appendix Table 5.2   Estimates of the net endogenous faecal calcium and coefficient of absorption of dietary calcium for sheep.

Diet	No. of animals	No. of estimates	Sex	Age (m)	Live weight (kg)	Intake (mg/kg liveweight per day)	Endogenous	Coefficient of absorption	Reference
Lucerne hay, hulls, starch &c.	6	1	M	18	36–46	82	58	0.49	Tillman & Brethour (1958a)
Lucerne meal, hulls, soya meal &c.	6	1	M	12	34–36	106	43	0.63	Tillman & Brethour (1958b)
Hay, concentrates	9	1	M	—	38	121	44	0.45	Schroder & Hansard (1958)
Hulls, maize meal, soya meal, lucerne	4	7	F	—	37	42	33	0.42	Thompson et al (1959)
Hay, concentrates	6	10	F	36	39–57	140–204	16–29	0.27–0.39	Braithwaite et al (1969)
Hay, concentrates	10	10	M	48	45–71	94–167	14–22	0.16–0.27	Braithwaite et al (1970)
Hay, concentrates, ammonium chloride	42	8	M	2–70	15–92	76–193	14–25	0.14–0.39	Braithwaite & Riazuddin (1971)
Hay, concentrates	12	5	M	12–24	51–67	97–105	13–15	0.16–0.45	Braithwaite (1972)
Hay, concentrates	8	3	M	24	50–60	106	16–18	0.11–0.19	Braithwaite (1974)
Straw, concentrates	4		M	24	50–60	30	9	0.34	
Hay, concentrates	6	5	M	8	36–44	111–137	19–22	0.21–0.35	Braithwaite (1975a)
Straw, concentrates	4	6	M	5	25–35	25–506	13–15	0.02–0.27	Braithwaite (1975b)
Straw, concentrates, calcium carbonate	4		M	60	80–90	51–498	9–11	0.01–0.20	

Appendix Table 5.3   Estimates of the coefficient of absorption of dietary calcium for dairy cows.

No. of animals	No. of balances	Live weight (kg)	Milk Ca (kg/day)	Dietary Ca (mg/kg live weight per day)	Coefficient of absorption	References
		Range of values				
6	18	359–486	10–17	18–54	−0.04–0.29	Forbes et al (1916)
6	12	399–530	16–24	40–88	0.23–0.44	Forbes et al (1917)
6	12	381–508	16–28	89–120	0.11–0.32	Forbes et al (1918)
10	18	440–594	0–27	64–130	0.08–0.38	Forbes et al (1922)
3	12	497–624	12–21	74–83	0.28–0.55	Hart et al (1922a)
3	33	509–560	9–27	26–97	0.16–0.75	Hart et al (1922b)
3	9	570–633	18–30	80–82	0.35–0.52	Hart et al (1927)
2	14	595–655	20–28	135–148	0.20–0.36	Hart et al (1926–27)
3	26	420–518	12–19	73–151	0.13–0.34	Turner et al (1927)
5	18	490–620	3–30	33–141	0.18–0.56	Turner & Hartman (1929)
3	15	520–592	18–22	154–166	0.09–0.19	Hart et al (1929a)
3	21	545–616	20–28	89–90	0.16–0.39	Hart et al (1929b)
3	24	482–540	16–25	83–90	0.32–0.44	Hart et al (1930)
5	134	445–550	0–44	26–119	0.27–0.90	Ellenberger et al (1931)
4	94	516–564	0–35	33–108	0.14–0.75	Ellenberger et al (1932)
14	180	425–660	0–31	21–147	0.08–0.71	Forbes et al (1935)
2	30	564	0–31	35–96	0.25–0.89	Newlander et al (1936)
8	18	370–483	4–20	15–41	−0.06–0.93	Owen (1948)
157	157	397–640	0–26	10–147	−0.60–1.50	Paquay et al (1968)

Coefficient = (I − Fo + Em)/I, where I and Fo represent intake and faecal excretion of Ca and Em the endogenous faecal Ca excretion (0.016 × live weight (kg)).

Appendix Table 5.4   Estimates of net endogenous faecal phosphorus and coefficient of absorption of dietary phosphorus for cattle.

Diet	No. of animals	No. of estimates	Sex	Age (m)	Live weight (kg)	Intake (mg/kg live weight per day)	Endogenous (mg/kg live weight per day)	Coefficient of absorption	References
Hay, concentrates	2	2	F	>24	427, 430	28–87	23–33	0.36–0.64	Kleiber et al (1951)
Synthetic milk	2	2	—	0.5, 08	27, 36	22, 23	4, 4	0.93–0.94	Lofgreen et al (1952)
Low P diet, phosphoric acid	6	6	M	14	227	45	18	0.76	Tillman & Brethour (1958a)
Low P diet, dicalcium phosphate	6	6	M	14	227	45	16	0.75	
Hay, casein, urea, dicalcium phosphate	4	4	F	6	190*	49	10–21	0.71–0.83	Brüggemann et al (1959)
Hay, casein, urea, P fertiliser	4	4	F	6	190*	47–49	12–16	0.82–0.88	
Low P diet, dicalcium phosphate	4	1	M	12	182	33, 44	13, 14	0.68, 0.73	Tillman et al (1959)
Milk	3	1	M	12	182	45	15	0.78	
Milk, hay, concentrates	3	1	M	2	75	140	2	0.98	Guéguen (1963)
Milk, mineral mixture	3	1	M	2	75	149	25	0.73	
Straw, concentrates	1	1	M	1.5	60	205	5	0.88	Guéguen (1964)
Straw, concentrates	1		F	48	500	32	11	0.31	Manston (1964)
Straw, concentrates	1	2	F	48, 60	500	53, 63	22, 24	0.43, 0.53	Symonds et al (1966)
Hay, concentrates	2	2	F	96	500	80	33, 39	0.45, 0.38	Manston (1966)

* Live weight estimated to be 190 kg from data on Black Pied cattle (Witt et al 1971).

Appendix Table 5.5   Estimates of net endogenous faecal phosphorus and coefficient of absorption of dietary phosphorus for sheep.

Diet	No. of animals	No. of estimates	Sex	Age (m)	Live weight (kg)	Intake (mg/kg liveweight per day)	Endogenous (mg/kg liveweight per day)	Coefficient of absorption	Reference
					Range of values				
Lucerne hay	5	5	M	L*	31–36	69–75	39—60	0.81–0.96	Lofgreen & Kleiber (1953)
Lucerne hay	4	4	M	12	36–40	62–67	39–42	0.93–096.	Lofgreen & Kleiber (1954)
Hay, concentrate	2	2	F	A	—	—	—	0.56–0.82	Wright (1955)
Low-P diet, monosodium phosphate	6	1	M	18	—	—	—	0.70	Tillman & Brethour (1958b)
Semipurified	6	—	M	12	—	—	—	0.81	Tillman & Brethour (1959c)
Low-P diet, monosodium phosphate	6	1	M	13	41–45	43	23	0.89	Tillman & Brethour (1958d)
Low-P diet, sodium metaphosphate	6	—	M	13	41–45	43	28	0.87	
Low-P diet, acid sodium pyrophosphate	6	1	M	13	41–45	43	23	0.90	
Hay, concentrates	18	2	M	L	38–39	74	58, 62	0.78, 0.83	Schroder & Hansard (1958)
Hay	2	2	M	24	40–50	59	40, 38	0.63, 0.65	Brüggemann et al (1959)
Hay, dicalcium phosphate	2	2	M	24	40–50	133	63, 66	0.47, 0.49	
Rye straw, low P concentrate	7	7	M	24	40–50	67	34–55	0.76–0.81	
Cottonseed hulls, maize meal, lucerne meal	12	6	M	L	33–41	42–58	34–52	0.75–0.88	Thompson et al (1959)
Hay, straw, beet pulp, P supplements	5	5	M	A	—	—	—	0.14–0.69	Lofgreen (1960)
Hay, concentrates, monosodium phosphate	15	3	M	6	—	—	—	0.56–0.64	Lueker & Lofgreen (1961)
Wheat, bran, concentrate	4	—	M	6	—	328	35	0.25	Ellis & Tillman (1961)
Hay	3	3	F	42–48	24–27	35–39	22–26	0.52–0.58	Gueguen (1962)
Molasses, beet pulp, maize grits	8	1	M	7	85–90	25	11	0.81	Preston & Pfander (1964)
Starch, dicalcium phosphate	8	—	M	7	25–33	33	15	0.84	
	8	1	M	7	25–33	73	40	0.86	
Beet pulp, starch &c.	11	6	M	6	—	—	—	0.54–0.80	Young et al (1966)
Beet pulp, straw, concentrates sodium dihydrogen phosphate	9	9	M	L	45	38–42	22–38	0.55–0.70	Compère (1966)
	9	9	M	L	45	72–82	17–56	0.58–0.91	
	9	9	M	L	45	92–118	36–61	0.66–0.81	
Milk	3	3	M	0.75	8–10	180–207	2–3	0.99–1.00	Compère et al (1967)
Milk, concentrates	1	1	M	2	14	68	3	0.95	
	1	1	M	2	17	97	3	0.88	
	1	1	M	2	19	142	8.5	0.87	
Hay, concentrates	3	3	M	5	34–44	75–97	25–37	0.64–0.78	Compère (1969)
Low-P diet (straw, cassava meal, beet pulp)	2	2	M	A	62	10	20, 20	0.47, 0.58	
Low-P diet, ammonium dihydrogen phosphate	2	2	M	A	62	48	31, 31	0.69, 0.64	
Low-P diet, diammonium hydrogen phosphate	2	2	M	A	62	43, 49	35, 36	0.77, 0.74	
Basic diet (hay, beet pulp, cassava meal)	3	3	M	L	32	59–77	42–53	0.59–0.69	
Basic diet, ammonium dihydrogen phosphate	3	3	M	L	32	208–225	90–125	0.59–0.77	
Basic diet, diammonium hydrogen phosphate	2	2	M	L	32	179–209	77, 86	0.68, 0.78	
Basic diet, P-Ca-Mg-supplement	3	3	M	L	32	200–225	103–120	0.70–0.77	

*L = < 12m, A = > 12m.

Appendix Table 5.6   Estimates of coefficient of absorption of dietary phosphorus for lactating dairy cows.

No. of animals	No. of balances	Live weight (kg)	Milk yield (kg/day)	Range of values	Coefficient of apparent absorption	Coefficient of absorption	References
				P intake (mg/kg live weight per day)			
6	18	359–486	10–17	46–82	0.26–0.49	0.39–0.69	Forbes et al (1916)
6	12	399–530	16–24	71–117	0.29–0.48	0.43–0.58	Forbes et al (1917)
6	12	381–508	16–28	92–140	0.24–0.40	0.31–0.49	Forbes et al (1918)
9	11	440–594	1–27	48–144	0.13–0.44	0.25–0.59	Forbes et al (1922)
3	33	509–560	9–27	60–149	−0.34–0.40	0.00–0.52	Hart et al (1922a)
2	6	570–633	18–30	93–97	0.28–0.36	0.38–0.46	Hart et al (1927)
26	12	420–518	12–19	90–262	0.09–0.34	0.18–0.39	Turner et al (1927)
2	24	490–620	3–30	179–269	0.11–0.24	0.17–0.28	Turner & Hartman (1929)
3	5	483–540	16–25	109–133	0.33–0.47	0.41–0.55	Hart et al (1930)
121		445–550	2–44	66–226	0.12–0.41	0.22–0.47	Ellenberger et al (1931)
4	81	516–564	3–35	81–173	0.14–0.41	0.24–0.52	Ellenberger et al (1932)
14	138	425–660	2–31	41–131	0.11–0.56	0.29–0.70	Forbes et al (1935)
	Overall mean	516	16.2	1.0	0.31	0.41	

Coefficient of absorption = $(I - F + Em)/I$, where I and F represent the dietary intake and faecal excretion of P(g/day), respectively, and Em the daily minimum endogenous faecal P (0.010 g/day × liveweight (kg)).

Chapter 6

Trace Elements

Introduction

Since the first edition of this publication many more investigations have been made of the effects of trace element deficiency and excess on ruminants. Most of these have studied the influence of two or three dietary concentrations of a single trace element on performance or clinical condition, and on the content of that trace element in one or two tissues. When considered individually, none of these reports presents an adequate quantitative description of trace element availability, absorption or retention. In the instance of copper and zinc, we have assessed data from a wide range of sources and have presented estimates of requirement based on a factorial approach which must be regarded as provisional until further information is available.

For the other trace elements, observations illustrating the influence of differing dietary concentrations on clinical response or performance were considered and estimates of the minimum gross requirement were subsequently based on an admittedly subjective analysis of such information.

Sufficient information is rarely available to indicate how gross requirement may be influenced by rate of growth, by pregnancy or by lactation. Furthermore, although with most of the trace elements other dietary components are known to affect true availability, only for the effects of molybdenum and sulphur on availability of copper has it proved possible to present even an approximate quantitative estimate of the influence of such antagonists. Nevertheless, we have presented as complete a picture as possible of the nature of the antagonists and their action on trace element availability to indicate circumstances in which deviations from the normal estimates of requirements may arise.

Finally, attention is drawn to the sections emphasizing the toxicity of copper to sheep and the toxicity of selenium to all mammalian species. When it is borne in mind that our knowledge of dietary factors which influence the availability of copper and selenium is meagre, it will be particularly evident that care must be exercised in correcting dietary deficits of these elements by the information we have presented.

Copper

The derivation of the copper requirements of ruminants is hampered by the scarcity of studies of relationships between copper intake and health or productivity and of the quantitative effects of common dietary antagonists on copper utilization. While there are ample data indicating that low copper concentrations in blood and liver of sheep and, particularly, of cattle are occurring with increasing frequency in the UK (e.g. Todd 1972, Davies & Baker 1974), the intakes of copper and its antagonists giving rise to these situations are rarely recorded. Aware of the need for realistic statements of copper requirements, we have opted for estimates based on a factorial analysis of the need for copper to meet endogenous losses and the demands for growth, pregnancy and lactation, while recognizing that the experimental work on which such estimates could be based is limited.

Two points relating to the practical use of these estimates require emphasis.

(*i*) Availability. It is apparent from both experimental and field studies that the availability of copper in ruminant diets can vary widely in differing and often ill-defined dietary circumstances. Where possible, guidance has been given on the selection of appropriate "availability factors" to use in deriving gross requirements from estimates of net requirement; this guidance must be regarded as provisional and likely to be modified as further knowledge becomes available. In view of the suscep-tibility of sheep to copper intoxication, the use of estimates based on a low availability factor is justifiable for sheep only when there is unequivocal evidence of high dietary contents of components known to reduce availability of copper (e.g. sulphur, molyb-denum, zinc, cadmium, iron or soil).

(*ii*) Liver stores. Unlike many other essential nutrients, sufficient copper may be deposited in the foetal liver during late pregnancy or subsequently retained in the liver during periods of luxus intake to offer a long period of protection to the ruminant against the adverse effects of a low copper intake. For example, veal calf rations frequently have low copper contents but the extent of depletion achieved before animals are slaughtered is usually insufficient to induce clinical signs of deficiency. Conversely, the early development of copper deficiency in some single-suckled calves may be a consequence of inadequate hepatic reserves at birth.

1. Factorial computation of net requirements of sheep for copper

(a) *Endogenous loss*

The daily output of copper in bile from mature ewes of normal copper status has been estimated to be 0.075 mg/day and that in pancreatic juice 0.065 mg/day (I. Bremner and R. N. B. Kay, unpublished results). The total output of 0.140 mg Cu/day is similar to the estimate of the metabolic faecal loss of 0.127 ± 0.019 mg Cu/day from ewes of low copper status derived by Suttle (1974a); urinary losses in the experiment of Suttle (1974a) were 0.047 ± 0.018 mg/day and thus total endogenous losses were estimated to be 3.8 μg Cu/kg liveweight per day. An endogenous loss of 4.8 μg Cu/kg live weight per day was predicted by Suttle & Field (1974) from the period of protection afforded by a single copper injection to ewes of low copper status. A constant value of 4 μg Cu/kg live weight per day has, therefore, been used as the maintenance requirement of sheep in the estimates presented in Table 6.3, although it is acknowledged that endogenous losses may vary with the copper status of the animal.

(b) *Growth*

In the absence of information on the body copper content of sheep, minced carcasses of 10 newborn lambs, 8 six-week-old lambs (shorn) and 10 mature ewes (shorn, all Scottish Blackface) have been analysed for copper. The regression of total body copper (*y*, mg) on fresh carcase weight (*x*, kg) was $y = 3.74 + 1.15 (\pm 0.08)x$ (*P* < 0.001, r = 0.94; N. F. Suttle, unpublished data). Each kg of carcass gain, therefore, required about 1.15 mg copper and at liveweight gains of 100–200 g/day, 0.11 to 0.22 mg copper would be required daily.

(c) *Pregnancy*

By estimating the copper content of the foetus, the placenta and the foetal fluids at 34, 90 and 136 days of gestation, Moss et al (1974) were able to calculate that copper accumulated in the total products of conception at rates of 15, 85 and 186 μg/day during the first, second and third trimesters. Since the liver normally provides about 50% of the total copper in the foetus at birth (Pryor 1964), the estimate of requirement for pregnancy will be determined largely by the concentration in the foetal liver. As normal concentrations (44 μg/g fresh liver) were present in the foetuses studied by Moss et al (1974), their estimates are taken as the best available values of the net requirement for pregnancy.

(d) Lactation

The net copper requirement for milk production is taken to be the product of milk yield and the copper concentration of milk. The three most extensive studies of the copper content of milk, taken from four different breeds, show that concentrations of 0.24–0.41 mg Cu/kg are found in early lactation and 0.11–0.32 mg/kg in late lactation (Beck 1941, Ashton & Yousef 1966, A. C. Field & G. Wiener, personal communication). The concentration of copper in colostrum is probably much higher (Underwood 1977). Concentrations of 0.32 and 0.22 mg Cu/kg for early and late lactation have been used in computing the net requirements for lactation.

(e) Wool production

The average copper concentration in wool is of the order of 5 mg/kg DM (Dick 1954, Suttle & Field 1968). The net requirement for wool production over the whole year will vary from 5 mg in low-yielding breeds such as the Welsh Mountain to 18 mg in high-yielding breeds such as the Merino; the corresponding daily copper requirements would be 0.014 and 0.050 mg.

2. Factorial computation of net requirements of cattle for copper

(a) Endogenous loss

There have been no direct measurements of metabolic faecal losses of copper by cattle and estimates of urinary output have varied from 0.5 (Joshi & Talapatra 1968) to 0.2 mg Cu/day (Hartmans 1970). These data are insufficient to permit derivation of a direct estimate of the endogenous copper loss of cattle. Because of the flexibility of the factorial approach we have decided to use an indirectly derived value for endogenous loss based on the relationship:

$$Cu_E = [Cu_I(A) + \Delta Cu_H - Cu_G (\Delta W)]/W, \qquad [6.1]$$

where Cu_E = endogenous loss of Cu (mg/kg per day), Cu_I = Cu intake (mg/day), A = fractional absorption of dietary Cu, Cu_H = hepatic Cu loss (mg/day), Cu_G = Cu content of weight gain (1.06 mg/kg gain; see p. 222), ΔW = change in live weight (kg/day) and W = live weight (kg).

Each component in the equation involves certain assumptions.

(i) Absorption (A) was predicted from the relationship between dietary molybdenum and sulphur concentrations and utilization of copper derived for sheep by Suttle & McLauchlan (1976) on the further assumptions that these were the major determinants of (A) and that cattle responded similarly to sheep. The validity of these assumptions is considered further on p. 226.

Table 6.1   Derivation of provisional estimates of endogenous Cu loss of growing cattle assuming that sulphur and molybdenum content of diets are primary determinants of Cu availability.

Authors	Nature of ration	Dietary Cu (mg/kg)	Mo (mg/kg)	S (g/kg)	Predicted coefficient of absorption (A) of Cu*	Available Cu (mg/kg DM)	Change in hepatic Cu (mg/day)	Derived factor for endogenous loss (μg/kg live wt)
Hartmans & Van der Grift (1964)	Grass silage, hay potatoes, concentrates	11.6 11.6	0.7 0.7	2.7 3.6	0.041 0.035	0.48 0.40	−0.27 −0.96	8.2 9.4
Lesperance & Bohman (1963)	Lucerne hay Grass hay	9.7 6.5	0.6 0.6	2.1 2.0	0.047 0.048	0.46 0.31	+0.18 +0.03	8.3 5.3
Mills, Burridge & Greenhalgh (unpublished observations)	Grass, zero grazed Free grazing at low stocking rate	6.7 6.3	2.03 2.42	1.99 2.30	0.044 0.038	0.29 0.24	−0.12 −0.38	3.6 7.1
Mills, Dalgarno & Wenham (1976)	Semisynthetic urea, starch	8 15	0.08 0.08	2.4 2.4	0.046 0.046	0.37 0.69	−0.50 +0.07	7.8 7.0

Mean 7.1 (±0.65) μg/kg live weight

* Predicted from equation $\log y = -1.153 - 0.076S - 0.013\ SMo$, where y = Cu availability, S = sulphur content of ration (g/kg DM) and Mo = molybdenum content (mg/kg DM).

(ii) Cu_H was predicted from initial and final liver copper concentrations and live weight by using a relationship relating liver DM (g) to live weight (kg), where liver $DM = xW$; selection of the appropriate value for x, which lay between 5.3 and 3.8 and was inversely related to live weight and to rate of gain, was based on carcass analysis data (M. Kay, J. F. D. Greenhalgh & C. F. Mills, unpublished).

(iii) For Cu_G (ΔW) it is assumed that the copper content of the extrahepatic component of live weight gain is a constant.

(iv) Using W as the divisor implies that Cu_E is linearly related to live weight.

Four studies involving nine groups of animals yielded data that was suitable for use in the above equation and the details are given in Table 6.1. The relatively uniform values obtained for Cu_E (mean $7.1 \pm SD\ 1.8$) suggest that the assumptions made may not have introduced serious errors. The derived value may not be applicable to animals on diets of very low copper status. Two further studies involving cattle with extremely low initial (Clawson et al 1972) or final liver copper content (Mills et al 1976) yielded much lower estimates of 1.8 and 0.8 μg/kg live weight for endogenous loss. These low values may reflect a real adaptation to copper deprivation. Alternatively, an increased efficiency of absorption may have occurred during deprivation, giving an apparently low endogenous loss. It was decided to ignore these two low values in deriving our mean value on the grounds that it was dangerous to allow for the maintenace of such a low copper status in the factorial calculation. Any adaptation occurring in the naturally deprived animal would provide a safety margin not included elsewhere in our recommendations.

(b) Growth

In the absence of comprehensive data on organ weights and copper concentrations in the bovine carcass, the growth requirement of 1.06 mg Cu/kg live weight gain estimated for sheep has been adopted for cattle.

(c) Pregnancy

The copper content of the foetus has been estimated from the data for four "mid-term" foetuses published by Pryor (1964). Values for older foetuses were obtained by extrapolating the relationship between foetal copper (y, mg) and foetal weight (x, kg) for the younger foetuses where $y = 3.92x - 1.1$. Bovine placental and uterine tissues and the foetal fluids were assumed to contain the same copper concentrations as those of the sheep 1.8 mg/kg fresh weight (Moss et al 1974) and 0.15 mg Cu/litre (N. F. Suttle, unpublished results), and their weights at different stages of gestation were predicted from information given in Chapter 1. After summating the data for the products of conception, we have predicted daily increments in the copper content of the conceptus (Table 6.2). During the last month of pregnancy, the daily requirement for copper appears to increase by approximately 70% of the maintenance requirement. Indirect evidence for a marked increase in copper requirement during late pregnancy is given by the fall which occurs at that time in the dam's liver copper (Van der Grift 1955, Allcroft & Uvarov 1959, Pryor 1964).

(d) Lactation

There is less variation in the copper concentrations reported for cow's milk than for sheep's milk and Kirchgessner et al (1967) suggest a general value of 0.1 mg/kg. The

Table 6.2   Estimated copper content of the bovine gravid uterus.

Age (days)	Foetal weight (kg)	Foetal Cu (mg)	Cu in membranes, fluids and uterus§ (mg)	Total Cu in conceptus (mg)	Daily Cu increment in conceptus (mg)
112	0.9	3.3*	—	—	—
140	2.1	5.6*	6.3	11.9	0.61
168	5.1	19.7*	9.2	28.9	0.64
196	9.4	35.5*	11.3	46.8	1.06
224	15.6	60.0†	16.6	76.6	1.62
252	26.0	110.7†	21.2	121.9	2.14
281	40.0	155.5†	26.4	181.9	2.07

* Pryor (1964).
† Foetal Cu (y) predicted from relationship for the younger foetuses in which
$y = 3.92$ (foetal weight, kg) $- 1.10$.
§ Foetal weight, weights of placental fluids, membranes and uterus estimated
from relationships given in Chapter 1. Cu concentrations assumed to be 0.15 mg/litre in
fluids (N. F. Suttle, unpublished results) and 1.8 mg/kg fresh tissue as in the sheep
(Moss et al 1974).

net requirement for milk production thus increases from 0.5 to 3 mg/day as milk
yield rises from 5 to 30 kg/day.

3. Absorption coefficients for converting net to gross requirements for copper

Recent studies have revealed a number of factors, some physiological, some dietary
and some genetic in origin, which affect the efficiency with which sheep and cattle use
dietary copper. It is therefore necessary to detail the evidence on which the selection
of appropriate absorption coefficients for conversion of net to gross requirements can
be based.

(a) Changes in apparent absorption with age at the preruminant stage

The preruminant lamb absorbs copper as efficiently as monogastric species and
much more efficiently than the mature animal. During the preruminant stage of
development, the apparent absorption (A) of copper has been found to fall dramatically
in lambs maintained on liquid milk substitute diets from the high value of 0.8–0.9
shortly after birth to 0.2–0.3 at 6 weeks of age, the precise relationship being
(A)$=0.971-0.0118a$ (where a is age, days) ± 0.0035 (Suttle 1975a).

The preruminant calf also absorbs copper efficiently; thus from the increase in live
copper stores of milk-fed calves, Bremner & Dalgarno (1973a, b) estimated that
50–60% of the copper from a diet supplemented with cupric sulphate (5 mg Cu/kg
offered between 3 and 14 weeks of age was retained in the liver. If age-related change
in availability occur in the calf they are probably completed within the first four weeks
of life (P. Thivend & R. N. B. Kay, personal communication).

(b) Changes attributable to the transition from liquid to solid diets

The development of a fully functional rumen is associated with a decline in the
apparent absorption of Cu and in one study with sheep (Suttle 1975a), the decline to
0.08 induced by early weaning obscured the age-related changes described above.

decline also occurs in the calf also after the establishment of rumination (P. Thivend & R. N. B. Kay, personal communication).

Provision for these effects of age and weaning has been made in the presentation of tentative estimates of the gross requirements of the lamb and the calf for copper (Tables 6.3 and 6.4).

(c) Effects of dietary composition on availability of copper

(i) Concentrated diets

It is possible to obtain an approximate estimate of the coefficient of true absorption (A) from experiments in which the effects of dietary copper supplements on liver concentrations have been reported. Since endogenous losses of copper and other tissue copper concentrations are unaffected by intake (Todd, 1969), then Δtotal liver Cu/ΔCu intake is a minimum estimate of the (A) of the supplementary copper. The application of this "hepatic copper retention" technique indicates that (A) is of the order of 0.06 ($\pm$0.01 SE) in diets low in molybdenum and sulphur and based on concentrates (from Hill & Williams 1965, MacPherson & Hemingway 1965, Dick 1956, Hemingway & MacPherson 1967, Abdellatif 1968, Hogan et al 1968, Kline et al 1971). This value falls between two values obtained by isotope dilution technique (0.086; Smith et al 1968) and by a repletion technique (0.057 $\pm$ 0,005 for 8 semipurified diets of low sulphur content, 1 g/kg (Suttle 1974a, b). The repletion technique assesses (A) for copper in the diet by comparing the increase in plasma copper produced in initially hypocupraemic sheep with that achieved by intravenous infusion of known amounts of inorganic Cu^{2+}.

(ii) Fresh and conserved herbage

The scanty information on (A) for copper in fresh herbage given to ruminants suggests that it is less than that in concentrate diets. Suttle & Price (1976), using a repletion technique, obtained a value of 0.048 for sheep given cut herbage. Much lower values have been estimated from hepatic retention in grazing lambs (0.024 and 0.017 by Hemingway et al 1962, MacPherson & Hemingway 1968) and cattle (0.018 and 0.015, Hartmans 1969) dosed orally with inorganic $CuSO_4$.

Hartmans & Bosman (1970) have suggested that availability of copper increases as herbage matures and is higher in hay than in fresh herbage, and that the decline in hepatic reserves on turning cattle out to graze short grass is more rapid than when grass at a later stage of growth is consumed. They suggest that similar changes in availability are responsible for the improvement in copper status which commonly occurs in cattle in the Netherlands when they are housed in winter.

(iii) Sulphur and molybdenum content of diet

Since the first discovery that inorganic sulphate potentiates the adverse effect of molybdenum on utilization of copper by sheep (Dick, 1953a, b), continuing research has revealed a highly complex interrelationship in which each element affects the metabolism of the other. For example, sulphate accelerates the clearance of molybdate (Dick 1956) and molybdenum influences the ruminal metabolism of sulphide (Mills 1960, Gawthorne & Nader 1976). Progress towards the quantitative definition of antagonistic relationships between molybdenum and sulphate on copper availability has been retarded by delay in recognizing that organic sulphur and inorganic sulphate share common metabolic fates in the rumen (Anderson 1956) and similarly influence the output of soluble copper from that organ (Bird 1970) and interact with molybdenum to reduce copper availability (Suttle 1975b). Until recently, the scarcity of data on total sulphur intakes has precluded even approximate description of the individual or combined effects of molybdenum and sulphur on the copper status of ruminants. Suttle & McLauchlan (1976) have examined the data from 10 repletion experiments with sheep in which the true availability (A) for copper was measured in 32 diets differing in molybdenum and sulphur content within the ranges 0.5–15 mg Mo/kg and 0.8–4 g S/kg. This was highly correlated (r=0.93) with the dietary concentrations of sulphur (g/kg DM) and molybdenum (mg/kg DM), the relationship being described by the equation:

$$\log (A) = -1.153 - 0.076 \,(S) - 0.013 \,(S \times Mo). \qquad [6.2]$$

The relationship confirms earlier suggestions that sulphur exerts an independent effect on the availability of copper to sheep or cattle (Wynne & McClymont 1956, Allcroft & Lewis 1956, Goodrich & Tillman 1966, Hartmans & Van der Grift 1964) and indicates that the effect of sulphur is greater than the sulphur-dependent effect of molybdenum.

The validity of the above equation for predicting (A) for copper was examined using the data of Wynne & McClymont (1956) and Dick (1953b) for sheep given hay and concentrates or hay alone. A significant correlation (r=0.94; 6 d.f.) was found between hepatic copper retention and predicted values of (A). A separate equation could not be derived for cattle because of lack of data. When the validation procedure used for sheep was applied to data from six groups of growing Friesian or Hereford cattle consuming fresh herbage or hay, a similar relationship was obtained in which

$$y = -1.03 \,(\pm 0.17) + 2.86 \,(\pm 0.59)x \,(R = 0.86), \qquad [6.3]$$

where x=dietary Cu (mg/kg DM)×(A) and y=daily change in hepatic Cu content (mg). It would, therefore, appear that use of the above equation for cattle will in general not introduce serious errors.

The relationship was less satisfactory for cattle consuming rations based on silage or semisynthetic diets with low roughage content. These exceptions may reflect the fact that the potency of molybdeum is influenced by the nature of the diet. In early work on the copper-responsive "teart scours" in cattle grazing forage of high molybdenum content, diarrhoea was observed to cease abruptly when herbage became frosted or was made into hay: this response was associated with a decrease in the water-soluble molybdenum content of forage but little or no change in total molybdenum (Ferguson et al 1943). The above equation is unlikely to be valid for predicting (A) for copper when dietary molybdenum is very high (e.g., 40–50 mg/kg dietary DM), as in the "teart scours" syndrome. Dietary molybdenum contents of this order have been shown to result in a net increase in the ruminal concentration of sulphide (Mills 1960, Bryden & Bray 1972, Gawthorne & Nader 1976) and the quantitative effect of this on availability of copper is not known.

Accepting that some revision of the above prediction equation may prove necessary for cattle and for particular diets, we have used it to predict the effects of a range of dietary sulphur and molybdenum contents on the availability of copper (Fig. 6.1) This approach is more practical than the presentation of arbitrary statements of tolerable molybdenum and sulphur contents and is preferable to the use of Cu:Mo ratios (cf. Bingley & Carillo 1966, Miltimore & Mason 1971, Alloway 1973), in that it allows for the involvement of sulphur in the interaction.

Fig. 6.1   Relationship for sheep of coefficient of absorption of dietary copper to dietary content of total sulphur and molybdenum. (From Suttle & McLauchlan 1976.)

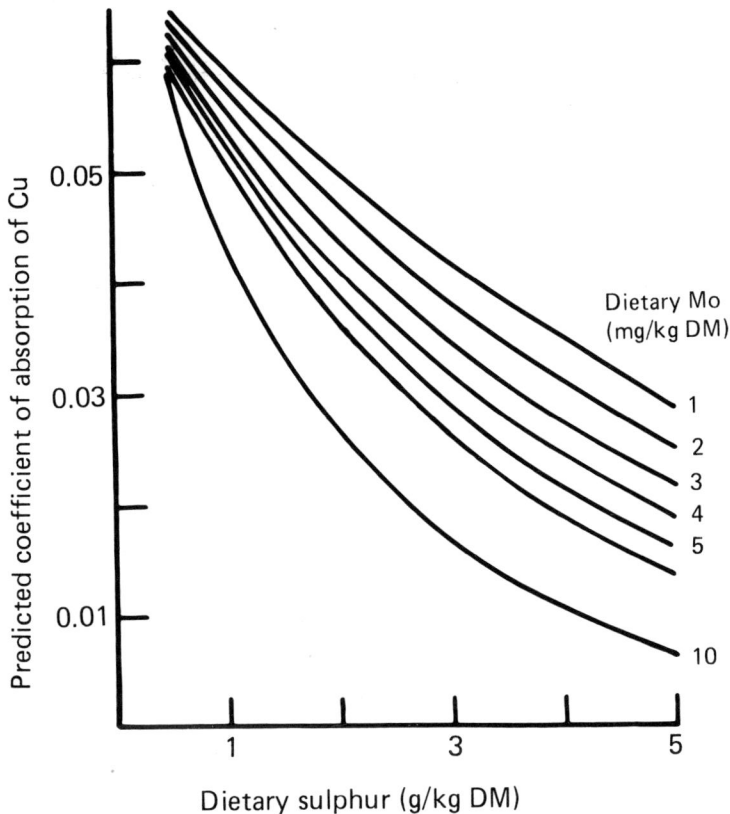

(iv) Soil ingestion

Grazing animals cannot avoid ingesting some soil as they graze. When soil structure is poor and herbage short, soil can constitute more than 10% of the DM intake (Field & Purves 1964, Healy 1967). The inclusion of 10% of three diverse soils in the diet of sheep reduced (A) for copper by at least 50% (Suttle et al 1975). Grazing cattle also ingest appreciable quantities of soil (Thornton 1974) and it is therefore possible that the soil which contaminates all herbage reduces the availability of copper in grazed herbage to ruminants generally.

(v) Calcium

Dick (1954), in experiments with sheep receiving 30 mg Cu/day, found that diets containing about 90 g calcium carbonate/kg halved the hepatic retention of copper; this response may have been attributable specifically to the carbonate salt, as a similar result was not obtained with dicalcium phosphate. Hemingway et al (1962) found no adverse effect on copper utilization of drenches providing 35 g CaCO₃/day to grazing sheep. In studies with cattle, Kirchgessner (1965) found that an increase in dietary calcium from 7 to 9.5 g/kg reduced copper retention. Others have failed to influence the copper status of cattle by wider adjustments of dietary calcium (e.g., 9–28 mg/kg,

Huber & Price 1971; 6–14 g/kg, Van Leeuwen & Van Kluijve 1971, Hartman 1970). These conflicting observations merit further investigation in view of the fact that the adverse effects of calcium or $CaCO_3$ occurred at intakes that could readily be attained by animals grazing herbage contaminated by calcareous soils.

(vi) Iron, cadmium and zinc

The data of Abdellatif (1968) suggest that iron intakes of 2.6–5.2 g/day by sheep reduced (A) for copper (estimated from hepatic retention) from 0.56 to 0.036. In attempting to simulate the harmful effects of irrigation with bore-water on grazing cattle, Campbell et al (1974) showed that liver copper reserves in calves at pasture were depleted by daily doses of iron (as $Fe(OH)_3$) equivalent to a dietary concentration of 1.4 g/kg DM.

Mills & Dalgarno (1972) found that cadmium at 7 mg/kg added to the diet of the pregnant ewe reduced the liver copper stores of her offspring and more recent work indicates that even 3 mg Cd/kg adversely affects copper retention and growth of lambs (J. K. Campbell, personal communication). Increasing the dietary concentration of zinc from 18 to 100 mg/kg DM has been shown to reduce the accumulation of copper in the liver of lambs on a semipurified diet (Mills 1974). Consumption of rations containing 40, 220 or 420 mg Zn/kg reduced the fraction of dietary copper retained by the liver of growing lambs during a four-week period from 0.04 to 0.02 or 0.015, respectively (Bremner et al 1976).

Although contamination of herbage with cadmium or zinc from industrial or natural environmental sources can lead to higher cadmium or zinc contents than this, it is difficult to make accurate allowances for the possible effects of environmental exposure to these heavy metals on copper requirements until more is known of the form and biological effectiveness of the contamination.

(d) Individual and genetic differences in copper utilization

Animals of similar age, breed and physiological state in a common environment can show marked differences in their efficiency of copper utilization. Thus (A) for copper varied from 0.042 to 0.112 in a group of 36 ewes on a semipurified diet (Suttle 1974a). There was, evidence from the work of Suttle & Price (1976) that such individual variation was much less on a herbage diet. In the grazing sheep Wiener & Field (1970) have reported hereditable differences among three breeds and their crosses in susceptibility to swayback and copper poisoning, the breed that was most susceptible to swayback being least susceptible to copper poisoning. Marked differences in the hepatic retention of dietary copper have been described for three other breeds of sheep by Lüke & Wiemann (1970).

Although individual differences exist in the severity of the clinical response of cattle of apparently identical copper status, genetic differences in the efficiency of absorption or hepatic retention of copper have not, as yet, been detected in this species. Since they have now been detected in sheep, the human infant (Danks et al 1972) and mice (Hunt 1974), it is improbable that the bovine will prove an exception. We are thus faced with the possibility of divergences both within and between breeds from the mean requirement, regardless of the steps taken to account for differences due to dietary or physiological conditions.

4. Derivation and validation of factorially-derived statements of copper requirement

(a) Sheep

Gross dietary requirements of sheep for copper from birth through to lactation are given in Table 6.3, together with the net requirements and availability factor

Table 6.3   Estimated requirements of sheep for copper*

	Live weight (kg)	Rate of gain, stage of pregnancy or milk yield	Net requirement (mg Cu/day)	Coefficient of absorption†	Dietary requirement (mg Cu/day)	Relative requirement (mg Cu/kg diet DM)‡
Growing lamb (castrate)	5	0.150 kg/day gain	0.19	0.90	0.21	1.0 (0.2)
	10	0.150	0.21	0.53	0.40	1.0 (0.4)
	20	0.150	0.25	0.22	1.1	1.7–1.9 (0.64–0.57)
	40	0.075	0.24	0.09	2.7	2.7–4.5 (1.0–0.6)
		0.150	0.33	0.09	3.7	2.6–4.6 (1.4–0.8)
		0.300	0.50	0.09	5.6	5.1 (1.1)
Adult	50	0	0.22	0.06	3.7	4.6–7.4 (0.8–0.5)
Pregnant ewe (twin foetuses)	75	Last 7 weeks	0.63	0.06	10.5	6.2–7.5 (1.7–1.4)
Lactating ewe	75	1 kg milk	0.52	0.06	8.7	4.6–5.8 (1.9–1.5)
		2 ,,	0.74	0.06	12.3	4.4–5.6 (2.8–2.2)
		3 ,,	1.24	0.06	20.7	5.6–8.6 (3.7–2.4)

* Calculated from the following components: endogenous loss, 4 μg Cu/kg LW/day; growth, 1.15 mg Cu/kg LW gain; milk, 0.32 mg Cu/kg (early lactation) to 0.22 mg Cu/kg (late lactation); pregnancy 0.32 mg Cu/day; wool, 5 mg Cu/kg DM.
† When dietary Mo and S contents are known, appropriate coefficients may be selected from Fig. 6.1. It is probable that the coefficient will be 0.03 or less when dietary Cd > 3 mg/kg, Zn > 200 mg/kg, Fe > 1 g/kg.
‡ Figures in parentheses are assumed DM intakes. For ruminating sheep, intakes are based on q ranges from 0.5 (or 0.6) to 0.7, for the adult from 0.4 to 0.6, for the pregnant ewe from 0.5 to 0.7, and for the lactating ewe from 0.5 to 0.6 (low yield) or 0.7 (high yield).

appropriate for the physiological or productive state of the animal. For preruminant lambs, true availability has been assumed to decrease from 0.90 to 0.22 in the first 6 weeks of life. The susceptibility of sheep to copper poisoning led us to give preference to a high factor, 0.06, for availability in computing requirements for more mature animals: this appears to be particularly appropriate for animals receiving mixed concentrate and roughage diets of relatively low sulphur content (<1.5 g/kg DM). It is noteworthy that late pregnancy and early lactation represent times of peak copper requirement. From the relationships given in Fig. 6.1 a factor of 0.04 would seem appropriate for herbage of low molybdenum (2 mg/kg DM) and normal sulphur content (2.5 g/kg) and 0.02 for herbage of high molybdenum and sulphur contents (5 mg Mo and 4 g S/kg DM). When heavy metal contamination from industrial or other sources results in dietary cadmium >3 mg/kg, zinc >200 mg/kg and iron >1 g/kg, a factor of 0.03 (i.e., a doubling of gross requirement) seems appropriate. As our knowledge of factors affecting copper availability becomes more complete it will no doubt become necessary to modify the list of conditions for which low availability factors are appropriate.

The recommendations do not take into account individual or genetic differences in availability and it is conceivable that extremely disparate animals could succumb to copper deficiency or toxicity on copper intakes close to those recommended. There is, however, evidence to suggest that the tabulated requirements provide a reasonable basis for assessing the copper nutriture of sheep. Regressions of copper balance (Murty 1957) and liver copper concentration (Abdellatif 1968) against intake on hay/concentrate diets of high expected availability (0.06) have indicated that the maintenance requirement of sheep is of the order of 3.5–4.0 mg/day as predicted in Table 6.3. The occurence of copper poisoning on similar diets providing 7–15 mg Cu/day (e.g., Hogan et al 1968, Suttle 1968, Adamson 1969) is not unexpected since the storage of 6% of the excess copper intake in the liver (of 300 g DM) would cause liver copper to increase by 8 to 15 μg/g DM per week. On the other had the presence of 10–20 mg Cu/kg DM in swayback pastures (e.g., Alloway 1973) would be regarded as marginal if availability was reduced to 0.02 by antagonists.

Table 6.4 Estimated requirements of cattle for copper*

Class of animal	Live weight (kg)	Rate of gain, stage of pregnancy, or milk yield	Absolute requirement (mg Cu/day)		Relative requirement (mg Cu/kg diet DM)‡
			Net	Dietary†	
Pre-ruminant calf	40	0.5 kg/day gain	0.81	1.16	1.2 (1.0)
Growing bullock, (medium sized breed)	100	0.5 1.0	1.24 1.76	31 44	8–14 (4.0–2.2) 10–16 (4.5–2.8)
	200	0.5 1.0	1.94 2.45	48 61	8–14 (6.0–3.5) 9–15 (6.5–4.0)
	300	0.5 1.0	2.64 3.16	66 79	8–15 (8.0–4.5) 9–15 (8.4–5.4)
Adult	500	0	3.50	88	12–19 (7.6–4.6)
Pregnant cow	500	7 months 9	4.56 5.64	114 141	13–17 (9.0–6.6) 13–20 (11.2–6.9)
Lactating cow	500	10 kg milk 20 30	4.50 5.50 6.50	113 138 163	10–14 (11.7–8.3) 8–11 (17.8–12.7) 8–11 (19.2–15.5)

* Calculated from the following components: endogenous loss, 7 µg Cu/kg LW per day; growth, 1 mg Cu/kg gain; milk, 0.1 mg Cu/kg; pregnancy (see Table 6.2).
† Coefficient of absorption assumed to be 0.7 for the 40 kg preruminant calf, and 0.04 for all ruminating cattle. When dietary S and Mo contents are known, appropriate coefficients for ruminating cattle may be selected from Fig. 6.1. It is probable that the coefficient will be 0.03 or less if dietary Cd is >3 mg/kg, Zn > 200 mg/kg, Fe > 1 g/kg.
‡ Figures in parentheses are assumed DM intakes. For ruminating cattle intakes are based on metabolizable energy requirements and likely values for metabolizability (q). For slow growth and pregnancy q = 0.4–0.6; for rapid growth and low or moderate milk yield, q = 0.5–0.65; for high milk yield, q = 0.6–0.7.

(b) Cattle

Estimates of the net and gross requirements of cattle for copper are presented in Table 6.4. The preruminant calf has been assumed to absorb copper with an efficiency of 0.70 and 0.50 in the first and third months of life, respectively. The gross requirements of weaned cattle have been derived by assuming a coefficient (A) of 0.04 on the basis of information obtained with sheep given predominantly grass or roughage diets. Situations in which the dietary content of copper antagonists justifies the use of low coefficients are similar to those given for sheep (p. 231) and are shown in a footnote to Table 6.4. It is interesting to note that the predicted requirements for cattle are approximately double those for sheep but relatively constant for the different classes of animal.

It is impossible to validate our estimates of requirements for cattle without further assumptions because all the complete sets of data were used in deriving the requirements. Assuming that DM intakes were the same as those in Table 6.4 and that sulphate generally constitutes one-third of the total sulphur in diets (e.g., Furrer 1966, Hartmans 1970), a further six studies provide comparable data for molybdenum and sulphur intakes (Havre, et al 1960, Havre & Dynna 1961, Todd, et al 1967, Thornton et al 1972, Bingley & Anderson 1972, Campbell et al 1974). In each case available copper concentrations in the diet were below the recommended value and in each case a decline in copper content of plasma or liver, the presence of clinical signs of deficiency or a response to copper therapy confirmed that the dietary supply was inadequate. We conclude that our tentative estimates of endogenous loss and availability, taken together, do not generally overestimate copper requirements. We are, however, aware of a few instances in which hay and concentrate diets of low copper content (<5 mg Cu/kg) have been offered to lactating or reproducing stock with little or no evidence of a reduction in body copper (Vanderveen & Keener 1964, Engel et al 1964). Such findings suggest that an absorption coefficient of ·0.06, similar to that used generally for sheep, might be appropriate for cattle given hay and concentrate diets and in some ill-defined circumstances might even overestimate requirements.

5. Copper toxicity

The occurrence of chronic copper toxicity is a major problem only in the intensive rearing of sheep. Calves also may develop copper toxicosis (Shand & Lewis 1957), but adult cattle are extremely tolerant of high copper intakes, only one possible cause of chronic copper toxicosis having been reported (Todd & Gribben 1965). It has been suggested that ewes are particularly susceptible to copper intoxication during late pregnancy (Simesen & Møller 1969, Suveges et al 1971).

Grazing sheep are rarely affected by copper toxicosis, except when pastures have been contaminated with copper, as from fungicides, or contain subterranean clovers of low molybdenum content (Bull 1951). There is some dispute as to whether the disposal of copper-rich pig slurries on grazing land constitutes a real hazard to sheep, as has been suggested by Van Ulsen (1972) and Feenstra & Van Ulsen (1973). Uptake of this copper by the plant is limited, but considerable contamination of grass or of soil surface residues may occur (Batey et al, 1972). Some of this copper can be absorbed by sheep (Dalgarno & Mills 1975, Price & Suttle 1975) and it was estimated that if slurry residues constituted 2% of their daily dry matter intake for a period of 6 months, a hazard might exist. However, in trials where sheep have grazed severely contaminated pastures for three successive years, no clinical case of toxicosis was reported (Woodside 1973) although there has been some evidence of tissue damage, judging by increased serum activity of aspartate aminotransferase and other enzymes indicative of tissue damage (Gracey et al 1976, Kneale & Howell 1974).

It is not possible to relate the incidence of copper toxicosis to the dietary copper content because of the dependence of availability of copper on dietary composition, and on the age, breed and physiological state of the animal. The onset of the terminal acute stage of the illness, when massive haemolysis occurs, can be precipitated by stress, change in diet and other unknown factors.

Copper poisoning has been reported in sheep on diets containing only 8 mg Cu/kg (Buck 1970), but Hemingway & MacPherson (1967) estimated that a daily intake of 38 mg copper for 16–20 weeks would be necessary to increase liver concentrations to 1000 mg/kg DM, and put animals at risk. This is in agreement with the report of a 35% incidence of toxicosis in feeder lambs on a barley-soya bean ration, with a daily copper intake of 40 mg for 16 weeks (27 mg Cu/kg diet) (Tait et al 1971). Van Adrichem (1965) recorded three cases of copper toxicosis within 21 weeks in five sheep receiving hay (8 mg Cu/kg) and concentrates (24–32 mg Cu/kg). Liver copper concentrations were 2150 mg/kg DM. Three of seven lambs on a barley-fish meal diet, containing 29 mg Cu/kg (largely from the inclusion of brewers' yeast and distillers' solubles) died within 22 weeks (Bremner et al 1976). Copper intoxication of growing lambs has also been reported with rations providing as little as 8 mg Cu/kg DM (Lüke & Marquering 1972). Such reports illustrate both the relative susceptibility of sheep to copper poisoning and our ignorance of factors which provoke its appearance.

Animals on milk-based rations may be at some risk from copper toxicosis, as hepatic retention may be equivalent to about 50% of the dietary copper intake in these animals (Bremner & Dalgarno 1973a, b, Suttle 1973). Addition of $CuSO_4$ to a milk-replacer diet for five calves to give copper concentrations of 50–300 mg/kg DM cause four deaths after 2–11 weeks (Weiss & Baur 1968). Possible reductions in weight gain have been reported in veal calves receiving a supplement of only 5 mg Cu/kg DM, without clinical signs of toxicosis (Bremner & Dalgarno 1973b). Similar effects of copper on weight gain have been reported in sheep on diets containing 30–40 mg Cu/kg (Hill & Williams 1965, Tait et al 1971), although in most experiments effects on growth and feed intake were apparent only in the terminal stages of the illness.

Control of copper toxicosis after onset of the haemolytic crisis has proved difficult. Several attempts have been made, using molybdenum and sulphate, without consistent success (e.g. Todd et al 1962). Long-term dietary supplementation with molybdenum (about 50 mg as ammonium molybdate) and sulphate (1 g) was shown to decrease liver copper accumulation and associated liver damage and prevent the onset of clinical toxicosis (Ross 1964 1966). More recently, liver copper concentration and serum aspartate aminotransferase activity were reduced by about 40% in growing lambs on a commercial diet (with 0.42% sulphate ion) when the molybdenum concentration was increased from 1.8 to 7.7 mg/kg (Harker 1976). Suttle (1977) has reported that the addition of only 4 mg molybdenum and 2.0 g sulphate/kg diet has protected growing lambs from copper poisoning, decreased liver copper content and prevented any increase in plasma aspartate aminotransferase activity; the basal diet contained 45 mg Cu, 0.7 mg Mo and 1.96 g S/kg. Increasing the dietary zinc concentration was found to be effective in reducing the incidence of copper toxicosis in growing lambs on a practical type of diet, with no serious side effect (Bremner et al 1976). Increase in dietary protein concentration from 10 to 20% has been shown to decrease liver copper retention and the incidence of toxicosis in sheep receiving 25 mg Cu/day (MacPherson & Hemingway 1965).

Iron

Interest in the iron requirements of ruminants centres primarily on the needs of the young animal maintained on milk or milk substitutes, such as artificially reared lambs and veal calves. With these animals, failure to meet iron requirements may retard growth and induce anaemia. Differing iron requirements for growth, for haemoglobin production and for muscle myoglobin formation demand careful regulation of iron supply to young stock and particularly to those offered diets based upon liquid milk or milk substitutes often of low iron content.

The definition of Fe requirements for older stock is less critical as instances of iron deficiency of nutritional origin are rare although evidence exists that utilization may be impaired during the later stages of copper deficiency, cadmium intoxication and possibly, as a secondary consequence of many infections.

1. Changes in tissue iron content during the development of iron deficiency

Substantial individual variations exist in hepatic iron reserves accumulated during the terminal stages of foetal development. Charpentier (1966a), in one study, found between 226 and 540 mg non-haem iron in neonatal calf liver but quoted other instances in which as little as 20–40 mg was found. The sequence of events subsequently arising during the development of iron deficiency has been described by Charpentier (1966a) in a study with Friesian calves maintained on liquid diet providing approximately 5 mg Fe/kg DM for 1½ to 3 months. Despite a substantial decline in blood haemoglobin concentration and in the concentration of myoglobin and associated haem pigments, resulting in the production of a pale musculature, the total pool of iron in blood increased by as much as 750 mg and that of muscle by 200 mg. Much of this increase in extrahepatic tissue iron is met by withdrawal of iron from the liver (about 450 mg), the balance presumably being accounted for by utilization from the diet which in these experiments provided 498 mg iron during the period. These observations are summarized in Table 6.5 with data derived from two other studies which, together, illustrate the extent to which hepatic iron is mobilized and redistributed during early postnatal life and thus may influence the interpretation of studies of the requirement for dietary iron. No comparable data are available to show whether this situation applies to the newborn lamb.

Table 6.5   Estimated changes in the body content and tissue distribution of iron in calves fed on liquid diets differing in iron content.

Live weight (W, kg):	40†	approx. 100			
Fe intake (mg) from 40 to 100 kg:	—	375	1120	3100	10350
Total organ content of Fe (mg)					
Liver	415	33	34	53	65
Muscle	325	330	455	530	1012
Blood	1645	985	1340	2200	2300
Estimated body Fe content (mg Fe/kg W)	60	13	18	28	34
Reference	(1)	(2)	(3)	(3)	(3)

References from which estimates derived:
1. Charpentier (1966a, b).
2. Roy et al (1964).
3. Bremner & Dalgarno (1973b); McDougall et al (1973).
† birth weight

Influence of iron supply on growth

Progressively decreasing growth rates have been noted in calves offered fresh milk diets unsupplemented with iron (Blaxter et al 1957). Reduced growth associated with a decline in feed consumption occurred after animals had been so treated for more than 8–12 weeks, by which time a moderate anaemia had developed (haemoglobin about 70 g/litre).

Adverse effects on growth rate or efficiency of feed utilization by calves have been noted in other studies in which diets had the following contents of iron (mg/kg DM): 10 (Bremner & Dalgarno 1973b), 14 (Bremner et al 1976, Donnelly & Bremner, unpublished data) and 21 (Eeckhout et al 1969a).

In an experiment with newly-weaned lambs offered a semisynthetic diet Lawlor et al (1965) found that, whereas 25 or 70 mg Fe/kg DM was enough to maintain normal growth rate, 10 mg Fe/kg rapidly induced growth failure and poor appetite.

The impression is gained from these studies and from those of Roy et al (1964), Matrone et al (1957) and Kirchgessner et al (1971) that the minimum dietary iron content necessary to meet requirements for growth may be about 25 mg/kg dietary dry matter. However, this conclusion is based on experiments in which a high proportion of the dietary iron has been in the form of soluble iron salts and only the investigation of Matrone et al continued for longer than three months.

The influence of iron supply on blood haemoglobin and muscle myoglobin

The incidence of iron deficiency in newborn calves has been examined by Hibbs et al (1963), who found that 30% of calves born to dams of normal haematological status had haemoglobin (Hb) values of less than 90 g/litre with a range of 50–150 g/litre. Injection of 500 mg iron as iron dextran was effective in raising haemoglobin values but had no influence on the subsequent weight gain of animals.

Calves consuming whole milk alone have long been known to develop anaemia which can be prevented or overcome by the administration of iron either by injection or orally. Blaxter et al (1957) found that Ayrshire calves became severely anaemic when offered only whole milk containing about 0.3 mg Fe/kg milk. A daily supplement of 20 mg iron as ferric citrate was sufficient to prevent all but a mild anaemia (Hb 79–84 g/litre at 18 weeks). Although in a field trial a supplement of 35 mg Fe/day reduced the proportion of calves with Hb<80 g/litre, that was not adequate to maintain the blood Hb of all calves above this figure. It was estimated that the net iron requirement of a calf gaining 1 kg/day for the synthesis of haemoglobin, myoglobin and iron-containing enzymes is about 50 mg/day. In a similar experiment with Friesian calves (Matrone et al 1957), a daily supplement of 30 or 60 mg iron was sufficient to maintain normal red cell volumes and blood haemoglobin concentrations (100–130 g/litre) or to increase the Hb of anaemic calves from 54 to 80 g/litre during 16–20 weeks. There was no difference in blood haemoglobin concentration between calves receiving 30 mg and those receiving 60 mg Fe/day.

The effects of intramuscular injections of a total of 1500 mg iron as iron dextran during the first three weeks of life on the performance and haematological status of Ayrshire calves reared on whole milk were examined by Roy et al (1964). Iron supplementation reduced the decline with age of packed cell volume, haemoglobin, mean corpuscular volume and mean corpuscular haemoglobin concentration. Red cell count also declined with age but was unaffected by iron supplementation.

In the commercial rearing of calves for veal the diet usually consists of a fat-supplemented skimmed milk preparation. Such diets produce greater weight gains than whole milk and might therefore be expected to increase the rate of development of anaemia (Roy et al 1964). The influence of variations in the iron supply to Ayrshire bull calves reared on fat-supplemented skimmed milk for 14 weeks was studied by Bremner & Dalgarno (1963b). Calves were given either an unsupplemented diet containing 10 mg Fe/kg DM or diets supplemented with ferrous sulphate to contain either 40 or 100 mg Fe/kg DM. Even 100 mg Fe/kg was insufficient to prevent an initial decline in blood Hb concentration and in packed cell volumes during the experiment. The final Hb concentrations obtained were, however, similar to those reported for adult cattle (110–120 g/litre; Underwood 1977) and there was little difference in the final Hb concentrations attained when the dietary iron was either 40 or 100 mg/kg diet. At no time were the mean Hb concentrations in these calves indicative of anaemia (i.e., less than 80 g/litre blood). Calves receiving 10 mg Fe/kg diet became progressively more anaemic as the experiment continued, with blood Hb concentrations of 66 g/litre ultimately attained. Reduced myoglobin and haem pigment concentrations in skeletal muscle and cytochrome concentrations in heart were found with either 10 or 50 mg Fe/kg diet (McDougall et al 1973, Bremner & Dalgarno 1973a). It was concluded that a dietary intake of 40 mg Fe/kg DM appears to be sufficient to prevent all but a very mild anaemia provided the iron is present in a soluble form, but that muscle myoglobin content (and thus flesh colour) increases if the dietary content of iron is increased from 40 to 100 mg/kg.

In a similar experiment with Friesian calves (Bremner et al 1976, Donnelly & Bremner, unpublished) a dietary iron content of 24 mg/kg DM was adequate to prevent all but a very mild anaemia from developing over a 14-week period. Again there were differences in the response of blood Hb and muscle pigment concentrations to iron intake. Although Hb values in calves receiving 24 mg Fe/kg were significantly higher than in those receiving 14 mg/kg, the myoglobin contents of the muscles were low in both groups and not significantly different, which suggests that Hb synthesis takes priority over myoglobin synthesis for available iron supplies. Muscle myoglobin content in the Friesian calf was almost doubled by increasing the dietary iron content from 24 to 44 mg/kg and increased further at 104 mg/kg diet.

Similar observations illustrating different requirements for Hb synthesis and muscle pigment formation have been made by Eeckhout et al (1969a, b). A dietary iron content of 26 mg/kg DM was sufficient to maintain normal Hb concentrations but, as in Ayrshire cattle, appreciable increases in muscle haem pigment concentration became evident only when the dietary iron was 50–60 mg/kg. The initial iron status of the calf (as judged by initial blood Hb concentrations) had a marked effect on the relationship of dietary iron concentration to the haem pigment concentration of muscle observed at the end of 14 weeks of treatment.

4. Availability of dietary iron

Several estimates have been made of the availability of soluble iron sources added as supplements to milk or milk substitute diets for calves. Most are based on the calculation of the coefficient of absorption of ingested iron for Hb synthesis. Few take adequate account of the probability suggested by Charpentier's data (1966a, b) that individual variations in the extent of tissue iron depots at birth can influence the utilization of dietary iron for Hb synthesis and the synthesis of haem pigments in muscle during the first 10–12 weeks of postnatal life.

Matrone et al (1957) estimated that the absorption coefficient of soluble iron in the form of supplements of ferric chloride was 0.60 when the diet of the young calf provided 30 mg iron and 0.30 when the diet provided 60 mg daily. Bremner & Dalgarno (1973a), again using estimates of Hb synthesis as the criterion, calculated that the absorption coefficient of the iron in a diet containing 10 mg/kg was 0.72, declining to 0.43 when the soluble iron sources, ferrous sulphate or ferric citrate, provided a total iron content of 40 mg/kg diet. In this instance utilization was estimated in calves 9–12 weeks old and thus the estimates are unlikely to be vitiated by the early postnatal utilization of hepatic iron. Thus the true availability of iron decreases as its dietary concentration increases.

A number of studies have been undertaken on the availability of iron from a variety of sources, with Hb response, change in tissue myoglobin content or tissue iron retention used as criterion of comparative availability. Kirchgessner et al (1971) found no detectable difference in the availability of iron from iron fumarate or sulphate. The availability of iron from ferrous sulphate, ferric citrate and the ferric complex of ethylenediaminetetra-acetic acid for Hb production in calves was found to be about 0.43 when the total dietary iron was 40 mg/kg. Iron from a ferric phytate complex was used less efficiently, 0.33 (Bremner & Dalgarno 1973a).

Using the faecal excretion and tissue retention of ^{59}Fe as the basis for comparing availability, Ammerman et al (1967) found that the iron of ferric chloride was 3 to 4 times as available as that of ferric oxide (Fe_2O_3) to iron-depleted calves. In a second experiment, although incomplete recovery of ferric oxide was obtained in the faeces, no ^{59}Fe activity was detected in tissues which suggests that, in this preparation, the iron was completely unavailable. In studies with sheep maintained on a roughage diet, Ammerman et al (1967) found no significant difference in the tissue retention or rate of faecal excretion of ^{59}Fe administered as sulphate, chloride or carbonate, while a similar study by Georgi (1964) suggested that the fractional absorption and tissue retention of ^{59}Fe administered as ferrous sulphate into the abomasum of the sheep might be as high as 0.16. It was less than 0.04 when radioactive iron was given in the form of haemoglobin.

5. Iron requirements   *(a) Calves*

Table 6.6 summarizes observations on the influence of variations in dietary iron content on the performance of calves offered liquid milk or milk substitute diets supplemented with soluble iron salts. For purposes of comparison a live weight of 100 kg has been assumed and, where necessary, the author's original data have been interpolated. Comparison of effects on flesh characteristics at slaughter has been on the basis of haematin contents of *longissimus dorsi* muscle, this being the only relevant measurement common to several studies.

Diets providing 25–40 mg Fe/kg DM frequently result in a depletion of the haem pigment content of skeletal muscle. Between these limits only a transient decline in blood Hb concentration is likely and only one study (Kirchgessner et al 1971) of those reviewed suggests that adverse effects on growth are possible. A depressed growth rate appears likely if the diet contains less than about 25 mg Fe/kg DM.

The gross iron requirement for growth and the maintenance of normal Hb concentration of blood and haem pigment content of muscle in calves is met only by diets providing at least 66 mg Fe/kg DM and studies with Friesian and Ayrshire breeds suggest it may be as high as 100 mg Fe/kg DM.

Table 6.6   Summary of observations on the influence of iron intake of calves weighing c. 100 kg on growth, changes in blood haemoglobin concentration and muscle pigmentation.

Dietary Fe concentration (mg Fe/kg DM)	Estimated gross Fe intake (mg/day)	Response to diet			
		Growth	Haemoglobin	Muscle pigmentation at slaughter	Reference
<14	<20	D	D	D	see text
	c. 23		D		1
	c. 38		D		1
21	34	D	D	D	2
	30		SD	?	3
24	48		SD	D	4
36	57		SD	SD	2
	60		SD	?	3
40	69		SD	D	5, 6
44	83		N	SD	4
51	82		N	SD	2
40	58	D	D	SD	7
66	106		N	N	2
100	172		N	N	5, 6
100	168		N	N	7
104	197		N	N	4

References
1. Blaxter et al (1957).
2. Eeckhout et al (1969a).
3. Matrone et al (1957a).
4. Donnelly & Bremner (unpublished).
5. Bremner & Dalgarno (1973b).
6. McDougall et al (1973).
7. Kirchgessner et al (1971).

Legend: D = depressed rate of gain, low Hb or muscle pigmentation.
SD = slightly depressed gain, Hb or pigmentation.
N = normal.
For haemoglobin: D = <90 g/litre.
SD = 90–110 g/litre.
N = >110 g/litre.
for muscle pigmentation: N = >50 mg haematin/kg fresh longissimus dorsi.
SD = 50–40 mg   ,,   ,,
D = <40 mg   ,,   ,,
,,   ,,   ,,

Matrone et al (1957) suggested that the net requirement for Fe for 1 kg daily weight gain is about 16 mg Fe/day and that the demand to replace the iron of Hb lost by erythrocyte turnover in a 200-kg calf is 1.2 mg Fe/day. Higher estimates of the iron content of 1 kg gain are presented in Table 6.5 for calves of about 100 kg live weight, 28 and 34 mg Fe/kg at the iron intakes found adequate to prevent anaemia and to maintain muscle haem pigments, respectively. Estimates derived from data in iron-deficient calves suggest that if body iron retention is only 13 mg/kg live weight (from Roy et al 1964) or 18 mg/kg (from Bremner & Dalgarno 1973b), growth may be retarded.

If the incremental tissue retention of iron for 1 kg gain is taken to be about 30 mg, this net requirement could be met by assuming that, in the studies summarized in Table 6.6, the true availability of iron from soluble sources added to a liquid diet declines from 0.40 to 0.15 as the iron content of the diet is increased from 40 to 100 mg/kg.

From the data of Matrone et al (1957) it seems probable that the iron requirement of the milk-fed calf may decline with increasing age. Thus at about 100 kg live weight an estimated iron intake of 32 mg/day (0.32 mg/kg live weight) was insufficient to prevent a transient but significant decline in blood Hb concentration. In contrast, as the same calves grew between 150 and 270 kg on a constant total iron intake, blood Hb concentration rose progressively. Over this period the iron intake expressed on a body weight basis declined from 0.22 to 0.13 mg/kg.

The requirement of weaned, ruminating cattle has not been assessed and the suggestion in the first edition that 30 mg Fe/kg dietary DM should be adequate has been neither verified nor contradicted.

The requirement for iron during the terminal stages of pregnancy to achieve a tissue iron content of the calf of 60 mg/kg weight at birth (see Table 6.5) may be greater. Even if the increasing iron content of the adnexa is ignored the iron content of the developing foetus will impose an additional net demand of 15, 20 and 30 mg daily during the last 3 months of development. If the true availability of iron from a roughage diet is assumed to be 0.20 (see data for sheep of Hoskins & Hansard, 1964), the gross requirement during the period may be 40–50 mg Fe/kg diet if dry matter intake is about 8.5 kg/day.

(b) Lambs and mature sheep

Although an anaemia responsive to the addition of about 14 mg Fe/kg dietary DM has been reported to develop in artificially reared lambs offered a liquid diet of skimmed milk plus fat (Brisson & Bouchard 1970), the iron requirements of the milk-fed lamb have not been studied experimentally. Milk substitute diets with iron content similar to that of ewe's milk (0.77±0.05 mg/litre; Ashton & Yousef 1966) have given satisfactory growth of lambs but the haematological status of such animals has not been investigated (E. R. Ørskov, personal communication). A content of 24 mg Fe/kg semisynthetic diet was found to be adequate to promote good growth and a normal haematological status in weaned lambs by Lawlor et al (1965) but 10 mg Fe/kg was grossly inadequate.

From balance experiments Hoskins & Hansard (1964) have estimated the gross requirement of ewes to be at least 34 mg/day (about 20 mg Fe/kg diet DM) during the final stages of pregnancy. In this study the true availability of iron from a ration of maize, soya bean oilmeal and cottonseed hulls was estimated to be 0.21.

Table 6.7 summarizes provisional estimates of iron requirements.

Table 6.7   Provisional estimates of iron requirements of ruminants.

		Dietary Fe (mg/kg DM)	Remarks
Cattle	Calves <150 kg	40	Adequate for growth
		100	Adequate for growth, haemoglobin and myoglobin production
	Calves >150 kg and mature animals	c. 30	
	Cattle, pregnant 6–9½ months	c. 40	
Sheep		c. 30	

. Iron toxicity

In cows pronounced loss of weight and an increased incidence of scouring resulted from the consumption of irrigated pasture high in iron (intake 73.5 g/day) (Coup & Campbell 1964). As part of the same investigation Coup & Campbell gave milking cows grazing on unirrigated pasture 15, 30 or 60 mg Fe/day as $Fe(OH)_3$. The minimum dose of iron depressing milk yield and herbage digestibility was less than 30 g/day and for butterfat and live weight below 60 g/day.

The effects of lower iron intakes on the performance of bullocks were examined by Standish et al (1969, 1971). A basal diet containing 77 to 100 mg Fe/kg was given to 24 bullocks of about 200 kg liveweight, with or without supplements of 400 to 1600 mg Fe/kg as $FeSO_4$. Average daily feed intake and weight gain were decreased by more than 500 mg Fe/kg diet and with 1000 to 1600 mg Fe/kg, plasma copper was reduced and liver and plasma phosphorus fell. The iron contents of liver, spleen, kidney and heart increased and in the liver, copper and zinc contents decreased.

Koong et al (1970) found that <2500 mg Fe/kg diet depressed the rate of gain o Friesian calves. Adverse effects on phosphorus metabolism were detectable when the diet contained >1000 mg Fe/kg and abnormally high blood Hb concentration resulted from dietary contents >500 mg Fe/kg.

We suggest that 500 mg Fe/kg diet should be regarded as the maximum tolerabl by ruminants.

Cobalt

The effects of cobalt deficiency in cattle and sheep are those of vitamin B_{12} deficienc and range from mild deficiency with an ill-defined and sometimes transient unthrif tiness with no clear clinical sign, to moderate or severe deficiency with appetite failure emaciation and listlessness accompanied by a characteristic pallor of the skin an mucous membranes caused by a progressively increasing anaemia. Viability of calve and lambs and milk yield are reduced by a moderate or severe deficiency.

The essentiality of cobalt in ruminant nutrition reflects the importance of its rol as an essential component of vitamin B_{12} (5,6-dimethylbenzamidazolyl cobamid cyanide). As cobalt intake declines, the synthesis of vitamin B_{12} by rumen microor ganisms is inhibited, tissue concentrations of vitamin B_{12} fall as reserves are consume and, ultimately, the activity of metabolic processes dependent on enzymes havin vitamin B_{12} as a cofactor declines. Particularly important among the enzymes i methylmalonyl coenzyme A mutase involved in the initial steps during the utilizatio of propionic acid as an energy source.

Although knowledge of the role of cobalt in ruminants has extended since the firs edition, quantitative aspects of the processes of cobalt utilization have received littl attention. Thus when attempting to define cobalt requirements we have continued t rely mostly on experimental and field observations describing the effects of a rang of dietary cobalt intakes on performance and health but, wherever the informatio is available, we have also considered their effects on tissue contents of vitamin B_{12}.

1. Utilization of dietary cobalt for vitamin B_{12} synthesis

The parenteral administration of vitamin B_{12} or vitamin $B_{12(b)}$ (its hydroxoanalogue is fully effective in preventing or promoting recovery from the effects resulting i ruminants of cobalt-deficient diets. From such studies it has been estimated that th minimum net requirement of growing sheep for vitamin B_{12} is 11 μg/day (Smith & Marston 1970). This demand is normally met on adequate diets by ruminal synthesi of vitamin B_{12} but it has been estimated that when the cobalt content in rumen flui falls below about 20 μg/litre the rate of vitamin B_{12} synthesis by rumen microorga nisms fails to meet demands of the host (Marston et al 1961).

The efficiency with which dietary cobalt is converted into vitamin B_{12} is inversel proportional to cobalt intake. Thus Smith & Marston (1970) reported $13\pm5\%$ con version in sheep on a cobalt-deficient diet but only about 3% when intake wa adequate. This effect is partly accounted for by diversion of the synthetic activity o rumen microorganisms towards the production of an analogue of vitamin B_{12} (2 methyl adenyl cobamide) which is neither readily absorbed nor physiologically activ (Gawthorne 1970).

Although adequate quantitative data are not available it is probable that th balance of production between vitamin B_{12} and its inactive analogues is influenced b other dietary variables. Thus the work of Smith & Marston (1970), Hine & Dawbar (1954) and Dawbarn et al (1957) has clearly shown that the rumen contents o pasture-fed animals have significantly higher proportions of vitamin B_{12} in a physio logically active form than the rumen contents of animals maintained on experimental diets based on chopped hay or straw with similar contents of cobalt. Elliot et al (1971) found a lower concentration of active vitamin B_{12} in sheep given a pelleted barley, soya bean, white-fish meal and molasses diet than those given pelleted grass meal. Furthermore, Sutton & Elliot (1972) found that increasing proportions of ground maize in a maize and hay ration significantly reduced vitamin B_{12} production. Unfortunately, the amounts of cobalt provided in most of these studies were such that cobalt deficiency was unlikely to arise and thus it is not possible to show whether such observations are strictly relevant to situations where cobalt intake is marginal or low. Nevertheless, they suggest that a larger margin of safety may be appropriate when considering cobalt supplementation of high production rations.

Sutton & Elliot (1972) found a positive correlation between vitamin B_{12} production in the rumen and the intake of digestible dry matter. On the other hand, Smith & Marston (1970) found no relationship between feed intake and the efficiency of conversion of dietary cobalt to vitamin B_{12}.

At present, there is little evidence that variations in the inorganic composition of ruminant diets influence the availability of cobalt. Spais et al (1966) obtained increased rates of weight gain by administering cobalt supplements to lambs grazing pastures with a high content of inorganic sulphate (>10 g SO_4^{2-}/kg DM) even though the cobalt content of herbage was 0.18 mg/kg and previously believed to be adequate. The possibility that variations in the dietary content of sulphur sources might influence utilization of cobalt has not been investigated further.

2. Relationship of cobalt intake to weight gain and tissue vitamin B_{12} content

In the previous edition the conclusion that approximately 0.1 mg Co/kg dietary DM probably met the requirements of ruminants for this element was based largely on consideration of a range of *ad hoc* observations made during the course of field studies on the incidence of cobalt deficiency. A later survey by Andrews (1965) led him to conclude that pasture cobalt contents of 0.11 mg/kg DM or higher are probably adequate and that mean values approaching 0.08 mg Co/kg may be regarded as marginal. The validity and limitations of these conclusions have become apparent since the publication of several reports describing the influence of variations in cobalt intake of sheep on their tissue contents of vitamin B_{12} which, incidentally, provide a more realistic guide to the adequacy of cobalt status than estimates of tissue cobalt content (for review see Underwood 1977).

Dawbarn et al (1957), Andrews & Stephenson (1966) and Somers & Gawthorne (1969) have suggested that serum vitamin B_{12} concentrations below 0.20 μg/litre are strongly indicative of cobalt deficiency and that 0.25–0.30 μg/litre may be marginal. From similar studies of the vitamin B_{12} content of liver, which provides a better indication than serum of the inadequacy of tissue vitamin B_{12} reserves, it has been concluded (Andrews & Isaacs, 1964; Marston, 1970) that the following criteria are applicable (units: mg vitamin B_{12}/kg fresh liver: severe deficiency, <0.07; moderate deficiency, 0.07–0.10; mild deficiency, 0.11–0.19; adequate Co status, >0.19.

Marston (1970) made an extensive study of the cobalt requirements of growing sheep. Although he concluded that the minimum requirement for the maintenance of body weight was about 0.07 mg Co/day it is apparent from his report that a group receiving a total of 0.11 mg Co/day attained a final weight about 10% greater than when 0.07 mg/day was given. Higher intakes of cobalt produced no additional increment of gain.

The serum vitamin B_{12} content of sheep receiving 0.07 mg Co/day was low and never significantly different from that of groups receiving less cobalt and growing slowly or losing weight but there was an obvious trend throughout the last 50 weeks of the experiment towards higher serum B_{12} concentrations in animals receiving

0.07 mg Co/day compared with those receiving 0.03–0.05 mg/day in which serum B_{12} content fell well below the suggested critical threshold of 0.3 μg/litre. Intakes of 0.11 and 0.13 mg Co/day produced significantly higher vitamin B_{12} concentrations in serum (about 1.8 mg/litre) for the last 30 weeks of the experiment.

The above findings suggest that while 0.07 mg Co/day was adequate for satisfactory growth it maintained serum vitamin B_{12} concentrations only just above the critical threshold. Furthermore, it was apparent that to establish a satisfactory reserve of vitamin B_{12} in the liver a cobalt intake of 0.11–0.13 mg/day was needed.

3. Cobalt requirements of ruminants

Table 6.8 summarizes previous experimental findings and field observations relevant to the assessment of cobalt requirements of cattle and sheep.

Table 6.8   Experiments on the cobalt requirement of cattle and sheep

Species	Diet	Dietary Co (mg/kg DM)	Observations	Reference
Sheep (mature)	Semisynthetic	0.07–0.08	Maintains normal clinical condition but serum B_{12} marginal	1
Sheep/cattle	Pasture	0.07–0.08	Maintains normal clinical condition and adequate growth rate	2, 3, 4
Sheep (34–60 kg)	Wheaten hay, wheat gluten	0.11–0.13	Maintains adequate serum and liver B_{12} contents and adequate for growth	5
Sheep/cattle	Pasture	0.08–0.11	0.08 regarded as marginal 0.11 as adequate	{ 6 { 7
Sheep	Pasture (high sulphate)	not less than 0.18	—	8
Sheep	Hay, barley silage, urea	not less than 0.20	—	9

References
1. Somers & Gawthorne (1969).   5. Marston (1970).
2. Filmer & Underwood (1937).   6. Andrews (1965).
3. Marston (1952).   7. Andrews et al (1966).
4. McNaught (1938).   8. Spais et al (1966).
                                                    9. Modjanov et al (1962).

From these data and the previous discussion of factors believed to influence the conversion of dietary cobalt to physiologically active vitamin B_{12} we conclude that pasture or diets of conserved roughage providing 0.11 mg Co/kg DM will meet the requirements of cattle and sheep in most circumstances. We point out that clinical signs of deficiency may not arise when diets provide only 0.08–0.1 mg Co/kg but question whether such diets will maintain a liver store of vitamin B_{12}.

Observation that an increased synthesis of inactive vitamin B_{12} analogues takes place in the rumen of sheep offered diets of high metabolizable energy content suggest that the above provision may not be fully adequate in such circumstances. However all the studies of this situation were made with diets with a relatively high cobalt content and, since signs of deficiency were not apparent, they preclude estimation of the additional increment that may be needed to meet requirements on high-energy diets.

4. Cobalt toxicity

The few studies on the chronic toxicity of cobalt all suggest it has a low order of toxicity. Becker & Smith (1951) suggest that 3 mg Co/kg live weight is tolerated by sheep but intakes of 4–10 mg/kg result in appetite failure, anaemia and, at the higher intakes, death. Young cattle may be less tolerant. Thus Keener et al (1949) and Ely et al (1948) reported that about 1 mg Co/kg live weight given by mouth produced a decline in appetite. From such results it appears reasonable to suggest that the cobalt content of ruminant rations should not exceed about 30 mg/kg DM.

Selenium and Vitamin E

1. Terminology and units of vitamin E activity

The term vitamin E is used in this report to describe the total biological activity of all tocol and tocotrienol derivatives. The biopotency of isomers of these compounds differs greatly. Thus the biopotency of α-tocopherol is about one hundred times that of γ-tocopherol. Many studies report only the content of total tocopherols in diets; these are quoted only where there are good grounds for the belief that the α-isomer accounts for most of the total content of tocopherols.

One International Unit (IU) of vitamin E activity is defined as the biopotency of 1 mg dl-α-tocopheryl acetate; 1 mg d-α-tocopherol has a biopotency of 1.5 IU vitamin E activity.

2. Selenium and vitamin E in ruminants

Situations can now be recognised where productive loss or pathological conditions in ruminants can be attributed to: (i) selenium deficiency alone with no response to vitamin E prophylaxis, (ii) disorders usually arising from an excessive intake of polyunsaturated fatty acids (PUFA) and responsive only to vitamin E prophylaxis or therapy and (iii) disorders, usually involving myopathy, which are frequently responsive to either vitamin E or selenium unless the selenium intake is very low.

Vitamin E and selenium are closely and mutually involved in a variety of metabolic processes, and it is often impracticable to consider them in isolation when defining nutrient requirements. Although these nutrients are often considered to have a mutual 'sparing' action, evidence for this concept is slender. For chicks (Noguchi et al 1973) both are essential nutrients in their own right, and although such detailed studies have not yet been made with ruminants, it is probable that the same applies.

The manifestations of selenium and vitamin E deficiencies in ruminants reflect their similar but independent roles in protecting tissue membranes against damage arising from the end products of some oxidative processes (Hoekstra 1975, Diplock 1976). Thus, selenium is a component of an enzyme, glutathione peroxidase, which destroys potentially toxic peroxides while vitamin E, in close physical association with cell membranes, is believed to act as a "scavenger" of any peroxides that escape destruction.

With such closely related functions it would be surprising if the severity of adverse responses to an inadequate supply of selenium was not influenced by differences in vitamin E supply. The converse may also apply. Few studies with ruminants have taken these possibilities into account. Although the effectiveness of vitamin E in protecting against the myopathies induced by high intakes of PUFA has been demonstrated (Blaxter 1962b, Van Gils & Zayed 1966), the possibility that the prophylactic effect of vitamin E is modified by the selenium status of animals has not been examined.

The limited evidence suggesting that at "marginal" intakes of selenium the dietary content of vitamin E may influence the severity of the response of the animal will be considered later. Metabolic studies of the fate of selenium or vitamin E in singly or doubly deficient animals provide no clear evidence that either nutrient influences the efficiency of absorption or metabolism or the other (Hidiroglu 1969 1970).

3. Vitamin E and selenium in relation to diseases in ruminants

Enzootic nutritional muscular dystrophy (NMD), white muscle disease or, in lambs, "stiff-lamb" disease is the most common manifestation of selenium/vitamin E deficiency in ruminants. Calves and young cattle and lambs up to 2 months of age are most frequently affected but congenital forms giving rise to dead foetuses or death of the newborn within the first 24 hours are known. Affected animals are weak and reluctant to stand and show a stiff gait. Muscular tremor is common. Respiratory distress, cardiac arrhythmia and collapse precede death in the acute forms of the disease in which cardiac muscle has undergone extensive degeneration. A severe diarrhoea frequently develops in chronic forms of the disease and myoglobinuria resulting from kidney damage is often apparent at a late stage.

In older animals the syndrome designated "paralytic myoglobinuria" appears more typical. In yearling cattle this is again characterized by muscular stiffness and weakness but accompanied by a generalized rather than the intense focal myopathy seen in neonatal forms of the disorder. Recent studies (Anderson et al 1976, Allen et al 1975) indicate the prophylactic or therapeutic effectiveness of combined administration of selenium and vitamin E in this condition. A similar diffuse myopathy in sheep 9 to 12 months old has been described by Hartley & Dodd (1957) in New Zealand. Supporting studies again suggest that a combined selenium/vitamin E deficiency may be involved in its aetiology. A similar condition was described by Buchanan-Smith et al (1969) in adult ewes offered diets low in selenium and vitamin E; increasing the dietary content of either nutrient reduced the incidence of the disease but combined supplementation was essential for complete prevention.

Trinder et al (1969), investigating claims by Boškov (1962), found a high incidence of retained placenta at parturition in dairy cows maintained on feeds or herbage of low selenium content. Selenium or, better, combined selenium/vitamin E administration reduced or completely abolished the incidence of this disorder.

Poor growth of cattle and sheep attributable to a low dietary content of selenium has been recognized in several areas despite the absence of clear pathognomonic indications of such deficiency. Such conditions, generally referred to as selenium-responsive "ill-thrift", have been described in sheep by Hartley & Grant (1961) and Blaxter (1963) and in cattle by Jolly (1960), Wilson (1964) and Dodson & Judson (1973). Other conditions sometimes found to be responsive to an increase in selenium intake or selenium therapy are infertility associated with early foetal resorption in sheep (Hartley 1963, Godwin et al 1970), decreased wool production (Andrews et al 1968, Gabbedy 1971), periodontal disease (Cousins & Cairney 1961) and, in cattle "sawdust liver" syndrome (Todd & Krook 1966) and abortion late in pregnancy (Mace et al 1963). A comprehensive review of these and other selenium/vitamin E deficiency syndromes has been prepared by Lannek & Lindberg (1975).

4. Dietary composition and the incidence of selenium/vitamin E deficiency syndromes

Tables 6.9 and 6.10, respectively, summarize observations on the relationships between dietary selenium content or both selenium and vitamin E contents and the development of clinical or subclinical lesions indicative of deficiency of these nutrients. These tables are restricted to trials in which the prophylactic or therapeutic effectiveness of selenium or vitamin E has been demonstrated.

The data presented in Table 6.9 suggest that as the selenium content of ruminant diets falls below about 0.08 mg/kg DM, the frequency with which clinical or sub-clinical signs of selenium-responsive disorders occur increases progressively. The most frequent observations reported are clinical lesions of muscular dystrophy or indirect evidence of myodegeneration from elevated plasma contents of the enzymes creatine phosphokinase, lactic dehydrogenase or aspartate aminotransferase.

Also included is a summary of observations on the selenium-responsive "ill-thrift" in growing sheep ("hogget ill-thrift") described by Hartley & Grant (1961; Table

Table 6.9   Relationship of dietary selenium content to clinical or subclinical signs of selenium/tocopherol deficiency in sheep and cattle. (data from studies in which dietary tocopherol content not reported)

Diet		Species	Observations		Reference
Se content (mg/kg DM)	Type		Clinical/subclinical	Prophylactic response to:	
<0.01	Hay	Ewes/lambs	"High" incidence NMD*, lambs (17 flocks)		1
d.009	Hay	Ewes/lambs	6 to 30% incidence NMD, lambs (6 flocks)	Se	2
0.01	Hay	Ewes/lambs	"Occasional" NMD, lambs (7 flocks)		1
c. 0.01	Hay/silage	Cows/calves	NMD, 3/48 calves	Se + E	3
0.01–0.002	Hay	Ewes/lambs	NMD, 9/33 lambs	Se	12
0.014	Hay	Cows/calves	NMD, 14/25 calves	Se	4
0.015–0.025	Lucerne hay	Ewes/lambs	NMD, 65/121 lambs		12
0.018	Hay	Ewes/lambs	NMD, 12/30 lambs		4
c. 0.02	Hay/"corn"	Ewes/lambs	NMD, 3/80 lambs; high plasma AAT 9/11	(E, to dam, ineffective)	5
0.022	Pasture	Ewes/lambs	NMD, 4/38 lambs	Se	12
0.02–0.03	Hay/barley/silage	Cows	Placental retention at parturition	Se + E	6
0.02–0.03	Pasture	Ewes/lambs	Neonatal and delayed NMD	Se (Se partially effective)	7
<0.03	Pasture	Lambs	Poor growth	Se	7
0.03	Lucerne hay	Ewes/lambs	NMD, 40/82 lambs	(E ineffective) E or Se (ewes or lambs)	11
0.03–0.05	Hay/corn concentrates	Ewes/lambs	No clinical signs		5
0.03–0.14	Pasture	Lambs	No clinical signs		7
0.05–0.05	Pasture	Cows	Placental retention at parturition	Se + E	6
0.05	Hay	Ewes/lambs	NMD incidence 1/8 lambs		4
0.05–0.15	Grass silage/maize	Ewes/lambs	NMD in 4/13 lambs	Se	9
	Hay/or hay/concentrates	Ewes/lambs	No clinical signs 0/98		8
0.16–0.52	Grass silage/hay/corn	Ewes/lambs	No clinical signs		9
0.23	Lucerne hay	Ewes/lambs	No clinical or subclinical signs		12
0.24	—	Ewes/lambs	No clinical or subclinical signs		10
0.39	Lucerne hay	Ewes/lambs	No clinical or subclinical signs, 0/16		12

* for key see Table 6.10.

References
1. Havre & Steinnes (1968); 2. Mikkelsen & Hansen (1967); 3. Jacobsson et al (1970); 4. Jenkins et al (1974); 5. Lindberg & Jacobson (1970); 6. Trinder et al (1969); 7. Hartley & Grant (1961); 8. Mikkelsen & Hansen (1968); 9. Paulson et al (1968); 10. Pedersen et al (1972); 11. Whanger et al (1976); 12. Whanger et al (1972).

Table 6.10   Relationship of dietary tocopherol and selenium contents to development of nutritional muscular dystrophy or other selenium/tocopherol responsive disorders in sheep and cattle.

Diet			Species	Observations		Reference
Tocopherol* content (mg/kg DM)	Se content (mg/kg DM)	Type		Clinical/subclinical	Prophylactic response to:	
0.5–0.7	0.03–0.05	Hay	Ewes/lambs	No NMD but high plasma AAT 12/49		2
0.6–0.7	0.004–0.01	Hay	Ewes/lambs	NMD, 5/18 lambs		2
0.5–0.8	0.004–0.03	Hay	Ewes/lambs	NMD, 1/14 lambs; high plasma AAT 27/49 lambs	Se; E, to dam, ineffective	2
(<1 estd)	<0.01	Semisynthetic	Ewes	NMD, 9/12 mortality	Se, partial; E partial	1
8–9 (α)	0.04	Hay/barley	Rams Cattle (6–20 mo)	NMD, 3/3 mortality Myopathy/myoglobinuria, high plasma CPK	(Se + E) effective Se + E	5
9–1 (α)	0.03–0.04	Hay/barley/silage	Cattle (6.12 mo)	No degenerative myopathy		5
>13	0.084	Hay + Se	Cows/calves	No NMD	Se	7
15	0.02	Hay	Cows/calves	NMD, 21/32 calves		3
16	0.024	Hay/concentrate	Ewes/lambs	NMD, 2/34 lambs; high plasma AAT 31/34 lambs		4
17	c. 0.025–0.03	Grain/silage	Lambs	Subclinical NMD in lambs elevated serum LDH	Se (to dams)	8
>17	0.11	Hay + Se	Ewes/lambs	No NMD 0/30		7
>17	0.14	Hay + Se	Ewes/lambs	No NMD 0/8		7
c. 20	0.034	Grass silage	Ewes	Subclinical NMD in lambs elevated serum LDH	Se (to dams)	8
20 (α)	0.08	Maize/silage	Cattle	No NMD		6
20 (α)	0.18	Maize/silage	Cattle	No NMD 0/12		6
20 (α)	0.28	Maize/silage	Cattle	No NMD 0/12		6
27	<0.05	Clover hay	Ewes/lambs	NMD 11/15 lambs		9
28	<0.05	Clover hay	Ewes/lambs	NMD 15/32 lambs		9
37	0.035	Grass hay/grass silage	Ewes/lambs	No clinical NMD		9
43	>0.25	Grass/lucerne hay	Ewes/lambs	No clinical NMD 0/18 subclinical 1/18		9

* as total tocopherols unless otherwise stated.
NMD = gross or microscopic indications of nutritional muscular dystrophy with confirmation that condition responsive to prophylaxis or therapy with Se ± α tocopherol.
AAT = aspartate aminotransferase.
CPK = creatine phosphokinase.

References
1. Buchanan-Smith et al (1969); 2. Mikkelsen & Hansen (1968); 3. Jenkins et al (1970); 4. Mikkelsen & Hansen (1967); 5. Allen et al (1975);
6. Perry et al (1976); 7. Jenkins et al (1974); 8. Paulson et al (1968); 9. Schubert et al (1961).

6.9, reference 7) and Cousins & Cairney (1961), which is responsive to oral or subcutaneous administration of selenium but not to vitamin E. The absence of evidence of myodegeneration in this condition, the ineffectiveness of vitamin E prophylaxis and the fact that its incidence is high in animals at pasture where vitamin E intake is likely to be adequate, all suggest that a deficiency of vitamin E is unlikely to be involved in its aetiology and that it arises from an uncomplicated, and often severe, dietary deficiency of selenium.

There are indications, particularly from the work of Whanger et al (1972), that the severity of the response to diets low in selenium may be influenced by dietary composition. In several of their studies the incidence of NMD in lambs was much higher when their dams had received lucerne hay during pregnancy rather than a grass hay ration of similar selenium content (0.01–0.025 mg/kg DM). In the same series of studies a significantly lower incidence of clinical and subclinical muscular dystrophy occurred in lambs born of ewes grazing pastures containing 0.02 mg Se/kg DM than when grass hay of almost identical selenium content was offered. The factors responsible for these differences in response have not been clearly identified.

Table 6.10 summarizes data from the relatively few studies with ruminants in which information on the dietary content of both selenium and vitamin E is provided. There are indications that even when the total tocopherol content of the diet is as high as 15–17 mg/kg DM, with α-tocopherol probably accounting for more than 90% of the total (Schubert et al 1961, Brown 1953, Paulson et al 1968), an appreciable incidence of clinical or subclinical muscular dystrophy can arise in lambs or calves if the diet of the dam has contained less than about 0.025 mg Se/kg DM (e.g. Jenkins et al 1970, Mikkelsen & Hansen 1967). At higher dietary contents of selenium, in the range 0.035–0.04 mg/kg DM, the reports of Schubert et al (1961), Mikkelsen & Hansen (1968) and Allen et al (1975) on the incidence of muscular dystrophy in lambs or delayed myopathy with myoglobinuria in growing cattle at differing vitamin E intakes provide some indication that the severity of clinical or subclinical lesions may be inversely related to the dietary content of vitamin E. Whanger et al (1972) suggest, without presenting analytical evidence, that the higher vitamin E content of pasture herbage than of hay may be responsible for their observation that the incidence of muscular dystrophy is frequently lower in lambs from grazing than from hay-fed ewes even when the selenium intake of such animals is almost identical.

Thus, while there are suspicions that the adverse effects of a "marginal' dietary content of selenium (say 0.03–0.05 mg/kg DM) may be modified by the dietary content of vitamin E, this possibility requires confirmation. As yet, in contrast to studies with poultry, there is no indication from experimental work that such an interrelationship is of practical significance if the content of selenium in the diet of the ruminant is less than 0.03 mg/kg DM.

It is particularly important to obtain clarification of the circumstances in which changes in dietary vitamin E content may modify the effects of a marginal selenium intake. Thus the storage of moist grain, even with preservatives, is known to accelerate vitamin E destruction, and such losses during barley storage, resulting in one typical report in a product containing only 0.7–3.5 mg α-tocopherol/kg DM, have been implicated as a primary cause of the development of paralytic myoglobinuria in growing cattle in areas where the selenium content of grain is marginal or low (0.05–0.01 mg/kg DM) (Allen et al 1974). A similar influence of retarded moisture loss on vitamin E losses during haymaking (Thafvelin & Oksanen 1966) may have been associated with an increase in incidence of muscular dystrophy in lambs born of ewes offered rain-damaged hay. It is not yet clear whether suboptimal selenium intake is involved in the aetiology of all these incidents. If it is, the frequency with which low selenium contents arise in such materials must be higher than is currently appreciated.

5. Absorption and utilization of selenium

Most of the selenium in common ruminant feeds and forages is in the selenium analogues of the sulphur amino acids. No difference in the uptake and retention of selenium from selenocystine, selenomethionine or selenite was found in sheep (Jenkins & Hidiroglu 1967). Later studies by the same authors (Jenkins et al 1971) show contradictory results which, in general, support the observation of Ehlig et al (1967) that ^{75}Se ingested as selenomethionine is more readily absorbed and retained by weaned lambs than selenium from inorganic selenite. Until studies have been made on the influence of different dietary forms of selenium on the metabolic and clinical response of selenium-depleted ruminants no firm conclusion can be drawn from the above studies.

Field investigations of a high incidence of muscular dystrophy in lambs born of ewes offered a mixed lucerne and grass hay relatively high in both selenium (<0.25 mg/kg DM) and sulphur led Schubert et al (1961) to suggest an antagonistic relationship between dietary selenium and sulphur. Hintz & Hogue (1964) found that sulphate providing up to 2.5 g S/kg in diets for ewes and lambs increased the incidence of muscular dystrophy, and in other work sulphate increased the incidence of cardiac muscle lesions in one experiment but not another (Whanger et al 1969, 1970). Hidiroglu et al (1965) failed to detect any association between dietary sulphur content and the incidence of myopathic disorders in cattle and Paulson et al (1966) did not detect any influence of variations in the dietary content of inorganic sulphate on ^{75}Se metabolism in the rumen of sheep. Allaway & Hodgson (1964) found an inconsistent relationship between the sulphur content of forages and the incidence of selenium deficiency in ruminants. The conflicting nature of these findings suggests that an additional variable may influence the effect of sulphur on selenium utilization, a situation which is reminiscent of early studies of the influence of dietary sulphur compounds on copper utilization by ruminants.

6. Absorption and utilization of vitamin E

The coefficient of apparent absorption of α- and γ-tocopherols by calves fed on milk replacer was in the range 0.61–0.71 when these vitamin E isomers were supplied at a rate of 0.5–1.5 mg/kg live weight per day, but may have been over-estimated owing to oxidation of faecal tocopherols between excretion and sampling (Chatterton et al 1961). Alderson et al (1971) suggested that the extent to which ingested vitamin E is destroyed in the rumen is directly related to the concentrate content of the ration. Pre-intestinal destruction of vitamin E (determined as total tocopherols) increased from 8 to 42% of the orally administered dose as the maize content of a ration for bullocks was increased from 200 to 800 g/kg.

It has been claimed repeatedly that the administration of vitamin E preparations to pregnant ruminants is ineffective in controlling the development of myopathy in their offspring (e.g., Muth et al 1958, Lindberg & Jacobson 1970). The inference is that placental transfer of vitamin E is inadequate to protect the developing foetus. A reappraisal of this possibility by Whanger et al (1976) has shown, firstly, that the effectiveness of vitamin E prophylaxis has varied with the stability of the preparation used and, secondly, that encapsulated vitamin E given by mouth to lambs can protect them against muscular dystrophy providing the dose rate is adequate (100 increasing to 700 mg tocopherol per week during the first six weeks of life). The prophylactic effectiveness of selenium administered by mouth to pregnant ewes or their lambs has rarely been questioned and the doses required are more closely related to the normal dietary intake than are the quantities of vitamin E required for oral prophylaxis. Thus, while a low vitamin E status of the dam cannot be considered irrelevant to the subsequent development of myopathy in her offspring, it is probable that variations in selenium status play a greater part in determining the degree of risk than do variations in tissue reserves of vitamin E in the dam or her offspring.

7. Production of vitamin E-responsive disorders by polyunsaturated fatty acids of dietary origin

High dietary intakes of PUFA induce skeletal and cardiac myopathies in calves (Blaxter et al 1953a, Blaxter & McGill 1955, Van Gils & Zayed 1966) and lambs (Boyd 1973). An antagonistic interaction between PUFA and vitamin E is indicated by experiments demonstrating the prophylactic effectiveness of vitamin E preparations (for reviews see Blaxter 1974b, Lannek & Lindberg 1975). It is probable that the selenium status of the ruminant has little bearing on the aetiology of myopathies arising from a high PUFA intake, since the administration of selenium to ruminants receiving diets high in PUFA is either ineffective (Blaxter 1962b) or only partly effective (Oksanen 1965) as a prophylactic or therapeutic measure.

Myopathies resulting from a high PUFA intake have been induced in calves consuming diets containing (g/kg diet DM) lard 280 (Thomas & Okamoto 1956); lard 160–220 (Van Gils & Zayed 1966); lard 280 and cod liver oil 20 (Blaxter et al 1952) or cod liver oil 37–149 (Blaxter et al 1954b). Boyd (1973) induced myopathy in lambs with diets containing cod liver oil 50 mg/kg but not with 5 mg/kg.

From a review of experiments on effects of dietary PUFA on vitamin E metabolism in a wide variety of species, Harris & Embree (1963) have suggested that if the ratio of dietary vitamin E (mg) to PUFA (g) is less than 0.6, signs of vitamin E deficiency are likely to develop particularly on diets of high fat content. The validity of this "critical ratio" of 0.6 for ruminant diets is difficult to assess, firstly because of a possibly variable degree of degradation of vitamin E in the rumen and secondly because of partial hydrogenation of PUFA by rumen microorganisms. While tentative estimates of the α-tocopherol and PUFA intakes of calves offered liquid diets supplemented with cod liver oil suggest that low α-tocopherol/PUFA ratios may induce muscular dystrophy, there are also some indications that the young ruminant may be particularly sensitive to high PUFA intake, as judged by the report of Blaxter & McGill (1955) that as little as 15–19 ml cod liver oil daily is sufficient to render ineffective the protective action of a daily supplement of 50 mg dl-α-tocopherol acetate (Blaxter & Brown 1952). The α-tocopherol/PUFA ratio for animals in this experiment is tentatively estimated to be about 3.

It has been the practice to regard the production of muscular dystrophy by high PUFA intakes as being an accidental circumstance arising from excess PUFA-rich dietary components low in vitamin E. However, the variations that can occur in the α-tocopherol and PUFA contents of common dietary ingredients are not well defined. Until additional data on the intake of these components and their relationship to the development of myopathic conditions become available it is unwise to assume that PUFA have such a limited relevance in the aetiology of these disorders.

8. Summary of selenium and vitamin E requirements

For both cattle and sheep, the data presented in Tables 6.9 and 6.10 suggest that when the selenium content of the diet is 0.03 mg/kg DM or less the incidence of disorders is increased, and contents of less than 0.03 mg/kg are therefore regarded as inadequate. The interpretation of dietary contents of selenium within the range 0.03–0.05 mg/kg is uncertain, because in earlier investigations 0.05 mg/kg was about the minimum detectable by the analytical methods used. Additionally, there is evidence from several investigations that subclinical muscle damage, as assessed from serum creatine phosphokinase, aspartate aminotransferase or lactic dehydrogenase, can be rectified by selenium supplementation of diets containing 0.03–0.05 mg/kg. For these reasons and the fact that clinical evidence of muscle degeneration preventable by selenium administration has arisen in two studies in which the dietary selenium content is reported as being 0.05 and 0.08 mg/kg (Jenkins et al 1974, Paulson et al 1968), it is probably appropriate to regard dietary selenium contents within the range 0.03–0.05 mg/kg as being "marginal".

Monitoring the selenium status of animals by assay of the selenium-containing enzyme, glutathione peroxidase, may ultimately permit the expression of selenium requirements with greater precision. Oh et al (1976) have already shown, for example, that increasing the selenium content of the diet of milk-fed lambs from 0.01 to 0.05 mg/kg DM substantially increases the glutathione peroxidase activity of many organs, and that maximal enzyme activity is not obtained until the diet provides 0.1 mg Se/kg. Better understanding of the relationship of changes in the activity of this enzyme to tissue damage resulting from selenium deficiency will be particularly useful in clarifying the range over which selenium intakes must be regarded as "marginal".

Only a provisional estimate can be made of the requirements of vitamin E for cattle and sheep during growth and reproduction, because of the infrequency with which both selenium and vitamin E contents of diets have been measured. The comprehensive study of myopathy and myoglobinuria in yearling cattle made by Allen et al (1975) presents adequate data (see Table 6.9, reference 5) to conclude that vitamin E contents in excess of about 10 mg/kg dietary DM may afford protection against this condition when the dietary selenium content is marginal (about 0.03 mg/kg). Diets providing more than 13 mg vitamin E/kg and adequate in selenium that have been offered to cows and their calves, or ewes and their lambs, have, in several instances, been shown to provide effective protection against clinical or subclinical myopathy (Table 6.10, references 6, 7, 9). On the basis of such slender information it is provisionally suggested that the minimal requirement for vitamin E in the diet of growing or pregnant sheep and cattle lies between 10 and 15 mg/kg dietary DM. Nevertheless, it is re-emphasized that substantially higher concentrations of vitamin E within the range 15–28 mg/kg may afford inadequate protection against the development of clinical or subclinical myopathy if the dietary content of selenium is low (see Table 6.10, references 3, 4, 9). Until more data are available on the quantitative relationship between PUFA intake of ruminants and its effects on vitamin E metabolism we are unable to make a reliable statement as to the additional provisions of vitamin E necessitated by the use of diets high in fat and PUFA in ruminant feeding.

9. Selenium toxicity

Decisions to introduce selenium supplementation of rations to correct a deficit revealed by analysis must be conditioned by the fact that selenium is a relatively toxic element. Selenium toxicity in ruminants and other animals has been comprehensively reviewed by Underwood (1971), Martin (1973) and Moxon & Olson (1974). Briefly, three types of selenium poisoning can be recognized. Acute poisoning results in death from respiratory failure and is preceded by a watery diarrhoea, elevated temperature and pulse rate, prostration and extensive tissue haemorrhage and oedema. Such effects can arise from sudden exposure to high selenium intakes as, for example, from poor mixing of mineral supplements containing selenium or the consumption of selenium-accumulator plants in large quantity.

One manifestation of chronic selenium poisoning, a form usually arising from prolonged consumption of the organoselenium compounds present in selenium accumulator plants growing on selenium-rich soils, is the development of muscular weakness and poor coordination when walking, impaired vision and loss of appetite. Additional features of this syndrome are paralysis and respiratory failure. The second form of chronic intoxication known to occur from the prolonged consumption of feeds containing >5 but usually <40 mg Se/kg is associated with loss of vitality and appetite, the development of a rough hair coat, hair loss and lameness. On the basis of field studies Munsell et al (1936) suggested that selenium toxicity may develop if the dietary selenium concentration exceeds 3–4 mg/kg DM while others (Anderson et al 1961) suggest that the danger level in forage is approximately 5 mg/kg DM.

With non-ruminant species the tolerance to selenium depends markedly on the form ingested. Thus organic selenium compounds present in vegetable crops appear to be poorly tolerated and more rapidly retained by the body than selenium administered as selenite. High dietary protein levels appear to offer some protection against selenium toxicity. The influence of such variations in diet composition on the tolerance of selenium by ruminants has not been explored adequately and thus, while provisionally we suggest that a concentration of 3 mg Se/kg diet DM should be regarded as the maximum tolerable, this may well be influenced by diet composition.

Iodine

1. Effects of iodine deficiency

The sole function of iodine is for the synthesis of the thyroid hormones, tetraiodothyronine (thyroxine) and triiodothyronine, which are intimately involved in the control of oxidative metabolism. The effects of inadequate thyroid hormone synthesis and release are numerous but reduced growth rate, reproductive failure and low milk production are probably most important economically (Underwood 1977). Reproductive problems include irregular oestrus, failure to conceive, loss of libido in the male, and abnormal development at any stage during the foetal period.

Clinically the main signs of iodine deficiency are thyroid enlargement, goitre, and the occurrence of stillborn offspring often with a poor growth of hair. Goitre is not always apparent in the live animal and postmortem determination of thyroid weight or other tests of thyroid function may be required for diagnosis of iodine deficiency.

2. Methods of assessing adequacy of iodine intake

Although increases in thyroid weight have often been used to detect suboptimal iodine intakes (e.g., Sinclair & Andrews 1961, Loosmore et al 1962, Hernandez et al 1972) it is clear from the work of Loosmore et al (1962) and Newton et al (1974) that this approach has limitations. The assessment of iodine status and thyroid function by determination of plasma protein-bound iodine concentration (PBI) has also been criticized (Kiesel & Burns 1960, Aschbacher 1968), particularly because PBI is not highly correlated with the flux of circulating iodine-containing hormones (Premachandra et al 1958, Sørensen 1958, Robertson & Falconer 1961a, Mixner et al 1962).

More satisfactory estimates of iodine requirement can be based on measurement of the influence of dietary iodine supply on thyroid hormone secretion rates (TSR) using radioactive iodine and we have particularly relied on the work of Henneman et al (1955) and Sørensen (1958) in the data presented below.

The recent development of radioimmunoassays for the thyroid hormones and for thyrotropin has provided alternative methods for assessing thyroid status (Hopkins et al 1975). Since circulating thyrotropin controls the activity of the thyroid gland and is elevated during periods of inadequate thyroid hormone output (Purves 1964), it may prove more appropriate for assessing the adequacy of iodine intake than direct estimates of thyroid hormone concentrations in plasma.

Even when the iodine requirement can be adequately defined, measurement of intake can present problems. Alderman & Stranks (1967) have, however, shown the following linear relationship between the iodine content of bulk milk from a herd (x, μg/litre) and the calculated average iodine intake of the herd (y, mg/cow per day):

$$y = 0.37x + 0.05 \qquad\qquad [6.4]$$

Measurement of iodine concentrations in milk may therefore give a convenient method for estimating iodine intakes. A similar relationship appears to hold for ewes where increase in milk iodine content was linearly correlated with increase in dietary iodine supplementation over a range both above and below requirement (Mason 1976). Unfortunately basal iodine intake from the diet was not estimated and an equation comparable with that above cannot be calculated.

3. Estimates of requirement and factors affecting dietary requirements for iodine

Because thyroid hormones affect basal metabolic rate, Mitchell & McClure (1937) suggested that iodine requirement should be related to metabolic body size and that such a relationship could be used to predict requirements for other species from experimental data obtained with man and rats. On this basis, they estimated the daily iodine requirement of sheep (50 kg) to be 50–100 μg (0.07–0.14 mg/kg DM) and for lactating cows (450 kg) 400–800 μg (0.03–0.07 mg/kg DM). These figures were for nonpregnant animals and made no allowance for loss of iodine in milk. The authors noted that their estimate for sheep agreed well with minimum iodine intakes in goitre-free areas reported by Orr & Leitch (1929). The minimum intakes for cows were substantially higher (3000 μg/day) than would be predicted on the basis of energy requirements.

Recent estimates based on indices of production have included the work of Shatalina & Veretennikova (1968), who found that in conditions of acute environmental iodine deficiency supplements of 8–10 μg I/kg live weight (0.2 mg/kg DM) improved beef production. Odynets & Asanbekov (1970) suggested that the iodine requirements of lactating ewes assessed on the basis of weight gain, wool production and lamb weight were met by 0.25 mg I/kg diet DM. This finding is supported by a recent study (R. Hill, personal communication) in which no adverse consequence arose in ewes or their lambs when diets providing 0.4 and 0.23 mg I/kg DM were offered in two successive reproductive cycles. On the other hand, recent reports from the USSR suggest optimal growth and reproductive performance of both nonpregnant sheep and cattle when their diets were supplemented with 0.5–0.75 mg I/kg dietary DM (Rizaev 1956b, Kovalskiĭ et al 1972).

In none of the above investigations was thyroid function directly monitored. Tables 6.11 and 6.12 present estimates of iodine requirement based on measurement of TSR in cattle and sheep. To deduce iodine requirement from TSR, the proportion of circulating iodide taken up by the thyroid gland (U) must be estimated. Since U increases when dietary intake of iodine is limiting (Robertson & Falconer 1961b, it is not possible to define a requirement without first agreeing a satisfactory value for U. When dietary supply of iodine is adequate, values of U measured in both sheep

Table 6.11   Dietary iodine requirements of cattle as estimated by measurements of thyroid hormone secretion rate (TSR).

Animal	Body weight (kg)	TSR (mg thyroxine I/day)	Dietary iodine requirements (mg/kg DM)	Reference
Cattle				
Calves	40	0.4	0.86	Sorensen, (1958)
Calves, heifer	180*	1.2	0.52	Lodge et al (1957)
Castrate	400	0.5	0.16	Sorensen (1958)
Heifers, nonpregnant	400	1.3	0.44	,, ,,
Heifers, pregnant	450	1.5	0.47	,, ,,
Cows, lactating				
10 litres/day†	500	1.6	0.36	,, ,,
25 litres/day†	500	4.1	0.47	,, ,,
Cows, lactating				
2-yr-old‡	500	4.4	0.75	Premachandra et al (1958)
4-yr-old	500	3.8	0.65	,, ,,
4-yr-old§	500	1.4	0.12	,, ,,

*, Assumed values;   †, Milk containing 4% butterfat;   ‡, No milk yield given;
§, TSR estimated during summer.

Table 6.12   Dietary iodine requirements of sheep and goats as estimated by measurements of thyroid hormone secretion rate (TSR).

Animal	Body weight (kg)	TSR (mg thyroxine I/day)	Dietary iodine requirements (mg/kg DM)	Reference
Sheep				
Lambs, ewe§	35	0.04	0.08	Singh et al (1956)
ram§	40*	0.03	0.08	,, ,,
wether§	35*	0.02	0.05	,, ,,
Lambs, ewe	15	0.15	0.47	Falconer & Robertson (1961)
ewe	35	0.47	1.14	Falconer & Robertson (1961)
Ewe				
1-yr-old	54	0.46	1.25	Falconer & Robertson (1961)
5–7-yr-old	67	0.31	0.81	Falconer & Robertson (1961)
lactating‡ 4-yr-old	65*	0.21	0.22	Robertson & Falconer (1961a)
Ewe				
2-yr-old	50	0.13	0.36	Henneman et al (1955)
2-yr-old§ pregnant	50	0.03	0.07	,, ,,
2-yr-old lactating	50	0.16	0.42	,, ,,
2-yr-old§	50	0.21	0.21	,, ,,
4-yr-old	60	0.14	0.13	,, ,,
Ewe, lactating‡	60*	0.11	0.11	Falconer (1963)
Goats				
Mature				
2-yr-old	70*	0.33	0.65	Flamboe & Reineke (1959)
4-yr-old	70*	0.19	0.38	,, ,,
4-yr-old§	70*	0.10	0.20	,, ,,

Calculations and footnotes as for Table 6.11.

and cattle appear to fall between 20% and 50% (Sørensen 1958, Falconer & Robertson 1961, Falconer 1963). A value of 33.3% was chosen since this results in the calculated requirement for iodine being twice the TSR (Sørensen 1958).

Breed differences in iodine utilization have been reported for both sheep and cattle but data are inadequate to make specific allowance for this possibility when considering iodine requirements (Long et al 1952, Allcroft et al 1954, Hennemar et al 1955, Lodge et al 1957, Sørensen 1958, Mixner et al 1962). Ageing in mature animals has been shown to decrease TSR in both cattle and sheep (Henneman et al 1955, Sørensen 1958, Premachandra et al 1958, Falconer & Robertson 1961) and in both species castrate males had a significantly lower TSR than intact females (Singh et al 1957, Sørensen 1958).

In the first edition, iodine requirements were assumed to increase during pregnancy. Recent investigations with both cattle and sheep have failed to show any significant increase in the TSR of the maternal thyroid during pregnancy (Henneman et al 1955, Søresen 1958, Robertson & Falconer 1961b). None of these estimates, however, make allowance for the demands of the foetus which will probably increase dietary requirements during the later stages of gestation.

Lactation has been shown to increase TSR in both cattle and sheep (Henneman et al 1955, Swanson 1972) and Sørensen (1958) showed that the dietary iodine requirement of a 500 kg cow for thyroid hormone secretion (y, mg/day) was related to butterfat production (x, kg/day) by the equation:

$$y = 7.9x + 0.07 \qquad [6.5]$$

No such relationship has yet been derived for the sheep. The quantitative importance of iodine losses in milk appears to differ between sheep and cattle. For example the above equation has been used to predict the dietary iodine intake of the cow needed to balance TSR, and the equation [6.4] of Alderman & Stranks (1967) used to predict milk iodine concentration from iodine intake. At milk yields of 10 to 25 litres per day containing 4% butterfat the iodine output in milk was estimated to be between 3 and 6% of the dietary requirement to maintain thyroid activity. Thus milk losses of iodine in cows are insignificant compared with the demand for iodine for thyroid hormone secretion. Additionally, when cows are subject to low dietary iodine intakes milk iodine concentration falls before TSR is affected (Swanson 1972). In contrast, lactating sheep fed on rations marginal in iodine show a high fractional uptake of iodine by the mammary gland at the expense of a fall in U for the thyroid (Falconer 1963). Ewe's milk is much richer in iodine than that of the cow, values of 80–90 μg/litre being associated with adequate iodine intake (Mason 1976). Losses of iodine in milk with ewes giving 0.5 and 1.5 litres per day therefore represent between 10% and 50% of the dietary iodine requirement for thyroid hormone secretion depending on which estimate of TSR is used, and can be a substantial proportion of the animal's iodine turnover. Blincoe (1975) has suggested that these species differences arise from the ability of the ewe's mammary gland to concentrate iodine from plasma at least 50-fold whereas the cow has little if any ability to concentrate iodine in the milk. Estimates of requirement based on TSR can probably be taken as adequate for cattle but must be increased, possibly doubled, when they relate to lactating ewes. A situation similar to that in the ewe appears to occur in the lactating goat (Flamboe & Reineke 1959).

The iodine requirements of an animal are influenced by its environment. Sørensen (1958) and Yousef & Johnson (1966) have shown that the rate of secretion of thyroid hormones in cattle is inversely related to environmental temperature and the substantial decrease in TSR found in cattle, sheep and goats during the summer is probably a consequence of this effect of temperature (Tables 6.11, 6.12) (Henneman et al 1955, Premachandra et al 1958, Flamboe & Reineke 1959).

The other main source of variation in dietary requirement for iodine is the presence of goitrogenic substances in feeds. These are of two types, namely the cyanogenetic goitrogens and those of the "thiouracil" type, most of the latter possessing a $(-NH-C=(S)-NH-)$ group. The former were initially found in leguminous plants such as white clover (Flux et al 1956). They impair iodide uptake by the thyroid gland and in general their effects can be overcome by iodine supplementation. In contrast the thiouracil goitrogens, which were first identified in brassica seeds (Astwood et al 1949), inhibit iodination of tyrosine residues in the thyroid gland and their effects are much less susceptible to reversal by iodine supplementation. Goitrogens have been found in a wide range of plant species (Sinclair & Andrews 1961, Allcroft & Salt 1961, Rodel 1972, Josefsson 1970) and there is evidence that the goitrogen content of many crops may be influenced by fertilizers.

Where the goitrogens were mainly of the cyanogenetic type, Roden (1971) found that adverse effects on the development of foetal lambs were prevented by giving ewes 135 mg iodine as KIO_3 twice during pregnancy. Goitrogenic effects of kale given to ewes have been overcome by dosing with 110 mg iodine as KI at weekly intervals for six weeks during pregnancy, or with 220 mg iodine as KI or KIO_3 twice, once at the beginning of the fourth and once at the beginning of the fifth month of pregnancy (Sinclair & Andrews 1954, 1958). However, when the goitrogens originated from rape seed and were probably of the thiouracil type, iodine supplementation (amounts unspecified) failed to prevent thyroid enlargement in fattening bulls (Iwarsson et al 1973). Similarly dietary iodine contents of 1 mg/kg DM did not prevent thyroid enlargement in calves from cows given goitrogenic soya bean concentrates during pregnancy (Hemken et al 1971).

Other dietary components have been reported to alter iodine requirement (Underwood 1977). Marginal intakes of manganese by sheep have been associated with increased weight of the thyroid gland and low iodine concentration in the gland (Rizaev 1965a). Both low and high intakes of cobalt have been found to reduce the iodine content of the thyroid gland of rats and rabbits (Koval'skiĭ & Gunstun 1966, Zak 1968) and the former authors have shown a direct correlation between iodine and cobalt contents of the thyroid glands of sheep in different regions of the USSR. Finally, an experiment by Hartmans (1971) suggested that increasing the calcium content of the rations of dairy cows from 4.9 to 10.9 g/kg may reduce iodine output in milk.

4. Iodine requirements

Published estimates of requirement differ considerably (see paragraph 3 and Tables 6.11, 6.12), but most suggest that even during pregnancy and lactation dietary iodine concentrations of 0.5 mg/kg DM are adequate for cattle and sheep when goitrogens are absent from their diet.

Few studies of the iodine requirements of cattle and sheep have been made during the summer period when TSR is low. These, separately identified in Tables 6.11 and 6.12, indicate a substantial reduction in iodine requirements during summer, such that demands may well be met by diets providing about 0.15 mg I/kg DM. It should be noted that, to date, only one estimate of iodine requirement of lactating cows has been made in these conditions (Premachandra et al 1958); the conclusion that 0.12 mg I/kg dietary DM meets requirements needs to be verified.

Not enough is known of the quantitative effect of dietary goitrogens on iodine utilization to allow us to estimate the increased requirements for iodine which will arise from goitrogen consumption. In general it may be advisable to increase the dietary iodine content to 2 mg/kg DM when the presence of substantial quantities of goitrogens is suspected.

Estimates of requirements derived from experimental observations made during winter suggest that the content of iodine needed per kg dietary DM may often exceed the iodine content of pasture grasses. Alderman & Jones (1967) found mean pasture iodine contents of 0.2–0.3 mg/kg DM although these varied considerably with season and tended to be lowest in summer when dietary requirements are probably also reduced. Hartmans (1974) analysed a wide range of grasses and clovers and found them to have between 0.06 and 0.17 mg/kg DM. Both groups found that nitrogen fertilization tended to decrease iodine concentration in the grasses although the increased yield of dry matter per acre resulted in a greater total iodine uptake from unit area. These reports suggest that particular care may be needed to ensure that the iodine intake of wintered stock is sufficient to meet requirements when conserved herbage forms a substantial proportion of their diet.

5. Methods of supplementation

Most investigations of requirement have used potassium iodide as a standard source of iodine. In practical conditions it suffers from extensive loss of iodine by oxidation and volatilization or by leaching. Iodine supplied as iodate appears to be readily available to sheep (Wright & Andrews 1955) and cattle (Miller, J. K. et al 1968) and is much more stable than iodine in KI. Ethylenediamine dihydriodide also appears to be a satisfactory source for cattle (Miller & Swanson 1973). Potassium or calcium iodate has been included in licks but they are subject to leaching when exposed to outdoor weather. Pentacalcium orthoperiodate, a compound of calcium iodate and lime, appears to offer a stable and available source of iodine for inclusion in licks (Miller, J. K. et al 1968). Another compound which has been suggested for inclusion in licks, 3,5-diiodosalicylic acid, was found to be a relatively poor source of iodine for ruminants (Miller, J. K. et al 1964, Aschbacher 1968). Iodine has been administered successfully as iodized poppyseed oil which contains 49% iodine by weight. When given intramuscularly to ewes once during pregnancy, it overcame the goitrogenic effects of kale (Sinclair & Andrews 1961). Similar preparations have been used to supply iodine during pregnancy in ewes feeding on pastures containing goitrogenic white clover (George 1966).

6. Toxicity of iodine

Forbes et al (1932) found that calves (50–100 kg) showed no adverse effect when given 10 mg I/day as iodized linseed oil (8 mg/kg DM) but grew less well when given 30 mg/day (24 mg/kg DM). Newton et al (1974) suggested that the minimum toxic concentration of iodine as calcium iodate for calves was 50 mg/kg DM (200–300 mg/day). Fattening sheep (30 kg) showed only marginally poorer growth on 33 mg I/day iodine as iodized linseed oil (42 mg/kg DM) (Forbes et al 1932). Malan et al (1935) found no effect of 15 mg I/day iodine as KI (20 mg/kg DM) on the reproduction of sheep (30 kg) but 38 mg/day (50 mg/kg DM) caused some loss of weight and irregularity of reproduction (Malan et al 1940). With lactating cows Blaxter (1943) found that 3 g I/day in protein of low thyroidal activity (200 mg/kg DM) produced signs of iodine poisoning but Forbes et al (1932) suggested that cows could tolerate, although not without severe signs of iodine excess, 0.65–2.3 g I/day as iodized linseed oil (40–150 mg/kg DM). The figures given above in brackets are estimates of equivalent dietary concentrations and suggest that the maximum safe amounts of iodine in diets lie between 8 and 20 mg/kg DM.

Zinc

1. Effects of zinc deficiency

Severe zinc deficiency in ruminants results in growth failure, poor feed intake, the development of skin lesions of parakeratosis or hyperkeratosis, brief periods of excessive salivation, deterioration of hair and wool texture followed by loss of hair and wool, reduced testicular development or atrophy, defective spermatogenesis and delayed wound healing following physical damage or infection of epidermal tissues. All these defects have been observed in experimental studies and in outbreaks of zinc deficiency arising in a genetically defective strain of Friesian calves (see below). In these conditions, clinical lesions usually appear after plasma zinc has declined to 0.5 mg/litre or less.

The appearance of zinc-responsive disorders in most field conditions differs in several respects from the above. With the exception of those of Legg & Sears (1960) and Pierson (1966), few reports describe as wide a variety of lesions as have been obtained in experimental studies. Other reports have described situations where the response to zinc administration is confined to an increased growth rate. In these, the deficiency may have been less severe but few of the reports present data on tissue zinc contents and this possibility cannot be evaluated.

2. Experimental studies of zinc deficiency and requirements

Table 6.13 summarizes 27 sets of clinical and chemical observations made with ruminants during experimental studies of the effects of dietary zinc concentrations between 1 and 100 mg/kg DM. All were made with semisynthetic diets and most used urea as the primary source of nitrogen. In only five instances were relatively

Table 6.13   Experimental studies of the influence of Zn content of semisynthetic diets on growth and clinical condition of ruminants.

Species	No.	Age at start of expt.	Dietary Zn (mg/kg DM)	Plasma Zn (mg/litre)	Summary of clinical effects*	Reference
Sheep, lambs	4	5–6 wk	1.20 (0.8–1.9)	0.35	gr, pk, hwl, app, hk, sal.	1
Cattle	4	4 wk	1.20 (0.8–1.9)	0.2	gr, pk, hwl, app, hk, sal.	1
Sheep, ram lambs	5		2.4		gr, pk, hwl, app, hk, sal. ta.	2
Cattle, bull calves	7		2.7–3.3		gr, pk, hwl, app, hk, sal, ta.	3
Sheep, lambs	6		2.7	0.4	gr, pk, hwl, sal.	4
Sheep, lambs	6		3	0.3	gr, pk, hwl, sal.	5
Sheep, lambs	4	5–6 wk	2.7	0.4	gr, pk, hwl, sal.	1
Cattle, calves	4	4 wk	3.1	0.2	gr, pk, hwl, sal.	1
Cattle, bull calves	4		4		gr, pk, hwl, ta, rwh.	6
Goat, kids		10 wk	5.3		gr, pk, hwl, ta, rwh.	7
Goats, lactating	3		6–7		gr, pk, hwl.	8
Goats, male			6–7		gr, pk, hwl.	8
Sheep, lamb	4	5–6 wk	7.1	0.5		1
Cattle, calves	4	4 wk	8	0.4		1
Cattle, bull calves	5		8.6			9
Cattle, calf		4 wk	14	0.9–1,2		1
Sheep, lambs	4	5–6 wk	15	0.9		1
Sheep, ram lambs	5		17		ta.	2
Sheep, lambs	6		18	1.0		5
Cattle, calves	6	7 wk	32			10
Sheep, ram lambs	5		32			2
Sheep, lambs	6		33	1.3		5
Cattle, bull calves	5		40–43			3
Cattle, bull calves	4		40			6
Goat, kids			40			7
Goat, kids	3		80			8
Sheep, lambs	6		100	1.2–1.5		4

* Key to clinical observations

    gr = failure or retardation of growth   hk = defective keratin of hoof or horn
    pk = parakeratosis   sal = excessive salivation (usually transient)
    hwl = hair or wool loss   ta = atrophy or reduced development of testes
    app = reduced feed consumption   rwh = retardation of wound healing
NOTE: In several of the above experiments in which diets containing <7 mg Zn/kg were offered, treatments were terminated abruptly to avoid death of animals. (Refs. 1, 3).

References
1. Mills et al (1967).   6. Miller et al (1965).
2. Underwood & Somers (1969).   7. Miller W. J. et al (1964).
3. Miller & Miller (1962).   8. Groppel & Hennig (1971).
4. Ott et al (1964).   9. Miller et al (1963).
5. Ott et al (1965).   10. Miller J. K. et al (1962).

uncharacterized sources of roughage (in the form of hay or straw) used and then only as minor components of the diet. From these results the conclusion may be drawn that providing the zinc content of such diets exceeds 7 mg/kg DM the risks of clinical signs of zinc deficiency are remote. Only one study (Ott et al 1965) suggested that a higher zinc content (18 mg/kg) may be still insufficient to meet requirements for rapid growth.

Requirements for pregnancy and lactation have not been established experimentally and only one investigation has been made of the influence of dietary zinc concentration on spermatogenesis in male ruminants. In a study with rams Underwood & Somers (1969) found that 17 mg Zn/kg dietary DM was insufficient to maintain normal spermatogenesis even though other clinical signs of deficiency were absent. They concluded that the zinc requirement for normal testicular function may be between 17 and 32 mg/kg diet. Weight is added to this suggestion by the finding that the male rat has a higher zinc requirement for normal spermatogenesis than for other functions (Swenerton & Hurley 1968). Furthermore, field observations with rams (Chernova 1969, Chernova & Aslanyan 1971) suggest that spermatogenesis may be sensitive to a reduced supply of zinc in the absence of other signs of deficiency.

Although the data in Table 6.13 indicate that 7 mg Zn/kg in semisynthetic rations may be sufficient to prevent the development of clinical signs of deficiency, it is also apparent that this is insufficient to raise blood zinc to within the range 0.9–1.5 mg/litre tentatively considered to be normal for ruminant animals receiving diets of adequate zinc content. This may well suggest that the minimum zinc requirement in these circumstances is not less than 14 mg/kg diet but there is, as yet, no clear evidence from experimental studies that the plasma concentrations attained when semisynthetic diets provide 7–14 mg Zn/kg are accompanied by an increased frequency of metabolic or gross clinical defects.

3. Factorial estimates of zinc requirement

Data on which to base factorial estimates of zinc requirement are scarce and thus our estimates derived from this approach must be regarded as provisional.

(a) Endogenous losses of zinc

The only published estimate of endogenous faecal output of zinc is derived from a study with pregnant heifers and suggests a faecal loss of 0.033 mg Zn/kg body weight (Hansard & Mohammed 1968). An estimated endogenous faecal loss of 0.053 mg Zn/kg body weight has been derived for sheep from regression analysis of faecal zinc output on diets providing between 20 and 730 mg Zn/day (N. D. Grace, personal communication). While we have adopted these estimates for the factorial derivation of requirements we emphasize that radioisotope balance studies with cattle and rats (e.g., Miller et al 1966) strongly suggest that endogenous faecal losses of zinc are influenced by the nature of the diet and are greatly reduced when zinc intake is low.

Estimates of endogenous urinary zinc loss range from 0.004 to 0.019 mg/kg live weight per day for cattle (Miller, W. J. et al 1966, 1968, 1970; Hansard et al 1968, Schwarz & Kirchgessner 1975a) and 0.023 mg/kg live weight per day for sheep (Grace 1975 and unpublished data).

From these results we have adopted a value of 0.045 mg Zn/kg live weight per day for the total endogenous zinc loss of cattle and, for sheep, 0.076 mg Zn/kg per day.

(b) Tissue growth

Estimates based on data of Kirchgessner & Neesse (1976) and Miller, W. J. et al (1966, 1970) for cattle suggest that between 16 and 31 mg zinc may be incorporated into body tissue for each kg gained during growth. Recent data (N. F. Suttle, unpublished) indicated that a value of 24 mg Zn/kg live weight gain may be appropriate for both cattle and sheep. This value has been used in our factorial estimates.

(c) Wool growth

Reports of the zinc content of the wool of normal sheep range from 77 (Pierson 1966) to 120 mg Zn/kg (Ott et al 1964). We have adopted a value of 115 mg Zn/kg wool on the basis of an extensive series of analyses made by Burns et al (1964) and Healy & Zieleman (1966).

Table 6.14   Factorial estimates of the zinc requirements of sheep*.

	Live weight (kg)	Rate of gain (kg/day), number of foetuses or yield (kg/day)	Net requirement (mg/day)	Assumed coefficient of absorption	Dietary requirement† (mg/day)	Dietary requirement† (mg/kg DM)
Growing lamb (castrate)	5	0.15	4.48	0.55	8.1	41
	10	0.15	5.36	0.55	9.7	24
	20	0.15	6.12	0.3	20	32–36
	40	0.075	5.84	0.2	29	29–49
		0.15	7.64	0.2	38	27–48
		0.30	11.24	0.2	56	51
Adult (nonproductive)	50	0	4.8	0.2	24	30–48
Pregnant ewe (late pregnancy)	75	2 foetuses	10.14	0.3	34	20–24
Lactating ewe	75	1 kg milk/day	13.9	0.3	46	24–31
		2 ,,	21.1	0.3	70	25–32
		3 ,,	28.3	0.3	94	25–39

* Calculated from the following components: endogenous loss, 76 µg Zn/kg LW per day; growth, 24 mg Zn/kg LW gain; pregnancy (twins), 3.44 mg Zn/day; lactation, 7.2 mg Zn/kg milk; wool growth, 115 mg Zn/kg. † Dry matter intake as indicated in Table 6.3.

Table 6.15   Factorial estimates of the zinc requirements of cattle.

	Live weight (kg)	Rate of gain, stage of pregnancy or milk yield	Net Zn requirement (mg/day)	Coefficient of absorption	Dietary requirement (mg/day)	Dietary requirement (mg/kg DM)†
Growing calf (liquid diet)	40	0.5 kg gain/day	13.8	0.5	28	28
Growing calf (solid diet)	80	0.5	15.6	0.4	39	26
		0.1	27.4	0.4	69	34
Growing bullock, medium-sized breed	100	0.5	16.5	0.3	55	14–28
		1.0	28.5	0.3	95	21–34
	200	0.5	21.0	0.3	70	12–20
		1.0	33.0	0.3	110	17–28
	300	0.5	25.5	0.2	128	16–28
		1.0	37.5	0.2	188	22–35
Adult	500	0	22.5	0.2	113	15–25
Pregnant cow	500	90–180 days	23.6	0.2	118	13–18
		180–270	28.8	0.2	144	13–21
Lactating cow	500	10 kg milk/day	62.5	0.3	208	18–25
		20 ,,	102.5	0.3	341	19–27
		30 ,,	142.5	0.3	475	25–31

† See Table 6.4 for assumed dry matter intakes.

(d) Pregnancy

Estimates based on results of Hansard et al (1968a) for pregnant cows suggest that the rate of increase in the zinc content of uterus plus foetus may be 1.1 mg Zn/day and 6.3 mg Zn/day at mid and late pregnancy, respectively. Corresponding estimates for pregnancy in ewes are 0.28 and 1.5 mg Zn/day (estimates from Hansard & Mohammed 1968b). While we have adopted these estimates we emphasize that data of Williams & Bremner (1976) indicate that the demand for each foetal lamb may rise to 2.5 mg Zn/day during the last days of intrauterine development.

(e) Lactation

Estimates of the mean content of zinc in cow's milk range from 3.4 ± 0.3 mg/litre (Osis et al 1972) to 5.8 ± 0.25 mg/litre (Schwarz & Kirchgessner 1975b). The zinc content of milk changes with that of the diet and may fall to 2–3 mg/litre in zinc-deficient animals. We have assumed a content of 4 mg Zn/litre milk in estimating the zinc requirements of lactating cattle. A higher value, 7.2 mg Zn/litre, was used for ewe's milk on the basis of recent analyses of 61 samples from two breeds (N. F. Suttle, unpublished).

(f) Availability of dietary zinc

Young ruminants offered milk or liquid milk-substitute diets absorb zinc very efficiently. An estimate derived from data of Miller & Cragle (1965) suggests a coefficient of absorption of at least 0.5 for milk-fed calves. The efficiency of absorption may be lower when milk substitutes include components high in phytic acid (e.g., soya bean meal) which, although readily degraded by animals with a fully functioning rumen, remains active in reducing availability of zinc to the preruminant calf or lamb (Van Leeuwen & Van der Grift 1969, Miller 1967). A 50% reduction in the apparent absorption of radioactive zinc resulted from the inclusion of soya bean protein in milk replacer given to calves in one study (Miller 1967).

Estimates of the coefficient of absorption for zinc in young ruminating cattle (70 to 150 kg live weight) range from 0.16 to 0.51 (Miller & Cragle 1965, Miller, W. J. et al 1968, 1970). The variables responsible for these differing estimates are not known. Estimates based on data of Grace (1975 and unpublished) from sheep given fresh herbage or lucerne hay give mean values of 0.31 and 0.25, respectively.

We have used a coefficient of absorption of 0.3 in estimating the gross requirements of young growing ruminants for zinc and, for mature animals, a value of 0.2 based on data of Miller & Cragle (1965), Hansard et al (1968a) and Hansard & Mohammed (1968).

4. Comparison of factorial estimates of zinc requirement with experimental and field observations

Observations on field cases of disorders found to be responsive to the provision of supplementary zinc are summarized in Table 6.16 together with data on the zinc content of the respective feeds. Comparison of factorially derived estimates of requirement with these observations shows that, in many instances, the feed provided less zinc than the estimated demands. In other instances (e.g., references 5–10, Table 6.16) dietary zinc appears to have been adequate as judged either by factorially-derived estimates of requirement or by experimental observation of responses to differing dietary contents of zinc. The reasons for the appearance of zinc-responsive syndromes in these situations are not understood.

In the factorial derivation of zinc requirements some uncertainty exists as to the validity of coefficients of absorption used to convert net to gross requirements. The "preferred" values we have used were, of necessity, derived from a small number of studies and only one of these used rations based on fresh or conserved forages or root crops. There are suspicions that both ration type and its absolute zinc content may modify the efficiency with which dietary zinc can be used. Such considerations may be relevant to the conclusion, derived from the data of Table 6.13, that as little as 8 mg Zn/kg dietary DM appears to be adequate for growth of cattle and sheep maintained on semisynthetic diets in experimental conditions and to the reported existence of deficiency in field conditions despite substantially higher intakes of zinc

Table 6.16 Field observations on clinical syndromes responsive to increases in Zn intake.

Species	No.	Age	Dietary Zn (mg/kg DM)	Type of ration*	Plasma Zn (mg/litre)	Summary of clinical effects**	Reference
Sheep, wethers	5		9.5	HRC		gr.	1
Cattle		1 mo	8–16	HEC		gr.	2
Cattle	37		20–30	CG	<0.65	gr, pk, hwl, d.	13
Cattle, calves and cows		1 mo–2 yr	19–42	G		gr, pk, hwl, d.	3
Cattle,	20		mean 25	HC	0.9	gr, pk, hwl, d.	4
calves	4				(0.3–1.5) 0.4	gr, pk, hwl, d.	4
Cattle,			30	G		rwh.	5
lactating cows			30–35	G		rwh.	5
(195 farms)			50	G			5
Cattle, bull calves	12	6 mo	30–56	RFC	0.8 (0.6–1.0)	rwh.	6
Cattle, calves and lactating cows			28–48	G		pk, hwl, ta.	7
Cattle, calves			35	SC		gr, pk.	4
Cattle	c. 30	young	37	G		gr, pk, hwl.	8
Sheep, rams		18–24 mo	45	HCE		ta.	9
Cattle,					0.6		10
calves	8	1 mo	>25	MHC	(0.2–1.75)	gr, pk, hwl, d.	10
	22	1 mo	>25	MHC	(55% below) 0.75	gr, pk, hwl, d.	11
Sheep, lambs	55	6–8 wk		G		gr, pk (?), hwl, ta, (d), sal. app.	12

* Key to ration components
C = concentrates (unspecified).
S = soya bean meal.
M = milk (or milk substitutes)
H = hay.
R = roots.
G = grass or other forage crops.
E = silage.

** Key to clinical observations
gr = failure or retardation of growth.
pk = parakeratosis or associated lesions of skin.
hwl = hair or wool loss.
rwh = retardation of wound healing or recovery from epidermal infection.
ta = atrophy or reduced development of testes.
d = death.
sal = excessive salivation (usually transient).
app = reduced feed consumption.

References
1. Gutkovich (1970a).
2. Gutkovich (1970b).
3. Legg & Sears (1960).
4. Van Leeuwen & Van der Grift (1969).
5. Bonomi (1964).
6. Demertzis & Mills (1973).
7. Haaranen (1963).
8. Dynna & Havre (1963).
9. Chernova (1969).
10. Chernova & Aslanyan (1971).
11. Van Adrichem et al (1970).
12. Pierson (1966).
13. Papasteriadis (1973).

There are indications that diets high in calcium adversely affect zinc utilization. Haaranen (1963), from field observations on the incidence of zinc-responsive skin disorders in cattle, suggested that consumption of herbage containing more than 3 g calcium/kg DM may increase zinc requirements, and Mills & Dalgarno (1967) found that the skin lesions induced in sheep offered diets marginal or low in zinc content were exacerbated by high-calcium diets (12 or 18 g Ca/kg DM).

Work with rats and sheep has shown that some bacterial and viral infections induce a marked decrease in plasma zinc content and an accompanying disturbance of zinc homeostasis, while studies with human subjects have illustrated that decreases in food intake or tissue trauma increase endogenous losses of zinc. The significance of such responses in ruminants has not been assessed but experimental studies of zinc depletion in growing calves and sheep indicate the rapidity with which gross clinical lesions of deficiency appear once endogenous losses of zinc exceed dietary intake. Experimentally induced zinc deficiency in cattle and goats has been found to retard the rate of epidermal wound healing (Miller, W. J. et al 1964, 1965). This response was no longer apparent in heifer calves given a hay and concentrate ration containing 30 mg Zn/kg DM but, later, Demertzis & Mills (1973) showed that recovery from infectious pododermatitis ("foul of foot") in bull calves receiving hay and roots rations providing up to 30–56 mg Zn/kg DM could be achieved solely by increasing the intake of zinc. In this last study the spontaneous outbreaks of zinc-responsive disorder occurred solely in bull calves but not in heifers or castrates, suggesting, in accord with observations with pigs, that the entire male may have a high requirement for zinc From an extensive field survey of the incidence of an apparently identical disorder (zoppina) in Italy, Bonomi (1964) concluded that pastures containing <30 mg Zn/kg DM were associated with a very high incidence and that 30–50 mg Zn/kg herbage DM must be regarded as marginal if other (unidentified) environmental circumstances favour the establishment of infection.

Genetic anomalies may influence zinc requirements, as has been shown by Krone man et al (1975) who identified a strain of Friesian cattle (strain A46) whose requirement for zinc may exceed 100 mg Zn/kg diet. Failure to meet the high zinc requirement of calves of this strain results in high mortality, a pronounced suscepti bility to infection and the development of skin lesions characteristic of Zn deficiency

5. Conclusions

While there is good evidence from studies with ruminants housed and maintained in experimental conditions that requirements for zinc will be fully met by rations providing approximately 30 mg Zn/kg DM, less definite conclusions can be derived from observations in field conditions.

The factors responsible for this situation may be associated with differences in the availability of dietary zinc, with incompletely identified interactions between zinc and other essential nutrients and with particularly high demands arising as a consequence of high rates of production or exposure to infection. Nevertheless, the factorial estimates of zinc requirement presented in Tables 6.14 and 6.15 are regarded as providing a realistic indication of the changing demands for this element during the productive life of ruminants under most practical circumstances.

6. Zinc toxicity

In contrast to many of the other trace elements, the tolerance of ruminants to zinc is relatively high. Nevertheless it must be emphasized that with the exception of some experiments in progress (J. K. Campbell, personal communication) studies of the chronic toxicity of zinc have been of relatively short duration. None exceeded 12 weeks of exposure.

Ott et al (1965) found that the weight gain and efficiency of feed conversion of lambs were adversely affected when the dietary content of zinc exceeded 1 g/kg DM Similar responses occurred in Hereford calves when dietary zinc exceeded 900 mg/kg When rations containing 3 g Zn/kg were offered for 8 weeks deaths occurred in lambs and calves developed a depraved appetite. Hemingway et al (1964) reported that the daily oral administration of 2 g zinc (as zinc sulphate solution) to pregnant ewes at pasture had no adverse effect on reproductive performance even though three of eight ewes receiving this treatment died before lambing.

None of the studies investigated the relationship of zinc tolerance to the dietary content of other essential trace elements (e.g., copper and iron) against which zinc is known to exert antagonistic effects. A recent study (J. K. Campbell, personal communication) has illustrated that 750 mg Zn/kg diet offered to ewes throughout pregnancy severely affects the viability of lambs after birth if the copper intake of the ewe is insufficient to permit the hepatic storage of copper. Typical results from this work are that of eight lambs born to ewes receiving a "marginal" copper intake, one was stillborn and seven survived and subsequently maintained a rapid growth rate When a high zinc intake (750 mg/kg diet) was superimposed on a marginal copper intake in two groups of ewes only 2 of 10 lambs in one group and 1 of 10 of the other survived beyond 24 hours after birth. No adverse effect was noted from 150 mg Zn/kg diet DM.

While investigating the efficacy of increased dietary supplementation with zinc in controlling copper intoxication of sheep Bremner et al (1976) found that 420 mg Zn/kg diet produced a small but statistically significant decline in blood haemoglobin concentration and packed cell volume, and lower plasma iron concentration when compared with diets providing 48 or 220 mg Zn/kg.

Thus, it appears that 150 mg Zn/kg diet DM is well tolerated but that mild adverse effects may become apparent within the range 150 to 420 mg Zn/kg. Pregnant animals may be less tolerant of zinc than nonpregnant.

Manganese

1. Manganese metabolism and effects of deficiency

Impaired growth, the development of skeletal abnormalities, poor reproductive performance and ataxia of the newborn are the most significant features of manganese deficiency in ruminants, judging by observations made in the course of experimental studies involving the use of semisynthetic diets with very low contents of manganese. Most of the pathological changes that develop appear to be related to defective synthesis of mucopolysaccharides and glycoproteins and probably reflect the importance of the role of manganese in several glycosyltransferase enzymes involved in their synthesis (Leach 1974).

Manganese is widely distributed throughout the body, the highest concentrations being found in the liver, hair and skeleton. The storage capacity of the liver for manganese is limited compared with its capacity to retain copper and iron and although the skeleton normally accounts for about one quarter of the total body content of manganese this reserve is not readily used when dietary intake is low.

Homeostatic control of manganese retention is almost entirely achieved by regulation of faecal endogenous secretion, primarily in bile and to a lesser extent in pancreatic and intestinal secretions (Miller et al 1972, Sansom et al 1976b). The coefficient of absorption of manganese from the intestinal tract of cattle was found to be nearly constant at 0.01 over a wide range of manganese intakes during a recent study by Sansom et al (1976b).

While some studies with ruminants (e.g., Rojas et al 1965) suggest that the development of clinical signs of manganese deficiency is accompanied by a small decline in the manganese content of whole blood, plasma or serum, lack of agreement between the analytical results of individual workers makes it impossible to establish criteria by which blood manganese analysis can be used to assess the adequacy of manganese intake (for review see Hartmans 1972).

Determination of the manganese content of hair or wool has been suggested as an alternative. Thus Anke (1966) found 14 mg Mn/kg hair in calves offered diets providing 23 mg Mn/kg DM but only 4 mg Mn/kg hair when diets low in manganese were offered, and Lassiter & Morton (1968) reported a decline in the manganese content of wool from 19 to 6 mg/kg during the induction of manganese deficiency in lambs. However, there are again serious discrepancies in reports of the manganese content of hair and wool of normal animals and clear indications of variability between individuals on the same diets and differences related to the degree of pigmentation and hair length (Hartmans 1972, Binot et al 1968).

In the absence of suitable biochemical criteria to assess changes in manganese status resulting from diets differing in manganese content, the only basis on which manganese requirements can be derived at present is from consideration of the reported effects of differing intakes on growth, performance and the presence or absence of clinical defects.

2. Influence of dietary manganese on development of skeletal lesions

In an experiment with lambs Lassiter & Morton (1968) found that a stiff gait, attributable to joint abnormalities, developed after 7 weeks exposure to a semisynthetic diet providing approximately 0.8 mg Mn/kg DM. Tibia length and breaking strength were much reduced. These defects did not appear in lambs receiving 30 mg Mn/kg diet. Similar defects were reported by Anke (1966) in calves offered 5.1 mg Mn/kg semisynthetic diet for 16 weeks.

Diets providing 6 mg Mn/kg offered for one year to goats during pregnancy and lactation induced tarsal bone deformities in 20% of the dams and 30% of their offspring (Groppel & Anke 1971) and deformities were so severe in some instances that paralysis or ataxia developed. Marked individual differences in the time at which such deformities developed were commented on. Provision of 20 mg Mn/kg diet virtually eliminated skeletal defects. Rojas et al (1965) offered diets based on barley straw, barley grain, cottonseed meal and urea and containing 16 or 17 mg Mn/kg DM to cows throughout pregnancy. All calves born exhibited enlarged joints and deformities of long bones. Humeral length was decreased by 15% and breaking strength halved compared with that of calves born to dams receiving 25 mg Mn/kg diet and exhibiting no skeletal defect. Howes & Dyer (1971) reported abnormalities of gait in calves born of dams receiving 13 or 14 mg Mn/kg diet but not when 21 mg Mn/kg was provided.

High dietary concentrations of calcium have been shown to decrease the availability of manganese and exacerbate the clinical signs of manganese deficiency in rats and chicks. In contrast, Rojas et al (1965) found no increase in the severity of skeletal lesions resulting from manganese deficiency in calves 1 to 7 days old when dietary calcium of their dams was increased from approximately 5 to 15 g/kg DM.

Earlier claims that leg deformities in cattle in the Netherlands were probably attributable to manganese deficiency, even though the pasture content of manganese in affected areas was at least 44 mg/kg (Grashuis et al 1953), have been refuted. In a reinvestigation of this claim Hartmans (1972) found no clinical sign of manganese deficiency in cattle maintained on rations providing 16 to 21 mg Mn/kg throughout one or two lactations even when dietary calcium was increased from 7.5 to 12 g/kg and phosphorus from 3.5 to 8.0 g/kg to investigate suggestions that high intakes of these major elements may be involved in the aetiology of manganese deficiency.

3. Manganese requirements for growth

Hawkins et al (1955) suggested that as little as 1 mg Mn/kg dietary DM was probably adequate for growth of young calves maintained on liquid feed. However, it is evident from a study of Lassiter & Morton (1968) with lambs that a similar content of manganese in solid semisynthetic diets (0.8 mg Mn/kg DM) was grossly inadequate to maintain an adequate rate of growth or to maintain voluntary intake of feed for longer than 16 weeks, during which period it must be assumed that unidentified tissue reserves of manganese were being used.

Bentley & Phillips (1951) concluded that 10 mg Mn/kg diet met the requirements of young heifers for growth and Rojas et al (1965) and Howes & Dyer (1971) found no adverse effect on weight gain of pregnant heifers from diets providing 13–16 mg Mn/kg DM. Other results that are relevant are those of Anke (1966) showing that 5.1 mg Mn/kg DM was adequate for goats during pregnancy to maintain normal birth weight in their offspring and of Hartmans (1972) that 16–21 mg Mn/kg DM was sufficient for pregnant and lactating cattle and for the subsequent growth of their calves.

While from these reports it may appear that manganese requirements of ruminants for growth may be met when diets provide approximately 10 mg/kg DM, there are indications from the studies considered in the preceding section that this concentration of manganese will probably be inadequate to meet demands for the development of a normal skeletal matrix in young stock.

4. Manganese and fertility in ruminants

Several experiments have shown that low dietary contents of manganese adversely affect reproductive performance in ruminants. Delayed or irregular oestrus and poor conception rates have resulted when heifers or mature cows have received diets with the following contents of Mn (mg/kg DM) before service: 7 to 10 (Bentley & Phillips 1951), 11 or 16 (Munro 1957), 16 or 17 (Rojas et al 1965). Anke & Groppel (1970) found irregular oestrus and poor foetal survival in goats given 6 mg Mn/kg diet.

Studies in similar experimental conditions showed that 25 (Rojas et al 1965), 13 to 21 (Howes & Dyer 1971) or 16 to 21 mg Mn/kg DM (Hartmans 1972) gave good reproductive performance in cattle.

Thus, although some uncertainty exists about the effects of dietary contents between 13 and 17 mg/kg on reproductive performance, it is probable that requirements for manganese are met, judging by experimental studies, if rations provide more than 20 mg Mn/kg DM. Evidence that requirements for reproduction are increased by high dietary contents of calcium or phosphorus is equivocal (Hignett 1956, Littlejohn & Lewis 1960). The possibility may merit further investigation in view of the claim of J. G. Wilson (1966) that at least 80 mg Mn/kg DM may be required for optimum fertility in cattle, judging by the results of an extensive field study which, nevertheless, did not include investigation of the effect of graded levels of manganese supplementation.

5. Summary of requirements for manganese

While the requirement of manganese for growth may be met by rations providing 10 mg/kg DM, about 20–25 mg/kg DM may be needed to permit optimum skeletal development. Experimental results so far available suggest that this should be adequate to meet requirements for reproduction. The above conclusions are based on studies with cattle. In the absence of comprehensive studies with sheep we must assume that similar conclusions apply.

6. Manganese toxicity

In a series of experiments to determine the toxicity of manganese (as manganese sulphate) to weaned calves, Cunningham et al (1966) found that more than 2.6 g Mn/kg diet depressed feed intake and rate of weight gain; 0.82 g Mn/kg had no adverse effect. In a second study 2 g Mn/kg also depressed blood haemoglobin concentration, suggesting that excessive intakes of manganese interfere with iron utilization. Further evidence of an antagonistic interaction between manganese and iron has been obtained in studies of the manganese tolerance of lambs (Hartman et al 1955). Thus 1 g Mn/kg DM greatly reduced serum iron and prevented haemoglobin regeneration in anaemic lambs, and Matrone et al (1959) have estimated that as little as 40 mg Mn/kg diet can have adverse effects on iron utilization in lambs with low tissue reserves of iron.

Fluorine Toxicity

Most instances of fluorine intoxication of ruminants in the United Kingdom have resulted from contamination of herbage and water supplies in relatively small and well defined areas, in the vicinity of industrial activity resulting in the emission of fluoride-containing dusts. Less frequently, the situation has arisen from inadequate monitoring of the adventitious fluorine contamination of diets resulting from the incorporation into mineral supplements of products derived from phosphate rock or phosphatic limestone which, when obtained from some geological sources, may have high fluorine contents.

Suttie (1969) reviewed much earlier work on the effects of chronic exposure of cattle to diets differing in fluorine content. From these field investigations and experimental studies he suggested that the following responses are likely to arise in cattle exposed for at least 2–3 years to the differing dietary fluoride concentrations (mg F/kg DM) quoted:

20–30, mild dental mottling sometimes observed;

>40, dental enamel hypoplasia and periosteal hyperostoses of long bones;

>50, significant incidence of lameness with increasing risks of adverse effects on milk production; skeletal fluoride content rising to >5000 mg/kg in 5-year exposure.

Burns & Allcroft (1964) and Allcroft et al (1965) point out that dairy cattle are appreciably more susceptible to fluorosis than beef cattle and other species and suggest that the maximum safe concentration of fluorine from highly soluble sources is probably between 30 and 50 mg/kg diet, whereas from relatively insoluble sources (e.g., as fluoride in rock phosphates), it may be 60–100 mg/kg DM.

These and other workers emphasize that the susceptibility to damaging effects arising from the ingestion of fluorine is particularly great during periods of rapid tooth and skeletal development. They stress that the consumption of diets with appreciably higher contents than the above may prove harmless if exposure is brief and emphasize that pregnancy and infection may well increase the susceptibility to fluorine intoxication. These and other aspects relevant to the determination of minimum tolerable fluorine content in diets are adequately considered by Phillips et al (1960) and Underwood (1977).

From these reports it appears probable that continuous exposure to dietary contents of fluorine of <35 or 40 mg/kg diet DM would produce no adverse effect in lactating ruminants and that up to 50 mg F/kg may be tolerated by non-lactating animals. Fluorine in concentrations up to 60 to 100 mg/kg may be tolerable if present in a relatively insoluble form such as in rock phosphate or phosphatic limestone.

Estimates of the minimum tolerable content of water-borne fluoride differ widely. In one group of studies the consumption of water containing between 7 and 20 mg F/litre produced no adverse effect whereas in a second group of studies, adverse effects have been reported from 2, 4, 5, 10 or 20 mg F/litre. Such reports are discussed in greater detail by Underwood (1977).

Although work with rats and mice suggests that fluorine may be an essential nutrient (Schwarz 1974, Messer et al 1974), adverse effects from a dietary deficiency of fluorine have not yet been demonstrated in ruminant animals.

Lead Toxicity

A variety of figures is given for the dietary contents of lead which cattle and sheep can tolerate. Allcroft & Blaxter (1950) gave cattle and sheep 5–6 mg Pb/kg live weight per day for 3 years without effect though this dose may be toxic on longer exposure (Allcroft 1951). No clinical or haematological sign was observed in sheep given diets providing 20 mg Pb/kg live weight per day as lead acetate for 12 weeks (Fick et al 1976) or 11 or 20 mg metallic Pb/kg per day (Carson et al 1973). Cattle were unaffected by 2 mg Pb/kg per day as chromate (Dinius et al 1973) but showed haematological changes after 42 days and polyuria after 200 days when they were given 15 mg Pb/kg per day as acetate. Studies with nonruminants have shown that tolerance is greatly influenced by the form in which lead is ingested. This possibility has not been adequately investigated in work with ruminants.

Low dietary contents of calcium, phosphorus or sulphur increase the susceptibility of sheep to lead intoxication. Quarterman et al (1977) and Morrison et al (1977) found that sheep given 4 or 8 mg Pb/kg live weight per day survived only 6 weeks on diets low in calcium (3 g/kg DM) or low in sulphur (1 g/kg DM). Mean survival time on a ration low in phosphorus (2.3 g/kg DM) was 30 weeks compared with 40 weeks on a ration containing Cu 8.3, P 4.8 and total S 3.7 g/kg DM.

Hormonal differences may also affect the susceptibility to lead. Male sheep were more susceptible than castrates (Quarterman et al 1977) and pregnant ewes than nonpregnant (Allcroft & Blaxter 1950). Such effects have not been adequately studied in cattle but Hammond & Aronson (1964) found that milking cows given 6–7 mg Pb/kg live weight per day died in 2 months, and Christian & Tryphonas (1971) showed that pregnant heifers died after receiving on average 2 or 6 mg Pb/kg per day. Adaptation to continued ingestion of lead also occurs so that blood and tissue concentrations cease to rise and even begin to decrease after 6 to 8 months (Kelliher et al 1973, Quarterman et al 1977).

Carson et al (1974) gave ewes 2 mg metallic Pb/kg live weight per day during pregnancy. At 10–15 months of age their lambs showed impaired visual discrimination compared with lambs of ewes given no lead. Allcroft & Blaxter (1950) reported that when pregnant ewes were offered a diet providing lead as lead acetate 1 mg/kg live weight per day for 50 days, 3 of 4 aborted.

Poisoning associated with pasture lead contents of 160–760 mg/kg DM has been described in Derbyshire, the northern Pennines and the Scottish Borders (Clegg & Rylands 1966, Stewart & Allcroft 1956, Butler et al 1957). It is characterized by stiffness of gait, fractures, osteoporosis and hydronephrosis.

From the above reports it appears possible that a risk of lead intoxication in ruminants may arise if the intake of lead exceeds about 2 mg/kg live weight per day although the appearance of clinical manifestations of toxicity from such a lead intake will be greatly influenced by diet composition and other variables. Expressed as a dietary concentration, such an intake may be achieved if the lead content of the ration exceeds about 60–100 mg Pb/kg DM.

Chapter 7

Requirements for Fat-Soluble Vitamins

Vitamin A

1. Basis of assessment *(a) Function*

Vitamin A, or more precisely, retinol, has many functions in the body. One of the earliest to be recognized was the prevention of night blindness with which the pigment rhodopsin present in the retina of the eye is concerned. Retinol is also necessary for the maintenance of the epithelial cells which line the surfaces of the cavities of the body. For this reason a wide range of histological and physiological changes may result from vitamin A deficiency, including keratinization of epithelial tissue, changes in bone structure and increase in the cerebrospinal fluid (CSF) pressure.

It is therefore to be expected that the signs of deficiency are many and varied. They include xerophthalmia, papilloedema, convulsive seizures and a staggering gait. Owing to changes in the gastrointestinal tract, in the kidneys and in the reproductive organs, the deficient animal may be affected by lack of appetite, diarrhoea and loss of body weight. Several problems in reproduction may be associated with this deficiency and, in cattle, these may include short gestation periods, a high incidence of retained placenta and the birth of incoordinated, blind or dead calves.

(b) Vitamin A and its provitamins

The International Union of Nutritional Sciences (1970) have suggested that "The term vitamin A should be used as the generic descriptor for all β-ionone derivatives other than the provitamin A carotenoids, exhibiting qualitatively the biological activity of retinol. The term provitamin A carotenoid should be used as the generic descriptor for all carotenoids exhibiting qualitatively the biological activity of β-carotene". A difficulty with this definition is that, by usage, the term vitamin A deficiency means not only deficiency of the vitamin as retinol but also of the provitamins. Retinol is exclusively of animal origin, but it can also be manufactured by the animal from the provitamin carotenoids which are widely distributed in plants. One of these, β-carotene, is by far the most important provitamin A.

The various forms of retinol can be determined chemically and related to biopotency. In the case of the provitamin A carotenoids the simple assay methods used are not always adequate to distinguish the frequently small structural differences which may have a marked influence upon biopotency. For example, 80 naturally occurring carotenes and carotenoids are known but only 15 have vitamin A activity and these can vary more than twofold. In fact, the routine procedures for the determination of carotene while separating active from inactive compounds do not differentiate between β-carotene and its isomers. The biopotency may vary between species; for instance, cats are unable to use carotene.

There have been a number of claims that carotene itself can exert an effect. One of these associates lack of carotene in the corpus luteum of the bovine and poor reproduction (Lotthammer et al 1974). However, an earlier review by Goodwin (1952) casts doubt on a specific effect of carotene on reproductive function, and further work by Lotthammer et al (1976) suggests that only in extreme conditions of low carotene intake by cattle would an effect be shown.

(c) Units

There are now four ways of expressing the requirement: in moles, in metric weights, in retinol equivalents or in international units. Before the chemical nature of the vitamin was known, it was necessary to express vitamin A activity in biological units; later, when pure retinol became available, it was found that one international unit (i.u.) of vitamin A was equivalent in activity to 0.3 μg of retinol. Crystalline preparations of vitamin A have been available since the early 1940s and it is now accepted that the results of physico-chemical analyses should be expressed in terms of metric weights and that the i.u. should be used only to express the results of biological assays.

There are two main difficulties with the use of the international unit which are most evident when the diet contains both the vitamin and the provitamin: a) it is impossible to tell how much of the activity comes from each source, and b) unless the conversion factors are stated, it is impossible to know what weight of the provitamin is present. For instance, if the evidence is of American origin then 8 μg β-carotene is taken to have the same activity as 1 μg retinol or 3.33 i.u., whereas in the UK it was accepted (largely from the same evidence) that 5 μg has the same activity as 1 μg retinol. As it is clearly impossible to ascribe a single conversion factor for β-carotene, quantitative measures of preformed vitamin A and of provitamin A are better expressed in mass units of retinol or β-carotene, respectively, rather than in international units. In this report all quantities will be expressed in metric weights.

(d) Conversion factors

Theoretically 1 μg β-carotene can give rise to 1 μg retinol but it is well known that the activity of β-carotene decreases as the amounts given are increased. Even at the low doses used in biological tests with rats, 2 μg β-carotene has the same activity as 1 μg retinol. In practice, at the marginal intakes needed to meet the requirement of cattle and sheep (ARC 1965, NRC 1971), 5–8 μg β-carotene has been stated to have the same activity as 1 μg retinol. Phillips et al (1966) have tabulated the various weight ratios of β-carotene and retinol that have from time to time been used or recommended to indicate vitamin A equivalence. Except for poultry, for which the ratios varied from 2 to 6:1, and for the fox and mink which were as high as 12:1, for all domestic animals and man the ratio varied from 4:1 to 10:1 with a mean of 7:1.

Jones et al (1962) studied the distribution of retinol in various organs of cattle given daily carotene or retinol between 50 and 500 μg/kg live weight and found 6 times as much retinol in the liver of those animals receiving the same weight of retinol as in those given β-carotene. Faruque & Walker (1970b) found, in lambs given milk containing 69 to 2200 μg β-carotene or 14 to 440 μg retinyl palmitate per kg live weight daily as water-miscible preparations, that the relative weights of β-carotene and retinol needed to produce the equivalent concentrations of retinol varied in the serum from 5:1 to 25:1 and in the liver from 3:1 to 9:1. Experiments in the USA at Storrs (Grifo et al 1960), using plasma concentrations as the criterion, suggested that the equivalence of β-carotene to retinol at marginal intakes (88 μg β-carotene/kg live weight) ranged from 3:1 to 8:1, varying with the carrier used. Using the minimum amount of vitamin A activity needed to prevent an increase in CSF pressure as a criterion, Eaton et al (1964) found that 106 μg β-carotene had the same effect as 20 μg retinol (Eaton et al 1972), suggesting a 5 or 6:1 conversion ratio. From the mass of other evidence, there are no precise data to contradict this assessment for normal diets and it is recommended that, with present knowledge, a ratio of 6 of β-carotene to 1 of retinol should be used in assessing the requirement of cattle and sheep.

(e) Criteria of adequacy

In the pioneering work with cattle of Guilbert & Hart (1935) night blindness was used as the criterion of vitamin A deficiency. These investigations suggested that 30 μg carotene/kg live weight daily was adequate. The work of Rousseau et al (1954) and of Eaton et al (1964) using cerebrospinal fluid pressure increased Guilbert's original estimate of 30 μg to 106 μg carotene/kg live weight daily. As a further example of the effect of the criterion used, Lewis & Wilson (1945) reported that calves given retinol needed 8 μg/kg live weight daily to cover minimum requirement (some growth), 16 μg for maximum growth and 32 μg for some liver storage. The latter is very similar to 29 μg retinol/kg live weight daily found by Eaton et al (1972) as the amount needed by the calf to prevent CSF pressure from rising above 76 mm.

Other criteria that have been used to distinguish adequate from inadequate intakes are appetite and reproductive performance. Abrams et al (1969) felt that depression of feed consumption may be the first objective sign of hypovitaminosis A. More specific criteria are histological changes such as squamous metaplasia of the parotid duct, ocular papilloedema and changes in renal function. As examples of the effect of difficulties of using these different criteria to determine requirement, it has been found that plasma retinol must be at least 270 μg/litre to maintain maximum renal clearance (Richards et al 1970), that squamous metaplasia is absent if the plasma retinol is above 200 μg/litre and that liver retinol must be over 8 mg/kg to prevent both squamous metaplasia and ocular papilloedema (Eaton et al 1970). The normal method of assessing deficiency in experimental conditions is to determine the retinol content of plasma, or less commonly to take liver biopsy samples. Unfortunately neither of these criteria is of much value in determining standards because both plasma and liver contents are more closely related to age than to immediate dietary intake.

Since the measurement of CSF pressure is now an established research procedure, and is known to be fairly closely related to carotene or retinol intake in young cattle, it is accepted that this criterion of adequacy should be used for assessing standards as far as evidence on it allows. It should, however, be recognised that CSF pressure is a physiological parameter and is not a direct measurement of requirement to sustain growth, reproduction and health. In cattle it appears to have the merit of reasonable sensitivity and repeatability at intakes of retinol or carotene somewhat above those which lead to overt signs of vitamin A deficiency.

2. Factors affecting utilization

(a) Dietary factors

Factors affecting the utilization of carotene and retinol include intake, dietary protein content, tocopherol, nitrate or nitrite, high-energy diets and the amounts of certain trace elements in the diet.

Calves achieve normal plasma retinol concentration and small liver reserves of retinol when reared on whole milk providing daily about 44 μg retinol/kg live weight (Roy et al 1964), that is 2.5 times the requirement of 16μg/kg previously recommended in the first edition (ARC 1965). However, when other feeds are introduced, it is important to ensure that the intake of vitamin A remains adequate, particularly as Spratling et al (1965) have shown that during periods of deficiency, appetite is reduced and the resultant loss in weight cannot subsequently be made good. Grifo et al (1960) observed that in calves maintained for 16 weeks on a diet low in carotene (26 μg/kg live weight) the subsequent utilization of carotene was reduced.

The utilization of carotene from pasture (McDowall & McGillivray 1963) and from artificially dried grass (Thompson 1970) decreases as the herbage matures. It is noteworthy that the proportion of unsaturated fatty acids in the lipids of grass decreases at about the same time (Hawke 1963).

Whether the carotene of forages is well or poorly utilized depends on the criterion used. For instance, the carotene from both grass and maize silage is well utilized as judged by milk and plasma retinol contents, but poorly as judged by liver stores (Thompson 1959, Miller, R. W. et al 1970, Smith et al 1964, Hatfield et al 1961). Several groups of workers using cattle (Weichenthal et al 1963, Davison et al 1964, Cameron 1966) and sheep (Goodrich et al 1964, Hoar et al 1968) have investigated the possibility that the nitrate content of forages and silage may be responsible for these effects, but the evidence overall does not support this (Jones et al 1966). Another explanation is that maize silage exerts its effect through its ethanol content (Miller et al 1969), as ethanol causes increased plasma concentrations and decreased liver stores of retinol (Moore 1957) in other species.

The trend towards the use of diets of high nutrient density for fattening cattle has led to an increase in the proportion of animals with night blindness and liver and kidney damage. The evidence on the association between the liver lesions and retino status is somewhat obscure (Quarterman & Mills 1964, Rowland 1970). There are clear indications that to maintain liver stores of retinol, animals given rations consisting mainly of barley require larger amounts of vitamin A (possibly 50% more) than comparable animals receiving traditional diets (Quarterman 1966, Topps et al 1966). There is recent evidence (J. I. D. Wilkinson, personal communication 1977) that losses of retinol in the rumen of sheep receiving diets of high nutrient density are of the order of 40–50%. It is interesting to note that in drought conditions, beef cows in Australia had large liver reserves of retinol despite the lack of pasture (Gartner & Alexander 1966). The evidence for both the higher requirement of extensively reared farm animals and the larger stores of retinol in the liver of cattle and sheep kept in drought conditions, when intakes of carotene are negligible, indicates that the requirement for vitamin A may in a general way be related to the rate of live weight increase. Moreover, in drought conditions, the very low protein intake will hinder the mobilization and/or transport of liver retinol. In less extreme conditions, Klosterman et al (1964) found no difference in the depletion rates of liver stores of fast and slow growing bullocks.

Trace elements may play a role in the utilization of vitamin A, and specifically zinc has been shown to have an effect in rats (Smith et al 1973) and lambs (Saraswa & Arora 1972).

(b) Animal factors

Moore (1957) has pointed out that the vitamin A status of the animal may be affected by a wide range of diseases and nutritional disorders. Those which are likely to be of most direct importance to retinol absorption and storage concern the gastrointestinal tract and the liver. Since the major site of conversion of carotene to retinol is the intestinal wall, damage due to diseases such as parasitic gastroenteritis, mucosal disease, Johne's disease and salmonellosis, which are often accompanied by diarrhoea, may not only reduce the efficiency of conversion of carotene to retinol but because of an increased rate of passage may also reduce the time available for absorption (Keener et al 1942, Blakemore et al 1948, Kon et al 1955).

Hepatic malfunction causing reduced retinol storage may arise, for example, from the alkaloids of ragwort (*Senecio jacobaea*) or from the hepatotoxin (aflatoxin) present in some strains of the fungus *Aspergillus flavus* in defective groundnuts.

It should also be borne in mind that the previous vitamin A status may affect requirement. Thus Grifo et al (1960) and Jordan et al (1963) have found that retinol-deficient cattle did not use carotene very efficiently, in agreement with an earlier observation of Byers et al (1955) that cattle conditioned to a continuous low carotene intake for 3–5 years lack the ability to effect storage even if massive doses of retinol are given. Furthermore, Helmboldt et al (1953) showed that the administration of carotene would not cause complete reversal of the metaplastic changes in the parotid duct or bring much fall of the raised CSF pressure in deficient animals. Retinol readily reversed those effects.

3. Hypervitaminosis A

Moderate overdosing of retinol (0.7 mg/kg live weight) as used during the last 13 weeks of pregnancy (Walker et al 1949) caused no harmful effect to dairy cows and their calves.

It has long been known that large oral doses of retinol can be toxic (cf. Moore 1957), and it is of interest that Hazzard et al (1964) observed with calves that when more than 7.3 mg retinol/kg live weight was given appetite was reduced and when more than 3.2 mg retinol/kg live weight was given daily, CSF pressure fell below normal. The studies of Grey et al (1964) and Hurt et al (1967) suggest that excess retinol reduces CSF pressure by making tissues more permeable and they lend support to the postulated role of retinol in regulating membrane stability (Lucy 1969). Although more than 3.2 mg retinol/kg live weight was needed to reduce CSF pressure, Eaton (1969) found that bone lesions were discernible when 1.33 mg retinol/kg live weight was given. It appears that the onset of bone lesions is a more sensitive indication of retinol excess than a decrease in CSF pressure. Bone lesions have been observed in cats given too much retinol (Lucke et al 1968). Fortunately the gap between the requirement and the dose causing the least discernible bone lesion is nearly 100-fold. It would seem that 30 times the requirement could be set as an upper limit since this would allow for very generous allowances and yet would ensure an adequate safety margin against toxicity. Obviously this recommendation applies only when the dose is given orally, as Moore (1957) has reported that large doses of retinol can be injected into rats without harmful effect.

In certain conditions of herd or flock management, when stock may be handled only a few times a year and doses of retinol must be given which are large enough to last from one treatment to the next, an effective way of giving the dose is by intramuscular injection. This reduces the risk of retinol toxicity because of the slower absorption by this route than from an oral dose. It also avoids losses in the gut due to incomplete absorption (Hale et al 1962).

Intramuscular injections of about 6 mg retinol/kg live weight to bullocks caused no deleterious effect on performance as judged by weight gain (Kirk et al 1970), and Roberts et al (1965) found that retinol given by the intramuscular route promoted greater hepatic storage of retinol than that administered by injection into the rumen. A similar observation has been made by Martin (1968).

4. Requirements of cattle

Since our previous assessment of requirements for vitamin A by ruminants some evidence has been published which suggests that amounts larger than those recommended for growing and adult cattle in the first edition may have beneficial effects on health and reproduction. Unfortunately, precise data are often not available, the retinol and carotene contents of the basal diets used are often not given and the requirement for maintaining normal health is difficult to assess because pregnancy or lactation has intervened.

(a) Calves

Before 1965 much of the work on calves had been done with varying amounts of β-carotene. Since then a number of experiments have been reported in which calves have been given differing amounts of retinol, following a depletion period on a vitamin A deficient diet (Eaton et al 1972, Mikkilineni et al 1973).

A reevaluation of the minimum requirement of carotene for calves was published by Eaton et al (1964) and on the basis of data from 108 vitamin A depleted calves receiving carotene in amounts ranging from 22 to 528 μg/kg live weight daily, the authors concluded that the CSF pressure was liable to increase when the intake of carotene was below 106 μg/kg live weight.

The critical work with calves at Storrs, in which retinol was used (Eaton et al 1972, Mikkilineni et al 1973), involved widely differing doses of vitamin A but with no dose between 16 and 64 μg retinol/kg live weight daily, within which range there appear to be particularly marked changes in CSF pressure. While no increase in CSF pressure was recorded in calves receiving over 16 μg retinol/kg live weight daily, interpolation of the data by the workers concerned suggested that the critical dose, with the small number of calves used, was about 29 μg retinol/kg live weight daily.

From the long series of experiments on the vitamin A requirement of calves at Storrs it seemed that a basic requirement of about 20 μg retinol or 100–110 μg β-carotene/kg live weight daily could be recommended. Even in the carefully controlled conditions of depletion used in these experiments, there appeared to be considerable variation in requirements for individual calves which must be added to the variation which may result from dietary factors discussed above.

(b) Adult cattle

There is little published evidence to add to that reviewed in the first edition to determine the basic requirement for growing and adult cattle. This is partly because long periods of depletion with substantial numbers of cattle would be required to obtain useful results. Much of the work that has been done has involved the complication of pregnancy and lactation and it appears to be impossible in the ruminant to establish changes in the basic requirement for vitamin A with age. For this reason it is necessary to interpolate from the older evidence with growing and adult cattle (Jones et al 1943, Boyer et al 1942, Page et al 1958) where criteria of deficiency were often much less precise than those for calves. In general this evidence suggests that requirements for growing and adult cattle which are not pregnant and not lactating are similar to those for calves on a body weight basis.

(c) Conception

Reduced fertility and low conception rates, in particular, have sometimes been associated with low intakes of vitamin A. Kuhlman & Gallup (1942) reported 1.99

services per conception in 21 cows receiving less than 86 μg carotene/kg live weight daily in the 90-day period before service, and substantially better conception at higher carotene intake. Many reports have followed in which various forms of infertility have been associated with low, but often unspecified, intakes of vitamin A.

It has been known for many years that retinol may inhibit oestrogen-induced cornification of the vaginal epithelium in the rat and Kahn (1954) showed that, in this respect, the relationship between retinol and α-oestradiol may follow a dose-response curve. Hayes (1969) sought to explain this from a morphological point of view and suggested that the antagonistic effect of oestrogen on retinol might be involved in reproductive failure. So far there appears to be no detailed physiological work relating vitamin A supplies to the reproductive cycle which could help to specify dietary supplies necessary for normal conception in cattle.

(d) Pregnancy

The requirement for pregnant cattle is here defined as that amount which will result in a calf which, although viable and healthy, lacks sufficient body stores of retinol to carry it many days into postnatal life. The work of Byers et al (1956), and of Converse & Meigs (1938), involving small numbers of cattle but fairly precise methods of determining deficiency, pointed to a requirement of 150 μg carotene/kg live weight daily. The less critical but more extensive work of Ronning et al (1959) indicated that normal pregnancies usually resulted when Holstein cows were supplied with 130 to 140 μg carotene/kg live weight daily, and in Channel Island cows when receiving 160–170 μg carotene/kg live weight daily.

In a recent experiment in the USSR (Burdina 1973) in which a 70% increase was made in the vitamin A supply to 32 cows in late pregnancy, udder trouble was reduced and less digestive upsets were recorded in the calves. Since the control animals appeared to be receiving substantially more carotene than that traditionally recommended it is difficult to reconcile the results with those from the United States quoted above.

(e) Lactation

In considering the vitamin A requirement for lactation two factors are of practical importance. The first is the amount of vitamin A required to maintain the health of the cow and her normal rate of milk production. The second is the variable requirement to enhance to specified extents the vitamin A potency of the milk produced, for the benefit of the calf or of the human consumer.

So far as the lactating cow herself is concerned the extensive data obtained by Ronning et al (1959) and Swanson et al (1968) indicate that the dietary amount of carotene which supports maintenance, growth and normal health is close to that amount which will, in addition, maintain milk yield. Consequently in the United States the requirement of lactating cattle is considered to be the same as that of nonlactating cattle of the same weight (NRC 1971). The amount of retinol in milk fat rarely reaches zero even on diets very low in vitamin A. For example, at the National Institute for Research in Dairying, the lowest values recorded have been 2 or 3 mg retinol/kg butterfat. For the 500-kg cow yielding 1 kg butterfat daily this output represents 5 μg retinol/kg live weight daily, which must be obtained from dietary sources. On this basis it is recommended that 5 μg live weight be added to the normal intakes for maintenance and growth, to provide for normal lactation.

(f) Suckled calves

In the first edition, an attempt was made to calculate the vitamin A supply to the newborn calf in relation to its maternal endowment. These assessments were of necessity speculative and little has since been published to add precision. The summary of evidence on body reserves of retinol of newborn calves indicated that these were usually very small except where substantial supplied of retinol were given to the dam in late pregnancy (Walker et al 1949). Branstetter et al (1973) have recently provided evidence on the relative importance of the dam's liver reserves and dietary supplies of vitamin A for placental or mammary transfer to the calf. Although both routes are of importance it seems that on ordinary farm diets the milk-fed calf will be very dependent on the potency of the colostrum and milk and hence on the dietary supply of the cows producing the milk. The evidence presented in the first edition of the relationship between the lactating cow's dietary intake of carotene and the calf's requirement for vitamin A suggested that, at normal intakes of milk by the calf, the dietary consumption of the cow should be about 400 μg carotene/kg live weight daily. In feeding conditions normal to the United Kingdom, the lactating cow will usually receive considerably higher intakes of carotene but on a diet of straw, poor hay, roots and cereals the milk secreted may supply insufficient retinol or carotene to meet the calf's requirement.

5. Requirements of sheep

There is evidence that the CSF pressure increases at an early stage in vitamin A deficiency in sheep (Eveleth et al 1949). There are no data comparable with those obtained for the calf to indicate with precision the point at which the CSF pressure begins to rise with decreasing carotene or retinol intake.

The use of overt signs of vitamin A deficiency such as night blindness is liable to lead to recommedations for somewhat lower requirements. It is difficult to discount the levels suggested by Peirce (1945, 1954) based on long studies in Australia where depletion times in sheep receiving no more than 10 μg carotene/kg live weight daily extended to several months before plasma retinol receded to a low concentration. Reproduction over 2 or 3 pregnancies appeared to be normal where carotene intake was on average no more than 50 μg/kg live weight per day. Some support for a requirement in young lambs of about 50 μg carotene/kg live weight daily is provided by Faruque & Walker (1970b), where 14 μg retinyl palmitate or 69 μg β-carotene/kg live weight daily permitted the establishment of a small but steady liver reserve of retinol.

Other Australian evidence (Franklin et al 1955, Bayfield et al 1972) has confirmed the slow depletion rates as measured by plasma vitamin A and the tendency for sheep to die as the result of anorexia and inanition without typical signs of vitamin A deficiency on diets which are low in energy as well as vitamin A.

There is little evidence that there is a substantial requirement for vitamin A for lactation although, by analogy with the cow, a small allowance could be made for the traces of retinol in milk even on diets low in carotene or retinol. It is also reasonable to assume that the newborn lamb, because it will normally have small body stores of vitamin A, may be quickly dependent on retinol supplies from its mother's colostrum and milk and, for this purpose, substantial increases in the mother's vitamin A supply would be needed.

Evidence on the requirement for the later stages of pregnancy is sparse. Guilbert et al (1940) suggested that it should be three or four times that for maintenance and growth, which, from their data, would amount to about 20 μg of retinol or 120 μg of carotene/kg live weight daily. This is not greatly different from the 100 μg carotene/kg live weight daily suggested by Peirce (1954).

There is fairly good agreement that in the ruminant lamb and in sheep the relative values of carotene and retinol at marginal intakes of vitamin A are 5 or 6:1 on a weight for weight basis (Faruque & Walker 1970b, Myers et al 1959). The equivalence depends partly on the criteria of deficiency and, at higher rates of intake, carotene becomes relatively much less efficient as a source of vitamin A. The data on sheep and cattle are probably not sufficiently precise to differentiate between the two species in this respect and in terms of basic requirement at marginal intakes a 6:1 ratio would seem to be appropriate. In practical conditions where the intake is rarely marginal a wider ratio should be used.

6. Use of recommended requirements in practice

It will be clear from the previous paragraphs that both the requirements for vitamin A and its supply to the ruminant are affected by powerful extraneous factors. In practice many of these would be unknown and some could be revealed only after complex and expensive analyses.

One factor which will determine the allowance needed in practice is the size of the body store, since when the intake exceeds the body's immediate requirement the excess retinol is stored in the liver. For instance, after a period of grazing pasture when the vitamin intake has produced substantial liver stores, it may be of economic importance to forecast the period of time an animal may perform optimally on a level in the diet below the immediate requirement.

In bullocks, initial liver stores of 20–40 mg/kg were exhausted in 3–4 months (Kohlmeier & Burroughs 1970). Six months was needed to exhaust higher stores of 112 mg/kg (Hale et al 1962), and Frey & Jensen (1947) observed that liver retinol stores were depleted by one half every 40 days in animals given a fattening ration, presumably low in vitamin A activity. The depletion rates did not differ between high and low yielding cows (Swanson et al 1968) or between fast and slow gaining bullocks (Klosterman et al 1964).

These findings suggest that the liver stores decline at a rate which is independent of the needs of the body. It is therefore unwise to assess by calculation how long an animal will subsist on its liver stores. Adequate performance can be maintained at intakes which are about a third of those needed to promote storage (Guilbert et al 1937).

Experiments by Anderson et al (1962) and Mitchell (1967) suggest that the rates of depletion of liver stores in sheep are similar to those observed in cattle. Perhaps the higher concentration of retinol (Moore 1957) in the liver of sheep (120 mg/kg) than of cattle (45 mg/kg) may explain why sheep are reputed to thrive for longer periods than cattle when receiving low retinol intakes. In addition, the effects of higher intakes of vitamin A than those which are thought to be necessary to prevent overt signs of deficiency may be of importance to health and fertility. Reliable evidence on this is still meagre and contradictory. Even in well controlled experiments with very similar animals there seems to be substantial difference between individuals in the precise requirement.

In these circumstances the interpretation of requirement into practical allowances may require a greater safety margin than that commonly recommended for most other nutrients; for instance, there is evidence of a higher requirement for cattle on high nutrient density rations. This is particularly important where much of the supply of vitamin A is in the form of β-carotene, since the ratio of potency of β-carotene to retinol recommended in this bulletin is arbitrary and would apply only to marginal intakes, in controlled conditions.

In practical situations at marginal intake it is recommended that 6 μg β-carotene is assumed to have the same vitamin A activity as 1 μg retinol, or 3.33 international units.

New recommendations for maintenance and growth (Table 7.1) are higher than those in the first edition, 120 compared with 80 μg β-carotene/kg live weight daily for cattle, and 60 compared with 50 μg/kg live weight daily for sheep.

Table 7.1   Proposed dietary requirements for vitamin A (μg/kg live weight per day).

	Retinol	β-carotene
Cattle:		
Maintenance and growth	20	120
Conception and pregnancy	30	180
Lactation—minimum	25	150
Lactation—to provide for the suckled calf	65	390
Sheep:		
Maintenance and growth	10	60
Pregnancy	20	120
Lactation	15	90

Vitamin D

It has been generally accepted that vitamin D_2 (ergocalciferol) and vitamin D_3 (cholecalciferol) are equally potent for mammals although not so for birds. Thus ergocalciferol has 100% activity in the rat but only 10% in the chick (Chen & Bosmann 1964). The experiments of Bechdel et al (1938), in which irradiated yeast and cod liver oil were used as the sources of the two forms of the vitamin, support this view for the calf. In a trial with sheep, Ewer (1950) found no difference in prophylactic efficiency of vitamin D_2 or vitamin D_3 but considered that it was not possible to draw a firm conclusion concerning this aspect in view of the limited number of sheep in the experiment.

In this report ergocalciferol is assumed to be as effective as cholecalciferol and both are measured in terms of μg crystalline cholecalciferol. (The antirachitic activity of 0.025 μg crystalline cholecalciferol = 1 International Unit (i.u.) = 1 USP unit.)

Recent studies of cholecalciferol have shown that it is metabolized to 25-hydroxy-cholecalciferol in the liver, and in this form is the major circulating metabolite. Conversion of this metabolite to dihydroxy derivatives (1,25; 24,25 and 25,26 dihydroxy) takes place in the kidney and the first-mentioned of these derivatives is considered to be the ultimate biologically active vitamin D_3 metabolite, at least regarding intestinal absorption of calcium and probably in remodelling of bone, in concert with parathyrin and calcitonin (Avioli & Haddad 1973, Olsen & DeLuca 1973). The 25-hydroxy derivative is the major storage form of the vitamin. Avioli & Haddad (1973) presume that ergocalciferol undergoes similar transformations. Greenbaum (1973) considers that, while hydroxy forms may have a particular use in the treatment of certain disorders, on the basis of cost cholecalciferol is likely to remain the material of choice for nutritional supplementation, at least in the near future.

It is relevant to note that in the past, measurements of vitamin D were based on biological assay techniques and were difficult. Now that chemical assays for 25-hydroxycholecalciferol are done routinely, the reserves of vitamin in the body are more easily assessed.

1. The importance of dietary calcium, phosphorus and energy in relation to vitamin D requirements

It is now well established that calcium, phosphorus and vitamin D are closely interrelated in the process of normal skeletal development in the body. There is evidence in the literature that a deficiency of either phosphorus or vitamin D in the presence of adquate intakes of the other two gives rise to rachitic lesions in ruminants. Thus a deficiency of vitamin D in the presence of sufficient dietary calcium and phosphorus has been shown to produce rickets in calves (Gullickson et al 1935, Thomas & Moore 1951, Thomas 1952) and in growing lambs (Andrews & Cunningham 1945). Similarly, a deficiency of dietary phosphorus in the presence of sufficient calcium and of vitamin D produced typical and extensive rachitic lesions in growing heifers (Theiler 1934). Benzie et al (1960) in Scotland, using growing lambs fed on a diet deficient in phosphorus but adequate in calcium and vitamin D, confirmed Theiler's findings.

Evidence in the literature suggests that ruminant animals given calcium deficient diets adequate in phosphorus and vitamin D, though showing poor skeletal development, do not develop rachitic lesions. Thus Auchinachie & Fraser (1932) found that with growing lambs supplementation of a calcium- and vitamin D-deficient diet with vitamin D (cod liver oil) prevented rickets and allowed satisfactory gains in weight over a period of 5 months. Poor skeletal growth but no evidence of rickets in sheep fed on a calcium-deficient but otherwise adequate diet is reported by Franklin (1934–35). Work by Benzie et al (1960), referred to later, supports these findings.

The experiments of Benzie et al (1960) throw considerable light on the interactions of different calcium, phosphorus, vitamin D and energy intakes on the growth and development of growing wether lambs. They gave a diet deficient in all four components and studied the effects of all combinations of supplements of these components on live weight gain, blood composition and skeletal development. The results (a) confirmed the findings of Duckworth et al (1943) in relation to the energy intake of growing animals and their vitamin D requirements; (b) showed that the addition of vitamin D to a vitamin D- and calcium-deficient diet stimulated body growth but not the mineralization of the skeleton, and that, nevertheless, there was no manifestation of rickets in any of these animals, a finding which agrees with the observations of Auchinachie & Fraser (1932), referred to above, that although no rickets occurred on a diet deficient in calcium and vitamin D supplemented with cod liver oil and on which the rate of gain was satisfactory, the animal had very thin bones compared with the bones of those receiving both cod liver oil and calcium; (c) showed that calcium and vitamin D supplementation of diets deficient in these two components resulted in a much greater improvement in mineral deposition in the skeleton than supplementation with either given individually, but that the effect on growth rate was that expected from the results of either given individually; (d) showed that the only major significant interaction between dietary phosphorus and vitamin D was that revealed in the radiological score (a measure of bone "quality") of the right limb radius. Though the supplementation of a phosphorus- and vitamin D-deficient diet with vitamin D gave a marked improvement in the score, vitamin D supplementation of the diet when adequate in phosphorus had no effect.

The basal diet used by Benzie et al (1960) was severely deficient in calcium and vitamin D and moderately deficient in phosphorus and it was intended that the supplements should provide adequate amounts. If it is assumed that the sheep averaged 30 kg live weight (Blackface wether lambs, 5–11 months), then on the basis of the requirements of calcium and phosphorus now proposed, the phosphorus intakes of all those on the diet intended to be phosphorus-deficient provided adequate phosphorus and the same was true for calcium. In all but the vitamin D-deficient diet, 2.5 ml cod liver oil was given daily assuming 2.13 μg vitamin D/ml would give a daily supplement of 7.8 μg, i.e. 0.26 μg vitamin D/kg W per day which is adequate.

It thus appears that in the growing ruminant, as in the chick or pig, the requirement for vitamin D is small provided that adequate amounts of calcium and phosphorus are present. As the calcium intake decreases, the requirement for vitamin D increases. The same probably applies to the dietary phosphorus, but the evidence is not so clear. Furthermore, it can be seen that the higher the growth rate the greater the requirement for vitamin D.

Wallis (1941) has shown that additional calcium and phosphorus for normal adult cows on a vitamin D-deficient diet did not prevent the deficiency syndrome from appearing nor delay its appearance. The addition of calcium and phosphorus to the ration of cows suffering from avitaminosis D did not relieve the condition.

2. Storage of vitamin D in the body

From the evidence available it appears that the ruminant like other domesticated animals is able to store vitamin D. Thus Duckworth et al (1943) concluded from their experiments that the lambs they studied had about a 6-week reserve of vitamin D in the body when housed in November.

Calves which after birth had received vitamin D only from the milk given during their first 10 days of life, but which were otherwise adequately fed, developed rickets within 60 days (Thomas & Moore 1951). In calves deprived of vitamin D rickets developed within 30 days (Duncan & Huffman 1936). In calves given adequate amounts of vitamin D before a vitamin D-deficient diet the deficiency syndrome developed in 90–100 days (Wallis et al 1935). Further evidence of the limited ability of the calf to store vitamin D is provided by Guerrant et al (1938), who analysed the vitamin D potency of the liver and blood of a number of calves that had been reared for up to 7 months on known intakes of vitamin D. While total daily intakes of the vitamin ranged from 0.08 to 1.2 μg/kg live weight, the contents of the vitamin in the blood and liver at slaughter ranged from 25 to 66 μg and 7.1 to 14.5 μg, respectively.

According to Wallis (1944), allowing for some variation due to storage of the vitamin, high yielding cows (27–30 kg milk/day) will reveal signs of vitamin D deficiency within 4 months of being deprived of a source of vitamin D whereas cows giving 11–18 kg milk/day may not reveal any signs for 6–8 months.

Eaton et al (1947) have shown that in cattle, storage of the vitamin in the liver and blood of the developing foetus is not significantly affected by giving 2.5 mg of vitamin D daily to the cow for a period of 8 weeks before calving.

3. Avitaminosis D in farm livestock in practical conditions

Only a limited number of investigators have reported the occurrence of rickets in young livestock reared in practical farm conditions. Leslie (1935) found a high incidence of rickets in growing lambs in the South Island of New Zealand, during autumn and winter. Fitch and Ewer (1944) and Franklin (1953) observed that in certain conditions pertaining in practice, vitamin D supplements given to grazing sheep during the winter improved weight gain and wool growth. Favourable responses were obtained both in sheep grazing winter oats, a crop which appears to possess rachitogenic properties, and in those grazing ordinary pastures. Ewer (1950) reported rickets in ewes and wether lambs wintered outside and fed on a diet which included green oats, old pasture and concentrates but little hay. Dunlop (1946, 1954) reported rickets in ram hoggs grazing low ground pasture in Ayrshire during winter, which was cured by the oral administration of a dose of 50 mg vitamin D. Crowley (1961) reported an outbreak of rickets in November-born lambs and ascribed it to a high intake of carotene at a time when intake of vitamin D was low. Nisbet et al (1966) observed rickets among 9–12 month old ewe and wether lambs born on two hill and arable farms in south Scotland. The occurrence of rachitic conditions in young calves reared in farm conditions in various parts of the USA has been reported by Rupel et al (1933), Gullickson et al (1935) and Hibbs et al (1945). Rickets has also been reported among calves being fattened on barley in north-east Scotland (Seawright 1966); signs of vitamin A deficiency also were present and treatment comprised intramuscular injection of 0.12 g of retinol and of 10 mg of vitamin D with the inclusion of bone meal in the feed. Vitamin D deficiency has also been reported in beef cattle reared intensively in the Ukraine (Danilenko et al 1972).

It is perhaps not surprising that so few instances of avitaminosis D in farm livestock have been reported. As discussed later, if dietary calcium and phosphorus are adequate, vitamin D requirements are small. Furthermore, the body has a capacity to store the vitamin. Vitamin D is synthesized within the body on exposure to sunlight and hence most ruminant livestock have an abundant supply during the summer months. In winter most ruminants receive some form of sun-cured fodder in which the vitamin D activity may well be high. Adult animals are thus unlikely to suffer from vitamin D deficiency. In young growing stock the vitamin D requirement is high only when the energy intake is sufficient to permit growth. In many conditions in which a dietary deficiency of vitamin D is likely, a concomitant shortage of dietary energy may prevent the overt signs of rickets. It must also be remembered that compared with the young chick, which has a rapid rate of skeletal growth and accordingly a high vitamin D requirement, the young ruminant after the suckling stage has a relatively much lower rate of skeletal development.

4. The minimum requirement for vitamin D to prevent rickets in calves

Experimental data have been collected in Table 7.2. Lack of experimental detail makes it difficult to assess the accuracy of the conclusions of Huffman & Duncan (1935) that 0.02 to 0.025 μg/kg live weight is sufficient to allow for normal growth. Nevertheless, those authors were careful to point out that this applies to vitamin D supplied in milk. It may be that the more efficient retention of the calcium and phosphorus supplied in milk may reduce the requirement of the calf for the vitamin. The experiments of Bechdel et al (1938) are those generally quoted as fixing the vitamin D requirement of the calf at 0.17 μg/kg live weight. This value is somewhat higher than that suggested by the work of Thomas & Moore (1951) and Thomas (1952), namely 0.1 and 0.08 μg/kg live weight, respectively. The latter amount protected fully against rickets only in the presence of more than adequate dietary calcium. The value of 0.1μg/kg W suggested by the work of Thomas & Moore is the arithmetical mean of two doses, one of which was adequate and the other not. It is probably too high, for at 0.08 μg/kg live weight these workers were unable to find any visible signs of rickets or evidence of the deficiency syndrome at post mortem examination.

With reference to the value of 0.17 μg/kg suggested by the data of Bechdel et al (1938), it will be noted from Table 7.2 that although the calcium intake was high, that of phosphorus was inadequate by National Research Council (1971) standards. While intakes of vitamin D of 0.073 and 0.098 μg/kg live weight (Table 7.2) were inadequate to prevent the development of rickets as judged by all criteria, that of 0.11 μg gave no visible sign of rickets and bone ash values were normal. Although X-ray analyses revealed that the closure of the epiphyseal cartilage in calves receiving this amount of vitamin D was not quite as satisfactory as in those on higher intakes of the vitamin, the major criterion of inadequacy was the low blood inorganic phosphorus concentration. To what extent the low dietary phosphorus was responsible for the low blood values is not certain but it is noteworthy that Bechdel et al (1938) point out that a higher dietary supply of phosphorus might have changed the whole picture at this rate of vitamin D intake.

Smith (1958) has suggested that milk-fed calves have a higher vitamin D requirement than calves given a dried skimmed milk and grain mixture from 28 days of age. His conclusion is based on the fact that although daily intakes of 0.11–0.18 μg/kg live weight were sufficient to maintain serum calcium levels in 5–10 week old calves, utilization of calcium was considerably increased when the daily intakes of vitamin D were raised as high as 23 μg/kg live weight. It will be noted that lower intakes of vitamin D than 0.11–0.18 μg/kg live weight per day may have been sufficient to prevent the decline in serum calcium content.

Table 7.2 Data concerning the minimum amount of vitamin D needed to prevent rickets in calves †

Reference	Breed and age (months)	Average Ca:P ratio	Diagnostic tests for vitamin D deficiency	Source of vitamin D	Amount supplied daily (µg/kg)	Average live weight gain (kg/day)	Remarks concerning adequacy for protection against rickets
Huffman & Duncan (1935)	Holstein (0–5)	—	Blood Ca and P; bone ash contents	Milk	—	Not given	Conclude that 0.02–0.025 µg/kg body weight adequate if vitamin D given in milk
Bechdel et al (1938)	Holstein, all but one female (2–7)	2.8:1	Blood Ca and P; X-ray analysis test and analyses of bone	Grain and hay	0.07	0.34	Inadequate for all criteria
				Grain and hay	0.10	0.35	Inadequate for all criteria
				Grain and hay + irradiated yeast or codliver oil.	0.11	0.44	Inadequate for blood P; borderline X-ray
				Grain and hay + irradiated yeast + codliver oil.	0.20	0.40	Adequate for all criteria
				Grain and hay + irradiated yeast + codliver oil.	0.27	0.49	Adequate for all crieria
Moore et al (1948)	Holstein and Jersey males (2–5)	—	Physical condition; bone ash content	Barn cured hay	0.08	0.75	Borderline
				Wilted silage	0.10	0.77	Adequate
				Field cured hay	0.12	0.69	Adequate
				Barn cured hay	0.14	0.77	Adequate
				Wilted silage	0.17	0.75	Adequate
				Barn cured hay	0.20	0.78	Adequate
				Wilted silage	0.24	0.79	Adequate
				Field Cured hay	0.24	0.73	Adequate
				Field cured hay	0.37	0.66	Adequate
Thomas & Moore (1951)	Jersey males except for two (3–8)	1.4→1.8:1	Blood Ca, P and phosphatase; X-ray analysis; bone ash	Early cut lucerne (barn dried)	x0.08	0.52	Borderline for plasma Ca, P + bone ash
		1.7→2.0:1		Medium cut lucerne (barn dried)	0.12	0.57	Adequate for all criteria
		1.4→1.8:1		Late cut lucerne (barn dried)	0.24	0.54	Adequate for all criteria
		Not given 1.3→1.9:1		Lucerne silage	0.33	0.60	Adequate for all criteria
				Field cured lucerne	0.60	0.53	Adequate for all criteria
		1.4→2.3:1		Viosterol	1.65	0.67	Adequate for all criteria
Thomas (1952)	Holstein and Jersey males (2–5)	1.2→1.4:1	Blood Ca, P, phosphatase; X-ray analysis; bone ash content; physical condition	Dried lucerne	0.03	0.67	Inadequate for all criteria
		1.5→1.9:1			0.03	0.67	Inadequate for all criteria
		2.0→2.7:1			0.03	0.67	Inadequate for all criteria
		1.5→1.9:1			0.05	0.67	Inadequate for all criteria
		2.0→2.7:1			0.05	0.67	Inadequate for all criteria
		1.2→1.4:1			0.08	0.68	Inadequate for bone ash
		2.0→2.7:1			0.08	0.68	Adequate for all criteria

† all Ca and P intakes were adequate as judged by current estimates of requirement for these elements (see Chapter 5).

From a consideration of the data in Table 7.2 it is suggested that the requirement for vitamin D for the prevention of rickets in calves receiving adequate amounts of dietary calcium and phosphorus is 0.1 μg/kg live weight per day.

Though the requirement for vitamin D to prevent rickets is obviously small, it may well be that higher intakes would ensure improved growth and general health. (The toxicity of very high single doses of vitamin D is referred to in a separate section; such doses may be of the order of 30 000 times the minimum daily dose to protect against rickets.) In this regard, it is the generally accepted practice in the USA to fortify calf starters with from 55 to 220 μg vitamin D/kg (Wallis 1952). Thus a 50-kg calf receiving 0.7 kg of calf starter per day and an 80-kg calf receiving 1.1 kg will have a vitamin D intake of 0.78–3.0 μg/kg live weight per day. Keyes (1944) reported that increasing by 1.1 μg/kg the daily vitamin D intake of calves already receiving a standard vitamin D-fortified calf starter improved growth and the general health of the calves.

It is also to be noted that in studies on coppper deficiency in calves fed on semisynthetic diets formulated to meet Agricultural Research Council (1965) standards for calcium, phosphorus and vitamin D, Mills et al (1976) and Suttle & Angus (1977) reported evidence of vitamin D deficiency, viz. low plasma calcium and phosphorus (Mills et al 1976) or 25-hydroxycholecalciferol concentrations (Suttle & Angus, 1978).

5. Vitamin D requirement of growing cattle

Only one experiment on the requirements of growing cattle for vitamin D has been traced. Bechtel et al (1936) reared heifer calves indoors on rations in which almost the only source of vitamin D was maize silage, the vitamin D potency of which was determined by biological assay with rats. The data are shown in Table 7.3 and indicate that the minimal requirement appears smaller than that necessary to prevent rickets in calves. Although in certain instances the daily intakes of calcium and phosphorus were inadequate, in no instance were they markedly so. Thus in the first edition it was suggested that a value of 0.06 μg/kg live weight per day be accepted as the minimal requirement of young cattle. In this revised edition it is proposed that in the absence of more complete data the minimal requirement be taken as that for the calf, 0.1 μg/kg live weight per day.

Table 7.3   Amounts of vitamin D adequate to allow normal growth in growing heifers (From Bechtel et al 1936)†.

Grade Holstein heifer	Age (months)	Average weight (kg)	Average Ca:P ratio	Intake of vitamin D (μg/kg live weight per day)	Average live weight gain (kg/day)	Remarks
No. 167 {	8–11	191	1.63:1	0.06	0.72	Adequate
	13–16	306	1.62:1	0.06	0.71	Adequate
	17–19	378	1.59:1	0.05	0.29	Adequate
No. 169	12–14	187	1.62:1	0.06	0.73	Adequate to cure severe rickets in 67 days and then to allow normal growth
No. 195	18–22	323	1.81:1	0.07	0.44	Adequate to cure severe rickets and allow normal growth

† see footnote to Table 7.2.

6. The requirement of vitamin D for pregnancy and lactation in dairy cows

There is so little information on the requirements of vitamin D for pregnancy and for lactation that these two reproductive processes have been considered jointly.

Conclusive evidence that vitamin D is essential for maintenance, reproduction and lactation in cows was obtained by Wallis (1938). The few data available on low and yet apparently adequate intakes of vitamin D for pregnancy and lactation are shown

Table 7.4   Amounts of dietary vitamin D adequate to protect against rickets in pregnant and lactating cows (From Bechtel et al 1936)†.

Grade Holstein	Age (months)	Months in calf	Vitamin D intake (µg/kg live weight per day)	Remarks
No. 167	⌈ 26–28	2–4	0.06‡	⎫ Adequate to protect against rickets
	{ 31–33	6–8	0.07	⎭ and produce normal calf
	⌊ 35–38	—	0.08	Adequate to enable cow to give daily yield of 20.4 kg milk
No. 195	25–29	4–8	0.06§	

† see footnote to Table 7.2.
‡ Vitamin D intake from 8 months of age did not exceed 6.3 µg/100 kg live weight per day (see Table 7.3).
§ Vitamin D intake from 18 months of age did not exceed 6.75 µ/100 kg live weight per day (see Table 7.3).

in Table 7.4. They suggest that the minimum lies below 0.08 µg/kg live weight per day.

Wallis (1941) on the basis of further, although incomplete, work suggests that the physiological minimum daily requirement approximates to 75 µg. This for a 500-kg cow would be equivalent to a daily intake of 0.15 µg/kg live weight per day.

In fixing a satisfactory minimum intake, two additional factors should perhaps be considered: first, that the doses of vitamin D likely to induce toxicity are almost 10^5 times the minimum requirement and, secondly, that the higher the vitamin D intake by the cows the greater will be the content of the vitamin in the milk produced (Wallis 1941). For these reasons it is suggested, in the absence of more complete data, that the minimal requirement for lactation is 0.25 µg/kg live weight per day. It is suggested that the requirement for pregnancy is the same.

With reference to milk as a source of dietary vitamin D for man, Krauss et al (1932) showed that the concentration in milk is positively correlated with the amount given to the cow.

7. The requirement of vitamin D for sheep

The only data intended expressly to determine the vitamin D requirement for growing sheep are those of Andrews & Cunningham (1945). Growing Corriedale lambs (sex

Table 7.5   Data concerning the requirements for vitamin D of growing lambs†.

Breed and age (months)	Ca:P ratio	Criteria used in judging deficiency	Vitamin D intake (µg/kg live weight per day)	Weight gain (kg/day)	Remarks
Halfbred wethers (7–13)	0.36:1	Physical appearance: serum Ca and blood inorganic P	0.47‡	0.08	Adequate
	1.43:1		0.47‡	0.08	Adequate
Halfbred wethers (6–11)	0.85:1	Physical appearance; serum Ca and blood inorganic P	0.15‡	0.07	Adequate
Corriedale lambs (6–11)	1.3 → 1.5:1	Presence or absence of rachitic lesions at slaughter	0.01	0.08	Inadequate
			0.02	0.09	Inadequate
			0.10	0.10	Borderline

† compiled from Auchinachie & Fraser (1932), Duckworth et al (1943) and Andrews & Cunningham (1945). All intakes of Ca and P were adequate as judged by current estimates of requirement (see Chapter 5), except that the Ca intake of halfbred wethers on a diet with a Ca:P ratio of 0.36:1 was judged inadequate.
‡ calculated by assuming that cod liver oil contains vitamin D 21.3 µg/ml.

not given) were used and the results shown in Table 7.5 indicate that for such sheep the minimal requirement borders on 0.1 μg/kg live weight per day. The calcium:phosphorus ratio of the diet used in these studies was near to 1.5:1. It is thus probable that the requirement will provide a good margin of safety in rations of more normal calcium:phosphorus ratio. Auchinachie & Fraser (1932) and Duckworth et al (1943) used cod liver oil supplements to provide the vitamin. Unfortunately, these oils were not assayed for potency. Nevertheless, if the potency is assumed to have been 2.1 μg/ml, these results provide some evidence that the vitamin D requirement of growing sheep approximates to 0.1 μg/kg live weight per day (Duckworth et al used a dose rate of 2.5 ml cod liver oil daily).

In the absence of further information the requirement has been fixed at 0.13 μg/kg live weight per day.

8. Vitamin D and bovine fertility

As already mentioned, the experiments of Wallis (1938, 1944) reveal that avitaminosis D in pregnant cows gives rise to bone abnormalities in the calf at birth. In these experiments it was noted also that when receiving vitamin D-deficient diets, cows failed to have regular oestrous cycles. When the deficiency was alleviated, normal oestrus was again observed. Moller (1959) gave a dose of 75 mg vitamin D to over 1500 cows with a similar number of animals as controls. There was a slight and not significant advantage in percentage of non-returns after first service in favour of those given the injection. Cohen (1960) found that an intravenous injection of 125–250 mg vitamin D to cows showing anoestrus during 2–5 months after calving significantly improved the return to normal service. Ward et al (1971) have examined reproductive performance of 37 Holstein cows 2 to 6 years old during 58 reproductive cycles to compare the effects of two calcium intakes (100 g and 200 g/day) given in a ration containing 80–100 g phosphorus per day, each with and without 7.5 mg cholecalciferol weekly (orally, by capsule). For an average body weight of 600 kg, this weekly supplement of vitamin corresponds to a daily intake of 1.8 μg/kg live weight. Services per pregnancy and milk production (6368–7208 kg per lactation) did not vary between treatments. However, first recognisable post-partum oestrus occurred 16 days earlier ($P<0.06$) and conception occurred 37 days earlier ($P<0.025$) in the vitamin D supplemented cows. First ovulation occurred 6 days earlier in cows given the higher calcium intake (Ca:P ratio 2.1:1 to 2.5:1; the ratio for the lower calcium intake varied from 0.9:1 to 1.3:1) and was not influenced by cholecalciferol supplementation. Uterine involution was completed 8 days sooner ($P<0.05$) in the cows receiving the higher calcium intake when both diets were supplemented with vitamin D; results for the unsupplemented diets were intermediate.

It is difficult to assess the significance of much of the earlier work since little information is given concerning the amounts of vitamin (from food or resulting from exposure to sunlight) the cows were receiving before the administration of the vitamin. In the experiments of Wallis (1938, 1944), the return to normal oestrus from the anoestrus which developed during avitaminosis D was achieved through intakes of vitamin D which, when considered in conjunction with the time they gave protection against rickets, are by no means much above the suggested minimum daily requirement for lactating animals of 0.25 μg vitamin D/kg live weight. In that of Ward et al (1971) the ration was considered by the authors to have furnished about 0.33 mg vitamin D/day which is approximately twice this minimum daily requirement. Supplementation with orally administered cholecalciferol raised this 7–8 times the daily minimum requirement. It perhaps should be noted that, in the studies of Ward et al (1971), the higher rate of calcium supplementation approximates to somewhat more than twice that suggested as the minimal requirement for the element in the first edition.

9. Vitamin D and the prevention of milk fever

Massive daily doses of vitamin D (0.5–0.75 g) given orally for 3–6 days prepartum and 1–2 days post partum (maximum of 7 days) have been shown to be very effective in reducing the incidence of milk fever in dairy cows (Hibbs & Pounden 1955, Dell & Poulton 1958, Hibbs & Conrad 1960). Smaller doses gave proportionately lower degrees of protection. The beneficial effect of such doses of vitamin D is considered to be largely due to the greatly enhanced absorption of calcium and phosphorus from the digestive tract and simultaneously lowered endogenous losses of these elements (Conrad et al 1956). The last-mentioned workers quote evidence that the size of the parathyroid and the amount of hormone secreted are diminished by such treatments. The suggested effect of massive doses of vitamin D in relation to the parathyroid and milk fever is in contrast to that of parathyroid stimulation by feeding pre partum on diets low in calcium and high in phosphorus as a cure for milk fever (Boda & Cole 1954). Since protection against milk fever diminishes rapidly a day or two after the last dose, and continuation of massive dosage with vitamin D over long periods may have harmful effects, it is necessary to predict the calving date fairly accurately. Hibbs and his co-workers, and Dell & Poulton, emphasize that treatment must be terminated after a maximum of 8 days even if parturition has not occurred, and Dell & Poulton (1958) found a very high incidence of milk fever in cows which had completed the treatment 1–2 days before parturition.

It is of interest that Van Dijk (1966) found that a single intramuscular dose of vitamin D_3 (0.4 g) given up to 7 days before calving was both effective and innocuous in the prevention of milk fever. Cows that did not calve within 7 days of treatment were given a second intramuscular injection at the same dose rate on the eighth day. Gregorovic et al (1967) also reported that such a dose given intramuscularly 2–8 days before calving is an effective and safe method of preventing milk fever in cattle.

Recently Sansom et al (1976) have reported that the intramuscular administration of 250 μg 1-hydroxy D_3 within 2h after calving significantly reduced post-parturient hypocalcaemia and hypophosphataemia.

10. Toxicity of vitamin D

Cows

Hibbs & Pounden (1955) mention that oral doses of 0.75 g/day given for 20–30 days to newly calved cows induced definite signs of toxicity. Swan (1952) found that a dose of 0.25 g given orally daily for 10 days to a lactating cow resulted in no clinical signs of toxicity and when she was slaughtered two months later there was no indication of excessive calcification in the body. Double the dose for a similar period caused no sign of toxicity but calcified lesions were detected in the left heart and the aorta when the animal was slaughtered two months later. A dose of 0.25 g for 3 days before calving in another cow caused slight calcification of the posterior aorta; its identical twin used as a control showed no such calcification at slaughter.

Payne (1963) reported that cows receiving 0.25 g or 1.0 g vitamin D, the recommended or 4 times the recommended treatment for milk fever prevention (the author does not state the fact but presumably the dose was given intramuscularly), when slaughtered 7 months later and subjected to post-mortem examination were found to have extensive calcified lesions of heart, aorta and main arteries; the kidneys showed calcification of the tubules in the medulla. In subsequent studies (Payne & Manston 1967), no metastatic calcification was found when the 0.25 g dose was used. They attributed the difference between this and the earlier finding (Payne 1963) to the fact that in the later work the cows had a more balanced input of calcium, phosphorus and magnesium than the earlier animals; the change in mineral intake had been made to bring the feeding practice in line with the mineral recommendations of the first edition. These authors concluded that 0.25 g of cholecalciferol given intramuscularly is satisfactory but that 1.0 g can be fatal. Manston & Vagg (1964) showed that an intramuscular dose of 1.0 g cholecalciferol induced a pronounced fall in milk yield, while that of cows given 62.5 mg or 250 mg was not affected.

Calves

Blackburn et al (1957) found that oral administration of 25 mg cholecalciferol daily to calves led to depression in growth, diarrhoea, distress on exercise and stiffness. The calves were destroyed when the effects had become severe. The mean age at death of those receiving vitamin D was 47 days, of those receiving $CaCO_3$+vitamin D it was 17 days, and of those receiving $CaHPO_4$+vitamin D, 37 days. Post-mortem examination revealed gross metastatic calcification particularly of the vascular system. No abnormality was found in calves given the minerals alone.

Sheep

Grant (1955) refers to preliminary results of trials being made on the toxicity of massive single doses of vitamin D to sheep as indicating that whereas the healthy animal is well able to tolerate a single dose of 50 mg vitamin D, in the sick or unthrifty animal a dose of 25 mg has a toxic effect resulting in a loss of weight within 10 days. Unfortunately, Grant does not state how the vitamin was administered. W. A. Greig (private communication) found that a single dose of 250 mg given intravenously produced metastatic calcification in the subendocardium and in the subintimal region of the major arteries in sheep. Blackface ewes, 1 year old, given a single dose of 25 mg cholecalciferol by intramuscular injection showed patchy degeneration of the aorta and atrophy of smooth muscle after two months, becoming progressively worse; proliferation of the vascular epithelium also was present in some (Fell et al 1964).

11. Sunlight and vitamin D

With the data available at the present time it is almost impossible to make a valid assessment of the amount of vitamin D likely to be synthesized by domestic livestock exposed to sunlight and therefore to assess dietary needs of such animals.

Most information about the antirachitic properties of sunlight has been derived from work with rats or chicks fed on rachitogenic diets. In most instances the hair coats of the rats were removed before exposure to the sun. From such studies it is known that the sun's rays capable of effecting vitamin D synthesis within the body are in the short ultraviolet range (290–315 nm).

The antirachitic power of the sun is greatest at elevation of 90° and declines markedly as the sun's elevation falls. Tisdall & Brown (1929) showed that the sun's rays have very limited antirachitic power when the sun is at an elevation of 35° or less. Abrams (1952) from a consideration of independent data confirmed this fact. It would appear that in Great Britain which lies between latitudes 50°N and 60°N the antirachitic capacity of the sun during a considerable part of the year must be very limited. Tisdall & Brown (1929) calculated that in London the sun never attains an elevation of more than 35° for 5 months of the year, and in Glasgow the equivalent period is 6 months.

The experiments of Campion et al (1937) with lactating cows showed clearly that the increase in vitamin D potency in milk which takes place in summer is due to exposure of the cows to sun- and sky-shine. In the United Kingdom over the two years 1958–59, average values for vitamin D in milk ranged from 0.14 to 0.30 µg/kg (Thompson et al 1964), the low value being recorded for each two-month period in the span November to April, the high value in May to June. Surprisingly, the vitamin D potencies were generally higher in samples taken from northern depots than from those in the south despite the higher sunshine bonus in the south and may reflect the reduction in effective radiation in southern areas due to industrial pollution. Henry et al (1971) reported that average vitamin D contents of milk in New Zealand varied from 0.12 µg/kg in August to 0.77 µg/kg in January. These authors, using their own data and those of Thompson et al (1964), showed that in both the United Kingdom and New Zealand vitamin D values were highly correlated with hours of sunshine and, for both countries, with total radiation; the last-mentioned includes the diffuse radiation in "skyshine" and makes allowance for variation in sunshine intensity. The higher content of vitamin D in New Zealand milk fat in mid-season is attributed to a greater total radiation in that country.

Quarterman et al (1964) have shown that in grazing sheep, clipped in early summer, vitamin D in blood rose to a maximum of 10–22.5 µg/kg whole blood in August or September and thereafter fell steadily to a minimum of 1.25 µg/kg in April. Comparison of sheep at Aberdeen (latitude near 58°N) with sheep at Reading (latitude near 51°N) showed that although the latter had higher blood concentrations of vitamin D in summer, the values in the two groups were not significantly different in the succeeding late winter.

The work of Auchinachie & Fraser (1932), however, has indicated that winter sunshine in Scotland possesses some antirachitic power. Sheep maintained indoors on a vitamin D- and calcium-deficient diet developed rickets, while similar sheep maintained on the same rations but out of doors did not. Nevertheless, that the amount of vitamin D so provided was not completely adequate can be inferred from the improvement in the outdoor sheep caused by a cod liver oil supplement.

Abrams (1952) attempted to assess the vitamin D supply from sunlight for domestic livestock from astrophysical data, meteorological records and data concerning the antirachitic potentials of short-wave irradiation in rachitic rats without hair coats. A drawback to such estimates is the lack of quantitative data relating to the antirachitic powers of short-wave radiation falling on the natural hair or fleece covering of our farm livestock. That this is undoubtedly a major consideration is well demonstrated in the experiments of Quarterman et al (1961); in sheep clipped in May and grazing throughout the summer the vitamin D content of the blood had risen to a mean value of 1.7 µg/kg by August, whereas in unclipped sheep grazing the same pastures the mean in August was only 0.75 µg/kg.

12. Comparison of the requirements suggested here with requirements given in literature

The comparisons are given in Table 7.6.

Table 7.6
Vitamin D requirements of cattle and sheep proposed in this Technical Review and requirements given in the literature (All values expressed as µg/kg liveweight per day).

References	Sheep (all classes)	Calves	Growing cattle	Cows (pregnant and lactating)
This Technical Review	0.13	0.10	0.10	0.25
National Research Council (1970)	—	0.10	0.20[1]	—
National Research Council (1971)	—	0.17[2]	0.17[2]	—
National Research Council (1968)	0.14	—	—	—
Sperling (1951)	—	0.25	—	—

[1] For beef cattle from 150 to 400 kg live weight the requirement of vitamin D is taken as 6.9 µg/kg dry ration.
[2] For calves from 20 kg to 200 kg liveweight the requirement of vitamin D is taken as 0.17 µg/kg weight. At higher weights and for lactating cows, requirements are considered to be met from forages.

Essential Fatty Acids

Linoleic acid (9,12-octadecadienoic acid) and its metabolic derivatives are required as essential components of membrane lipids (Holman 1968) and as precursors of prostaglandins (Horton 1969). The derivatives, γ-linolenic acid (6,9,12-octadecatrienoic acid) and (5,8,11,14-eicosatetraenoic acid), are effective dietary substitutes for linoleic acid, though they are not usually present in the feed of herbivorous animals.

The dietary supply of linoleic acid for both young and adult nonruminants is normally assured and when it contributes 1 to 2% of total energy intake, the minimum requirement is met (Sewell & McDowell 1966, Leat 1970). The diet of adult ruminants, or of younger animals with a functioning rumen, commonly contains less than 1% of gross energy as linoleic acid (Leat 1970, Leat & Harrison 1972) and much of this is hydrogenated in the rumen. Bickerstaffe et al (1972) and Mattos & Palmquist (1977) measured the ruminal hydrogenation of octadecadienoic (18:2) acids present in diets in unusually high concentrations (6 and 2.7% of gross energy, respectively). The proportions hydrogenated were 86 and 68%, respectively; of the 18:2 acid reaching the duodenum 90–95% was identified as linoleic acid by Bickerstaffe et al (1972). Even after ruminal hydrogenation, the digesta of these diets contained linoleic acid equivalent to about 0.8% of gross energy intake.

Both young and mature ruminants appear to conserve and use linoleic acid and its derivatives more efficiently than do nonruminants. Though the young ruminant is born with very little linoleic acid in its tissue lipids, pronounced increases take place during the first few days after birth, despite the extremely low concentration of the acid in colostral and milk lipids (Leat 1966, Noble et al 1971); that which is absorbed is all retained (Noble & Moore 1971). In addition to linoleic acid of maternal origin, some may be available to the young ruminant as a result of microbial synthesis in the undeveloped rumen (Sklan et al 1971; Sklan & Budowski 1974). In adult ruminants there is evidence for a low turnover rate of linoleate (Lindsay & Leat 1977).

When milk-fed ruminants have been given diets containing little or no linoleate they have sometimes shown clinical signs of deficiency (Cunningham & Loosli 1954 a, b, Lambert et al 1954), although Sklan et al (1972) found no sign of deficiency in calves given diets to which linoleate contributed as little as 0.01% of total energy content. Nevertheless, for the milk-fed ruminant it is suggested that the dietary concentration of linoleate should be no less than that recommended for nonruminant animals (i.e., 1% of energy content, or 6 g linoleate per kg of a typical milk or milk substitute containing 25 MJ/kg DM).

Adult ruminants, or young animals with a functioning rumen, seem to be remarkably resistant to dietary deficiency of linoleate; we can find no well-authenticated case of essential fatty acid deficiency in such animals. Cunningham & Loosli (1954b) fed lambs on a dry semisynthetic diet containing no lipid for 240 days without inducing deficiency. Palmquist et al (1977) fed adult sheep parenterally on a lipid-free diet for at least 4 weeks without inducing in blood serum the production of eicosatrienoic acid that is considered to indicate essential fatty acid deficiency (Holman 1960). In Britain, dairy cattle have been fed exclusively for their complete lifespan on a diet containing about 5 g linoleate/kg (i.e., 1% of energy content) without the appearance of eicosatrienoic acid in serum lipids (G. A. Garton, W. R. H. Duncan & D. Gibson, 1976 unpublished results). Lactation should increase the requirement for linoleic acid although not to a large extent. The linoleic acid content of the milk of ruminants is a debatable quantity since much of the 18:2 acid commonly measured consists of positional and geometrical isomers of linoleic acid. If linoleic is assumed to constitute 70% of 18:2 acid or 1.3% of total fatty acid (Cuthbertson 1976), a cow yielding 15 kg milk per day would need to absorb a minimum of 8 g/day of linoleic acid. If 25% of dietary linoleic acid escaped rumen hydrogenation and was subsequently absorbed, the minimum dietary requirement for milk production would be 32 g/day, or about 2 g/kg feed DM consumed.

It must be emphasized that calculations of this kind are at present based on inadequate and sometimes conflicting evidence. For example, the cows of Mattos & Palmquist (1977) secreted 24 g of 18:2 fatty acid in 15.7 kg milk daily, which is about twice the quantity expected. Until more information becomes available it seems unwise to make specific recommendations on the essential fatty acid requirements of adult ruminants.

Chapter 8

The Vitamin B Complex

1. Introduction

The tissues of ruminant animals require all the vitamins of the B complex, but in considering dietary requirements it is necessary to consider separately the needs of the young milk-fed animal and those of older, ruminating animals. The preruminant animal depends on its diet to supply its needs for vitamins of the B complex, whereas the older animal with a functioning rumen usually receives adequate supplies as the result of microbial synthesis in the rumen.

2. Requirements of the Milk-fed Ruminant

Dietary needs for thiamin, riboflavin, vitamin B_6, pantothenic acid, nicotinic acid, vitamin B_{12}, folic acid and choline have been established in experiments in which young calves and lambs were given purified diets deficient in the vitamin being studied. The experimental proof of a requirement for biotin is less certain (Kon & Porter 1951).

Table 8.1   Amounts of B vitamins providing adequate intakes for the preruminant calf, and representative values for the B vitamin content of cow's milk and whey, and ewe's milk.

	Adequate intake (μg/kg live weight per day)	References	Cow's milk (mg/kg)					Ewe's milk (reference 12) (mg/kg)
			Whole liquid	Whole dried	Dried skimmed	Whey	Dried whey	
Thiamin	65–150	(1, 2)	0.45	3.0	4.5	4.0	4.0	1.2
Riboflavin	15–45	(3, 4)	1.8	13.5	20	0.80	12	4.3
Vitamin B_6	65	(5)	0.40	3.0	4.0	0.20	3.0	0.7
Pantothenic acid	195	(6)	3.2	25	32	3.2	48.5	5.3
Nicotinic acid	260	(7)	0.80	6.0	8.0	0.70	10.5	5.4
Vitamin B_{12}	0.4–0.8	(8)	0.0035	0.025	0.035	0.003	0.045	0.0098
Folic acid	5	(9)	0.06	0.45	0.60	0.05	0.75	0.054
Biotin	1.9	(10)	0.025	0.20	0.25	0.015	0.20	0.050
Choline	26 (mg)	(11)	130	1000	1300	130	2000	—

1. Johnson et al (1948).
2. Benevenga et al (1966).
3. Draper & Johnson (1952).
4. Brisson & Sutton (1951).
5. Johnson et al (1950).
6. Sheppard & Johnson (1957).
7. Hopper & Johnson (1955).
8. Lassiter et al (1953).
9. Wiese et al (1947).
10. Wiese et al (1946).
11. Johnson et al (1951).
12. Williams, A. P. et al (1976).

Intakes that prevented the onset of signs of deficiency are listed in Table 8.1. It must be stressed that the values quoted do not necessarily indicate specific requirements; only for riboflavin and vitamin B_{12} have attempts been made to assess dietary needs quantitatively. Thiamin requirements almost certainly depend on the carbohydrate content of the diet and it is noteworthy that the requirement was higher with a low-fat diet in which a large proportion of the dietary energy was supplied by carbohydrate (Benevenga et al 1966).

Present indications are that the suckled animal and the animal given milk or a milk substitute based on milk proteins receive adequate amounts of all the vitamins of the B complex. Although the nicotinic acid concentration in milk is below the likely requirement, milk proteins contain adequate amounts of tryptophan which can be converted in the body to nicotinic acid, and it is only with diets low in both tryptophan and nicotinic acid that signs of deficiency develop (Blaxter & Wood 1952).

Milk substitute diets based on or containing appreciable quantities of alternative protein sources such as soya bean, fish or microbes may require the addition of one or more B vitamins to raise their concentrations at least to those present in milk (Table 8.1). Some commercial calf diets contain added vitamin B_{12} about 30 $\mu g/kg$ dry matter, though there appears to be no experimental proof of the benefit of such addition.

3. Requirements of the Ruminating Animal

The earlier work providing evidence of the synthesis in the rumen of all the vitamins of the B complex was reviewed by Kon & Porter (1954) and Ling et al (1961). More recently a number of workers have reported studies of the effects of diet on the extent of ruminal synthesis of B vitamins and their findings generally confirm earlier conclusions that vitamins are synthesized in amounts adequate for normal metabolism in the host and secretion of normal amounts into milk. However, in certain conditions problems may arise in ensuring adequate supplies of thiamin and vitamin B_{12} (cyanocobalamin).

(a) Thiamin

Evidence of thiamin deficiency in young ruminants (lambs, calves and goats 2–6 months old) has been reported from several countries (cf. Edwin & Lewis 1971). Affected animals developed signs of a disease involving the central nervous system and called cerebrocortical necrosis (CCN) (Terlecki & Markson 1959) in Europe and Australasia, and polioencephalomalacia (PEM) (Jensen et al 1956) in the USA. Such animals became dull and were unable to coordinate their movements; within a few days they collapsed and, after convulsive seizures, passed into a coma and died. Post mortem the only consistent gross lesions were in the brain. Affected animals responded dramatically to intravenous or intramuscular injection of an aqueous solution of thiamin and recovery was complete when treatment was started before widespread neuronal damage had occurred.

A somewhat similar condition, molasses toxicity, observed in cattle given diets containing large amounts of molasses (Verdura & Zamora 1970) now appears not to be due to a deficiency of thiamin (Losada et al 1971).

The signs of CCN differ from those of frank thiamin deficiency and it seems likely that the disease is a modified deficiency arising from the presence in the gut of a thiaminase, probably of microbial origin (Shreeve and Edwin 1974). CCN has not been reproduced experimentally without the use of the thiamin antagonist amprolium (Markson et al 1974). Outbreaks of the disease in practical conditions have been sporadic and unpredictable; no clearly defined predisposing circumstances have been reported, though the use of diets high in concentrates may be a contributory factor. Such diets may encourage the production of thiaminase and at the same time increase the animal's need for thiamin to metabolize a high intake of carbohydrate.

(b) Vitamin B$_{12}$ (see also Cobalt p.240)

Cyanocobalamin and a number of related but biologically inactive vitamin B$_{12}$ analogues are synthesized by the rumen microorganisms in animals given cobalt. The amounts and relative proportion of cyanocobalamin and analogues synthesized are determined primarily by the cobalt content of the diet but they may also be influenced by the composition of the diet. Smith & Marston (1970) showed that 400–700 μg of cyanocobalamin was synthesized daily in the rumen of sheep given a diet high in roughage and adequate in cobalt. But the vitamin was poorly absorbed from the intestine and less than 5% of that synthesized entered the animal. Even so, the amount absorbed (20–30 μg) was in excess of the animal's requirement, estimated by Marston (1970) to be about 11 μg/day (i.e., about 0.3 μg/kg body weight per day).

Cyanocobalamin generally contributes 10–20% to the total vitamin B$_{12}$ activity synthesized by rumen microorganisms. The proportion of cyanocobalamin tends to be highest in the rumen of grazing animals, intermediate in stall-fed animals receiving diets high in roughage and lowest in animals given diets high in concentrates (Smith & Marston 1970, Dryden & Hartman 1971, Sutton & Elliott 1972).

The essential role of cyanocobalamin in the metabolism of propionate in ruminants is now well established (Smith & Marston 1971) and failure to metabolize propionate at the normal rate leads to a progressive loss of appetite in sheep deprived of vitamin B$_{12}$ (Marston et al 1972). It is not yet clear whether reduced synthesis of cyanocobalamin in animals given diets high in concentrates affects their ability to metabolize effectively the large amounts of propionate that may be produced and absorbed.

Chapter 9

Requirements for Water

Introduction The water requirements of ruminant animals are met from three sources:
  1. Water consumed voluntarily;
  2. Water contained in feed;
  3. Water formed within the animal's body as a result of metabolic oxidation.

This last source is important in the water economy of the animal since the catabolism of 1 kg of fat, carbohydrate or protein produces about 1070, 500 or 400 ml water, respectively (Roubicek 1969). These figures are slightly lower than those given in the first edition.

In this report the *estimated water intake* will be considered, defined as the sum of the water contained in the feed and the water drunk voluntarily. It does not include the metabolic water formed in the body. Although estimated water intake can be assumed to be the best approximation to water requirement, it must be stressed that true physiological requirements are not known with any precision and there is a need for more basic research on the physiological aspects of the water balance of ruminants. A valuable review of the water economy of farm animals (Leitch and Thomson 1944–45) summarized the information on water intake then available. The uncertainties and difficulties involved in arriving at reliable values for "apparent utilization" were fully discussed in that review.

In temperate conditions, the main need for water is to meet physiological requirements. The extent to which animals can tolerate dehydration, and the efficiency with which they can use water to rectify body imbalance, vary widely between species. For example, cattle have a limited ability in this respect compared with camels. Popovici et al (1973) reported that total water deprivation caused no significant reduction in milk yield for 24 hours, but caused severe reductions in the subsequent 24 hours even though water had again been made available. Konar & Thomas (1970) found a similar delayed effect on milk yield in goats, and both groups of authors reported significant increases in the total solids content of the milk. Little et al (1976) conducted a trial in which cows received only 60% or more of their normal water intake for a period of up to 6 days, the daily allowance being given in two equal portions. Milk yield was depressed only after the first 24 hours of deprivation, but there was little effect on the milk composition. The reduction in yield may have been due partly to the associated reduction in feed intake. Such reduction in feed intake occurred when nonlactating cattle (Balch et al 1953) and sheep (English 1966) were partially deprived of water, and complete deprivation had a similar effect on goats (Konar & Thomas 1970).

Little et al (1976) found that cows compensated for the restriction in water intake to a much greater extent by reducing the output of water in faeces than by reducing milk yield. Surprisingly they did not reduce significantly their output of urine. The animal's response to periods of deprivation longer than 6 days had not been assessed; until recently it was thought that daily loss of 6 kg water would become intolerable and a reduction in urine output, similar to that observed in nonlactating cows by Balch et al (1953) and in sheep by MacFarlane et al (1961), might occur.

Little et al (1978) measured voluntary water intakes of eight housed lactating cows, four of which were subsequently deprived of 40% of this water intake for 3 weeks. With the deprived cows there was a decrease in the mean intake of dry matter of 24% and in milk yield of 16%. These changes were apparent after 24 hours, but there were no further changes. Faecal water was significantly reduced and faecal dry matter concentration was significantly increased. Urine osmolality increased but the reduction in urine volume was not statistically significant. Concentrations of urea and sodium in serum and serum osmolality were significantly higher throughout the period of deprivation. The differences in blood composition increased slightly but significantly during each day from the time the animals were last offered water. The deprived cows lost weight during the first week of deprivation but showed no further change until rehydration, after which normal body weight was regained within 14 days. The deprived cows had consistently lower mean rates of respiration and rumination than the control cows throughout the period of deprivation but they showed no sign of distress. It appeared that the housed cows could have tolerated receiving only 60% of their normal voluntary water intake indefinitely. Little et al (1976) postulated that the measurement of the total ionic concentration (serum osmolality) might be useful in identifying herds of cows which are receiving insufficient water, and could be added to the routine metabolic profile test if water deficiency were suspected.

In this report we deal mainly with water intakes of livestock in temperate conditions such as those in the United Kingdom. We have therefore taken an environmental temperature of 30°C as the normal upper limit.

Factors Affecting Water Intake

1. Physiological condition and stage of growth of the animal

Water intake varies according to whether the animal is in a state of maintenance, growth, fattening, pregnancy or lactation.

(a) Cattle

Young calves

Because of the high water content of the feed, calves receiving liquid milk diets consume greater amounts of water in relation to the dry matter of their diet than older animals receiving dry diets. Atkeson et al (1934) showed that calves 1–5 weeks old receiving liquid milk diets consumed from 5.4 to 7.5 kg water/kg dry matter. This intake of water consisted mainly of the water contained in the diet, but small additional amounts of water were drunk. Burt & Bell (1962), using dry matter concentrations of 10, 14 and 20% in the diet, found that restricting the amount of water in the milk tended to reduce weight gain. Pettyjohn et al (1963) gave replacer diets of varying dry matter content to calves aged 4–8 weeks; on diets containing more than 15% dry matter (i.e., less than 5.7 kg water/kg dry matter) their efficiency of feed conversion was lower.

Thickett et al (1980) have demonstrated the value of providing water for young calves on an early-weaning system. When liquid milk diets were given at the rate of 4 kg/day the mean voluntary water intake of calves from 0 to 5 weeks was 65.1 kg with a coefficient of variation of 53%. There was a difference in average consumption of water according to calf weight. There were also significant correlations between weight gain, dry matter intake and water intake. It was found that for an increase in dry matter intake of 0.58 kg/day and an increase in weight gain of 0.39 kg/day, an extra 7 kg/day of water was consumed by the calf.

Pregnant cows

Little research has been done specifically on the water intake of pregnant cows. Roubicek (1969) found that heifers drank almost 50% less water on the day of oestrus than on the other days of the oestrus cycle, but the reason for this decrease is not known. The total body water content is increased in pregnant cattle by foetal tissues and associated embryonic fluids which require additional water intake. Piatkowski (1966) found that the voluntary water intake of heifers and cows in late pregnancy was, on average, 70 g water per kg live weight when offered water twice daily. Castle & Watson (1973) found that dry Ayrshire cows had a daily water intake of only 13.7 kg from July to September. They reported that the water intake was significantly reduced in proportion to the daily rainfall, and the major part of the daily water requirement of the animals was obtained from the herbage.

Lactating cows

Dairy cows in milk require the greatest amount of water in proportion to their weight or surface area because water constitutes 85–87% of the milk they yield. In addition, total body weight of most cows includes 55–65% of water; very fat cows may have less than 50%, and very thin cows as much as 70% of water (Reid et al 1955). In the previous edition it was considered that $0.87 \times$ the yield of milk (kg) should be added to the values quoted in Table 1 to compensate for the water secretion in milk. Recent work by Little et al (1978), in which two groups of four British Friesian$\times$Ayrshire cows were used, has produced a new relationship differing from that in the first edition. In experiments conducted over 7 consecutive days, water intake (I_w, kg/day), milk yield (M, kg/day), dry matter intake (I_d, kg/day) and the dry matter content of the hay and concentrates were measured. A multiple regression analysis gave the relationship:

$$I_w = 2.15\ I_d + 0.73\ M + 12.3 \qquad [9.1]$$

This relationship ignores the water content of the feed (c. 2 kg/day on average) but this would have only a small effect on the regression coefficients. The partial regression coefficient with milk yield (0.73) is similar to that recommended in the first edition (0.87) but that for dry matter intake (2.15) is considerably less than the value of the first edition (3.9). The new relationship also differs from the first edition in containing a constant (12.3); this is probably more realistic because it implies that if cows consume no dry matter and produce no milk they would nevertheless drink water. However, in order to provide adequately for the secretion of large amounts of water in the milk it is now suggested that for practical purposes the additional water allowance for lactation should be equal to the total weight of milk secreted.

The data recorded in Table 9.1 refer to water intakes determined principally with cattle receiving dry feed. The column headed water intake (kg/kg DM ingested) refers to the ratio:

(Water in feed + water drunk − water in milk)/DM intake.

Evidence assembled by Sykes (1955), MacLusky (1959–60), Calder et al (1964) and Paquay et al (1970a) suggests that cows fed on forages with a high moisture content (e.g., a fresh leafy sward) have a considerably increased total water intake. In these circumstances the total water intake may be higher than the true water requirement of the animal. The provision of a water supply to grazing dairy cattle has been studied by Castle (1972b) and by T. P. Thomas (1971). Castle (1972b) studied the normal

Table 9.1   Water intakes of cattle at various environmental temperatures.

Environmental temperature (°C)	Description of stock	Live weight (kg)	Water intake (kg/kg DM ingested)	Water intake (g/kg live weight per day)	References
−12 to +4	Various breeds	—	3.1	—	
4.4	of European	—	3.1	50*	
10.0	(*B. taurus*)	—	3.3	54*	
15.6	and Indian (*B.*	—	3.9	63*	Winchester & Morris (1956)
21.1	*indicus*) cattle	—	4.4	73*	
26.7		—	5.1	83*	
−18 to +4	Holstein-Friesian cows (lactating)	—	3.5	—	MacDonald & Bell (1958)
−18 to +4	Shorthorn heifers	—	3.5	—	
10		107	3.4	—	
26.7		79	6.6	—	
10		199	3.2	—	
26.7		133	8.5	—	Johnson et al (1958)**
10		289	3.8	—	
26.7		202	10.2	—	
10		359	3.7	—	
26.7		272	9.5	—	
—	Friesian, Guernsey and Jersey cows (lactating)	—	3.4	—	Blosser and Soni (1957)***
22	Dairy	620	4.7:5.0	—	
18	Shorthorn	645	4.4	—	
18	cows (dry)	571	4.2	—	Balch et al (1953)
21		568	5.0	—	
15 to 24	Holstein-Friesian	486	—	119	
32.2	cows (lactating)	482	—	155	Moody et al (1967)
—	Ayrshire cows (dry and nonpregnant)	525	—	45.5	Campling (1966)
8.2	British Friesian cows (lactating)	525	—	55.6	
		—	3.7	—	Castle & Thomas (1975)
—	British Friesian heifers	511	4.0	80	Owen et al (1968)
—	Friesian heifers (pregnant)	500	—	46	Enyedi & Illes (1967)
—	Friesian cows (dry, nonpregnant)	550	—	60	Paquay et al (1970ab)
22	E. African	198	—	51	
24	Zebu heifers	230	—	88	Wilson (1961)

* These values are for heifers of 450 kg and cows of 540 kg live weight.
** The moisture content of the hay and grain consumed was assumed to be 10%.
*** The silage dry matter content was taken as 28%.

intake of drinking water by dairy cows at grass, and found that a daily supply of 40 kg/cow would suffice on 98% of occasions. It was further suggested that there should be sufficient trough space to allow 10% of the whole herd to drink at any one time. T. P. Thomas (1971) has suggested that for a 100-cow herd at grass a flow rate of at least 655 kg/hour should be provided in a trough or troughs where up to six cows can drink at once. When the flow rate is too low, provision should be made for some form of reserve storage.

Castle & Watson (1970) compared three grass silages in a 12-week winter feeding experiment with 12 Ayrshire cows. Silages were eaten to appetite with a supplement of moist barley and groundnut cake. Silage A was made from unwilted herbage with no additive, silage B from identical herbage treated with formic acid, and silage C from the herbage after wilting for 28 hours. The total amounts of water consumed by the cows on the different treatments did not differ significantly. However, the amount of water drunk with silage C was significantly ($P < 0.01$) higher than with silages A and B and, conversely, the amount of water in the feed with silage C was significantly lower than with A and B. Water intake per kg of dry matter consumed did not differ significantly between A and B (i.e., between the untreated and the acid-treated silages) but was higher with silage C than with B ($P < 0.05$). The average temperature in the cowshed was 9°C.

Further work by Castle (1972a) showed that, on average, the consumption of water per unit of dry matter by cows was 18% higher with rations containing dried sugarbeet pulp than with rations containing barley. In a subsidiary experiment, it was concluded that when relatively large amounts of dried sugarbeet pulp are given, cows should be able to drink at will, but there is no apparent advantage in soaking the beet before feeding.

Castle & Thomas (1975) studied 14 mainly autumn-calving commercial British Friesian herds housed from November to April and concluded that 160 cm of trough available for drinking would be sufficient for either 50 milking cows being offered a ration of low dry matter content or 30 cows given a ration of high dry matter content. When water bowls are used, there should be one bowl per 10 cows on a low dry matter ration and 1 bowl per 6 cows on a high dry matter ration. The mean rate of drinking from bowls was 4.5 kg/minute whereas for herds with water troughs it ranged from 5.6 to 14.9 kg/min. This value is similar to the 5.0 kg/min recorded by Castle & Watson (1973). Consequently these rates suggest that the use of water bowls could limit water intake in some circumstances since cows can drink at the rate of 16 to 27 kg/min (MacLusky, 1959–60; Thomas, T. P., 1971). It appears that the optimum height of the top of the water trough is about 90 cm.

The water intake of Friesian heifers throughout their first lactation has been recorded by Owen et al (1968) (Table 9.2).

Table 9.2   Water intakes of heifers according to stage of lactation.†

Successive periods pre-and post-calving (weeks)	Free water intake (kg)	Water from feed (kg)	Total water intake (kg)	Total water intake (kg/kg DM)	Total water intake (g/kg live weight per day)
Pre-calving	2285	100	2385	4.2	83
0–8	3262	120	3382	5.0	126
9–16	3547	148	3695	4.4	131
17–24	3356	139	3495	4.4	120
25–32	2956	130	3086	4.2	101
33–40	2590	126	2716	3.8	85

† After Owen et al (1968).

(b) Sheep

Pregnant sheep

The intake of water by pregnant sheep increases rapidly as pregnancy advances. Head (1953) demonstrated an increase in the water intake of pregnant Cheviot ewes from early to late pregnancy. Similar findings were subsequently recorded by Forbes (1968) and both sets of data are summarized in Table 9.3 (a). Forbes also showed that these requirements could be modified according to litter size, as shown in Table 9.3 (b). Forbes attributed the increase in water intake with advancing pregnancy to a combination of increased heat production and urinary output, resulting from increased metabolic activity. The fact that the water requirements of heavily pregnant ewes are greatly in excess of those of nonpregnant ewes is of particular practical importance. The failure of a shepherd to meet these increased water requirements might depress feed intake (Gordon 1965). Such a depression in dry matter intake would predispose heavily pregnant ewes to pregnancy toxaemia (Forbes 1968).

Table 9.3   Intake of water by pregnant ewes.

(a) Effect of stage of gestation

Month of gestation	Water intake (kg/kg dry matter eaten)	
	(Head 1953)	(Forbes 1968)
1	1.9	—
2	2.6	—
3	3.3	2.2
4	3.8	3.3
5	4.3	5.2

(b) Effect of foetal burden (% of requirement for nonpregnant ewe)

Month of gestation	Water requirement for single-bearing ewe	Water requirement for twin-bearing ewe	
	(Forbes 1968)	(Head 1953)	(Forbes 1968)
3	106	150	125
4	125	180	152
5	138	220	212

Lactating sheep

Work by Forbes (1968) suggests that in the first 4 weeks of lactation, total water intakes per unit of dry matter intake are in excess of those given in the first edition which indicated that lactating ewes need 50% more water than dry ewes. If the water content of the milk is subtracted from the total water intake of the lactating ewes and the result then divided by dry matter intake, the result is still higher than the total water intake per unit dry matter intake of nonpregnant ewes. The difference may be due to the higher metabolic rate of the lactating ewe and the greater need for water in vaporization and excretion. From week 5 to week 7 of lactation, however, the discrepancy between the water intake of lactating ewes minus the water in milk, and the estimated water intake of dry ewes was not great (Table 9.4). Consequently it is suggested that the water intake of nonlactating, nonpregnant sheep be increased by 50% in early lactation and by 25% in late lactation.

Table 9.4   Water intake of lactating ewes (Forbes 1968).

Week of lactation	Actual total water intake		Intake less water in milk	
	As kg/kg DM intake†	As % of value for nonlactating ewes	As kg/kg DM intake	As % of value for nonlactating ewes
1	3.90 ± 0.34	175	—	—
2	4.20 ± 0.25	205	3.04	148
3	4.41 ± 0.21	227	3.19	164
4	3.71 ± 0.25	173	2.77	129
5	3.69 ± 0.22	141	2.58	99
6	3.40 ± 0.33	126	2.80	104
7	3.51 ± 0.33	139	2.90	115

† Mean for 15 ewes with standard error.

2. Ambient temperature   Increases in ambient temperature above 4°C increase the water intake per unit of dry matter consumed by cattle. The ratio of water intake to dry matter consumed appears to remain constant for environmental temperatures between –12°C and +4°C (Winchester & Morris 1956, MacDonald & Bell 1958) but then increases at an increasing

Table 9.5   Water intakes of sheep at various environmental temperatures.

Breed	Feed offered	Environmental temperature °(C)	Water intake (kg/kg DM ingested)	Water intake (g/kg live weight per day)	Reference
Halfbred × Down wethers (closely clipped)	Dried grass	8 18 28	1.4 1.4 3.0	— — —	} Blaxter et al (1959)
Merino wethers	Very poor quality hay	20 22 20	4.2 5.0 4.4	— — —	} Clark & Quin* (1947–49a)
	Lucerne hay	21 22 20	3.2 3.2 3.3	— — —	} Clark & Quin* (1947–49b)
	Maize endosperm and hay	22 26	1.8 2.1	— —	} Brauns (1930)** (see Leitch & Thomson, 1944–45)
Cheviot stores	Hay, oats and skimmed milk Hay, concentrates and water	— —	— —	70 57.2	} Cresswell et al (1968)
Herdwick wethers	Hay, oats and water Hay, oats, linseed and water	— —	— —	69 74	} Cresswell et al (1968)
Suffolk cross ewes	Dried lucerne	22	—	66.8	Anand et al (1966)
Cheviot wethers		20 −11	— —	49.4 20.8	} Bailey et al (1962)

* Hay assumed to have dry matter content of 90%; temperatures quoted are mean values.
** Diet assumed to have dry matter content of 87%.

Table 9.6   Water intake of nonlactating cattle (kg/kg DM ingested).

Class of stock	Environmental temperature (°C)			
	−17–+10	11–15	16–20	21–25
Calves (up to 6 weeks of age)	—	7.0	8.0	—
Cattle (over 100 kg, nonpregnant, nonlactating, nongrazing)	3.5	3.6	4.1	4.7
Cattle (over 100 kg, nonpregnant, nonlactating, grazing)	5.2	5.4	6.1	7.0
Cattle (over 100 kg, pregnant, non-lactating, nongrazing)	5.2	5.4	6.1	7.0

rate with elevation in temperature (Winchester & Morris 1956). Winchester & Morris (1956) further noted that the feed intake of lactating cows of European breeds began to decline at temperatures of 21–26°C. Tables 9.5 and 9.6 summarize these results for sheep and cattle, respectively.

The values reported by Blaxter et al (1959) in Table 9.5 relate to closely clipped sheep. The heat loss from the skin would be severely reduced in sheep carrying several months' growth of wool. In such cases it is likely that more heat would be lost from the respiratory tract, involving greater vaporisation of water and hence increased water intake. At environmental temperatures up to 15°C, the water intake of nonpregnant, nonlactating sheep is about 2.0 kg water/kg dry matter consumed; at temperatures of 15°C to 20°C it is about 2.5 kg/kg DM and at temperatures above 20°C, about 3.0 kg/kg DM.

It is suggested that for temperatures below 10°C the water intake of nonpregnant, nonlactating cattle should be 3.5 kg/kg dry matter ingested, and that for temperatures above 10°C the water intake should be as tabulated in Table 9.6.

3. Relative humidity, wind velocity and rainfall

Differences in relative humidity at moderate temperatures have little effect on water consumption. Increasing humidity at high temperatures results in decreased water consumption but increased frequency of drinking (Ragsdale et al 1953). It would seem that these changes reflect, in part, the lower intake of feed and the reduced vaporisation of moisture at high temperatures and humidities.

Increasing the wind velocity up to 45 m/s at air temperatures of 10–26.7°C result in a slightly lower consumption of water by cows of European breeds (*Bos taurus*) at 35°C, water consumption was about the same at velocities of 1.8 and 45 m/ (Brody et al 1954).

Castle (1972b) has postulated that the amount of rainfall per day also influence the intake of water, and that relative humidity and rainfall are of greater importance than temperature *per se*. Further work (Castle & Watson 1973; Castle & Watson 1975) with cows at grass has shown that the major factors influencing the amount of water drunk were rainfall, maximum air temperature, and the dry matter content of the herbage. Relative humidity and the hours of sunshine were also significantly correlated with water intake but, surprisingly, in these studies milk yield was not an important factor. However, the yields of milk were modest and greater milk output might be expected to have an effect on water intake.

4. Other factors

Water intake is subject to both diurnal and seasonal variation. In temperate conditions differences between water intake in summer and winter are not large, whereas in very hot environmental conditions the water intake of cattle can be increased by as much as 72% (Wilson 1961). Castle & Thomas (1975) have shown that peak water demand occurs between 1 and 3 hours after evening milking, while a second smaller peak demand occurs for 1–3 hours during the morning or early afternoon.

5. Quantity of dry matter consumed

Increasing intakes of feed dry matter are generally associated with increasing voluntary intakes of water (e.g. Ragsdale et al 1950, 1951, Winchester & Morris 1956, Kama et al 1962, Owen et al 1968, Utley et al 1970, Paquay et al 1970b), but for a given body size the water intake per unit of dry matter eaten is higher for low dry matter intake than for high (Leitch & Thomson 1944–45). In trials with Guernsey cattle in which wet material, in the form of cut grass, was given in excess of appetite for months, voluntary water intake was positively related to the percentage of the dry matter of the grass and the estimated average water intake rose to 141 g/kg live weight (Halley & Dougall 1962). Kay & Hobson (1963) observed that 2–4 kg of water was consumed per kg DM eaten. It appears that the "estimated water intake" of animals fed on dry diets agrees more closely with prediction than that of animals fed on low dry matter diets, when total water intake may often be in excess of apparent needs. This effect can be partially explained by an increase in the water loss in the faeces at higher feed intakes (Forbes 1968) but also by a higher rate of metabolism or by increased losses via the respiratory tract, followed by the need to excrete more waste products in the urine.

6. Composition of the diet

If large amounts of salt are introduced into the diet of an animal in stable water balance, the electrolyte concentrations in body fluids tend to increase while water consumption also increases. In general, the higher the proportion of minerals in the diet the greater the excretion of urine and accordingly the larger the water intake. This may not necessarily be true for calcium. Cattle drink more on a high-protein than on a low-protein diet, since the nitrogenous endproducts require a larger urine volume for excretion (Sykes 1955, Weeth & Haverland 1961). Whitlock et al (1975) observed in contrast that chronically sodium-deficient dairy cows exhibit signs of extreme thirst and polyuria. Diets high in pentosans and crude fibre result in increased losses of water in the faeces, and, therefore, in increased water intakes (Paquay et al 1970a).

If animals are to tolerate moderate or high intakes of salt in their diets then an adequate water supply is essential. Animals are more tolerant of salt in the dry feed (provided they have unlimited water) than they are of salt in the drinking water. This may be ascribed partly to adaptive changes in kidney function resulting in increased excretion of both sodium and chlorine (Potter 1966 1968).

(a) Effect of salt in feed dry matter

A. D. Wilson (1966) has shown that when sheep were given 10–20% sodium chloride in the feed, feed intake decreased and water intake increased in relation to salt concentration. Further experiments (Wilson & Hindley 1968) in which sheep were fed on diets containing 7.5 and 15.0% added sodium chloride, allowing the animals access to water only once daily resulted in a reduction in fluid intake which became more severe with increasing salt concentration.

(b) Effect of various salts in drinking water

Weeth & Haverland (1961) offered cattle drinking water containing various concentrations of sodium chloride for up to 90 days and concluded that cattle could tolerate 1.0% but not 1.2% of salt in the water. The water consumption of animals drinking water containing 1% salt increased by 53%. At this level of salt the animal's pulse rate tended to be depressed, serum potassium was significantly elevated and serum magnesium and urea were significantly depressed. Further work (Weeth & Lesperance 1965) showed that yearling heifers tolerated a 1% salt solution without adverse effect whilst water containing 1.5% sodium chloride resulted in a marked increase in the urinary excretion of urea. The effect of intermittent watering with saline water has also been studied (Weeth et al 1968). These workers found that watering once a day decreased the water consumption when compared with unlimited intake, and that watering every two days further decreased consumption of saline water. However, a solution of 2% sodium chloride is toxic, causing severe anorexia, weight loss and anhydraemia (Weeth et al 1960).

Peirce (1962) showed that sheep could tolerate drinking water containing as much as 0.3% calcium chloride in addition to 1.0% sodium chloride. He also showed that they could tolerate water containing up to 0.1% magnesium chloride in addition to 1.2% sodium chloride (Peirce 1959) and water containing 0.5% sodium sulphate in addition to 0.9% sodium chloride (Peirce 1960).

Blosser & Soni (1957), in experiments with lactating cows, found no difference in performance when the cows were offered hard or soft water. The contents of calcium plus magnesium in the two waters were respectively 33 mg/kg (hard) and 1 mg/kg (soft). Similar findings have been reported by Graf & Holdaway (1952) and by Allen et al (1958). Other elements such as iron, aluminium, zinc, manganese and strontium can contribute to the hardness of the water. Thus although hardness does not itself affect the performance of cattle, some of the elements causing water hardness might be toxic when present in unusual concentrations. On the other hand, Singh & Talapatra (1961), working with calves, suggested that the digestibility of crude fibre was significantly improved with hard water, an observation which has not been repeated by other workers.

7. Variation

(a) Variation between species

There are large differences between species in the rate of increase in water consumption with increasing ambient temperatures. Water consumption increases more slowly with rising temperatures for *Bos indicus* than for *Bos taurus* whether based on comparisons by body weight or body surface area (Ragsdale et al 1950 1951, Winchester & Morris 1956). In comparable conditions, the ratio of water drunk to hay eaten is significantly smaller for *Bos indicus* than for *Bos taurus*. The former drinks significantly less water than does the latter in relation to body weight (Horrocks & Phillips 1961).

(b) Individual animal variation

There are large variations in the amount of water drunk by similar animals maintained on the same feed and in the same environmental conditions. Up to 50% difference in water consumption has been noted among individual animals by Kelley (1945) and by Thompson et al (1949) with lactating cows, and by Evans (1957) with Blackface sheep. More recently T. P. Thomas (1971) has found that daily consumption of water by individual Friesian cows at pasture ranged from 18 to 64 kg/day, while in New Zealand the daily voluntary water intake of Jersey cows during a summer grazing period ranged from 6 to 52 kg/cow (Campbell 1958). Castle (1972b) has recorded a range of 4–44 kg/cow per day with data collected daily for the mean of 10 cows. Little & Shaw (1978) have shown in experiments with four groups of four lactating dairy cows in a metabolism house that individual water intakes range from 26.0 to 75.0 kg of water/day with a mean of 56.5 and a standard deviation of 7.22 when measured for seven consecutive days. Moreover, water intake, was significantly correlated with dry matter intake (range 4.6 to 14.4 kg/day) and milk yield (range 13.7 to 30.0 kg/day). There was no significant correlation with dry matter content of the feed (range 833 to 898 g/kg), body weight (range 400 to 620 kg) or mean air temperature (range 7 to 20°C).

The range of water intakes is even greater with calves, when mean free water intake over the period 0–5 weeks can have a coefficient of variation of 53% between individuals (Thickett et al 1960).

8. Temperature of drinking water

It was found by Ittner et al (1951) that in high environmental temperatures (29–30°C) cattle drank less when the water was kept cool than when it was allowed to get warm. This has been confirmed by Cunningham et al (1964) who found that the average daily intake of water by nonlactating dairy cattle increased with the temperature of the water. At ambient temperature −12°C, however, the temperature of the drinking water does not influence the amount of water consumed, so that in these conditions water need only be maintained unfrozen (Altman 1955, Bailey et al 1962).

9. Frequency and periodicity of watering

There is some evidence (Altman 1955) that dairy cows which have water continually available drink 18% more water, and yield more milk, than when watered only once a day. This finding is endorsed by Morrison (1956). Weeth et al (1968) found that watering once every day or once every two days reduced consumption 10% and 31% respectively when compared with unlimited drinking, while Altman (1955) observed that consumption was reduced by 11% and 32% by drinking once every two days or once every three days compared with once every day. Hancock (1953) has reported that when water is freely available to grazing cows they usually drink 2–5 times daily and probably not more than 7 times. This finding has been substantiated by Castle et al (1950), and also applies in tropical conditions (Wilson 1961). Weeth et al (1967 found that in a hot, dry climate cattle wholly deprived of water for 4 days had a depressed feed consumption of 45% on the first day, and about 50% of the preceding day's consumption on each of the following 3 days. Some animals refused to eat on the fourth day of deprivation.

Frequency of watering can also affect milk composition. Aschaffenburg and Rowland (1950) noted that, when dairy cattle were deprived of water overnight, they drank copiously when first let out to grass but seldom drank again before being rehoused in late afternoon. This intake of water in the morning was associated with increased water content of the evening milk, and it was concluded that excessive water intake over such a limited period may well have led to excess water secretion in the evening milk. This finding substantiates work done by Hillman et al (1950). Castle and Thomas (1975) have shown that both frequency and duration of drinking were related to type of roughage and drinking source available. Thus cows fed on silage drank less frequently than did those fed on dried grass, while water bowls were used more frequently than were troughs. Castle (1972b) has questioned whether a supply of water in the field is as important in temperate conditions as is generally supposed. The times of drinking are flexible so that there could well be periods during the grazing season when all the drinking water needed by dairy cows could be obtained either in or near the cowshed or parlour. It would be unwise to extrapolate these findings for hotter or drier environments and it should be remembered that, on occasion, ambient temperatures in the South of England can reach tropical levels in very hot summers, such as that experienced in 1976.

10. pH and toxicity of water

Most natural waters range in pH from 6 to 9. Water from springs sometimes has a pH above 9. It appears that water within a pH range of 6–9 is satisfactory for cattle (National Research Council 1974). It has long been known that cattle are sometimes poisoned when they drink lake water invaded by blue-green algae. Since at least six species have been identified as potential causes of toxicity (Gorham 1964), cattle should be prevented from drinking water with a heavy algal growth.

Conclusions

The water intake of ruminants is supplied in part from the water present in the feed, in part from water consumed voluntarily and to a smaller but nevertheless significant extent from the water formed within the body as a result of several oxidation processes. The estimated water intakes given here include only the first two (i.e., the water in the feed and the water consumed voluntarily).

There is considerable variation among individual animals, and factors such as the nature of the feed and ambient temperatures can affect water needs.

There is no clear advantage to be gained by restricting drinking water to ruminants, and no problems are normally encountered when water is consumed in excess of apparent needs. The provision of an unlimited supply of clean drinking water, with access to such a supply at least once daily, should therefore be recommended. In the case of lactating ruminants, or ruminants maintained in hotter environments, such an adequate supply should preferably be available several times each day. The water intakes estimated to meet requirements of cattle and sheep that are summarized in Tables 9.6, 9.7 and 9.8 are for livestock maintained in temperate conditions. In all these tables the estimates are calculated from the higher end of the ranges quoted in the literature and reviewed in this chapter. The estimates may be regarded as maximum rather than as mean estimates, but until further critical research data become available it is deemed prudent to use values unlikely to prove limiting in practice.

Table 9.7   Total water intake of lactating cows (kg/cow/day).

Milk yield (kg/day)	Live weight (kg)	Environmental temperature (°C)			
		−17–+10	11–15	16–20	21–25
10	600	78	81	92	105
	350	52	54	61	70
20	600	88	92	104	119
	350	62	65	73	84
30	600	99	103	116	133
	350	73	76	85	98
40	600	109	113	128	147
	350	88	92	104	119

Table 9.8   Water intake of sheep (kg/kg DM ingested).

Class of stock	Environmental temperature (°C)		
	<16	16–20	>20
Lambs (up to 4 weeks)	4.0	5.0	6.0
Sheep (growing or adult, nonpregnant, nonlactating)	2.0	2.5	3.0
Ewes			
Mid pregnancy, single bearing	2.5	3.1	3.7
Late pregnancy, single bearing	2.8	3.5	4.2
Mid pregnancy, twin bearing	3.3	4.1	4.9
Late pregnancy, twin bearing	4.4	5.5	6.6
Ewes			
Lactating, first month	4.0	5.0	6.0
Lactating, second month	3.0	3.7	4.5
Late lactation	2.5	3.1	3.7

Chapter 10

Nutrient Requirements of Ruminants Expressed as Dietary Concentrations

Nutrient requirements expressed as absolute quantities are converted into dietary concentrations by dividing by dry matter intake. The latter is determined by the energy requirement. For example a 600 kg Friesian cow giving 20 kg milk per day is estimated to require 48 g Ca/day (Table 5.5); if her diet has a metabolizability (q) of 0.6, her requirement for metabolizable energy (M_E, kg/day) is estimated as 155 MJ/day (Table 3.28). If the dry matter of the diet is assumed to have a heat of combustion of 18.4 MJ/kg, the quantity of dry matter required (I_T, kg/d) is calculated from the general equation:

$$I_T = M_E/18.4 \; q \qquad\qquad [10.1]$$

to be 14.0 kg/day. The required calcium concentration is therefore 48/14.0=3.4 g/kg DM.

If the metabolizability of the diet were lower (say q=0.5) the required concentration would be calculated to be lower (2.7 g/kg) and if it were higher (q=0.7) the required concentration would be higher (4.2 g/kg).

In calculations of this kind the common assumption made is that neither the coefficients of absorption of nutrients nor their endogenous losses from the animal will vary with the metabolizability of dietary energy. Such an assumption may be valid for feeds and diets differing little in metabolizability from those used in the derivation of coefficients of absorption and estimates of endogenous loss, but may not hold for more extreme diets. For example, one may question whether the protein of straw (q=0.4) that escapes rumen degradation is as efficiently digested and absorbed from the lower gut as is the protein of soya bean meal (q=0.7). Likewise, the application of coefficients of absorption for many mineral elements to low quality diets (q<0.5) must be viewed with suspicion. For this reason, calculations of nutrient requirements as dietary concentrations must be made with caution, especially when they lead to low concentrations for poor quality diets.

Introduction to the Tables

Requirements expressed as dietary concentrations are summarized in six tables (Tables 10.3–10.8), dealing respectively with growing, gestating and lactating cattle and sheep. These tables are based on those in earlier chapters but their scope has been somewhat restricted. In general, fewer combinations of animal size, breed and production rate have been included. There are several reasons for this simplification. One is the need to keep the tables a reasonable size for easy reference. Another is that requirements expressed as dietary concentrations differ less with animal production rate than do requirements expressed as absolute quantities. A third reason is that there may be insufficient data to justify a multiple classification. For example, although in Chapters 3 and 4, energy and protein requirements of growing cattle have been stated separately for the sexes, no such distinction was possible for mineral elements because insufficient data were available on sex differences in the mineral composition of gains; consequently no attempt has been made in this chapter to give separate dietary concentrations of minerals for male and female cattle.

A further restriction in the scope of the tables in this chapter is that the range in metabolizability is only from 0.5 to 0.7.

Dry Matter Intake

Each table begins with figures for the dry matter intake required from diets of q=0.5, 0.6 and 0.7 for the stated levels of production. These figures should be used in conjunction with the estimates of maximal intake given in Chapter 2. When the required dry matter intake is clearly in excess of appetite limits, it has been omitted from the tables. However, it should be remembered that appetite limits vary with the type of diet (coarse or fine) and — in the case of cattle — with the proportion of concentrates in the diet. Thus in Table 10.6 a 30 kg lamb gaining 0.2 kg/day on a diet with q=0.6 would be required to consume 1 kg DM/day. If a coarse diet (e.g. long roughage) were used, this would be in excess of the appetite limit of 0.73 kg DM/day (Table 2.2) but if a fine diet (e.g., pelleted roughage) were used, the required intake would be within the limit of appetite of 1.17 kg DM/day.

Protein

The tables give estimates of the minimum crude protein concentration required for diets with q=0.5, 0.6 and 0.7. These estimates apply only when the degradability of dietary protein in the rumen is optimal, and values for optimal degradability are included in the tables. The derivation of these figures is complex and is therefore explained in detail below.

1. Rumen-degradable protein

The requirement for RDP (g/day) can be predicted from the requirement for metabolizable energy (M_E, MJ/day) by the equation given in Chapter 4:
$$RDP = 7.8 \, M_E$$
and RDP (g/day) is converted to CP (g/day) by dividing by degradability (dg). Thus:

$$CP = 7.8 \, M_E/dg \qquad [10.2]$$

The concentrations of crude protein required to supply sufficient RDP can be stated as a single set of values for all sheep and cattle, regardless of age, sex, breed, form of production or rate of production (Table 10.1).

For many combinations of animals and diets, satisfaction of the requirement for RDP will automatically ensure satisfaction of the requirement for UDP. For example, the CP requirement of an animal requiring 400 g RDP can be met by 5 kg of a diet containing 100 g/kg CP, of degradability 0.8. This diet will also supply $5 \times 100 \times 0.2 = 100$ g UDP, and no additional UDP need be supplied if the animal's requirement for UDP is <100 g.

The need for additional UDP can be assessed from the optimal degradability of dietary protein for the animal in question (see below).

2. Requirements for undegraded protein: optimal degradability and concentration of crude protein

From Tables 4.13–4.27 data for RDP and UDP requirements can be combined to estimate the optimal degradability of dietary protein ($\hat{dg}$).

$$\hat{dg} = RDP/(RDP + UDP)$$ [10.3]

When degradability is optimal, the requirement for crude protein is the sum of requirements for RDP and UDP. When the animal's tissue protein requirement is small enough to be met entirely by rumen microbial protein, and there is no requirement for UDP, then optimal degradability = 1. As production rates, and hence tissue protein requirements, increase, the optimal degradability falls.

To use Tables 10.3 to 10.8, the probable degradability of protein in the diet being formulated must be compared with the optimal degradability. If the probable degradability is lower than the optimal degradability, the primary requirement will be for RDP and hence the simplified estimates of Table 10.1 may be employed. For example, Table 10.3 indicates that a 100 kg calf gaining 0.5 kg/day on a diet with q = 0.6 requires a minimum dietary protein concentration of 121 g/kg DM, with optimal degradability of 0.71. If the probable degradability is 0.6, reference to Table 10.1 shows the estimated requirement to be higher, at 144 g CP/kg DM. This last figure may also be calculated without reference to Table 10.1 by multiplying the requirement at optimal degradability (121 g/kg DM) by $\hat{dg}/dg$ (0.71/0.6).

If the probable degradability of protein is greater than optimal degradability, there will be a requirement for additional UDP. To supply this, the minimal requirement must be multiplied by $(1-\hat{dg})/(1-dg)$. To return to the example from Table 10.3, of a 100 kg calf gaining 0.5 kg/day on a diet with q = 0.6, if probable degradability is 0.8, the minimal requirement must be increased from 121 g CP/kg DM to $121 \times (1-0.71)/(1-0.8) = 175$ g CP/kg DM.

Table 10.2 gives correction factors for adjusting minimal crude protein requirements for the difference between probable and optimal degradability.

Calculations of the kind exemplified above emphasize the importance of protein degradability as a determinant of the protein requirement of ruminants. At the present time, degradability values for individual feeds are very limited in number, and their sensitivity to other dietary factors (e.g. associative effects between feeds, levels of feeding, etc.) is unknown. It may therefore be prudent to calculate crude protein requirements in relation to the likely range in degradability rather than to a specific value for degradability. From the examples given above, for which optimal degradability was 0.71, it can be shown that a crude protein concentration of 175 g/kg DM would be adequate for the degradability range 0.6–0.8.

Minerals

Tables 10.3–10.8 give requirements for Ca, P, Mg and Na. Potassium is not included because deficiency is rare. Trace elements are not included because requirements as dietary concentrations are given in Chapter 6.

Endogenous losses, and hence maintenance requirements, of mineral elements have in general been related to animal liveweight (see Chapter 5), whereas fasting metabolism, and hence maintenance requirements for energy, are related to liveweight raised to the power $W^{0.67}$ or $W^{0.75}$ (see Chapter 3). This means that for animals at maintenance, mineral requirements expressed as dietary concentrations are greater for heavier animals. In growing animals this trend is reversed, because as animals grow their energy requirement per unit of gain increases, whereas mineral requirements remain constant or decrease. A further complication with respect to phosphorus in diets for cattle is that the coefficient of absorption is lower for older animals (see Table 5.14), and this leads to higher dietary requirements at heavier weights.

Mineral concentrations for diets with metabolizability greater or less than 0.6 may be calculated from the dry matter intakes given in Tables 10.3–10.8. For example, in Table 10.6, a 30 kg lamb gaining 0.2 kg/day is estimated to require 2.1 g P/kg DM in a diet with q=0.6, and its dry matter intake would be 1.00 kg/day. For a diet with q=0.7, dry matter intake would be 0.75 kg/day, and the required phosphorus concentration would be $2.1 \times 1.00/0.75 = 2.8$ g/kg. When calculations of this kind are made the cautionary remarks included in the introduction to this chapter should be borne in mind.

Fat-Soluble Vitamins

In Chapter 7 requirements for vitamins A and D are related to body weight, without reference to rate of production (although, for cattle, distinctions are drawn between growing, pregnant and lactating animals). This means that requirements expressed as dietary concentrations fall as production rate, and hence dry matter intake, increase.

Table 10.1   Dietary crude protein concentrations required to meet the requirements of ruminants for rumen degradable protein (g/kg DM).

Metabolizability (q)	Degradability of dietary protein (dg)			
	0.6	0.7	0.8	1.0
0.5	120	103	90	72
0.6	144	123	108	86
0.7	168	144	126	100

Table 10.2   Factors for calculating the crude protein requirement of ruminants from their minimal protein requirement (Tables 10.3 to 10.8) when the actual rumen degradability of protein (dg) differs from the optimal degradability (d̂g)[1].

Optimal degradability (d̂g)	Actual degradability (dg)						
	0.55	0.60	0.65	0.70	0.75	0.80	0.85
0.55	1.00	1.12	1.29	1.50	1.80	2.25	3.00
0.60	1.09	1.00	1.14	1.33	1.60	2.00	2.67
0.65	1.18	1.08	1.00	1.17	1.40	1.75	2.33
0.70	1.27	1.17	1.08	1.00	1.20	1.50	2.00
0.75	1.36	1.25	1.15	1.07	1.00	1.25	1.67
0.80	1.45	1.33	1.23	1.14	1.07	1.00	1.33
0.85	1.55	1.42	1.31	1.21	1.13	1.06	1.00
0.90	1.64	1.50	1.38	1.29	1.20	1.12	1.06
0.95	1.73	1.58	1.46	1.36	1.27	1.19	1.12
1.00	1.82	1.67	1.54	1.43	1.33	1.25	1.18

[1] When $\hat{dg} > dg$, factor $= \dfrac{1 - \hat{dg}}{1 - dg}$

When $\hat{dg} < dg$, factor $= \dfrac{\hat{dg}}{dg}$

Table 10.3   Nutrient requirements of beef cattle, expressed as dietary concentrations.

Live weight (kg)	Live weight gain (kg/day)	Dry matter intake (kg/day) when q =			Crude protein requirement (g/kg DM) at optimal degradability (in parentheses) when q =			Mineral requirements (g/kg DM) when q = 0.6				Vitamin requirements (µg/kg DM) when q = 0.6	
		0.5	0.6	0.7	0.5	0.6	0.7	Ca	P	Mg	Na	A	D
100	0	1.8	1.4	1.2	72 (1.00)	86 (1.00)	100 (1.00)	1.7	1.1	1.2	0.50	1400	7.0
	0.5	2.8	2.1	1.7	93 (0.77)	121 (0.71)	150 (0.67)	5.5	2.9	1.4	0.75	1000	5.0
	1.0	—	3.2	2.4	—	130 (0.66)	169 (0.59)	6.7	3.4	1.3	0.75	600	3.0
200	0	3.0	2.3	1.9	72 (1.00)	85 (1.00)	100 (1.00)	2.0	1.3	1.5	0.65	1700	8.5
	0.5	4.3	3.3	2.6	72 (1.00)	90 (0.97)	110 (0.91)	4.3	2.4	1.5	0.70	1200	6.0
	1.0	6.4	4.7	3.6	72 (1.00)	94 (0.91)	121 (0.83)	5.2	2.7	1.3	0.65	900	4.5
300	0	3.9	3.1	2.5				2.3	1.5	1.7	0.75	1900	9.5
	0.5	5.6	4.3	3.4				4.0	2.2	1.5	0.70	1400	7.0
	1.0	8.3	6.0	4.7				4.3	2.3	1.3	0.65	1000	5.0
	1.5	—	8.8	6.4				4.0	2.1	1.0	0.55	700	3.5
400	0	4.8	3.8	3.1				2.5	2.2	1.9	0.80	2100	10.5
	0.5	6.8	5.2	4.2	72 (1.00)	86 (1.00)	100 (1.00)	3.6	2.8	1.6	0.75	1500	7.5
	1.0	10.0	7.3	5.6				3.8	2.9	1.3	0.65	1100	5.5
	1.5	—	10.6	7.7				3.6	2.6	1.0	0.50	800	4.0
500	0	5.5	4.4	3.6				2.7	2.3	2.0	0.85	2300	11.5
	0.5	7.9	6.1	4.9				3.5	2.7	1.7	0.75	1600	8.0
	1.0	11.6	8.5	6.6				3.5	2.7	1.3	0.65	1200	6.0
	1.5	—	12.2	9.0				3.3	2.4	1.0	0.50	800	4.0
600	0	6.3	5.0	4.1				2.8	2.5	2.1	0.90	2400	12.0
	0.5	8.9	6.9	5.5				3.4	2.7	1.7	0.75	1700	8.5
	1.0	13.1	9.6	7.4				3.5	2.6	1.4	0.65	1200	6.0
	1.5	—	13.8	10.1				3.1	2.3	1.0	0.50	900	4.5

Table 10.4 Nutrient requirements of pregnant cattle, expressed as dietary concentrations.[1]

Breed etc.	Weeks before term	Dry matter intake (kg/day) when q =			Mineral requirements (g/kg DM) when q = 0.6				Vitamin requirements (µg/kg DM) when q = 0.6	
		0.5	0.6	0.7	Ca	P	Mg	Na	A	D
Standard (40 kg calf)	12	6.4	5.1	4.2	3.0	2.5	1.8	0.8	2900	24
	8	7.1	5.7	4.7	3.1	2.6	1.7	0.8	2600	22
	4	8.2	6.6	5.5	3.2	2.6	1.6	0.8	2300	19
	0	10.2	8.3	6.9	3.1	2.4	1.4	0.8	1800	15
Friesian (42 kg calf)	12	7.2	5.8	4.7	3.1	2.6	2.0	0.9	3100	26
	8	7.9	6.3	5.2	3.2	2.7	1.8	0.9	2900	24
	4	9.1	7.3	6.1	3.3	2.6	1.7	0.8	2500	20
	0	11.2	9.0	7.6	3.2	2.5	1.4	0.8	2000	17
Ayrshire (32 kg calf)	12	6.2	5.0	4.1	2.9	2.5	1.9	0.8	3000	25
	8	6.8	5.4	4.5	3.0	2.5	1.8	0.8	2800	23
	4	7.7	6.2	5.1	3.1	2.5	1.6	0.8	2400	20
	0	9.3	7.5	6.2	3.1	2.4	1.4	0.8	2000	17
Jersey (24 kg calf)	12	5.3	4.2	3.4	2.8	2.3	1.8	0.8	2400	20
	8	5.7	4.5	3.7	2.9	2.4	1.7	0.8	2200	18
	4	6.4	5.1	4.2	2.9	2.4	1.6	0.8	2000	16
	0	7.5	6.1	5.1	2.9	2.3	1.4	0.7	1600	14

[1] for crude protein requirements, see Table 10.1 (i.e. requirements are for rumen degradable protein).

Table 10.5   Nutrient requirements of lactating cows, expressed as dietary concentrations.

Breed etc.	Milk yield (kg/day)	Dry matter intake (kg/day) when q =			Crude protein requirement (g/kg DM) at optimal degradability (in parentheses) when q =			Mineral requirements (g/kg DM) when q = 0.6				Vitamin requirements (μg/kg DM) when q = 0.6	
		0.5	0.6	0.7	0.5	0.6	0.7	Ca	P	Mg	Na	A[1]	D
Friesian; 600 kg live weight; 36.8 g fat/kg milk	0	6.3	5.0	4.1	72 (1.00)	87 (1.00)	101 (1.00)	2.8	2.5	2.1	0.8	3000	30
	10	12.0	9.4	7.7	73 (0.97)	94 (0.93)	112 (0.89)	3.3	3.0	1.9	1.1	1600	16
	20	17.8	14.0	11.4	87 (0.83)	109 (0.79)	133 (0.76)	3.4	3.1	1.8	1.2	1100	11
	30	—	18.8	15.3	—	116 (0.74)	142 (0.71)	3.4	3.1	1.7	1.2	800	8
Ayrshire; 500 kg liveweight; 38.6 g fat/kg milk	0	5.5	4.4	3.6	72 (1.00)	87 (1.00)	101 (1.00)	2.7	2.3	2.0	0.8	2800	28
	10	11.4	9.0	7.3	78 (0.92)	98 (0.88)	120 (0.84)	3.2	3.0	1.8	1.0	1400	14
	20	17.4	13.7	11.2	91 (0.79)	114 (0.76)	139 (0.72)	3.4	3.1	1.7	1.1	900	9
	30	—	18.6	15.1	—	120 (0.71)	147 (0.68)	3.4	3.2	1.7	1.2	700	7
Jersey; 400 kg live weight; 49.0 g fat/kg milk	0	4.8	3.8	3.1	72 (1.00)	87 (1.00)	101 (1.00)	2.5	2.2	1.9	0.7	2600	26
	10	11.4	9.0	7.3	85 (0.85)	106 (0.81)	130 (0.77)	3.3	3.2	1.4	0.9	1100	11
	20	18.4	14.5	11.8	95 (0.76)	119 (0.72)	146 (0.69)	3.5	3.5	1.4	0.9	700	7

[1] multiply by 2.6 for cows suckling calves.

Table 10.6   Nutrient requirements of castrate lambs on solid diets, expressed as dietary concentrations.

Live weight (kg)	Live weight gain (kg/day)	Dry matter intake (kg/day) when q =			Crude protein requirement (g/kg DM) at optimal degradability (in parentheses) when q =			Mineral requirements (g/kg DM) when q = 0.6				Vitamin requirements (μg/kg DM) when q = 0.6	
		0.5	0.6	0.7	0.5	0.6	0.7	Ca	P	Mg	Na	A	D
20	0	0.39	0.31	0.25	72 (1.00)	86 (1.00)	100 (1.00)	1.5	1.2	1.2	1.8	650	8.5
	0.1	0.67	0.50	0.40	85 (0.84)	112 (0.77)	138 (0.73)	3.9	2.3	1.2	1.4	400	5.0
	0.2	—	0.76	0.57	—	117 (0.65)	154 (0.65)	4.5	2.5	1.1	1.1	250	3.5
30	0	0.54	0.43	0.35	72 (1.00)	86 (1.00)	100 (1.00)	1.7	1.3	1.4	2.0	700	9.0
	0.1	0.89	0.67	0.53	72 (1.00)	90 (0.97)	111 (0.90)	3.3	2.0	1.2	1.4	450	6.0
	0.2	1.44	1.00	0.75	72 (1.00)	94 (0.91)	121 (0.82)	3.7	2.1	1.0	1.1	300	4.0
	0.3	—	—	1.02	—	—	123 (0.82)	3.7	2.1	0.9	1.0	250	3.0
40	0	0.67	0.53	0.44	72 (1.00)	86 (1.00)	100 (1.00)	1.8	1.4	1.5	2.1	750	10.0
	0.1	1.10	0.83	0.66	72 (1.00)	86 (1.00)	100 (1.00)	2.9	1.8	1.2	1.5	500	6.5
	0.2	1.77	1.23	0.93	72 (1.00)	86 (1.00)	103 (0.97)	3.2	1.8	1.0	1.1	350	4.0
	0.3	—	1.80	1.26	—	86 (1.00)	102 (0.98)	3.0	1.7	0.8	0.8	200	3.0

Table 10.7   Nutrient requirements of pregnant ewes, expressed as dietary concentrations.

Live weight (kg)	No. of foetuses	Weeks before term	Dry matter intake (kg/day) when q =			Crude protein requirement[1]	Mineral requirements (g/kg DM) when q = 0.6				Vitamin requirements (µg/kg DM) when q = 0.6	
			0.5	0.6	0.7		Ca	P	Mg	Na	A	D
40	1	12	0.80	0.64	0.53		1.6	1.6	1.1	2.0	1250	8.0
		8	0.90	0.72	0.60		1.9	1.8	1.1	1.8	1100	7.0
		4	1.02	0.83	0.69		2.6	2.0	1.1	1.5	950	6.5
		0	1.18	0.96	0.80		3.2	2.0	1.1	1.2	850	5.5
40	2	12	0.91	0.74	0.61		1.4	1.5	1.0	1.7	1100	7.0
		8	1.07	0.86	0.72		1.9	1.8	1.0	1.6	950	6.0
		4	1.27	1.03	0.87		2.9	2.1	1.0	1.2	800	5.0
		0	1.53	1.25	1.05		3.5	2.0	1.0	1.0	650	4.0
75	1	12	1.28	1.03	0.85		1.8	1.9	1.3	2.3	1450	9.5
		8	1.42	1.15	0.95		2.1	2.0	1.3	2.0	1300	8.5
		4	1.62	1.30	1.09		2.8	2.2	1.3	1.7	1150	7.5
		0	1.86	1.51	1.26		3.3	2.1	1.3	1.5	1000	6.5
75	2	12	1.45	1.17	0.97		1.7	1.7	1.2	2.1	1300	8.5
		8	1.68	1.36	1.13		2.0	2.0	1.1	1.8	1100	7.0
		4	1.98	1.61	1.35		3.0	2.2	1.1	1.5	930	6.0
		0	2.37	1.93	1.63		3.6	2.1	1.1	1.2	800	5.0

[1] for crude protein requirements the estimates in Table 10.1 apply, provided degradability <0.90.

Table 10.8   Nutrient requirements of lactating ewes, expressed as dietary concentrations.

Live weight (kg)	Milk yield (kg/day)	Dry matter intake (kg/day) when q =			Crude protein requirement (g/kg DM) at optimal degradability (in parentheses) when q =			Mineral requirements (g/kg DM) when q = 0.6				Vitamin requirements (µg/kg DM) when q = 0.6	
		0.5	0.6	0.7	0.5	0.6	0.7	Ca	P	Mg	Na	A	D
40	0	0.61	0.49	0.40	72 (1.00)	86 (1.00)	100 (1.00)	1.9	1.9	1.4	2.3	1200	10.5
	1.0	1.49	1.18	0.96	95 (0.76)	118 (0.73)	144 (0.70)	2.8	2.6	1.4	1.3	500	4.5
	2.0	2.41	1.90	1.54	102 (0.71)	128 (0.67)	156 (0.64)	2.9	2.8	1.4	1.1	300	2.5
75	0	1.00	0.80	0.65	72 (1.00)	86 (1.00)	100 (1.00)	2.2	2.2	1.7	2.7	1400	12.0
	1.0	1.87	1.48	1.20	83 (0.86)	104 (0.82)	128 (0.79)	2.8	2.6	1.6	1.7	750	6.5
	2.0	2.76	2.18	1.77	94 (0.76)	118 (0.73)	145 (0.70)	3.0	2.8	1.5	1.4	500	4.5
	3.0	3.68	2.90	2.36	99 (0.73)	125 (0.69)	152 (0.66)	3.0	2.9	1.5	1.2	400	3.5

References

For explanation of references beginning (S), see p.159.

Abdellatif, A. M. M. (1968) *Versl. landbouwk. Onderzoek.* No. 709.

Abrams, J. T. (1952) *Vet. Rec.* **64**, 151.

Abrams, J. T.; Barnett, K. C.; Bridge, P. S.; Palmer, A. C.; Spratling, F. R.; Sharman, I. M. (1969). *Int. J. Vitam. Nutr. Res.* **39**, 416.

Adamson, A. H.; Valks, D. A.; Appleton, M. A.; Shaw, W. B. (1969) *Vet. Rec.* **85**, 368.

Agricultural Research Council (1965) The Nutrient Requirements of Farm Livestock. No. 2, Ruminants. London; HMSO.

Aines, P. D.; Smith, S. E. (1957) *J. Dairy Sci.* **40**, 682.

Aitchison, T. E.; Mertens, D. R.; McGilliard, A. D.; Jacobson, N. L. (1976) *J. Dairy Sci.* **59**, 2056.

Aitken, F. C. (1976) *Tech. Comm. Commonw. Bur. Nutr.* No. 26.

Albert, W. W.; Garrigus, U. S.; Forbes, R. M.; Norton, H. W. (1956) *J. Anim. Sci.* **15**, 559.

Alderman, G.; Jones, D. I. H. (1967) *J. Sci. Fd Agric.* **18**, 197.

Alderman, G.; Stranks, M. H. (1967) *J. Sci. Fd Agric.* **18**, 151.

Alderman, G.; Morgan, D. E.; Lessells, W. J. (1970) In *Energy metabolism of farm animals:* 5th symp., Vitznau, p. 81. (Eds A. Schürch & C. Wenk). Zurich: Juris Druck. (*Publs Eur. Ass. Anim. Prod.* **13**).

Alderson, N. E.; Mitchell, G. E.,Jr; Little, C. O.; Warner, R. E.; Tucker, R. E. (1971) *J. Nutr.* **101**, 655.

Alexander, G. (1962a) *Aust. J. agric. Res.* **13**, 82.

Alexander, G. (1962b) *Aust. J. agric. Res.* **13**, 144.

Allaway, W. H.; Hodgson, J. F. (1964) *J. Anim. Sci.* **23**, 271.

Allcroft, R. (1951) *Vet. Rec.* **63**, 583.

Allcroft, R.; Blaxter, K. L. (1950) *J. comp. Path.* **60**, 209,

Allcroft, R.; Lewis, G. (1956) *Landbouwk. Tijdschr.'s-Grav.* **68**, 711.

Allcroft, R.; Salt, F. J. (1961) In *Advances in Thyroid Research* p. 4 (Ed. R. Pitt-Rivers) Oxford: Pergamon.

Allcroft, R.; Uvarov, O. (1959) *Vet. Rec.* **71**, 797.

Allcroft, R.; Scarnell, J.; Hignett, S. L. (1954) *Vet. Rec.* **66**, 367.

Allcroft, R.; Burns, K. N.; Hebert, C. N. (1965) *Fluorosis in cattle-2. Development and alleviation: experimental studies.* London: HMSO.

Allden, W. G. (1970) *Nutr. Abstr. Rev.* **40**, 1167.

Allen, N. N.; Ausman, D.; Patterson, W. N.; Hays. O. E. (1958) *J. Dairy Sci.* **41**, 688.

Allen, W. M.; Parr, W. H.; Bradley, R.; Swannack, K.; Barton, C. R. Q.; Tyler, R. (1974) *Vet. Rec.* **94**, 373.

Allen, W. M.; Bradley, R.; Berrett, S.; Parr, W. H.; Swannack, K.; Barton, C. R. Q.; McPhee, A. (1975) *Br. vet. J.* **131**, 292.

Alloway, B. J. (1973) *J. agric. Sci., Camb.* **80**, 521.

Allsop, T. F.; Rook, J. A. F. (1972) *Proc. Nutr. Soc.* **31**, 65A.

Altman, L. B. (1955). *U.S. Dep. Agric. Leafl.* No. 395.

American Dry Milk Institute Inc., Chicago (1971) Standards for grades of dry milks including methods of analysis. *Bull.* No. 916.

Amir, S.; Aitken, J. N.; Kay, M. (1968) *Proc. Nutr. Soc.* **27**, 35A.

Ammerman, C. B.; Wing, J. M.; Dunavant, B. G.; Robertson, W. K.; Feaster, J. P.; Arrington, L.R. (1967) *J. Anim. Sci.* **26**, 404.

Ammerman, C. B.; Chicco, C. F.; Loggins, P. E.; Arrington, L. R. (1972) *J. Anim. Sci.* **34**, 122.

Anand, R. S.; Parker, A. H. & Parker, H. R. (1966) *Am. J. vet. Res.* **27**, 899.

Anderson, C. M. (1956) *N.Z. Jl Sci. Technol. A.* **37**, 379.

Anderson, M. S.; Lakin, H. W.; Beeson, K. C.; Smith, F. F.; Thacker, E. (1961) *U.S. Dep. Agric. Handbk* no. 200.

Anderson, P. H.; Berrett, S.; Patterson, D. S. P. (1976) *Vet. Rec.* **99**, 316.

Anderson, T. A.; Hubbert, F. Jr; Roubicek, C. B.; Taylor, R. E. (1962) *J. Nutr.* **78**, 341.

Andrews, E. D. (1965) *N.Z. Jl. agric. Res.* **8**, 788.

Andrews, E. D.; Cunningham, I. J. (1945) *N.Z. Jl. Sci. Technol.* **27**, 223.

Andrews, E. D.; Isaacs, C. E. (1964) *N.Z. vet. J.* **12**, 147.

Andrews, E. D.; Stephenson, B. J. (1966) *N.Z. Jl agric. Res.* **9**, 491.

Andrews, E. D.; Stephenson, B. J.; Isaacs, C. E.; Register, R. H. (1966) *N.Z. vet. J.* **14**, 191.

Andrews, E. D.; Hartley, W. J.; Grant, A. B. (1968) *N.Z. vet J.* **16**, 3.

Andrews, R. P.; Ørskov, E. R. (1970a) *J. agric. Sci., Camb.* **75**, 19.

Andrews, R. P.; Ørskov, E. R. (1970b) *J. agric. Sci., Camb.* **75**, 11.

Andrews, R. P.; Escuder-Volonte, J.; Curran, M. K.; Holmes, W. (1972) *Anim. Prod.* **15**, 167.

Andrews, R. P.; Kay, M.; Ørskov, E. R. (1969) *Anim. Prod.* **11**, 173.

Anke, M. (1966). *Arch. Tierernähr.* **16**, 199.

Anke, M.; Groppel, B. (1970) In *Trace element metabolism in animals:* WAAP/IBP int. symp., Aberdeen, 1969, p. 133. (Ed C. F. Mills). Edinburgh: Livingstone.

Annison, E. F.; Armstrong, D. G. (1970) In *Physiology of digestion and metabolism in the ruminant:* 3rd int. symp. Cambridge, 1969, p. 422. (Ed. A. T. Phillipson). Newcastle upon Tyne: Oriel Press.

Archibald, J. G.; Bennett, E. (1935) *J. agric. Res.* **51**, 83.

Armsby, H. P. (1917) *The Nutrition of Farm Animals.* New York; Macmillan.

Armstrong, D. G. (1964) *J. agric. Sci., Camb.* **62**, 399.

Armstrong, D. G. (1969) In *Handbuch der Tierernährung,* vol. 1, p. 385. (Eds W. Lenkeit, K. Breirem & E. Crasemann) Hamburg: Paul Parey.

Armstrong, D. G.; Annison, E. F. (1973) *Proc. Nutr. Soc.* **32**, 107.

Armstrong, D. G.; Blaxter, K. L. (1957a) *Br. J. Nutr.* **11**, 247.

Armstrong, D. G.; Blaxter, K. L. (1957b) *Br. J. Nutr.* **11**, 413.

Armstrong, D. G.; Blaxter, K. L. (1961) In *2. Symposium on energy metabolism: methods and results of experiments with animals* p. 187. (Eds E. Brouwer & A. J. H. Van Es). Wageningen: Lab. Anim. Physiol (*Publs. Eur. Ass. Anim. Prod.* **10**)

Armstrong, D. G.; Blaxter, K. L. (1965) In *Energy metabolism:* 3rd symp., Troon, 1964, p. 59. (Ed. K. L. Blaxter). London Academic Press. (*Publs. Eur. Ass. Anim. Prod.* **11**)

Armstrong, D. G.; Blaxter, K. L.; Graham, N. McC. (1957) *Br. J. Nutr.* **11**, 392.

Armstrong, D. G.; Blaxter, K. L.; Graham, N. McC.; Wainman, F. W. (1958). *Br. J. Nutr.* **12**, 177.

Armstrong, D. G.; Blaxter, K. L.; Graham, N. McC. (1960). *Proc. Nutr. Soc.* **19**, xxxi.

Armstrong, D. G.; Blaxter, K. L.; Graham, N. McC. (1961) *Br. J. Nutr.* **15**, 169.

Arnold, G. W. (1975) *Aust. J. agric. Res.* **26**, 1017.

Arnold, G. W.; Dudzinski, M. L. (1967) *Aust. J. agric. Res.* **18**, 349.

Arthur, G. H. (1969) *J. Reprod. Fert. Suppl.* No. 9, p. 45.

Aschaffenburg, R.; Rowland, S. J. (1950) *Chemy Ind.* p. 636.

Aschbacher, P. W. (1968) *J. Anim. Sci.* **27**, 127.

Ashton, W. M.; Yousef, I. M. K. (1966) *J. agric. Sci., Camb.* **67**, 77.

Ashton, W. M.; Owen, J. B.; Ingleton, J. W. (1964) *J. agric. Sci., Camb.* **63**, 85.

Astwood, E. B.; Greer, M. A.; Ettlinger, M. G. (1949) *J. biol. Chem.* **181**, 121.

Atkeson, F. W.; Warren, T. R.; Anderson, G. C. (1934) *J. Dairy Sci.* **17**, 249.

Auchinachie, D. W.; Fraser, A. H. H. (1932) *J. agric. Sci. Camb.,* **22**, 560.

Avioli, L. V.; Haddad, J. G. (1973) *Metabolism* **22**, 507.

Ayala, H. J. (1974) Ph.D. Thesis, Cornell University (*Diss. Abstr. int. B.* **34**, 5752).

Bailey, C. B.; Hironaka, R.; Slen, S. B. (1962) *Can. J. Anim. Sci.* **42**, 1.

Balch, C. C.; Line, C. (1957) *J. Dairy Res.* **24**, 11.

Balch, C. C.; Balch, D. A.; Johnson, V. W.; Turner, J. (1953) *Br. J. Nutr.* **7**, 212.

Baldwin, R. L.; Lucas, H. L.; Cabrera, R. (1970) In *Physiology of digestion and metabolism in the ruminant:* 3rd int. symp. Cambridge, 1969, p. 319. (Ed. A. T. Phillipson). Newcastle upon Tyne: Oriel Press.

Barnicoat, C. R.; Murray, P. F.; Roberts, E. M.; Wilson, G. S. (1957) *J. agric. Sci., Camb.* **48**, 9.

Bartley, E. E.; Deyoe, C. W.; Behnke, K. C.; Griffel, G. W.; Meyer, R. M. (1973) *J. Anim. Sci.* **37**, 336.

Batey, T.; Berryman, C.; Line, C. (1972) *J. Br. Grassld Soc.* **27**, 139.

Bath, D. L.; Ronning, M.; Meyer, J. H.; Lofgreen, G. P. (1965) *J. Dairy Sci.* **48**, 374.

Bath, D. L.; Ronning, M.; Lofgreen, G.P.; Meyer, J. H. (1966) *J. Dairy Sci.* **49**, 830.

Bayfield, R. F.; Saville, D. G.; Falk, R. H. (1972). *Aust. J. exp. Agric. Anim. Husb.* **12**, 19.

Bechdel, S. I.; Hilston, N. W.; Guerrant, N. B.; Dutcher, R. A. (1938). *Bull. Pa agric. Exp. Stn* No. 364.

Bechtel, H. E.; Huffman, C. F.; Duncan, C. W.; Hoppert, C. A. (1936) *J. Dairy Sci.* **19**, 359.

Beck, A. B. (1941) *Aust. J. exp. Biol. med. Sci.* **19**, 145.

Becker, D. E.; Smith, S. E. (1951) *J. Anim. Sci.* **10**, 266.

Becker, R. B.; Arnold, P. T. D.; Marshall, S. P. (1950) *J. Dairy Sci.* **33**, 911.

Beever, D. E. (1969) Ph.D. Thesis, University of Newcastle upon Tyne.

Beever, D. E.; Thomson, D. J.; Pfeffer, E.; Armstrong, D. G. (1971a) *Br. J. Nutr.* **26**, 123.

Beever, D. E.; Thomson, D. J.; Harrison, D. G. (1971b) *Proc. Nutr. Soc.* **30**, 86A.

Beever, D. E.; Coelho da Silva, J. F.; Prescott, J. H. D.; Armstrong, D. G. (1972) *Br. J. Nutr.* **28**, 347.

Beever, D. E.; Harrison, D. G.; Thomson, D. J.; Cammell, S. B.; Osbourn, D. F. (1974a) *Br. J. Nutr.* **32**, 99.

Beever, D. E.; Thomson, D. J.; Harrison, D. G. (1974b) In *Programme Handbk, 4th Int. Symp. Ruminant Physiol., Sydney, Australia*, p. 123.

Benevenga, N. J.; Baldwin, R. L.; Ronning, M. (1966) *J. Nutr.* **90**, 131.

Ben-Ghedalia, D.; Tagari, H.; Bondi, A.; Tadmor, A. (1974). *Br. J. Nutr.* **31**, 125.

Bensadoun, A.; Reid, J. T.; Van Vleck, L. D.; Paladines, O. L.; Van Niekerk, B. D. H. (1968) *Publs natn. Res. Coun., Wash.* No. 1598, p. 452.

Bentley, O. G.; Phillips, P. H. (1951) *J. Dairy Sci.* **34**, 396.

Benzie, D.; Boyne, A. W.; Dalgarno, A. C.; Duckworth, J.; Hill, R.; Walker, D. M. (1960) *J. agric. Sci., Camb.* **54**, 202.

Bickerstaffe, R.; Noakes, D. E.; Annison, E. F. (1972) *Biochem. J.* **130**, 607.

Bines, J. A. (1976) *Livestk Prod. Sci.* **3**, 115.

Bines, J. A.; Balch, C. C. (1973) *Br. J. Nutr.* **29**, 457.

Bines, J. A.; Suzuki, S.; Balch, C. C. (1969) *Br. J. Nutr.* **23**, 695.

Bingley, J. B.; Anderson, N. (1972) *Aust. J. agric. Res.* **23**, 885.

Bingley, J. B.; Carrillo, B. J. (1966) *Nature, Lond.* **209**, 834.

Binot, H.; Lomba, F.; Chauvaux, G.; Bienfet, V. (1968). *Annls Méd. vét.* **112**, 666.

Bird, P. R. (1970) *Proc. Aust. Soc. Anim. Prod.* **8**, 212.

Bird, P. R. (1972) *Aust. J. biol. Sci.* **25**, 1073.

Bird, P. R. (1973). *Aust. J. biol. Sci.* **26**, 1429.

Bird, P. R. (1974) *Aust. J. agric. Res.* **25**, 631.

Bitman, J.; Hawk, H. W.; Cecil, H. C.; Sykes, J. F. (1961) *Am. J. Physiol.* **200**, 827.

Black, J. L. (1971) *Br. J. Nutr.* **25**, 31.

Black, J. L.; Griffiths, D. A. (1975) *Br. J. Nutr.* **33**, 399.

Black, J. L.; Pearce, G. R.; Tribe, D. E. (1973) *Br. J. Nutr.* **30**, 45.

Blackburn, P. S.; Blaxter, K. L.; Castle, E. J. (1957) *Proc. Nutr. Soc.* **16**, xvi.

Blair-West, J. R.; Coghlan, J. P.; Denton, D. A.; Wright, R. D. (1970) In *Physiology of digestion and metabolism in the ruminant:* 3rd int. symp., Cambridge, 1969, p. 350. (Ed. A. T. Phillipson). Newcastle upon Tyne: Oriel Press.

Blakemore, F.; Davies, A. W.; Eylenburg, E.; Moore, T.; Sellers, K. C.; Worden, A. N. (1948) *Biochem. J.* **42**, xxx.

Blaxter, K. L. (1943) *Nature, Lond.* **152**, 751.

Blaxter, K. L. (1956) *Proc. Br. Soc. Anim. Prod.* p. 3.

Blaxter, K. L. (1961) In *Digestive physiology and nutrition of the ruminant*, p. 183. (Ed. D. Lewis). London: Butterworths. (*Proc. Easter Sch. agric. Sci. Univ. Nott.* 7, 1960).

Blaxter, K. L. (1962a) *The Energy Metabolism of Ruminants*. London: Hutchinson.

Blaxter, K. L. (1962b) *Proc. Nutr. Soc.* **21**, 211.

Blaxter, K. L. (1963) *Br. J. Nutr.* **17**, 105.

Blaxter, K. L. (1964) *Proc. Nutr. Soc.* **23**, 62.

Blaxter, K. L. (1976) George Scott Robertson Memorial Lecture, Queen's University, Belfast.

Blaxter, K. L. (1969) In *Energy metabolism of farm animals:* 4th symp., Warsaw, 1967, p. 21. (Eds. K. L. Blaxter, J. Kielanowski & G. Torbek). Newcastle upon Tyne: Oriel Press. (*Publs Eur. Ass. Anim. Prod.* **12**)

Blaxter, K. L. (1971) *Fedn Proc. Fedn Am. Socs. exp. Biol.* **30**, 1436.

Blaxter, K. L. (1972) *Festskrift til Knut Breirem* p. 19, Oslo: Mariendals Boktrykkeri.

Blaxter, K. L. (1974a) In *Energy metabolism of farm animals*; 6th symp., Hohenheim, 1973, p. 115. (Eds K. H. Menke, H.-J. Lantzsch and J. R. Reichl). Hohenheim: Universitäts Dokumentationsstelle. (*Publs Eur. Ass. Anim. Prod.* **14**)

Blaxter, K. L. (1974b) In *Disorders of Voluntary Muscle*, 3rd ed., p. 907 (Ed. J. N. Walton). Edinburgh: Churchill Livingston

Blaxter, K. L. (1977) In *Nutrition and the climatic environment* p. 1. (Eds W. Haresign, H. Swan and D. Lewis). Londo Butterworths. (*Proc. Nutr. Conf. Feed Mfrs Univ. Nott.* **10**, 1976)

Blaxter, K. L.; Boyne, A. W. (1970) In *Energy metabolism of farm animals*: 5th symp., Vitznau, p. 9 (Eds A. Schürch & C Wenk). Zurich: Juris Druck. (*Publs Eur. Ass. Anim. Prod.* **13**)

Blaxter, K. L.; Boyne, A. W. (1974). See Blaxter (1974a).

Blaxter, K. L.; Brown, F. (1952) *Nutr. Abstr. Rev.* **22**, 1.

Blaxter, K. L.; Clapperton, J. L. (1965) *Br. J. Nutr.* **19**, 511.

Blaxter, K. L.; Graham, N. McC. (1955) *J. agric. Sci., Camb.* **46**, 292.

Blaxter, K. L.; Joyce, J. P. (1963) *Br. J. Nutr.* **17**, 523.

Blaxter, K. L.; Martin, A. K. (1962) *Br. J. Nutr.* **16**, 397.

Blaxter, K. L.; McGill, R. F. (1955) *Vet. Revs Annot.* **1**, 91.

Blaxter, K. L.; McGill, R. F. (1956) *Vet. Revs Annot.* **2**, 35.

Blaxter, K. L.; Rook, J. A. F. (1953) *Br. J. Nutr.* **7**, 83.

Blaxter, K. L.; Rook, J. A. F. (1954) *J. comp. Path. Ther.* **64**, 176.

Blaxter, K. L.; Rook, J. A. F. (1957) *J. agric. Sci., Camb.* **48**, 194.

Blaxter, K. L.; Sharman, G. A. M. (1955) *Vet. Rec.* **67**, 108.

Blaxter, K. L.; Wainman, F. W. (1961) *J. agric. Sci., Camb.* **57**, 419.

Blaxter, K. L.; Wainman, F. W. (1964) *J. agric. Sci., Camb.* **63**, 113.

Blaxter, K. L.; Wainman, F. W. (1966) *Br. J. Nutr.* **20**, 103.

Blaxter, K. L.; Wood, W. A. (1951a) *Br. J. Nutr.* **5**, 29.

Blaxter, K. L.; Wood, W. A. (1951b *Br. J. Nutr.* **5**, 11.

Blaxter, K. L.; Wood, W.A. (1952). *Br. J. Nutr.* **6**, 56.

Blaxter, K. L.; Watts, P. S.; Wood, W. A. (1952) *Br. J. Nutr.* **6**, 125.

Blaxter, K. L.; Wood, W. A.; MacDonald, A. M. (1953a) *Br. J. Nutr.* **7**, 34.

Blaxter, K. L.; Brown, F.; Wood, W. A.; MacDonald, A. M. (1953b) *Br. J. Nutr.* **7**, 337.

Blaxter, K. L.; Rook, J. A. F.; MacDonald, A. M. (1954) *J. comp. Path.* **64**, 157.

Blaxter, K. L.; Sharman, G. A. M.; MacDonald, A. M. (1957) *Br. J. Nutr.* **11**, 234.

Blaxter, K. L.; Graham, N. McC.; Wainman, F. W.; Armstrong, D. G. (1959) *J. agric. Sci., Camb.* **52**, 25.

Blaxter, K. L.; Clapperton, J. L.; Martin, A. K. (1966a) *Br. J. Nutr.* **20**, 449.

Blaxter, K. L.; Clapperton, J. L.; Wainman, F. W. (1966b) *J. agric. Sci., Camb.* **67**, 67.

Blaxter, K. L.; Clapperton, J. L.; Wainman, F. W. (1966c) *Br. J. Nutr.* **20**, 283.

Blaxter, K. L.; Wainman, F. W.; Davidson, J. L. (1966d) *Anim. Prod.* **8**, 75.

Blincoe, C. (1975) *J. Anim. Sci.* **40**, 342.

Blosser, T. H.; Soni, B. K. (1957) *J. Dairy Sci.* **40**, 1519.

(S) Boaz, T. G.; Kirk, G.; Johnson, C. L. (1974) *J. agric. Sci., Camb.* **82**, 97.

Boda, J. M.; Cole, H. H. (1954) *J. Dairy Sci.* **37**, 360.

Boekholt, H. A. (1972) In *Proc. 'mini-symposium' on 'Protein requirements for growing and lactating ruminants', Wageninge* p. 1.

Bogart, R.; England, N. C. (1971) *J. Anim. Sci.* **32**, 420.

Bonomi, A. (1964) *Atti Soc. ital. Sci. vet.* **18**, 268.

Bonsma, F. N. (1939) *Publs Univ. Pretoria, Agric. Ser.* No. 48.

Borgida, L. P.; Durand, M.; Delort-Laval, J. (1976) *Annls Zootech.* **25**, 71.

Bose, P. K. (1955) Ph.D. Thesis, University of Aberdeen.

Boškov, L. (1962) *Vet. Glasn.* **16**, 800.

Bouchard, R.; Brisson, G. J. (1969) *Can. J. Anim. Sci.* **49**, 143.

Bouchard, R.; Conrad, H. R. (1973a) *J. Dairy Sci.* **56**, 1429.

Bouchard, R.; Conrad, H. R. (1973b) *J. Dairy Sci.* **56**, 1276.

Bouvier, J. C.; Vermorel, M. (1975) *Annls Zootech.* **24**, 697.

(S) Bowers, H. B.; Preston, T. R.; McDonald, I.; MacLeod, N. A.; Philip, E. B. (1965) *Anim. Prod.* **7**, 19.

Boyd, J. W. (1973) *Acta Agric. scand.* Suppl.19, 136.

Boyer, P. D.; Phillips, P. H.; Lundquist, N. S.; Jensen, C. W.; Rupel, I. W. (1942) *J. Dairy Sci.* **25**, 433.

Braithwaite, G. D. (1972) *Br. J. Nutr.* **27**, 201.

Braithwaite, G. D. (1974) *Br. J. Nutr.* **31**, 319.

Braithwaite, G. D. (1975a) *Br. J. Nutr.* **33**, 309.

Braithwaite, G. D. (1975b) *Br. J. Nutr.* **34**, 311.

Braithwaite, G. D.; Riazuddin, S. (1971) *Br. J. Nutr.* **26**, 215.

Braithwaite, G. D.; Glascock, R. F.; Riazuddin, S. (1969) *Br. J. Nutr.* **23**, 827.

Braithwaite, G. D.; Glascock, R. F.; Riazuddin, S. (1970) *Br. J. Nutr.* **24**, 611.

Branstetter, R. F.; Tucker, R. E.; Mitchell, G. E., Jr.; Boling, J. A.; Bradley, N. W. (1973) *Int. J. Vitam. Nutr. Res.* **43**, 142.

Bray, A. C.; Hemsley, J. A. (1969) *Aust. J. agric. Res.* **20**, 759.

Breirem, K. (1944) *Lantbr. Akad. Tidsskr.* **83**, 345.

Breirem, K. (1953) *Tidsskr. norske Landbr.* **60**, 25.

Breirem, K.; Homb, T. (1972) In *Handbuch der Tierernährung* vol.2 p.547. (Eds. W. Leinkeit, K. Breirem & E. Crasemann). Hamburg: Paul Parey.

Breirem, K.; Ender, F.; Halse, K.; Slagsvold, L. (1949) *Acta Agric. Suec.* **3**, 89.

Bremner, I.; Dalgarno, A. C. (1973a) *Br. J. Nutr.* **29**, 229.

Bremner, I.; Dalgarno, A. C. (1973b) *Br. J. Nutr.* **30**, 61.

Bremner, I.; Young, B. W.; Mills, C. F. (1976) *Br. J. Nutr.* **36**, 551.

Brett, D. J.; Corbett, J. L.; Inskip, M. W. (1972) *Proc. Aust. Soc. Anim. Prod.* **9**, 286.

Brisson, G. J.; Bouchard, R. (1970) *J. Anim. Sci.* **31**, 810.

Brisson, G. J.; Sutton, T. S. (1951) *J. Dairy Sci.* **34**, 28.

Brochart, M.; Larvor, P. (1959) *Cah. Elevage* no. 254.

Brockway, J. M.; Pullar, J. D.; McDonald, J. D. (1969) In *Energy metabolism of farm animals:* 4th symp., Warsaw, 1967, p. 423. (Eds. K. L. Blaxter, J. Kielanowski & G. Thorbek). Newcastle upon Tyne: Oriel Press. (*Publs Eur. Ass. Anim. Prod.* **12**).

Brody, S. (1945) *Bioenergetics and Growth.* New York: Reinhold.

Brody, S.; Ragsdale, A. C.; Thompson, H. J.; Worstell, D. M. (1954) *Res. Bull. Mo. agric. Exp. Stn.* No. 545.

Broster, W. H. (1972) *Dairy Sci. Abstr.* **34**, 265.

Broster, W. H. (1974) *Bienn. Rev. natn. Inst. Res. Dairy,* p. 14.

Broster, W. H.; Tuck, V. J.; Balch, C. C. (1964) *J. agric. Sci., Camb.* **63**, 51.

Broster, W. H.; Tuck, V. J.; Smith, T.; Johnson, V. M. (1969a) *J. agric. Sci., Camb.* **72**, 13.

Broster, W. H.; Broster, V. J.; Smith, T. (1969b) *J. agric. Sci., Camb.* **72**, 229.

Broster, W. H.; Broster, V. J.; Smith, T.; Siviter, J. W. (1975) *J. agric. Sci., Camb.* **84**, 173.

Brouwer, E.; Nijkamp, H. J. (1966) *Meded. LandbHogesch. Wageningen* **66**, 14.

Brouwer, E.; Van Es, A. J. H.; Nijkamp, H. J. (1965) In *Energy metabolism:* 3rd symp., Troon, 1964, p. 205. (Ed. K. L. Blaxter). London: Academic Press. (*Publs Eur. Ass. Anim. Prod.* **11**).

Brown, F. (1953) *J. Sci. Fd Agric.* **4**, 61.

Brown, L. D. (1966) *J. Dairy Sci.* **49**, 223.

Bruce, J.; Goodall, E. D.; Kay, R. N. B.; Phillipson, A. T.; Vowles, L. E. (1966) *Proc. R. Soc. B.* **166**, 46.

Brüggemann, J.; Bronsch, K.; Lörcher, K.; Seuss, H. (1959) *Z. Tierphysiol. Tierernähr.* **14**, 224.

Bryant, M. P. (1970) *Am. J. clin. Nutr.* **23**, 1440.

Bryden, J. McG.; Bray, A. C. (1972) *Proc. Aust. Soc. Anim. Prod.* **9**, 335.

Buchanan-Smith, J. G.; Nelson, E. C.; Osburn, B. I.; Wells, M. E.; Tillman, A. D. (1969) *J. Anim. Sci.* **29**, 808.

Buck, W. B. (1970) *J. Am. vet. med. Ass.* **156**, 1434.

Bull, L. B. (1951) *Proc. Br. Commonw. Spec. Conf. Agric., Australia* p. 300 Australia p. 300 (1949).

Bull, L. S. (1969) Ph.D. Thesis, Cornell University. *Diss. Abstr. int. B.* **30**, 918.

Bull, L. S.; Baumgardt, B. R.; Clancy, M. (1976) *J. Dairy Sci.* **59**, 1078.

Burdina, V. G. (1973) *Trudy mosk. vet. Akad.* **64**, 34.

Burns, K. N.; Allcroft, R. (1964) *Fluorosis in cattle–1. Occurrences and effects in Industrial Areas in England and Wales* 1954–57. London: HMSO.

Burns, R. H.; Johnston, A.; Hamilton, J. W.; McColloch, R. J.; Duncan, W. E.; Fisk, H. G. (1964) *J. Anim. Sci.* **23**, 5.

Burris, W. R.; Bradley, N. W.; Boling, J. A. (1974) *J. Anim. Sci.* **38**, 233.

Burroughs, S.; Raun, A.; Cheng, E.; Culbertson, C. C. (1956) *Iowa St. Coll. Anim. Husb. Leafl.* No.209.

Burroughs, W.; Fowler, M. A.; Adeyanju, S. A. (1970) *J. Anim. Sci.* **30**, 450.

Burroughs, W.; Nelson, D. K.; Mertens, D. R. (1975a) *J. Dairy Sci.* **58**, 611.

Burroughs, W.; Nelson, D. K.; Mertens, D. R. (1975b) *J. Anim. Sci.* **41**, 933.

Burt, A. W. A. (1957) *Dairy Sci. Abstr.* **19**, 435.

Burt, A. W. A.; Bell, E. O. (1962) *J. agric. Sci., Camb.* **58**, 131.

Burton, J. H.; Reid, J. T. (1969) *J. Nutr.* **97**, 517.

Burton, J. H.; Anderson, M.; Reid, J. T. (1974) *Br. J. Nutr.* **32**, 515.

Butler, E. J.; Nisbet, D. I.; Robertson, J. M. (1957) *J. comp. Path.* **67**, 378.

Buttery, P. J. (1976) In *Principles of cattle production,* p. 145. (Eds. H. Swan & W. H. Broster). London: Butterworths. (*Proc. Easter Sch. agric. Sci. Univ. Nott.* **23**, 1975).

Byers, J. H.; Weswig, P. H.; Bone, J. F.; Jones, I. R. (1955) *J. Dairy Sci.* **38**, 657.

Byers, J. H.; Jones, I. R.; Bone, J. F. (1956) *J. Dairy Sci.* **39**, 1556.

Calder, F. W.; Nicholson, J. W. G.; Cunningham, H. M. (1964) *Can. J. Anim. Sci.* **44**, 266.

Camalesa, N.; Dumitru, V.; Stoenescu, V. (1967) *Buletin şti. Univ. Craiova* **9**, 411.

Cameron, C. D. T. (1966) *Can. J. Anim. Sci.* **46**, 19.

Campbell, A. G.; Coup, M. R.; Bishop, W. H.; Wright, D. E. (1974) *N.Z. Jl. agric. Res.* **17**, 393.

Campbell, I. L. (1958) *Dairyfmg A.* p. 53.

Campbell, L. D.; Roberts, W. K. (1965) *Can. J. Anim Sci.* **45**, 147.

Campion, J. E.; Henry, K. M.; Kon, S. K.; MacKintosh, J. (1937) *Biochem. J.* **31**, 81.

Campling, R. C. (1966) *J. Br. Grassl. Soc.* **21**, 41.

Campling, R. C.; Freer, M.; Balch, C. C. (1962) *Br. J. Nutr.* **16**, 115.

Care, A. D. (1960) *Res. vet. Sci.* **1**, 338.

Care, A. D. (1965) *Proc. Nutr. Soc.* **24**, 99.

Care, A. D.; Ross, D. B. (1962) *Proc. Nutr. Soc.* **21**, x.

Carson, T. L.; Van Gelder, G. A.; Buck, W.B.; Hoffman, L. J.; Mick, D. L.; Long, K. R. (1973) *Clin. Toxicol.* **6**, 389.

Carson, T. L.; Van Gelder, G. A.; Karas, G. C.; Buck, W. B. (1974) *Arch. environ. Hlth* **29**, 154.

Castle, M. E. (1972a) *J. agric. Sci., Camb.* **78**, 371.

Castle, M. E. (1972b) *J. Br. Grassl. Soc.* **27**, 207.

Castle, M. E.; Thomas, T. P. (1975) *Anim. Prod.* **20**, 181.

Castle, M. E.; Watson, J. N. (1969) *J. Br. Grassld. Soc.* **24**, 187.

Castle, M. E.; Watson, J. N. (1970) *J. Br. Grassld Soc.* **25**, 651.

Castle, M. E.; Watson, J. N. (1973) *J. Br. Grassld Soc.* **28**, 73.

Castle, M. E.; Watson, J. N. (1974) *J. Br. Grassld Soc.* **29**, 101.

Castle, M. E.; Watson, J. N. (1975) *J. Br. Grassld Soc.* **30**, 1.

Castle, M. E.; Watson, J. N. (1976) *J. Br. Grassld Soc.* **31**, 191.

Castle, M. E.; Foot, A. S.; Halley, R. J. (1950) *J. Dairy Res.* **17**, 215.

(S) Castle, M. E.; Drysdale, A. D.; Waite, R.; Watson, J. N. (1963) *J. Dairy Res.* **30**, 199:

Chalmers, M. I.; Marshall, S. B. M. (1964) *J. agric. Sci., Camb.* **63**, 277.

Chalmers, M. I.; Synge, R. L. M. (1954) *J. agric. Sci., Camb.* **44**, 263.

Chalmers, M. I.; Cuthbertson, D. P.; Synge, R. L. M. (1954) *J. agric. Sci., Camb.* **44**, 254.

Chalmers, M. I.; Jaffray, A. E.; White, F. (1971) *Proc. Nutr. Soc.* **30**, 7.

Charpentier, J. (1966a) *Annls. Zootech.* **15**, 181.

Charpentier, J. (1966b) *Annls. Zootech.* **15**, 361.

Chatterton, R. T., Jr.; Hazzard, D. G.; Eaton, H. D.; Dehority, B. A.; Grifo, A. P., Jr.; Gosslee, D. G. (1961) *J. Dairy Sci.* **44**, 1061.

Chen, P. S., Jr.; Bosmann, H. B. (1964) *J. Nutr.* **83**, 133.

Chernova, Z. I. (1969) *Ovtsevodstvo* No. 9, 20.

Chernova, Z. I.; Aslanyan, M. M. (1971) *Khimiya v Sel'skom Khozaistve* **9**, 376.

Chiou, P. W. S.; Jordan, R. M. (1973) *J. Anim. Sci.* **37**, 581.

Chomyszyn, M.; Ziołecka, A.; Kużdowicz, M.; Bieliński, K. (1960) *Roczn. Nauk roln.* **75B**, 513.

Christian, R. G.; Tryphonas, L. (1971) *Am. J. vet. Res.* **32**, 203.

Chudy, A.; Schiemann, R. (1969) *Arch. Tiernernähr.* **19**, 231.

Clapperton, J. L. (1964a) *Br. J. Nutr.* **18**, 39.

Clapperton, J. L. (1964b) *Br. J. Nutr.* **18**, 47.

Clapperton, J. L.; Armstrong, D. G.; Blaxter, K. L. (1960) *Proc. Nutr. Soc.* **19**, xvii.

Clark, R.; Quin, J. I. (1947–49a) *Onderstepoort J. vet. Sci.* **22**, 335.

Clark, R.; Quin, J. I. (1947–49b) *Onderstepoort J. vet. Sci.* **22**, 345.

Clark, R. C.; Budtz-Olsen, O. E.; Cross, R. B.; Finnamore, P.; Bauert, P. A. (1973) *Aust. J. agric. Res.* **24**, 913.

Clark, R. M.; Holter, J. B.; Colovos, N. F.; Hayes, H. H. (1972) *J. Dairy Sci.* **55**, 257.

Clarke, E. M. W.; Ellinger, G. M.; Phillipson, A. T. (1966) *Proc. R. Soc. B* **166**, 63.

Clawson, W. J.; Lesperance, A. L.; Bohman, V. R.; Layhee, D. C. (1972) *J. Anim. Sci.* **34**, 516.

Clegg, F. G.; Rylands, J. M. (1966) *J. comp. Path.* **76**, 15.

Coelho da Silva, J. F. (1971) Ph.D. Thesis, University of Newcastle upon Tyne.

Coelho da Silva, J. F.; Seeley, R. C.; Beever, D. E.; Prescott, J. H. D.; Armstrong, D. G. (1972a) *Br. J. Nutr.* **28**, 357.

Coelho da Silva, J. F.; Seeley, R. C.; Thomson, D. J.; Beever, D. E.; Armstrong, D. G. (1972b) *Br. J. Nutr.* **28**, 43.

Cohen, P. (1960) *Tijdschr. Diergeneesk.* **85**, 1163.

Colovos, N. F.; Keener, H. A.; Davis, H. A. (1958) *J. Dairy Sci.* **41**, 676.

Colovos, N. F.; Holter, J. B.; Clark, R. M.; Urban, W. E., Jr.; Hayes, H. H. (1970) In *Energy metabolism of farm animals:* 5th symp., Vitznau, p. 89 (Eds. A. Schürch & C. Wenk). Zurich: Juris Druck. (*Publs Eur. Ass. Anim. Prod.* **13**).

Comar, C. L.; Monroe, R. A.; Visek, W. J.; Hansard, S. L. (1953) *J. Nutr.* **50**, 459.

Comberg, G. (1967) *Dt. tierärztl. Wschr.* **74**, 613.

Compère, R. (1966) *Bull. Rech. Agron. Gembloux* **1**, 190.

Compère, R. (1969) *Bull. Rech. Agron. Gembloux* **4**, 339.

Compère, R.; Vanuytrecht, S.; Fabry, J. (1965) *C.R. Soc. Biol.* **159**, 1258.

Compère, R.; Cava, R.; Maudoux, C. (1967) *Bull. Rech. Agron. Gembloux* **2**, 379.

Conrad, H. R.; Hansard, S. L.; Hibbs, J. W. (1956) *J. Dairy Sci.* **39**, 1697.

Conrad, H. R.; Pratt, A. D.; Hibbs, J. W. (1964) *J. Dairy Sci.* **47**, 54.

Converse, H. T. (1954) *U.S. Dep. Agric. Tech. Bull.* No. 1092.

Converse, H. T.; Meigs, E. B. (1938) *J. Dairy Sci.* **21**, 114.

Cook, C. W.; Mattox, J. E.; Harris, L. E. (1961) *J. Anim. Sci.* **20**, 866.

Coombe, J. B.; Christian, K. R. (1969) *J. agric. Sci., Camb.* **72**, 261.

Coop, I. E.; Drew, K. R. (1963) *Proc. N.Z. Soc. Anim. Prod.* **23**, 53.

Coppock, C. E.; Noller, C. H.; Wolfe, S. A.; Callahan, C. J.; Baker, J. S. (1972) *J. Dairy Sci.* **55**, 783.

Coppock, C. E.; Noller, C. H.; Wolfe, S. A. (1974) *J. Dairy Sci.* **57**, 1371.

Coppock, C. E.; Everett, R. W.; Belyea, R. L. (1976) *J. Dairy Sci.* **59**, 571.

Corbett, J. L. (1968) *Aust. J. agric. Res.* **19**, 283.

Coup, M. R.; Campbell, A. G. (1964) *N.Z. Jl agric. Res.* **7**, 624.

Cousins, F. B.; Cairney, I. M. (1961) *Aust. J. agric. Res.* **12**, 927.

Cowan, E. D.; Oliver, J.; Elliott, R. C. (1970) *Rhod. J. agric. Res.* **8**, 15.

Crabtree, R. M. (1976) Ph.D. Thesis, University of Aberdeen.

Cresswell, E.; Gill, J. C.; Fraser, C. (1968) *Anim. Prod.* **10**, 117.

Crowley, J. P. (1961) *Vet. Rec.* **73**, 295.

Cunningham, H. M.; Brisson, G. J. (1957) *Can. J. Anim. Sci.* **37**, 152.

Cunningham, H. M.; Loosli, J. K. (1954a) *J. Dairy Sci.* **37**, 453.

Cunningham, H. M.; Loosli, J. K. (1954b) *J. Anim. Sci.* **13**, 265.

Cunningham, G. N.; Wise, M. B.; Barrick, E. R. (1966) *J. Anim. Sci.* **25**, 532.

Cunningham, M. D.; Martz, F. A.; Merilan, C. P. (1964) *J. Dairy Sci.* **47**, 382.

Curran, M. K.; Holmes, W. (1970) *Anim. Prod.* **12**, 213.

Curran, M. K.; Wimble, R. H.; Holmes, W. (1970) *Anim. Prod.* **12**, 195.

Cuthbertson, W. F. J. (1976) *Am. J. clin. Nutr.* **29**, 559.

Czerkawski, J. W.; Blaxter, K. L.; Wainman, F. W. (1966) *Br. J. Nutr.* **20**, 349.

Daccord, R. (1970) Thesis, Dr. ès Sc. Swiss Fed. Poly. Zurich: Juris Druck.

Dalgarno, A. C.; Mills, C. F. (1975) *J. agric. Sci., Camb.* **85**, 11.

Daly, R. A.; Carter, H. B. (1955) *Aust. J. agric. Res.* **6**, 476.

Daneshvar, K.; Salter, D. N.; Smith, R. H. (1976) *Proc. Nutr. Soc.* **35**, 52A.

Danilenko, I. A.; Zhgun, N. B.; Kosmachev, N. V. (1972) *Dokl. vses. Akad. sel'.-khoz. Nauk* No. 5, 22 (see *Nutr. Abstr. Rev.* (1973) **43**, p. 77.

Danks, D. M.; Stevens, B. J.; Campbell, P. E.; Gillespie, J. M.; Walker-Smith, J.; Blomfield, J.; Turner, B. (1972) *Lancet* i, 1100.

Davies, D. A. R.; Owen, J. B. (1967) *Anim. Prod.* **9**, 501.

Davies, D. G.;Baker, M. H. (1974) *Vet. Rec.* **94**, 561.

(S) Davies, P. J. (1968) *Anim. Prod.* **10**, 311.

Davison, K. L.; Hansel, W.; Krook, L.; McEntee, K.; Wright, M. J. (1964) *J. Dairy Sci.* **47**, 1065.

Dawbarn, M. C.; Hine, D. C.; Smith, J. (1957) *Aust. J. exp. Biol. med. Sci.* **35**, 97.

Dawes, S. N. (1965) *N.Z. Jl Sci.* **8**, 161.

Deif, H. I.; el-Shazly, K.; Abou Akkada, A. R. (1968) *Br. J. Nutr.* **22**, 451.

Dell, J. C.; Poulton, B. R. (1958) *J. Dairy Sci.* **41**, 1706.

Demarquilly, C. (1973) *Annls Zootech.* **22**, 1.

Demertzis, P. N.; Mills, C. F. (1973) *Vet. Rec.* **93**, 219.

Deutsche Demokratische Republik (1970) *Das D.D.R. Futterbewertungs-system.* Berlin: V.E.B. Deutscher Landwirtschaftsverlag.

Devlin, T. J.; Roberts, W. K. (1963) *J. Anim. Sci.* **22**, 648.

Dick, A. T. (1953a) *Aust. vet. J.* **29**, 18.

Dick, A. T. (1953b) *Aust. vet. J.* **29**, 233.

Dick, A. T. (1954) *Aust. J. agric. Res.* **5**, 511.

Dick, A. T. (1956) In *Symposium on inorganic nitrogen metabolism: function of metallo-flavoproteins*, p. 445. (Eds W. D. McElroy & H. B. Glass). Baltimore, Md: Johns Hopkins Press.

Dinius, D. A.; Brinsfield, T. H.; Williams E. E. (1973) *J. anim. Sci.* **37**, 169.

Diplock, A. T. (1976) *CRC Crit. Rev. Toxicol.* **4**, 271.

Dodson, M. E.; Judson, G. J. (1973) *Aust. vet. J.* **49**, 320.

Donald, H. P.; Russell, W. S. (1970) *Anim. Prod.* **12**, 273.

Dowe, T. W.; Matsushima, J.; Arthaud, V. H. (1957) *J. anim. Sci.* **16**, 811.

Downey, N. E. (1960) *Vet. Rec.* **72**, 1023.

Drabkin, D. L. (1951) *Physiol. Rev.* **31**, 345.

Draper, H. H.; Johnson, B. C. (1952) *J. Nutr.* **46**, 37.

Drew, K. R. (1971) Ph.D. Thesis, Cornell University (*Diss. Abstr. int. B.* **32**, 5552).

(S) Dror, Y.; Tagari, H.; Bondi, A. (1970) *J. agric. Sci., Camb.* **75**, 381.

Drummond, J. C.; Baker, L. C. (1929) *J. Soc. chem. Ind., Lond.* **48**, 232.

Dryden, L. P.; Hartman, A. M. (1971) *J. Dairy Sci.* **54**, 235.

Duckworth, J.; Godden, W.; Thomson, W. (1943) *J. agric. Sci., Camb.* **33**, 190.

Dulphy, J. P.; Michalet, B. (1975) *Annls Zootech.* **24**, 757.

Duncan, C. W.; Huffman, C. F. (1936) *J. Dairy Sci.* **19**, 291.

Duncan, D. L. (1958) *Nutr. Abstr. Rev.* **28**, 695.

Dunlop, G. (1946) *Proc. Nutr. Soc.* **4**, 69.

Dunlop, G. (1954) *Nature, Lond.* **173**, 453.

Du Toit, P. J.; Malan, A. I.; Groenewald, J. W. (1934) *Onderstepoort J. vet. Sci. Anim. Ind.* **2**, 565.

Dynna, O.; Havre, G. N. (1963) *Acta vet. scand.* **4**, 197.

Eaton, H. D. (1969) *Am. J. clin. Nutr.* **22**, 1070.

Eaton, H. D.; Spielman, A. A.; Loosli, J. K.; Turk, K. L. (1947) *J. Dairy Sci.* **30**, 588.

Eaton, H. D.; Rousseau, J. E. Jr.; Woelfel, C. G.; Calhoun, M. C.; Nielsen, S. W.; Lucas, J. J. (1964) *Bull. Conn. Agric. Exp. Stn.* No. 383, 1.

Eaton, H. D.; Lucas, J. J.; Nielsen, S. W.; Helmboldt, C. F. (1970) *J. Dairy Sci.* **53**, 1775.

Eaton, H. D.; Rousseau, J. E. Jr.; Hall, R. C. Jr.; Frier, H. I.; Lucas, J. J. (1972) *J. Dairy Sci.* **55**, 232.

Edwin, E. E.; Lewis. G. (1971) *J. Dairy Res.* **38**, 79.

Eeckhout, W.; Casteels, M.; Buysse, F. (1969a) *Annls Zootech.* **18**, 249.

Eeckhout, W.; Casteels, M.; Buysse, F. (1969b) *Annls Zootech.* **18**, 263.

Ehlig, C. F.; Hogue, D. E.; Allaway, W. H.; Hamm, D. J. (1967) *J. Nutr.* **92**, 121.

Ekern, A. (1972a) *Meld. Norg. LandbrHøgsk.* **51** (31).

Ekern, A. (1962b) *Meld. Norg. LandbrHøgsk.* **51** (30).

Ellenberger, H. B.; Newlander, J. A.; Jones, C. H. (1931) *Bull. Vt. agric. Exp. Stn.* No. 331.

Ellenberger, H. B.; Newlander, J. A.; Jones, C. H. (1932) *Bull. Vt agric. Exp. Stn.* No. 342.

Ellenberger, H. B.; Newlander, J. A.; Jones, C. H. (1950a) *Bull. Vt. agric. Exp. Stn.* No. 558.

Ellenberger, H. B.; Newlander, J. A.; Jones, C. H. (1950b) *Bull. Vt agric. Exp. Stn.* No. 556.

Elliott, J. M.; Kay, R. N. B.; Goodall, E. D. (1971) *Life Sci. Part 1* **10**, 647.

Elliott, R. C.; Topps, J. H. (1963a) *Br. J. Nutr.* **17**, 539.

Elliott, R. C.; Topps, J. H. (1963b) *Nature, Lond.* **197**, 668.

Elliott, R. C.; Topps, J. H. (1964) *Br. J. Nutr.* **18**, 245.

Elliott, R. C.; Reed, W. D. C.; Topps, J. H. (1964) *Br. J. Nutr.* **18**, 519.

Ellis, L. C.; Tillman, A. D. (1961) *J. Anim. Sci.* **20**, 606.

Ellis, W. C.; Pfander, W. H. (1965) *Nature, Lond.* **205**, 974.

Ellis, W. C.; Garner, G. B.; Muhrer, M. E.; Pfander, W. H. (1956) *J. Nutr.* **60**, 413.

El-Sokkary, A. M.; Sirry, I.; Hassan, H. A. (1949) *J. agric. Sci., Camb.* **39**, 287.

Ely, R. E.; Dunn, K. M.; Huffmann, C. F. (1948) *J. Anim. Sci.,* **7**, 239.

Engel, R. W.; Hardison, W. A.; Miller, R. F.; Price, N. O.; Huber, J. T. (1964) *J. Anim. Sci.* **23**, 1160.

English, P. B. (1966) *Res. vet. Sci.* **7**, 233.

Enyedi, S.; Illés, A. (1967) *Állattenyésztés* **16**, 15.

Eskeland, B.; Pfander, W. H.; Preston, R. L. (1973) *Br. J. Nutr.* **29**, 347.

Eskeland, B.; Pfander, W. H.; Preston, R. L. (1974) *Br. J. Nutr.* **31**, 201.

Evans, R. E. (1960) *Rations for livestock*, 15th ed. London: H.M.S.O. (*Bull. Minist. Agric. Fish. Fd, Lond.* No. 48).

Evans, J. V. (1957) *Nature, Lond.* **180**, 756.

Eveleth, D. F.; Bolin, D. W.; Goldsby, A. I. (1949) *Am. J. vet. Res.* **10**, 250.

Evered, D. F. (1961) *Nature, Lond.* **189**, 228.

Ewer, T. K. (1950) *Vet. Rec.* **62**, 603.

Falconer, I. R. (1963) *J. Endocr.* **25**, 533.

Falconer, I. R.; Robertson, H. A. (1961) *J. Endocr.* **22**, 23.

Farnworth, A. J. (1956) *Aust. J. appl. Sci.* **7**, 233.

Farrell, D. J.; Leng, R. A.; Corbett, J. L. (1972a) *Aust. J. agric. Res.* **23**, 483.

Farrell, D. J.; Leng, R. A.; Corbett, J. L. (1972b) *Aust. J. agric. Res.* **23**, 499.

Faruque, O.; Walker, D. M. (1970a) *Br. J. Nutr.* **24**, 11.

Faruque, O.; Walker, D. M. (1970b) *Br. J. Nutr.* **24**, 23.

Federation of United Kingdom Milk Marketing Boards (1974) *United Kingdom Dairy Facts and Figures.*

Feenstra, P.; Van Ulsen, F. W. (1973) *Tijdschr. Diergeneesk.* **98**, 632.

Fell, B. F. (1972) *Wld Rev. Nutr. Diet.* **14**, 180.

Fell, B. F.; Boyne, R.; Quarterman, J. (1964) *J. comp. Path.* **74**, 514.

Fennessy, P. F.; Woodlock, M. R.; Jagusch, K. T. (1972) *N.Z. Jl agric. Res.* **15**, 795.

Ferguson, K. A. (1975) In *Digestion and metabolism in the ruminant*: 4th int. symp. Ruminant Physiol., Sydney, 1974, p. 448. (Eds. I. W. McDonald & A. C. I. Warner). Armidale: University of New England Publishing Unit.

Ferguson, W. S.; Lewis, A. H.; Watson, S.J. (1943) *J. agric. Sci., Camb.* **33**, 44.

Fick, K. R.; Ammerman, C. B.; Miller, S. M.; Simpson, C. F.; Loggins, P. E. (1976) *J. anim. Sci.* **42**, 515.

Field, A. C. (1959) *Nature, Lond.* **183**, 983.

Field, A. C. (1961) *Br. J. Nutr.* **15**, 349.

Field, A. C. (1962) *Br. J. Nutr.* **16**, 99.

Field, A. C.; Purves, D. (1964) *Proc. Nutr. Soc.* **23**, xxiv.

Field, A. C.; Suttle, N. F. (1967) *J. agric. Sci., Camb.* **69**, 417.

Field, A. C.; Suttle, N. F. (1969) *J. agric. Sci., Camb.* **73**, 507.

Field, A. C.; Suttle, N. F. (1970) *Proc. Nutr. Soc.* **29**, 34A.

Field, A. C.; McCallum, J. W.; Butler, E. J. (1958) *Br. J. Nutr.* **12**, 433.

Field, A. C.; Suttle, N. F.; Gunn, R. G. (1968) *J. agric. Sci., Camb.* **71**, 303.

Field, A. C.; Sykes, A. R.; Gunn, R. G. (1974) *J. agric. Sci., Camb.* **83**, 151.

Field, A. C.; Suttle, N. F.; Nisbet, D. I. (1975) *J. agric. Sci., Camb.* **85**, 435.

Filmer, J. F.; Underwood, E. J. (1937) *Aust. vet. J.* **13**, 57.

Fisher, L. J.; MacIntosh, A. I.; Carson, R. B. (1970) *Can. J. Anim. Sci.* **50**, 121.

Fishwick, G.; Hemingway, R. G. (1973) *J. agric. Sci., Camb.* **81**, 441.

(S) Fishwick, G.; Fraser, J.; Hemingway, R. G.; Parkins, J. J.; Ritchie, N. S. (1974) *J. agric. Sci., Camb.* **82**, 427.

Fitch, L. W. N.; Ewer, T. K. (1944) *Aust. vet. J.* **20**, 220.

Flamboe, E. E.; Reineke, E. P. (1959) *J. Anim. Sci.* **18**, 1135.

Flatt, W. P.; Coppock, C. E. (1963) *J. Dairy Sci.* **46**, 638.

Flatt, W. P.; Coppock, C. E.; Moore, L. A. (1965) In *Energy metabolism:* 3rd symp., Troon, 1964, p. 121. (Ed. K. L. Blaxter). London: Academic Press. (*Publs. Eur. Ass. Anim. Prod.* **11**).

Flatt, W. P.; Moe, P. W.; Van Es, A. J. H. (1968) *Proc. 2nd Wld Conf. Anim. Prod.* p. 399.

Flatt, W. P.; Moe, P. W.; Moore, L. A. (1969a) In *Energy metabolism of farm animals:* 4th symp., Warsaw, 1967, p. 123. (Eds. K. L. Blaxter, J. Kielanowski & G. Thorbek). Newcastle upon Tyne: Oriel Press. (*Publs. Eur. Ass. Anim. Prod.* **12**).

Flatt, W. P.; Moe, P. W.; Moore, L. A.; Van Soest, P. J. (1969b) In *Energy metabolism of farm animals:* 4th symp., Warsaw, 1967, p. 59. (Eds. K. L. Blaxter, J. Kielanowski & G. Thorbek). Newcastle upon Tyne: Oriel Press. (*Publs. Eur. Ass. Anim. Prod.* **12**).

Flatt, W. P.; Moe, P. W.; Moore, L. A.; Breirem, K.; Ekern, A. (1972) In *Handbuch der Tierernährung*, vol. 2, p. 341. (Eds. W. Lenkeit, K. Breirem & E. Crasemann). Berlin: Paul Parey.

Flux, D. S.; Butler, G. W.; Johnson, J. M.; Glenday, A. C.; Petersen, G. B. (1956) *N.Z. Jl Sci. Technol. A* **38**, 88.

Folin, O. (1905) *Am. J. Physiol.* **13**, 117.

Foot, J. Z. (1969) Ph.D. Thesis, University of Aberdeen.

Foot, J. Z.; Greenhalgh, J. F. D. (1970) *Anim. Prod.* **12**, 669.

Forbes, E. B. (1933) *Rec. Proc. Amer. Soc. Anim. Prod.* **25**, 32.

Forbes, E. B.; Beegle, F. M.; Fritz, C. M.; Morgan, L. E.; Rhue, S. N. (1916) *Bull. Ohio agric. Exp. Stn* No. 295.

Forbes, E. B.; Beegle, F. M.; Fritz, C. M.; Morgan, L. E.; Rhue, S. N. (1917) *Bull. Ohio agric. Exp. Stn* No. 308.

Forbes, E. B.; Halverson, J. O.; Morgan, L. E.; Schulz, J. A.; Mangels, C. E.; Rhue, S. N.; Burke, G. W. (1918) *Bull. Ohio agric. Exp. Stn* No. 330.

Forbes, E. B.; Hunt, C. H.; Schulz, J. A.; Winter, A. R.; Remler, R. F. (1922) *Bull. Ohio agric. Exp. Stn* No. 363.

Forbes, E. B.; Kriss, M.; Braham, W. W.; Jeffries, C. D.; Swift, R. W.; French, R. B.; Miller, R. C.; Smythe, C. V. (1927) *J. agric. Res.* **35**, 946.

Forbes, E. B.; Braman, W. W.; Kriss, M. (1928) *J. agric. Res.*, **37**, 253.

Forbes, E. B.; French, R. B.; Letonoff, T. V. (1929) *J. Nutr.* **1**, 201.

Forbes, E. B.; Braman, W. W.; Kriss, M.; Swift, R. W. (1931) *J. agric. Res.* **43**, 1003.

Forbes, E. B.; Karns, G. M.; Bechdel, S. I.; Williams, P. S.; Keith, T. B.; Callenbach, E. W.; Murphy, R. R. (1932) *J. agric. Res.* **45**, 111.

Forbes, E. B.; Black, A.; Braman, W. W.; Frear, D. E. H.; Kahlenberg, O. J.; McClure, F. J.; Swift, R. W.; Voris, L. (1935) *Bull. Pa agric. Exp. Stn* No. 319.

Forbes, J. M. (1968) *Br. J. Nutr.* **22**, 33.

Forbes, J. M. (1969a) *J. agric. Sci., Camb.* **72**, 119.

Forbes, J. M. (1969b) *Anim. Prod.* **11**, 263.

Forbes, J. M. (1970) *Br. vet. J.* **126**, 1.

(S) Forbes, T. J.; Robinson, J. J. (1967) *Anim. Prod.* **9** 521.

Fox, D. G.; Johnson, R. R.; Preston, R. L.; Dockerty, T. R.; Klosterman, E. W. (1972) *J. Anim. Sci.* **34**, 310.

Franke, E. R.; Weniger, J. H. (1958) *Arch. Tierernähr.* **8**, 81.

Franklin, M. C. (1934–35) 4th report of the Director, Institute of Animal Pathology, University of Cambridge, p. 111.

Franklin, M. C. (1953) *Aust. vet. J.* **29**, 302.

Franklin, M. C.; McClymont, G. L.; Briggs, P. K.; Campbell, B. L. (1955) *Aust. J. agric. Res.* **6**, 324.

) Fraser, C.; Ørskov, E. R. (1974) *Anim. Prod.* **18**, 75.

Freer, M.; Jones, D. B.; Christian, K. R. (1972) *J. agric. Sci., Camb.* **78**, 7.

Freney, M. R. (1934) *J. Soc. chem. Ind., Lond.* **53**, 151T.

Freney, M. R. (1940) *Bull. Coun. scient. ind. Res., Melb.* No. 130.

Frens, A. M.; Dijkstra, N.D. (1959) *Versl. landbouwk. Onderz.* No. 65.9.

Frey, P. R.; Jensen, R. (1947) *Science, N.Y.* **105**, 313.

Furrer, O. J. (1966) *Schweiz. landw. Mh.* **44**, 125.

Gabbedy, B. J. (1971) *Aust. vet. J.* **47**, 318.

Gaines, W. L.; Davidson, F. A. (1923) *Bull. Ill. agric. Exp. Stn.* No. 245.

Gaines, W. L.; Overman, O. R. (1938) *J. Dairy Sci.* **21**, 261.

Gardner, R. W.; Park, R. L. (1973) *J. Dairy Sci.* **56**, 390.

Garrett, W. N. (1958) Ph.D. Thesis, University of California.

Garrett, W. N. (1971) *J. Anim. Sci.* **32**, 451.

Garrett, W. N.; Hinman, N. (1969) *J. Anim. Sci.* **28**, 1.

Garrett, W. N.; Meyer, J. H.; Lofgreen, G. P. (1959) *J. Anim. Sci.* **18**, 528.

Garrett, W. N.; Rollins, W. C.; Tanaka, M.; Hinman, N. (1971) *Proc. Am. Soc. Anim. Prod. (Western Sect.)* **22**, 273.

Gartner, R. J. W.; Alexander, G. I. (1966) *Qd. J. agric. Anim. Sci.* **23**, 93.

Gawthorne, J. M. (1970) *Aust. J. exp. Biol. med. Sci.* **48**, 285; 293.

Gawthorne, J. M.; Nader, C. J. (1976) *Br. J. Nutr.* **35**, 11.

Geay, Y.; Robelin, J.; Jarrige, R. (1976) In *Energy metabolism of farm animals:* 7th symp., Vichy, p. 225 (Ed. M. Vermorel). Clermont-Ferrand: de Bussac. (*Publs. Eur. Ass. Anim. Prod.* **19**).

George, J. M. (1966) *Proc. Aust. Soc. Anim. Prod.* **6**, 403.

Georgi, J. R. (1964) *Am. J. vet. Res.* **25**, 952.

Gharaybeh, H. R.; McManus, W. R.; Arnold, G. W.; Dudzinski, M. L. (1969) *J. agric. Sci., Camb.* **72**, 65.

Gibb, M. J.; Penning, P. D. (1972) *Anim. Prod.* **15**, 177.

Glimp, H. A. (1972) *J. Anim. Sci.* **34**, 1085.

Glover, J.; Dougall, H. W. (1960) *J. agric. Sci., Camb.* **55**, 391.

Godden, W.; Puddy, C. A. (1935) *J. Dairy Res.* **6**, 307.

Godwin, K. O.; Kuchel, R. E.; Buckley, R. A. (1970) *Aust. J. exp. Agric. Anim. Husb.* **10**, 672.

Gonzalez-Jimenez, E.; Blaxter, K. L. (1962) *Br. J. Nutr.* **16**, 199.

Goodrich, R. D.; Tillman, A. D. (1966) *J. Anim. Sci.* **25**, 484.

Goodrich, R. D.; Emerick, R. J.; Embry, L. B. (1964) *J. Anim. Sci.* **23**, 100.

Goodwin, T. W.; (1952) *The comparative biochemistry of the carotenoids.* London: Chapman & Hall.

Gordon, F. J. (1977) *Anim. Prod.* **24**, 125.

Gordon, J. G. (1965) *J. agric. Sci., Camb.* **64**, 31.

Gorham, P. R. (1964) In *Algae & Man,* p. 307 (Ed. D. F. Jackson). New York: Plenum Press.

Goto, T.; Ohashi, S. (1969) *Jap. J. zootech. Sci.* **40**, 496.

Grace, N. D. (1975) *Br. J. Nutr.* **34**, 73.

Gracey, H. I.; Stewart, T. A.; Woodside, J. D.; Thompson, R. H. (1976) *J. agric. Sci., Camb.* **87**, 617.

Graf, G. C.; Holdaway, C. W. (1952) *J. Dairy Sci.* **35**, 998.

Graham, N. McC. (1962) *Proc. Aust. Soc. Anim. Prod.* **4**, 138.

Graham, N. McC. (1964a) *Aust. J. agric. Res.* **15**, 127.

Graham, N. McC. (1964b) *Aust. J. agric. Res.* **15**, 969.

Graham, N. McC. (1965) In *Energy Metabolism,* p. 231 (Ed. K. L. Blaxter) (*Publs. Europ. Assoc. Anim. Prod.,* **11**).

Graham, N. McC. (1966) *Proc. Aust. Soc. Anim. Prod.* **6**, 364.

Graham, N. McC. (1968) *Aust. J. agric. Res.* **19**, 821.

Graham, N. McC. (1969) *Aust. J. agric. Res.* **20**, 375.

Graham, N. McC.; Williams, A. J. (1962) *Aust. J. agric. Res.* **13**, 894.

Graham, N. McC.; Wainman, F. W.; Blaxter, K. L.; Armstrong, D. G. (1959) *J. agric. Sci., Camb.* **52**, 13.

Graham, N. McC.; Searle, T. W.; Griffiths, D. A. (1974) *Aust. J. agric. Res.* **25**, 957.

Grant, A. B. (1955) *Vet. Revs. Annot.* **1**, 115.

Grashuis, J. (1963) *Landbouwk. Tijdschr.* **75**, 1127.

Grashuis, J.; Lehr, J. J.; Beuvery, L. L. E.; Beuvery-Asman, A. (1953) *Mededeling, De Schothorst* No. S40.

Graves, R. R.; Dawson, J. R.; Kopland, D. V.; Watt, A. L.; Van Horn, A. G. (1938) *Tech. Bull. U.S. Dep. Agric.* No. 610.

Greenbaum, S. B. (1973) *Feedstuffs, Minneap.* April 16, p. 30.

Greenhalgh, J. F. D.; Reid, G. W. (1969a) *J. Br. Grassld Soc.* **24**, 98.

Greenhalgh, J. F. D.; Reid, G. W. (1969b) *J. agric. Sci., Camb.* **72**, 223.

Greenhalgh, J. F. D.; Reid, G. W. (1974) *Anim. Prod.* **19**, 77.

Greenhalgh, J. F. D.; Reid, G. W. (1975) *Proc. Br. Soc. Anim. Prod.* **4**, 111.

Greenhalgh, J. F. D.; Runcie, K. V. (1962) *J. agric. Sci., Camb.* **59**, 95.

Greenhalgh, J. F. D.; Reid, G. W.; Aitken, J. N.; Florence, E. (1966) *J. agric. Sci., Camb.* **67**, 13.

Greenhalgh, J. F. D.; Reid, G. W.; Aitken, J. N. (1967) *J. agric. Sci., Camb.* **69**, 217.

Gregorović, V.; Skusek, F.; Kesnar, F.; Beks, L. (1967) *Vet. Rec.* **81**, 161.

Grey, R. M.; Calhoun, M. C.; Rousseau, J. E., Jr.; Woelfel, C. G.; Hall, R. C., Jr.; Nielsen, S. W.; Eaton, H. D.; Lucas, J. J. (1964) *J. Dairy Sci.* **47**, 783.

Grieve, D. G.; Coppock, C. E.; Merrill, W. G.; Tyrrell, H. F. (1973) *J. Dairy Sci.* **56**, 218.

Grifo, A. P., Jr.; Rousseau, J. E., Jr.; Eaton, H. D.; Dehority, B. A.; Hazzard, D. G.; Helmboldt, C. F.; Gosslee, D. G. (1960) *J. Dairy Sci.* **43**, 1809.

Groppel, B.; Anke, M. (1971) *Arch. exp. Vet Med.* **25**, 779.

Groppel, B.; Hennig, A. (1971) *Arch. exp. Vet Med.* **25**, 817.

(S) Guada, J. A.; Robinson, J. J.; Fraser, C. (1975) *J. agric. Sci., Camb.* **85**, 175.

Guéguen, L. (1962) *Annls. Biol. anim. Biochim. Biophys.* **2**, 143.

Guéguen, L. (1963) *Annls. Biol. anim. Biochim. Biophys.* **3**, 243.

Guéguen, L. (1964) *C.R. hebd. Séanc. Acad. Sci., Paris*, **258**, 5985.

Guerrant, N. B.; Morck, R. A.; Bechdel, S. I.; Hilston, N. W. (1938) *Proc. Soc. exp. Biol. Med.* **38**, 827.

Guilbert, H. R.; Hart, G. H. (1935) *J. Nutr.* **10**, 409.

Guilbert, H. R.; Miller, R. F.; Hughes, E. H. (1937) *J. Nutr.* **13**, 543.

Guilbert, H. R.; Howell, C. E.; Hart, G. H. (1940) *J. Nutr.* **19**, 91.

Gullickson, T. W.; Palmer, L. S.; Boyd, W. L. (1935) *Tech. Bull. Minn. agric. Exp. Stn.* No. 105.

Gutkovich, Ya. L. (1970a) *Khim. sel'sk. Khoz.* **8**, 457.

Gutkovich, Ya. L. (1970b) *Khim. sel'sk. Khoz.* **8**, 788.

Haaranen, S. (1963) *Nord. VetMed.* **15**, 536.

(S) Hadjipieris, G.; Holmes, W. (1966) *J. agric. Sci., Camb.* **66**, 217.

Haecker, T. L. (1920) *Bull. Minn. agric. Exp. Stn.* No. 193.

Hagemeister, H.; Kaufmann, W. (1974) *Kieler milchw. ForschBer.* **26**, 199.

Hagemeister, H.; Pfeffer, E. (1973) *Z. Tierphysiol. Tierernähr. Futtermittelk.* **31**, 275.

Hagemeister, H.; Kaufmann, W.; Pfeffer, E. (1976) In *Protein metabolism and nutrition:* 1st int. symp., Nottingham, 1974, p. 425. (Eds. D. J. A. Cole, K. N. Boorman, P. J. Buttery, D. Lewis, R. J. Neale & H. Swan) London: Butterworths. (*Publs. Eur. Ass. Anim. Prod.* **16**).

Haigh, L. D.; Moulton, C. R.; Trowbridge, P. F. (1920) *Res. Bull. Mo. agric. Exp. Stn* No. 38.

Hale, W. H.; Hubbert, F., Jr; Taylor, R. E.; Anderson, T. A.; Taylor, B. (1962) *Am. J. vet. Res.* **23**, 992.

Hall, W. C.; Brody, S. (1933) *Res. Bull. Mo. agric. Exp. Stn* No. 180.

Hall, W. C.; Brody, S. (1934) *Res. Bull. Mo. agric. Exp. Stn* No. 208.

Halley, R. J.; Dougall, B. M. (1962) *J. Dairy Res.* **29**, 241.

Hamilton, T. S.; Robinson, W. B.; Johnson, B. C. (1948) *J. Anim. Sci.* **7**, 26.

Hammond, J. (1927) *The Physiology of Reproduction in the Cow.* Cambridge: University Press.

Hammond, P. B.; Aronson, A. L. (1964) *Ann. N.Y. Acad. Sci.* **111**, 595.

Hancock, J. (1953) *Anim. Breed. Abstr.* **21**, 1.

Hansard, S. L.; Mohammed, A. S. (1968) *J. Anim. Sci.* **27**, 807.

Hansard, S. L.; Comar, C. L.; Plumlee, M. P. (1954) *J. Anim. Sci.* **13**, 25.

Hansard, S. L.; Crowder, H. M.; Lyke, W. A. (1957) *J. Anim. Sci.* **16**, 437.

Hansard, S. L.; Mohammed, A. S.; Turner, J. W. (1968) *J. Anim. Sci.* **27**, 1097.

Harker, D. B. (1976) *Vet. Rec.* **99**, 78.

Harris, L. E.; Mitchell, H. H. (1941) *J. Nutr.* **22**, 167.

Harris, L. E.; Work, S. H.; Henke, L. A. (1943) *J. Anim. Sci.* **2**, 328.

Harris, L. T.; Loosli, J. K. (1944) *J. Dairy Sci.* **27**, 650.

Harris, P. L.; Embree, N. D. (1963) *Am. J. clin. Nutr.* **13**, 385.

Harrison, D. G. (1977) M.Sc. thesis, University of Newcastle upon Tyne.

Harrison, D. G.; Beever, D. E.; Thomson, D. J.; Osbourn, D. F. (1975) *J. agric. Sci., Camb.* **85**, 93.

Harrop, C. J. F. (1974) *J. agric. Sci., Camb.* **83**, 249.

Hart, E. B.; Humphrey, G. C.; Morrison, F. B. (1912) *J. biol. Chem.* **13**, 133.

Hart, E. B.; Steenbock, H.; Hoppert, C. A.; Humphrey, G. C. (1922a) *J. biol. Chem.* **53**, 21.

Hart, E. B.; Steenbock, H.; Hoppert, C. A.; Bethke, R. M.; Humphrey, G. C. (1922b) *J. biol. Chem.* **54**, 75.

Hart, E. B.; Steenbock, H.; Scott, H.; Humphrey, G. C. (1926–27) *J. biol. Chem.* **71**, 263.

Hart, E. B.; Steenbock, H.; Scott, H.; Humphrey, G. C. (1927) *J. biol. Chem.* **73**, 59.

Hart, E. B.; Steenbock, H.; Teut, E. C.; Humphrey, G. C. (1929a) *J. biol. Chem.* **84**, 359.

Hart, E. B.; Steenbock, H.; Teut, E. C.; Humphrey, G. C. (1929b) *J. biol. Chem.* **84**, 367.

Hart, E. B.; Steenbock, H.; Kline, O. L.; Humphrey, G. C. (1930) *J. biol. Chem.* **86**, 145.

Hartley, W. J. (1963) *Proc. N.Z. Soc. Anim. Prod.* **23**, 20.

Hartley, W. J.; Dodd, D. C. (1957) *N.Z. vet. J.* **5**, 61.

Hartley, W. J.; Grant, A. B. (1961) *Fedn. Proc. Fedn. Am. Socs. exp. Biol.* **20**, 679.

Hartman, R. H.; Matrone, G.; Wise, G. H. (1955) *J. Nutr.* **57**, 429.

Hartmans, J. (1965) *Versl. landbouwk. Onderz.* No. 664.

Hartmans, J. (1969) *Jaarb. Inst. biol. scheik. Onderz. LandbGewass.* No. 405, p. 45.

Hartmans, J. (1970) In *Trace element metabolism in animals:* WAAP/IBP int. symp., Aberdeen, 1969, p. 441. (Ed. C. F. Mills). Edinburgh: Livingstone.

Hartmans, J. (1971) *Proc. 8th Colloq. int. Potash Inst. (Uppsala)* p. 207.

Hartmans, J. (1972) *Landw. Forsch.* **27**, Suppl. p. 1.

Hartmans, J. (1974) *Neth. J. agric. Sci.* **22**, 195.

Hartmans, J.; Bosman, M. S. M. (1970) In *Trace element metabolism in animals:* WAAP/IBP int. symp., Aberdeen, 1969, p. 362. (Ed. C. F. Mills). Edinburgh: Livingstone.

Hartmans, J.; Van Der Grift, J. (1964) *Jaarb. Inst. biol. scheik. Onderz. LandbGewass.* No. 249, p. 145.

Hashizume, T.; Kaishio, Y.; Ambo, S.; Morimoto, H.; Masabuchi, T.; Abe, M.; Horii, S.; Tanaka, K.; Hamada, T.; Takahashi, S. (1962) *Bull. nat. Inst. agric. Sci., Japan*, **G21**, 213.

Hashizume, T.; Morimoto, H.; Masubichi, T.; Abe, M.; Hamada, T. (1965a) In *Energy metabolism:* 3rd symp., Troon, 1964, p. 111. (Ed. K. L. Blaxter). London: Academic Press. (*Publs. Eur. Ass. Anim. Prod.* **11**).

Hashizume, T.; Morimoto, H.; Masubichi, T.; Abe, M.; Hamada, T.; Horii, S.; Zitsukawa, Y.; Yokota, C.; Yamamoto, T.; Takahashi, K.; Toshino, M.; Saito, H. (1965b) *Spec. Rep. natn. Inst. Anim. Ind., Chiba* No. 6, p. 1.

Hatfield, E. E.; Smith, G. S.; Neumann, A. L.; Forbes, R. M.; Garrigus, U. S.; Ross, O. B. (1961) *J. Anim. Sci.* **20**, 676.

Havre, G. N.; Dynna, O. (1961) *Acta vet. scand.* **2**, 375.

Havre, G. N.; Steinnes, E. (1968) *Nord. VetMed.* **20**, 420.

Havre, G. N.; Dynna, O.; Ender, F. (1960) *Acta vet. scand.* **1**, 250.

Hawke, J. C. (1963) *J. Dairy Res.* **30**, 67.

Hawkins, G. E. Jr.; Wise, G. H.; Matrone, G.; Waugh, R. K.; Lott, W. L. (1955) *J. Dairy Sci.* **38**, 536.

Hayes, K. C. (1969) *Am. J. clin. Nutr.* **22**, 1081.

Hazzard, D. G.; Woelfel, C. G.; Calhoun, M. C.; Rousseau, J. E. Jr.; Eaton, H. D.; Nielsen, S. W.; Grey, R. M.; Lucas, J. J. (1964) *J. Dairy Sci.* **47**, 391.

Head, M. J. (1953) *J. agric. Sci., Camb.* **43**, 214.

Healy, W. B. (1967) *Proc. N.Z. Soc. Anim. Prod.* **27**, 109.

Healy, W. B.; Zieleman, A. M. (1966) *N.Z. Jl. agric. Res.* **9**, 1073.

Heaney, D. P.; Pritchard, G. I.; Pigden, W. J. (1968) *J. Anim. Sci.* **27**, 159.

Hedrick, H. B.; Thompson, G. B.; Krause, G. F. (1969) *J. Anim. Sci.* **29**, 687.

Helmboldt, C. F.; Jungherr, E. L.; Eaton, H. D.; Moore, L. A. (1953) *Am. J. vet. Res.* **14**, 343.

Helmer, L. G.; Bartley, E. E. (1971) *J. Dairy Sci.* **54**, 25.

Hemingway, R. G.; MacPherson, A. (1967) *Vet. Rec.* **81**, 695.

Hemingway, R. G.; Brown, N. A.; Inglis, J. S. S. (1962) *Res. vet. Sci.* **3**, 348.

Hemingway, R. G.; Inglis, J. S. S.; Brown, N. A. (1964) *Res. vet. Sci.* **5**, 7.

Hemken, R. W.; Vandersall, J. H.; Sass, B. A.; Hibbs, J. W. (1971) *J. Dairy Sci.* **54**, 85.

Henderickx, H. K.; Demeyer, D. I.; Van Nevel, C. J. (1972) In *Tracer studies on non-protein nitrogen for ruminants:* Proc. Panel, Vienna, 1971, p. 57. Vienna: I.A.E.A.

Henneman, H. A.; Reineke, E. P.; Griffin, S. A. (1955) *J. Anim. Sci.* **14**, 419.

Henry, K. M.; Hosking, Z. D.; Thompson, S. Y.; Toothill, J.; Edwards-Webb, J. D.; Smith, L. P. (1971) *J. Dairy Res.* **38**, 209.

Henseler, G.; Jentsch, W. (1973) *Arch. Tierenähr.* **23**, 567.

Henseler, G.; Jentsch, W.; Schiemann, R.; Wittenburg, H. (1973) *Arch. Tierernähr.* **23**, 353.

Hernandez, M. V.; Etta, K. M.; Reineke, E. P.; Oxender, W. D.; Hafs, H. D. (1972) *J. Anim. Sci.* **34**, 780.

Hibbs, J. W.; Conrad, H. R. (1960) *J. Dairy Sci.* **43**, 1124.

Hibbs, J. W.; Pounden, W. D. (1955) *J. Dairy Sci.* **38**, 65.

Hibbs, J. W.; Krauss, W. E.; Monroe, C. F.; Pounden, W. D. (1945) *J. Dairy Sci.* **28**, 525.

Hibbs, J. W.; Conrad, H. R.; Vandersall, J. H.; Gale, C. (1963) *J. Dairy Sci.* **46**, 1118.

Hidiroglu, M.; Carson, R. B.; Brossard, G. A. (1965) *Can. J. Anim. Sci.* **45**, 197.

Hidiroglu, M.; Hoffman, I.; Jenkins, K. J. (1969) *Can. J. Physiol. Pharmac.* **47**, 953.

Hidiroglu, M.; Jenkins, K. J.; Lessard, J. R.; Carson, R. B. (1970) *Br. J. Nutr.* **24**, 917.

Hignett, S. L. (1956) *Int. Congr. Anim. Reprod.* 3, Cambridge, p. 116.

Hignett, S. L. (1959) *Vet. Rec.* **71**, 247.

Hill, R.; Williams, H. L. (1965) *Vet. Rec.* **77**, 1043.

Hillman, H. C.; Provan, A. L.; Steane, E. (1950) *Chemy. Ind.* p. 333.

Hine, D. C.; Dawbarn, M. C. (1954) *Aust. J. exp. Biol. med. Sci.* **32**, 641.

Hintz, H. F.; Hogue, D. E. (1964) *J. Nutr.* **82**, 495.

Hjerpe, C. A. (1968a) *Cornell Vet.* **58**, 193.

Hjerpe, C. A. (1968b) *Am. J. vet. Res.* **29**, 143.

Hoar, D. W.; Embry, L. B.; Emerick, R. J. (1968) *J. Anim. Sci.* **27**, 1727.

Hobson, P. N.; Summers, R. (1972) *J. gen. Microbiol.* **70**, 351.

Hodge, R. W. (1973) *Aust. J. agric. Res.* **24**, 237.

Hodge, R. W.; Pearce, G. R.; Tribe, D. E. (1973) *Aust. J. agric. Res.* **24**, 229.

Hoekstra, W. G. (1975) *Fedn Proc. Fedn Am. Socs exp. Biol.* **34**, 2083.

Hoffmann, L.; Jentsch, W.; Wittenburg, H.; Schiemann, R. (1972) *Arch. Tierernähr.* **22**, 721.

Hoffmann, L.; Klippel, W.; Schiemann, R. (1967) *Arch. Tierernähr.* **17**, 441.

Hoffmann, L.; Schiemann, R.; Jentsch, W. (1971) In *Energetische Futterbewertung und Energienormen,* p. 118. (Eds. R. Schiemann, K. Nehring, L. Hoffmann, W. Jentsch and A. Chudy). Berlin: VEB Deutscher Landwirtschaftsverlag.

Hoffmann, L.; Schiemann, R.; Jentsch, W.; Henseler, G. (1974) *Arch. Tierernähr.* **24**, 245.

Hogan, A. G.; Nierman, J. L. (1927) *Res. Bull. Mo. agric. Exp. Stn.* No. 107.

Hogan, J. P. (1973) *Aust. J. agric. Res.* **24**, 587.

Hogan, J. P.; Weston, R. H. (1967a) *Aust. J. agric. Res.* **18**, 803.

Hogan, J. P.; Weston, R. H. (1967b) *Aust. J. agric. Res.* **18**, 973.

Hogan, J. P.; Weston, R. H. (1969) *Aust. J. agric. Res.* **20**, 339.

Hogan, J. P.; Weston, R. H. (1970) In *Physiology of digestion and metabolism in the ruminant:* 3rd int. symp., Cambridge, 1969, p. 474. (Ed. A. T. Phillipson). Newcastle upon Tyne: Oriel Press.

Hogan, J. P.; Weston, R. H. (1971) *Aust. J. agric. Res.* **22**, 951.

Hogan, J. P.; Connell, P. J.; Mills, S. C. (1972). *Aust. J. agric. Res.* **23**, 87.

Hogan, K. G.; Money, D. F. L.; Blayney, A. (1968) *N.Z. Jl. agric. Res.* **11**, 435.

Holman, R. T. (1960) *J. Nutr.* **70**, 405.

Holman, R. T. (1968) *Prog. Chem. Fats.* **9**, 275.

Holmes, C. W.; Davey, A. W. F. (1976) *Anim. Prod.* **23**, 43.

Holmes, C. W.; Davey, A. W. F.; McLean, N. A.; Jukes, G. C. (1975) *Proc. N.Z. Soc. Anim. Prod.* **35**, 36.

Holmes, C. W.; Stephens, D. B.; Toner, J. N. (1976) *Livestk. Prod. Sci.* **3**, 333.

Holmes, W. (1968) *Herb. Abstr.* **38**, 265.

Holmes, W.; Waite, R.; MacLusky, D. S.; Watson, J. N. (1956) *J. Dairy Res.* **23**, 1.

Holter, J. B.; Heald, C. W.; Colovos, N. F. (1970) *J. Dairy Sci.* **53**, 1241.

Holter, J. B.; Jones, L. A.; Colovos, N. F.; Urban, W. E., Jr. (1972) *J. Dairy Sci.* **55**, 1757.

Holzschuh, W. (1966) *Arch. Tierz.* **9**, 159.

Hopkins, P. S.; Wallace, A. L. C.; Thorburn, G. D. (1975) *J. Endocr.* **64**, 371.

Hopper, J. H.; Johnson, B. C. (1955) *J. Nutr.* **56**, 303.

Horrocks, D.; Phillips, G. D. (1964) *J. agric. Sci., Camb.* **63**, 359.

Horton, E. W. (1969) *Physiol. Rev.* **49**, 122.

Hoskins, F. H.; Hansard, S. L. (1964) *J. Nutr.* **83**, 10.

House, W. A.; Van Campen, D. (1971) *J. Nutr.* **101**, 1483.

Hovell, F. D. DeB. (1972) Ph.D. thesis, University of Aberdeen.

Hovell, F. D. DeB.; Greenhalgh, J. F. D. (1972) *Proc. Nutr. Soc.* **31**, 68A.

Hovell, F. D. DeB.; Greenhalgh, J. F. D.; Wainman, F. W. (1976) *Br. J. Nutr.* **35**, 343.

Howes, A. D.; Dyer, I. A. (1971) *J. Anim. Sci.* **32**, 141.

Huber, J. T. (1975) *J. Anim. Sci.* **41**, 954.

Huber, J. T.; Price, N. O. (1971) *J. Dairy Sci.* **54**, 429.

Hudson, L. W. (1969) *Diss. Abstr. int. B* **30**, 1970.

Hudson, L. W.; Glimp, H. A.; Little, C. O.; Woolfolk, P. G. (1970) *J. Anim. Sci.* **30**, 609.

Huffman, C. F.; Duncan, C. W. (1935) *J. Dairy Sci.* **18**, 432.

Huffman, C. F.; Duncan, C. W.; Robinson, C. S.; Lamb, L. W. (1933) *Tech. Bull. Mich. agric. Exp. Stn.* No. 134.

Huffman, C. F.; Conley, C. L.; Lightfoot, C. C.; Duncan, C. W. (1941) *J. Nutr.* **22**, 609.

Hume, I. D. (1970a) *Aust. J. agric. Res.* **21**, 297.

Hume, I. D. (1970b) *Aust. J. agric. Res.* **21**, 305.

Hume, I. D. (1974) *Aust. J. agric. Res.* **25**, 155.

Hume, I. D.; Bird, P. R. (1970) *Aust. J. agric. Res.* **21**, 315.

Hume, I. D.; Purser, D. B. (1975) *Aust. J. agric. Res.* **26**, 199.

Hungate, R. E. (1966) *The Rumen and its microbes.* New York: Academic Press.

Hunt, D. M. (1974) *Nature, Lond.* **249**, 852.

Hurt, H. D.; Eaton, H. D.; Rousseau, J. E., Jr.; Hall, R. C., Jr. (1967) *J. Dairy Sci.* **50**, 1941.

Hutchinson, J. C. D.; Morris, S. (1936) *Biochem. J.* **30**, 1682.

Hutton, J. B. (1963) *Proc. N.Z. Soc. Anim. Prod.* **23**, 39.

Hutton, J. B.; Jury, K. E.; Davies, E. B. (1965) *N.Z. Jl agric. Res.* **8**, 479.

International Union of Nutritional Sciences (IUNS) (1970) *Nutr. Abstr. Rev.* **40**, 395.

Isaacson, H. R.; Hinds, F. C.; Bryant, M. P.; Owens, F. N. (1975) *J. Dairy Sci.* **58**, 1645.

Ittner, N. R.; Kelly, C. F.; Guilbert, H. R. (1951) *J. Anim. Sci.* **10**, 742.

Iwarsson, K.; Ekman, L.; Everitt, B. R.; Figueiras, H.; Nilsson, P. O. (1973) *Acta vet. scand.* **14**, 610.

Jackson, H. M.; Kromann, R. P.; Ray, E. E. (1971) *J. Anim. Sci.* **33**, 872.

Jacobson, D. R.; Barnett, J. W.; Carr, S. B.; Hatton, R. H. (1967) *J. Dairy Sci.* **50**, 1248.

Jacobsson, S. O.; Lidman, S.; Lindberg, P. (1970) *Acta vet. scand.* **11**, 324.

Jagusch, K. T.; Mitchell, R. M. (1971) *N.Z. Jl agric. Res.* **14**, 434.

Jagusch, K. T.; Nicol, A. M. (1970) *Proc. N. Z. Soc. Anim. Prod.* **30**, 116.

Jagusch, K. T.; Norton, B. W.; Walker, D. M. (1970) *J. agric. Sci., Camb.* **75**, 273.

Jagusch, K. T.; Mitchell, R. M.; McConnell, G. R.; Fennessy, P. F.; Woodlock, M. R.; Jay, N. P. W. (1971) *Proc. N. Z. Soc. Anim. Prod.* **31**, 121.

Jahn, S. (1967) *Arch. Tierernähr.* **17**, 501.

Jakobsen, P. E. (1957) *Beretn. Forsøgslab. Kbh.* No. 299.

Jakobsen, P. E.; Sørensen, P. H.; Larsen, H. (1957) *Acta Agric. scand.* **7**, 103.

Jenkins, K. J.; Hidiroglu, M. (1967) *Canad. J. Anim. Sci.* **51**, 237.

Jenkins, K. J.; Hidiroglu, M. (1971) *Can. J. Anim. Sci.* **51**, 389.

Jenkins, K. J.; Hidiroglu, M.; MacKay, R. R.; Proulx, J. G. (1970) *Can. J. Anim. Sci.* **50**, 137.

Jenkins, K. J.; Proulx, J. G.; Hidiroglu, M. (1971) *Canad. J. Anim. Sci.* **51**, 237.

Jenkins, K. J.; Hidiroglu, M.; Wauthy, J. M.; Proulx, J. E. (1974) *Can. J. Anim. Sci.* **54**, 49.

Jenkinson, D. M.; Mabon, R. M. (1973) *Br. Vet. J.* **129**, 282.

Jensen, E.; Klein, J. W.; Rauchenstein, E.; Woodward, T. E.; Smith, R. H. *Tech. Bull. U.S. Dep. Agric.* No. 815.

Jensen, R.; Griner, L. A.; Adams, O. R. (1956) *J. Am. vet. med. Ass.* **129**, 311.

Johnson, B. C.; Hamilton, T. S.; Mitchell, H. H.; Robinson, W. B. (1942) *J. Anim. Sci.* **1**, 236.

Johnson, B. C.; Hamilton, T. S.; Nevens, W. B.; Boley, L. E. (1948) *J. Nutr.* **35**, 137.

Johnson, B. C.; Pinkos, J. A.; Burke, K. A. (1950) *J. Nutr.* **40**, 309.

Johnson, B. C.; Mitchell, H. H.; Pinkos, J. A. (1951) *J. Nutr.* **43**, 37.

Johnson, H. D.; Ragsdale, A. C.; Yeck, R. G. (1958) *Res. Bull. Mo. agric. Exp. Stn* No. 683.

Johnson, W. H.; Goodrich, R. D.; Meiske, J. C. (1970) *J. Anim. Sci.* **31**, 1003.

Johnson, W. L.; Trimberger, G. W.; Wright, M. J.; Van Vleck, L. D.; Henderson, C. R. (1966) *J. Dairy Sci.* **49**, 856.

Jolly, R. D. (1960) *N.Z. vet. J.* **8**, 11.

Josefsson, E. (1970) *J. Sci. Fd Agric.* **21**, 98.

Jones, I. R.; Weswig, P. H.; Bone, J. F. (1962) *J. Dairy Sci.* **45**, 683.

Jones, I. R.; Weswig, P. H.; Bone, J. F.; Peters, M. A.; Alpan, S. O. (1966) *J. Dairy Sci.* **49**, 491.

Jones, J. H.; Schmidt, H.; Dickson, R. E.; Fraps, G. S.; Jones, J. M.; Riggs, J. K.; Kemmerer, A. R.; Howe, P. E.; Black, W. H.; Ellis, N. R.; Marion, P. T. (1943) *Bull. Tex. agric. Exp. Stn* No. 630

Jordan, H. A.; Smith, G. S.; Neumann, A. L.; Zimmerman, J. E.; Breniman, G. W. (1963) *J. Anim. Sci.* **22**, 73.

Joshi, D. C. (1973) *Acta Agric. scand.* **23**, 5.

Joshi, D. C.; Talapatra, S. K. (1968) *Indian J. vet. Sci. Anim. Husb.* **38**, 665.

Journet, M.; Rémond, B. (1976) *Livestk. Prod. Sci.* **3**, 129.

Journet, M.; Poutous, M.; Calomiti, S. (1965) *Annls Zootech.* **14**, 5.

Joyce, J. P.; Blaxter, K. L. (1964) *Br. J. Nutr.* **18**, 5.

Joyce, J. P.; Rattray, P. V. (1970) *N.Z. Jl agric. Res.* **13**, 792.

Joyce, J. P.; Blaxter, K. L.; Park, C. (9166) *Res. vet. Sci.* **7**, 342.

Kahn, R. H. (1954) *Am. J. Anat.* **95**, 309.

Kalinin, V. V.; Kozmanishvili, A. B. (1970) *Ovtsevodstvo* **16**(8), 35.

Kamal, T. H.; Johnson, H. D.; Ragsdale, A. C. (1962) *Res. Bull. Mo. agric. Exp. Stn* No. 785.

Kataoka, K.; Nakae, T. (1971) *Jap. J. Dairy Sci.* **20**, A222.

Kaufmann, W.; Hagemeister, H. (1973) *Kraftfutter* **56**(12), 1.

Kay, M.; Houseman, R. A. (1975) In *Meat*, p. 85 (Ed. D. J. A. Cole and R. A. Lawrie). London: Butterworths.

(S) Kay, M.; Macdearmid, A. (1972) *Anim. Prod.* **14**, 367.

(S) Kay, M.; MacLeod, N. A.; McKiddie, G.; Philip, E. B. (1967) *Anim. Prod.* **9**, 197.

(S) Kay, M.; Bowers, H. B.; McKiddie, G. (1968) *Animl Prod.* **10**, 37.

(S) Kay, M.; Macdearmid, A.; Massie, R. (1970a) *Anim. Prod.* **12**, 419.

(S) Kay, M.; MacLeod, N. A.; McLaren, M. (1970b) *Anim. Prod.* **12**, 413.

(S) Kay, M.; MacLeod, N. A.; Andrews, R. P. (1972) *Anim. Prod.* **14**, 149.

Kay, R. N. B.; Hobson, P. N. (1963) *J. Dairy Res.* **30**, 261.

Keenan, D. M.; McManus, W. R.; Freer, M. (1969) *J. agric. Sci., Camb.* **72**, 139.

Keener, H. A.; Bechdel, S. I.; Guerrant, N. B.; Thorp, W. T. S. (1942) *J. Dairy Sci.* **25**, 571.

Keener, H. A.; Percival, G. P.; Morrow, K. S.; Ellis, G. H. (1949) *J. Dairy Sci.* **32**, 527.

Kehar, N. D.; Mukherjee, R.; Sen, K. C. (1943) *Indian J. vet. Sci.* **13**, 257.

Kellaway, R. C. (1973) *J. agric. Sci., Camb.* **80**, 17.

Kelley, J. T. (1945) *J. Dep. Agric. Vict.* **43**, 158.

Kelliher, D. J.; Hilliard, E. P.; Poole, D. B. R.; Collins, J. D. (1973) *Ir. J. agric. Res.* **12**, 61.

Kemp. A. (1964) *Neth. J. agric. Sci.* **12**, 263.

Kemp. A.; Deijs, W. B.; Hemkes, O. J.; Van Es, A. J. H. (1961) *Neth. J. agric. Sci.* **9**, 134.

Kennedy, P. M. (1974) *Aust. J. agric. Res.* **25**, 1015.

Kennedy, P. M.; Siebert, B. D. (1973) *Aust. J. agric. Res.* **24**, 143.

Kennedy, P. M.; Williams, E. R.; Siebert, B. D. (1975) *Aust. J. biol. Sci.* **28**, 31.

Keyes, E. A. (1944) *Bull. Pa agric. Exp. Stn* No. 446, Suppl. No. 2, p. 4.

) Khalifa, H. A.; Prescott, J. H. D.; Armstrong, D. G. (1973) *Anim. Prod.* **16**, 185.

Kielanowski, J. (1976) In *Protein metabolism and nutrition:* 1st int. symp., Nottingham, 1974, p. 207. (Eds. D. J. A. Cole, K. N. Boorman, P. J. Buttery, D. Lewis, R. J. Neale and H. Swan). London: Butterworths. (*Publs. Eur. Ass. Anim. Prod.* **16**).

Kielanowski, J.; Lassota, L. (1960) *Zesz. probl. Postęp. Nauk roln.* **22**, 173.

Kiesel, G. K.; Burns, M. J. (1960) *Am. J. vet. Res.* **21**, 226.

Kirchgessner, M. (1965) *Proc. Nutr. Soc.* **24**, 89.

Kirchgessner, M.; Neesse, K. R. (1976) *Z. Lebensmittelunters. u. -Forsch.* **161**, 1.

Kirchgessner, M.; Friesecke, H.; Koch, G. (1967) *Nutrition and the Composition of Milk*, p. 219. London: Crosby Lockwood.

Kirchgessner, M.; Grassmann, E.; Krippl, J.; Müller, H. L. (1971) *Zuchtungskunde* **43**, 336.

Kirchgessner, M.; Müller, H. L.; Neesse, K. R. (1976) *Z. Tierphysiol. Tierernähr. Futtermittelk.* **37**, 334.

Kirk, W. G.; Shirley, R. L.; Easley, J. F.; Peacock, F. M. (1970) *Bull. Fla agric. Exp. Stn* No. 741.

Kleiber, M.; Smith, A. H.; Ralston, N. P.; Black, A. L. (1951) *J. Nutr.* **45**, 253.

Kline, R. D.; Hays, V. W.; Cromwell, G. L. (1971) *J. Anim. Sci.* **33**, 771.

Klosterman, E. W.; Johnson, L. J.; Moxon, A. L.; Grifo, A. P. Jr. (1964) *J. Anim. Sci.* **23**, 723.

Kneale, W. A.; Howell, J. McC. (1974) *Proc. Br. Soc. Anim. Prod.* **3**, 98.

Kohlmeier, R. H.; Burroughs, W. (1970) *J. Anim. Sci.* **30**, 1012.

Kon, S. K.; Porter, J. W. G. (1951) *Rep. natn. Inst. Res. Dairy.* p. 83.

Kon, S. K.; Porter, J. W. G. (1954) *Vitams Horm.* **12**, 53.

Kon, S. K.; McGillivray, W. A.; Thompson, S. Y. (1955) *Br. J. Nutr.* **9**, 244.

Konar, A.; Thomas, P. C. (1970) *Br. vet. J.* **126**, 25.

Koong, L. J.; Wise, M. B.; Barrick, E. R. (1970) *J. Anim. Sci.* **31**, 422.

Koval'skii, V. V.; Gunstun, M. I. (1966) *Vest. sel'khoz. Nauki, Alma-Ata* **3**, 77.

Koval'skii, V. V.; Ladan, A. I.; Gribovskaya, I. F.; Blokhina, R. I. (1972) *Vest. sel'-khoz. Nauki, Alma-Ata* No. 10, 66.

Krauss, W. E.; Bethke, R. M.; Monroe, C. F. (1932) *J. Nutr.* **5**, 467.

Krebs. H. A. (1960) *Arzneimittel-Forsch.* **10**, 369.

Kromann, R. P. (1973) *J. Anim. Sci.* **37**, 200.

Kromann, R. P.; Ray, E. E. (1967) *J. Anim. Sci.* **26**, 1379.

Kroneman, J.; van der Mey, G. J. M.; Helder, A. (1975) *Zentbl. VetMed. A* **22**, 201.

Kuhlman, A. H.; Gallup, W. D. (1942). *J. Dairy Sci.* **25**, 688.

Laird, A. K. (1966) *Growth* **30**, 263.

Lambert, M. R.; Jacobson, N. L.; Allen, R. S.; Zaletel, J. H. (1954) *J. Nutr.* **52**, 259.

Langlands, J. P. (1972) *Proc. Aust. Soc. Anim. Prod.* **9**, 321.

Langlands, J. P.; Sutherland, H. A. M. (1968) *Br. J. Nutr.* **22**, 217.

Langlands, J. P.; Sutherland, H. A. M. (1969) *Br. J. Nutr.* **23**, 603.

Langlands, J. P.; Sutherland, H. A. M. (1973) *Br. J. Nutr.* **30**, 529.

Lannek, N.; Lindberg, P. (1975) *Adv. vet. Sci.* **19**, 127.

Large, R. V. (1964) *Anim. Prod.* **6**, 169.

Large, R. V. (1965) *Anim. Prod.* **7**, 325.

Large, R. V.; Penning, P. D. (1967) *J. agric. Sci., Camb.* **69**, 405.

Larvor, P. (1976) *Cornell Vet.* **66**, 413.

Lassiter, C. A.; Ward, G. M.; Huffman, C. F.; Duncan, C. W.; Webster, H. D. (1953) *J. Dairy Sci.* **36**, 997.

Lassiter, J. W.; Morton, J. D. (1968) *J. Anim. Sci.* **27**, 776.

Lawlor, M. J.; Smith, W. H.; Beeson, W. M. (1965) *J. Anim. Sci.* **24**, 742.

Leach, R. M. Jr. (1974) In *Trace element metabolism in animals—2.* 2nd int symp., Madison, Wis., 1973, p. 51 (Eds. Hoekstra, W. G., Suttie, J. W., Ganther, H. E. and Mertz, W.) Baltimore, Md; University Park Press.

Leat, W. M. F. (1966) *Biochem. J.* **98**, 598.

Leat, W. M. F. (1970) In *Physiology of digestion and metabolism in the ruminant:* 3rd int. symp., Cambridge, 1969, p. 211. (Ed. A. T. Phillipson) Newcastle upon Tyne: Oriel Press.

Leat, W. M. F.; Harrison, F. A. (1972) *Proc. Nutr. Soc.* **31**, 70A.

Leaver, J. D.; Campling, R. C.; Holmes, W. (1969) *Anim. Prod.* **11**, 11.

Legg, S. P.; Sears, L. (1960) *Nature, Lond.* **186**, 1061.

Leibholz, J. (1972a) *Aust. J. agric. Res.* **23**, 1073.

Leibholz, J. (1972b) *Aust. J. exp. Agric. Anim. Husb.* **12**, 561.

Leibholz, J.; Hartmann, P. E. (1972) *Aust. J. agric. Res.* **23**, 1059.

Leitch, I.; Thomson, J. S. (1944–45) *Nutr. Abstr. Rev.* **14**, 197.

Lengemann, F. W. (1965) *J. Dairy Sci.* **48**, 1718.

Leroy, A. M.; Zelter, S. Z. (1948) *Annls agron.* **18**, 194.

Leroy, F.; Zelter, S. Z. (1970) *Annls Biol. anim. Biochim. Biophys.* **10**, 401.

Leslie, A. (1935) Cited by Franklin (1953).

Lesperance, A. L.; Bohman, V. R. (1963) *J. Anim. Sci.* **22**, 686.

L'Estrange, J. L.; Axford, R. F. E. (1966) *J. agric. Sci., Camb.* **67**, 295.

L'Estrange, J. L.; Owen, J. B.; Wilman, D. (1967) *J. agric. Sci., Camb.* **68**, 173.

Levy, D.; Holzer, Z. (1971) *Anim. Prod.* **13**, 569.

Levy, D.; Holzer, Z.; Neumark, H.; Amir, S. (1974) *Anim. Prod.* **18**, 67.

Lewis, D.; Hill, K. J.; Annison, E. F. (1957) *Biochem. J.* **66**, 587.

Lewis, J. K.; Burkitt, W. H.; Willson, F. S. (1951) *J. Anim. Sci.* **10**, 1053.

Lewis, J. M.; Wilson, L. T. (1945) *J. Nutr.* **30**, 467.

Lindberg, P.; Jacobsson, S. O. (1970) *Acta vet. scand.* **11**, 49.

Lindsay, D. B.; Leat, W. M. F. (1977) *J. agric. Sci., Camb.* **89**, 215.

Lindsay, J. R.; Hogan, J. P. (1972) *Aust. J. agric. Res.* **23**, 321.

Line, C.; Head, M. J.; Rook, J. A. F.; Foot, A. S.; Rowland, S. J. (1958) *J. agric. Sci., Camb.* **51**, 353.

Lineweaver, J. A.; Hafez, E. S. E. (1969) *J. Dairy Sci.* **52**, 2001.

Ling, E. R.; Kon, S. K.; Porter, J. W. G. (1961) In *Milk: the Mammary Gland and its Secretion.* Vol. 2, p. 195. New York: Academic Press.

Little, C. O.; Mitchell, G. E., Jr. (1967) *J. Anim. Sci.* **26**, 411.

Little, W.; Manston, R. (1972) *J. agric. Sci., Camb.* **78**, 309.

Little, W.; Shaw, S. R. (1978) *Anim. Prod.* **26**, 225.

Little, W.; Sansom, B. F.; Manston, R.; Allen, W. M. (1976) *Anim. Prod.* **22**, 329.

Little, W.; Sansom, B. F.; Manston, R.; Allen, W. M. (1978) *Anim. Prod.* **27**, 79.

Littlejohn, A. I.; Lewis, G. (1960) *Vet. Rec.* **72**, 1137.

Lodge, G. A.; Heaney, D. P. (1970) In *Energy metabolism of farm animals:* 5th symp., Vitznau, p. 109. (Eds. A. Schürch & C. Wenk). Zurich: Juris Druck. (*Publs. Eur. Ass. Anim. Prod.* **13**).

Lodge, G. A.; Heaney, D. P. (1973) *Can. J. Anim. Sci.* **53**, 95.

Lodge, J. R.; Lewis, R. C.; Reineke, E. P. (1957) *J. Dairy Sci.* **40**, 209.

Lofgreen, G. P. (1960) *J. Nutr.* **70**, 58.

Lofgreen, G. P.; Garrett, W. N. (1968) *J. Anim. Sci.* **27**, 793.

Lofgreen, G. P.; Kleiber, M. (1953) *J. Anim. Sci.* **12**, 366.

Lofgreen, G. P.; Kleiber, M. (1954) *J. Anim. Sci.* **13**, 258.

Lofgreen, G. P.; Kleiber, M.; Luick, J. K. (1952) *J. Nutr.* **47**, 571.

Lofgreen, G. P.; Hull, J. L.; Otagaki, K. K. (1962) *J. Anim. Sci.* **21**, 20.

Lomba, F.; Paquay, R.; Bienfet, V.; Lousse, A. (1969) *J. agric. Sci., Camb.* **73**, 453.

Lomba, F.; Chauvaux, G.; Fumière, I.; Bienfet, V.; Paquay, R.; De Baere, R.; Lousse, A. (1970) *Z. Tierphysiol. Tierernähr. Futtermittelk.* **27**, 9.

Long, J. F.; Gilmore, L. O.; Curtis, G. M.; Rife, D. C. (1952) *J. Dairy Sci.* **35**, 603.

Lonsdale, C. R.; Tayler, J. C. (1971) *Anim. Prod.* **13**, 384.

Lonsdale, C. R.; Poutiainen, E. K.; Tayler, J. C. (1971) *Anim. Prod.* **13**, 461.

Loosli, J. K. (1952) *Feed Age* **2**, 44.

Loosmore, R. M.; Allcroft, R.; Salt, F. J.; Hughes, D. E.; Chou, J. T.-Y.; Wilson, D. C. (1962) *Nature, Lond.* **193**, 595.

Losada, H.; Dixon, F.; Preston, T. R. (1971) *Revta cubana Cienc. agric. (Eng. ed.)* **5**, 369.

Lotthammer, K. H.; Ahlswede, L.; Meyer, H.; Schultz, G.; Hofmann, B. (1974) *Z. Tierphysiol. Tierernähr. Futtermittelk.* **33**, 212.

Lotthammer, K. H.; Ahlswede, L.; Meyer, H. (1976) *Dt. tierärztl. Wschr.* **83**, 353.

Lucke, V. M.; Baskerville, A.; Bardgett, P. L.; Mann, P. G. H.; Thompson, S. Y. (1968) *Vet. Rec.* **82**, 141.

Lucy, J. A. (1969) *Am. J. clin. Nutr.* **22**, 1033.

Lueker, C. E.; Lofgreen, G. P. (1961) *J. Nutr.* **74**, 233.

Lüke, F.; Marquering, B. (1972) *Züchtungskunde* **44**, 56.

Lüke, F.; Wiemann,H. (1970) *Berl. Münch. tierärztl. Wschr.* **83**, 253.

) Lyons, T.; Caffrey, P. J.; O'Connell, W. J. (1970) *Anim. Prod.* **12**, 323.

Macdonald, D. C.; Care, A. D.; Nolan, B. (1959) *Nature, Lond.* **184**, 736.

MacDonald, M. A.; Bell, J. M. (1958) *Can. J. Anim. Sci.* **38**, 23.

MacDougall, D. B.; Bremner, I.; Dalgarno, A. C. (1973) *J. Sci. Fd Agric.* **24**, 1255.

Mace, D. T.; Tucker, J. A.; Bills, C. B.; Ferreira, C. J. (1963) *Bull. Calif. Dep. Agric.* No.1, p. 63.

MacFarlane, W. V.; Morris, R. J. H.; Howard, B.; McDonald, J.; Budtz-Olsen, O. E. (1961) *Aust. J. agric. Res.* **12**, 889.

MacLusky, D. S. (1959–60) *Agriculture, Lond.* **66**, 383.

MacPherson, A.; Hemingway, R. G. (1965) *J. Sci. Fd Agric.* **16**, 220.

MacPherson, A.; Hemingway, R. G. (1968) *J. Sci. Fd Agric.* **19**, 53.

MacRae, J. C.; Armstrong, D. G. (1969) *Br. J. Nutr.* **23**, 377.

MacRae, J. C.; Ulyatt, M. J. (1974) *J. agric. Sci., Camb.* **82**, 309.

MacRae, J. C.; Ulyatt, M. J.; Pearce, P. D.; Hendtlass, J. (1972) *Br. J. Nutr.* **27**, 39.

Mäkelä, A. (1956) *Suom. maatal. Seur. Julk.* No. 85.

Mäkelä, A. (1960) *Maataloust. Aikakausk.* **32**, 9.

Malan, A. I.; Du Toit, P. J.; Groenewald, J. W. (1935) *Onderstepoort J. vet. Sci. Anim. Ind.* **5**, 189.

Malan, A. I.; Du Toit, P. J.; Groenewald, J. W. (1940) *Onderstepoort J. vet. Sci. Anim. Ind.* **14**, 329.

Mangan, J. L. (1972) *Br. J. Nutr.* **27**, 261.

Manston, R. (1964) *Br. vet. J.* **120**, 365.

Manston, R. (1966) *Br. vet. J.* **122**, 443.

Manston, R.; Vagg, M. J. (1964) *Br. vet. J.* **120**, 580.

Markson, L. M.; Edwin, E. E.; Lewis, G.; Richardson, C. (1974) *Br. vet. J.* **130**, 9.

) Marsh, R. (1974) *Anim. Prod.* **18**, 201.

) Marsh, R. (1975) *Anim. Prod.* **20**, 345.

Marshall, S. P.; Smith, K. L. (1970) *J. Dairy Sci.* **53**, 1622.

Marshall, S. P.; Smith, K. L. (1971) *J. Dairy Sci.* **54**, 1064.

Marshall, S. P.; Smith, K. L. (1973) *J. Anim. Sci.* **37**, 833.

Marston, H. R. (1935) *J. agric. Sci., Camb.* **25**, 113.

Marston, H. R. (1948a) *Aust. J. scient. Res. B* **1**, 93.

Marston, H. R. (1948b) *Biochem. J.* **42**, 564.

Marston, H. R. (1952) *Physiol. Rev.* **32**, 66.

Marston, H. R. (1970) *Br. J. Nutr.* **24**, 615.

Marston, H. R.; Allen, S. H.; Smith, R. M. (1961) *Nature, Lond.* **190**, 1085.

Marston, H. R.; Allen, S. H.; Smith, R. M. (1972) *Br. J. Nutr.* **27**, 147.

Martin, A. K.; Blaxter, K. L. (1961) In *2. Symposium on energy metabolism: methods and results of experiments with animals* p. 200. (Eds. E. Brouwer & A. J. H. Van Es). Wageningen: Lab. Anim. Physiol. (*Publs Eur. Ass. Anim. Prod.* **10**).

Martin, F. H. (1968) Ph.D. thesis, Michigan State University (*Diss. Abstr. B.* **29**, 256).

Martin, J. L. (1973) In *Organic Selenium Compounds: their Chemistry and Biology* p. 663. (Eds. D. L. Klayman and W. H. H. Günther). New York: Wiley.

Mason, R. W. (1976) *Br. vet. J.* **132**, 374.

Mason, V. C.; Palmer, R. (1971) *J. agric. Sci., Camb.* **76**, 567.

Matěj, V. (1970) *Acta Univ. Agric. Fac. agroecon.* **6**, 365.

Mathers, J. C.; Horton, C. M.; Miller, E. L. (1977) *Proc. Nutr. Soc.* **36**, 37A.

Mathison, G. W.; Milligan, L. P. (1971) *Br. J. Nutr.* **25**, 351.

Matrone, G.; Conley, C.; Wise, G. H.; Waugh, R. K. (1967) *J. Dairy Sci.* **40**, 1437.

Matrone, G.; Hartman, R. H.; Clawson, A. J. (1959) *J. Nutr.* **67**, 309.

Mattos, W.; Palmquist, D. L. (1977) *J. Nutr.* **107**, 1755.

McAllan, A. B.; Smith, R. H. (1972) *Proc. Nutr. Soc.* **31**, 24A.

McCance, R. A.; Widdowson, E. M. (1946) *Spec. Rep. Ser. med. Res. Coun.* No. 235, 2nd ed.

McCarrick, R. B. (1966) *Int. Grassld Congr. 10, Helsinki* p. 575.

McCarrick, R. B. (1967) *Occ. Symp. Br. Grassld Soc.* No. 3, p. 121.

(S) McClelland, T. H.; Forbes, T. J. (1971) *Anim. Prod.* **13**, 643.

(S) McClelland, T. H.; Forbes, T. J. (1973) *Anim. Prod.* **16**, 165.

McClymont, G. L.; Wynne, K. N.; Briggs, P. K.; Franklin, M. C. (1957) *Aust. J. agric. Res.* **8**, 83.

McConnell, G. R.; Jagusch, K. T. (1972) *Proc. N.Z. Soc. Anim. Prod.* **32**, 50.

McCullough, M. E. (1974) *Optimum feeding of dairy animals for meat and milk,* 2nd ed. Athens, Ga: Univ. Georgia Press.

(S) McCullough, T. A. (1970) *J. agric. Sci., Camb.* **75**, 337.

McDonald, I. W. (1954) *Biochem. J.* **56**, 120.

McDonald, I. W.; Hall, R. J. (1957) *Biochem. J.* **67**, 400.

McDowall, F. H.; McGillivray, W. A. (1963) *J. Dairy Res.* **30**, 59.

McGilliard, A. D. (1961) Ph.D. thesis, Michigan State University.

(S) McLeod, D. S.; Wilkins, R. J.; Raymond, W. F. (1970) *J. agric. Sci., Camb.* **75**, 311.

McManus, W. R.; Reid, J. T.; Donaldson, L. E. (1972) *J. agric. Sci., Camb.* **79**, 1.

McMeniman, N. P. (1976) Ph.D. thesis, University of Newcastle upon Tyne.

McMeniman, N. P.; Ben-Ghedalia, D.; Elliott, R. (1976) *Br. J. Nutr.* **36**, 571.

McNaught, K. L. (1938) *N.Z. Jl Sci. Technol. A* **20**, 14.

Mehrez, A. Z. (1976) Ph.D. thesis, University of Aberdeen.

Mehrez, A. Z.; Ørskov, E. R. (1977) *J. agric. Sci., Camb.* **88**, 645.

Menke, K. H.; Sarban, M. S.; Oelschläger, W. (1969) *Landw. Forsch.* **22**, 229.

Mercer, J. R.; Annison, E. F. (1976) In *Protein metabolism and nutrition:* 1st int. symp., Nottingham, 1974, p. 397. (Eds. D. J. A. Cole, K. N. Boorman, P. J. Buttery, D. Lewis, R. J. Neale and H. Swan). London: Butterworths. (*Publs Eur. Ass. Anim. Prod.* **16**).

Mercik, L. (1970) *Zesz. Nauk. wyższ. Szk. roln. Kraków* **26**, 131.

Messer, H. H.; Armstrong, W. D.; Singer, L. (1974) In *Trace element metabolism in animals–2:* 2nd int. symp., Madison, Wis., 1973, p. 425. (Eds. W. G. Hoekstra, J. W. Suttie, H. E. Ganther and W. Mertz). Baltimore, Md: University Park Press.

Meyer, J. H.; Clawson, W. J. (1964) *J. Anim. Sci.* **23**, 214.

Meyer, J. H.; Weir, W. C. (1954) *J. Anim. Sci.* **13**, 443.

Meyer, J. H.; Weir, W. C.; Ittner, N. R.; Smith, J. D. (1955) *J. Anim. Sci.* **14**, 412.

Meyer, J. H.; Hull, J. L.; Weitkamp, W. H.; Bonilla, S. (1965) *J. Anim. Sci.* **24**, 29.

Mikkelsen, T.; Hansen, M. A. (1967) *Nord. VetMed.* **19**, 393.

Mikkelsen, T.; Hansen, M. A. (1968) *Nord. VetMed.* **20**, 402.

Mikkilineni, S. R.; Rousseau, J. E., Jr.; Hall, R. C., Jr.; Frier, H. I.; Eaton, H. D. (1973) *J. Dairy Sci.* **56**, 395.

Milford, R.; Minson, D. J. (1965) *Br. J. Nutr.* **19**, 373.

Milk Marketing Board (1973) *Tech. Div. Rep* No. 331.

Miller, E. L. (1973) *Proc. Nutr. Soc.* **32**, 79.

Miller, H. G.; Brandt, P. M.; Jones, R. C. (1924) *Am. J. Physiol.* **69**, 169.

Miller, H. G.; Yates, W. W.; Jones, R. C.; Brandt, P. M. (1925) *Am. J. Physiol.* **72**, 647.

Miller, J. K. (1967) *J. Nutr.* **93**, 386.

Miller, J. K.; Cragle, R. G. (1965) *J. Dairy Sci.* **48**, 370.

Miller, J. K.; Miller, W. J. (1962) *J. Nutr.* **76**, 467.

Miller, J. K.; Swanson, E. W. (1973) *J. Dairy Sci.* **56**, 378.

Miller, J. K.; Miller, W. J.; Clifton, C. M. (1962) *J. Dairy Sci.* **45**, 1536.

Miller, J. K.; Aschbacher, P. W.; Swanson, E. W. (1964) *J. Dairy Sci.* **47**, 169.

Miller, J. K.; Moss, B. R.; Swanson, E. W.; Aschbacher, P. W.; Cragle, R. G. (1968) *J. Dairy Sci.* **51**, 1831.

Miller, R. W.; Hemken, R. W.; Waldo, D. R.; Moore, L. A. (1969) *J. Dairy Sci.* **52**, 1998.

Miller, R. W.; Hemken, R. W.; Waldo, D. R.; Moore, L. A. (1970) *J. Anim. Sci.* **30**, 984.

Miller, W. J.; Clifton, C. M.; Cameron, N. W. (1963) *J. Dairy Sci.* **46**, 715.

Miller, W. J.; Pitts, W. J.; Clifton, C. M.; Schmittle, S. C. (1964) *J. Dairy Sci.* **47**, 556.

Miller, W. J.; Morton, J. D.; Pitts, W. J.; Clifton, C. M. (1965) *Proc. Soc. exp. Biol. Med.* **118**, 427.

Miller, W. J.; Blackmon, D. M.; Powell, G. W.; Gentry, R. P.; Hiers, J. M. Jr. (1966) *J. Nutr.* **90**, 335.

Miller, W. J.; Martin, Y. G.; Gentry, R. P.; Blackmon, D. M. (1968) *J. Nutr.* **94**, 391.

Miller, W. J.; Blackmon, D. M.; Gentry, R. P.; Pate, F. M. (1970) *J. Nutr.* **100**, 893.

Miller, W. J.; Neathery, M. W.; Gentry, R. P.; Blackmon, D. M.; Lassiter, J. W.; Pate, F. M. (1972) *J. Anim. Sci.* **34**, 460.

Mills, C. F. (1960) *Proc. Nutr. Soc.* **19**, 162.

Mills, C. F. (1974) In *Trace element metabolism in animals–2:* 2nd int. symp., Madison, Wis., 1973, p. 79. (Eds. W. G. Hoekstra, J. W. Suttie, H. E. Ganther and W. Mertz). Baltimore, Md: University Park Press.

Mills, C. F.; Dalgarno, A. C. (1967) *Proc. Nutr. Soc.* **26**, xix.

Mills, C. F.; Dalgarno, A. C. (1972) *Nature, Lond.* **239**, 171.

Mills, C. F.; Dalgarno, A. C.; Williams, R. B.; Quarterman, J. (1967) *Br. J. Nutr.* **21**, 751.

Mills, C. F.; Dalgarno, A. C.; Wenham, G. (1976) *Br. J. Nutr.* **35**, 309.

Milne, J. A. (1974) *J. agric. Sci., Camb.* **83**, 281.

Miltimore, J. E.; Mason, J. L. (1971) *Can. J. Anim. Sci.* **51**, 193.

Ministry of Agriculture and Fisheries (1925). *Report of the Departmental Committee on Rationing of Dairy Cows.* London: HMSO.

Ministry of Agriculture, Fisheries & Food (1972). *Nutrient Standards for Ruminants.* Reports of Working Parties.

Ministry of Agriculture, Fisheries & Food (1975). *Tech. Bull. Minist. Agric. Fish. Fd* No.33, London: HMSO.

Mitchell, G. E., Jr. (1967) *J. Am. vet. med. Ass.* **151**, 430.

Mitchell, H. H. (1962) *Comparative Nutrition of Man and Domestic Animals* vol. 1. New York: Academic Press, (pp. 131 or 225).

Mitchell, H. H.; Hamilton, T. S. (1940) *J. agric. Res.* **61**, 847.

Mitchell, H. H.; Hamilton, T. S. (1941) *J. Nutr.* **22**, 541.

Mitchell, H. H.; McClure, F. J. (1937) *Bull. natn. Res. Coun., Wash.* No. 99.

Mitchell, H. H.; Kammlade, W. G.; Hamilton, T. S. (1926) *Bull.Ill. agric. Exp. Stn.* No. 283.

Mitchell, H. H.; Kammlade, W. G.; Hamilton, T. S. (1928) *Bull. Ill. agric. Exp. Stn.* No. 314.

Mitchell, R. M.; Jagusch, K. T. (1972) *N.Z. Jl agric. Res.* **15**, 788.

Mixner, J. P.; Kramer, D. H.; Szabo, K. T. (1962) *J. Dairy Sci.* **45**, 999.

Modyanov, A. V.; Raetskaya, Yu. I.; Kholmanov, A. M. (1962) *Trudy vses. nauchno-issled. Inst. Zhivot.* **24**, 203.

Moe, P. W.; Flatt, W. P. (1969) *J. Dairy Sci.* **52**, 928.

Moe, P. W.; Reid, J. T.; Tyrrell, H. F. (1965) *J. Dairy Sci.* **48**, 1053.

Moe, P. W.; Tyrrell, H. F.; Flatt, W. P. (1970) In *Energy metabolism of farm animals:* 5th symp., Vitznau, p. 65 (Eds. A. Schürch and C. Wenk). Zurich: Juris Druck. (*Publs Eur. Ass. Anim. Prod.* **13**).

Moe, P. W ; Tyrrell, H. F.; Flatt, W. P. (1971) *J. Dairy Sci.* **54**, 548.

Moe, P. W.; Flatt, W. P.; Tyrrell, H. F. (1972) *J. Dairy Sci* **55**, 945.

Mohrenweiser, H. W.; Donker, J. D. (1968) *J. Dairy Sci.* **51**, 367.

Moir, R. J. (1970) In *Sulfur in Nutrition* p. 165 (Ed. O. H. Muth) Westport, Conn.: AVI Publ. Co.

Moir, R. J. (1974) In *Sulphur in Australasian Agriculture* (Ed. K. D. McLachlan). Sydney: University Press.

Moir, R. J.; Harris, L. E. (1962) *J. Nutr.* **77**, 285.

Moller, K. (1959) *N.Z. vet. J.* **7**, 126.

Møllgaard, H. (1929) *Futterungslehre des Milchviehs.* Hanover: M. & H. Schaper.

Monroe, C. F. (1924) *J. Dairy Sci.* **7**, 58.

Monroe, C. F.; Perkins, A. E. (1925) *J. Dairy Sci.* **8**, 293.

Monteiro, L. S. (1972) *Anim. Prod.* **14**, 263.

Monteith, J. L. (1973) *Principles of Environmental Physics.* London: Edward Arnold.

Monteith, J. L.; Mount, L. E. (Eds) (1974) *Heat loss from animals and man.* London: Butterworths. (*Proc. Easter Sch. agric. Sci. Univ. Nott.* **20**, 1973).

Moody, E. G.; Van Soest, P. J.; McDowell, R. E.; Ford, G. L. (1968) *J. Dairy Sci.* **51**, 969.

Moore, L. A.; Thomas, J. W.; Jacobson, W. C.; Melin, C. G.; Shepherd, J. B. (1948) *J. Dairy Sci.* **31**, 489.

Moore, T. (1957) *Vitamin A.* Amsterdam: Elsevier.

Moore, W. F.; Fontenot, J. P.; Tucker, R. E. (1971) *J. Anim. Sci.* **33**, 502.

Morris, J. G.; Gartner, R. J. W. (1971) *Br. J. Nutr.* **25**, 191.

(S) Morris, J. G.; O'Bryan, M. S. (1965) *J. agric. Sci., Camb.* **64**, 343.

Morrison, F. B. (1956) *Feeds & Feeding,* 22nd ed. Ithaca, N.Y.: Morrison Publ. Co.

Morrison, J. N.; Quarterman, J.; Humphries, W. R. (1977) *J. comp. Path.* **87**, 417.

Moss, B. R.; Madsen, F.; Hansard, S. L.; Gamble, C. T. (1974) *J. Anim. Sci.* **38**, 475.

Moulton, C. R.; Trowbridge, P. F.; Haigh, L. D. (1922) *Res. Bull. Mo. agric. Exp. Stn.* No. 55.

Mowat, D. N. (1963) Ph.D. thesis, Cornell University (*Diss. Abstr.* **24**, 3915).

Moxon, A. L.; Olsen, O. E. (1974) In *Selenium* p. 675 (Eds. R. A. Zingaro and W. C. Cooper). New York: Van Nostrand Reinhold.

Mukherjee, R.; Mitchell, H. H. (1951) *J. Anim. Sci.* **10**, 149.

Munro, I. B. (1957) *Vet. Rec.* **69**, 125.

Munsell, H. E.; DeVaney, G. M.; Kennedy, M. H. (1936) *Tech. Bull. U.S. Dep. Agric.* No. 534.

Murty, V. N. (1957) *J. scient. ind. Res.* **16**(c), 121.

Muth, O. H.; Oldfield, J. E.; Remmert, L. F.; Schubert, J. R. (1958) *Science, N.Y.* **128**, 1090.

Myers, G. S., Jr.; Eaton, H. D.; Rousseau, J. E., Jr. (1959) *J. Anim. Sci.* **18**, 288.

Nakanishi, T.; Tokita, F. (1957) *Tohoku J. agric. Res.* **8**, 155.

National Research Council (1958) *Nutrient Requirements of Sheep,* 4th edn. Washington, D.C.: National Academy of Sciences.

National Research Council (1970) *Nutrient Requirements of Beef Cattle,* 5th edn. Washington, D.C.: National Academy of Sciences.

National Research Council (1971) *Nutrient Requirements of Dairy Cattle,* 5th edn. Washington D.C.: National Academy of Sciences.

National Research Council (1974) *Nutrients and toxic substances in water for livestock and poultry.* Washington, D.C.: National Academy of Sciences.

National Research Council (1975) *Nutrient Requirements of Sheep,* 5th ed. Washington, D.C.: National Academy of Sciences.

Needham, J. (1931) *Chemical Embryology* Vol. 3. Cambridge: University Press.

Nehring, K.; Schiemann, R. (1956) In *Festschrift Anlässlich des 100 Jährigen Bestehens der Landwirtschaftlichen Versuchsstation Leipzig-Möckern* (ed. K. Nehring).

Nehring, K.; Schiemann, R. (1966) In *Vergleichende Ernährungslehre des Menschen und seiner Haustiere,* p. 581. (Ed. A. Hock). Jena: Gustav Fischer.

Nehring, K.; Schiemann, R.; Hoffmann, L.; Jentsch, W. (1961) *Arch. Tierernähr.* **11**, 359.

Nel, J. W.; Moir, R. J. (1974) *S. Afr. J. Anim. Sci.* **4**, 1.

Nelson, B. D.; Ellzey, H. D.; Morgan, E. B.; Allen, M. (1968) *J. Dairy Sci.* **51**, 1796.

Newlander, J. A.; Ellenberger, H. B.; Jones, C. H. (1936) *Bull. Vt agric. Exp. Stn* No. 406.

Newton, G. L.; Fontenot, J. P.; Tucker, R. E.; Polan, C. E. (1972) *J. Anim. Sci.* **35**, 440.

Newton, G. L.; Barrick, E. R.; Harvey, R. W.; Wise, M. B. (1974) *J. Anim. Sci.* **38**, 449.

Nicholson, J. W. G.; Sutton, J. D. (1969) *Br. J. Nutr.* **23**, 585.

Niedermeier, R. P.; Allen, N. N.; Lance, R. D.; Rupnow, E. H.; Bray, R. W. (1959) *J. Anim. Sci.* **18**, 726.

Nimrick, K.; Hatfield, E. E.; Kaminski, J.; Owens, F. N. (1970a) *J. Nutr.* **100**, 1293.

Nimrick, K.; Hatfield, E. E.; Kaminski, J.; Owens, F. N. (1970b) *J. Nutr.* **100**, 1301.

Nisbet, D. I.; Butler, E. J.; Smith, B. S. W.; Robertson, J. M.; Bannatyne, C. C. (1966) *J. Comp. Path.* **76**, 159.

Noble, R. C.; Moore, J. H. (1971) *Proc. Nutr. Soc.* **30**, 61A.

Noble, R. C.; Steele, W.; Moore, J. H. (1971) *Br. J. Nutr.* **26**, 97.

Noguchi, T.; Cantor, A. H.; Scott, M. L. (1973) *J. Nutr.* **103**, 1502.

Nolan, J. V.; Leng, R. A. (1972) *Br. J. Nutr.* **27**, 177.

Nolan, J. V.; Norton, B. W.; Leng, R. A. (1973) *Proc. Nutr. Soc.* **32**, 93.

Norton, B. W.; Walker, D. M. (1971) *Br. J. Nutr.* **26**, 7.

Norton, B. W.; Jagusch, K. T.; Walker, D. M. (1970) *J. agric. Sci., Camb.* **75**, 287.

O'Donovan, W. M.; Elliott, R. C. (1971a) *Rhod. J. agric. Res.* **9**, 65.

O'Donovan, W. M.; Elliott, R. C. (1971b) *Rhod. J. agric. Res.* **9**, 77.

Odynets, R. N.; Asanbekov, O. A. (1970) *Mikroélementy v zhivotnovodste i rastenievodstve, p. 29.* Frunze, Kirgizian S.S.R.: Izdatelstvo 'Ilim'.

Oh, S. H.; Sunde, R. A.; Pope, A. L.; Hoekstra, W. G. (1976) *J. Anim. Sci.* **42**, 977.

O'Hara, P. J.; Fraser, A. J. (1975) *N.Z. vet. J.* **23**, 45.

Oksanen, H. E. (1965) *Acta vet. scand.* **6**, Suppl. 2.

Olson, E. B. Jr; DeLuca, H. F. (1973) *Wld Rev. Nutr. Diet.* **17**, 164.

Oltjen, R. R.; Slyter, L. L.; Kozak, A. S.; Williams, E. E., Jr. (1968) *J. Nutr.* **94**, 193.

Onions, W. J. (1962) *Wool,* London: Benn.

Orr, J. B.; Leitch, I. (1929) *Spec. Rep. Ser. med. Res. Coun.* No. 123.

Ørskov, E. R. (1970) *Proc. Nutr. Conf. Feed Mfrs. Univ. Nott.* **4**, 20. (Eds. H. Swan & D. Lewis). London: Churchill.

Ørskov, E. R. (1975) *Tech. Note Scott. agric. Coll.* No. 7.

Ørskov, E. R. (1976) In *Protein metabolism and nutrition:* 1st int. symp., Nottingham, 1974, p. 457. (Eds. D. J. A. Cole, K. N. Boorman, P. J. Buttery, D. Lewis, R. J. Neale & H. Swan). London: Butterworths. (*Publs. Eur. Ass. Anim. Prod.* **16**).

Ørskov, E. R.; Allen, D. M. (1966a) *Br. J. Nutr.* **20**, 295.

Ørskov, E. R.; Allen, D. M. (1966b) *Br. J. Nutr.* **20**, 509.

Ørskov, E. R.; Allen, D. M. (1966c) *Br. J. Nutr.* **20**, 519.

Ørskov, E. R.; Fraser, C. (1969) *J. agric. Sci., Camb.* **73**, 469.

Ørskov, E. R.; Fraser, C. (1973) *Proc. Nutr. Soc.* **32**, 68A.

Ørskov, E. R.; Fraser, C. (1975) *Br. J. Nutr.* **34**, 493.

Ørskov, E. R.; Mehrez, A. Z. (1977) *Proc. Nutr. Soc.* **36**, 78A.

Ørskov, E. R.; Hovell, F. D.; Allen, D. M. (1966) *Br. J. Nutr.* **20**, 307.

Ørskov, E. R.; Flatt, W. P.; Moe, P. W. (1968) *J. Dairy Sci.* **51**, 1429.

Ørskov, E. R.; Flatt, W. P.; Moe, P. W.; Munson, A. W.; Hemken, R. W.; Katz, I. (1969a) *Br. J. Nutr.* **23**, 443.

Ørskov, E. R.; Fraser, C.; Kay, R. N. B. (1969b) *Br. J. Nutr.* **23**, 217.

Ørskov, E. R.; Fraser, C.; Corse, E. L. (1970) *Br. J. Nutr.* **24**, 803.

Ørskov, E. R.; Fraser, C.; Gill, J. C.; Corse, E. L. (1971a) *Anim. Prod.* **13**, 485.

Ørskov, E. R.; McDonald, I.; Fraser, C.; Corse, E. L. (1971b) *J. agric. Sci., Camb.* **77**, 351.

Ørskov, E. R.; Fraser, C.; McDonald, I. (1971b) *Br. J. Nutr.* **25**, 243.

Ørskov, E. R.; Fraser, C.; McDonald, I. (1972) *Br. J. Nutr.* **27**, 491.

Ørskov, E. R.; Fraser, C.; McHattie, I. (1974a) *Anim. Prod.* **18**, 85.

Ørskov, E. R.; Fraser, C.; McDonald, I.; Smart, R. I. (1974b) *Br. J. Nutr.* **31**, 89.

Osińska, Z.; Ziołecka, A. (1972) *Anim. Prod.* **14**, 119.

Osis, D.; Kramer, L.; Wiatrowski, E.; Spencer,H. (1972) *Am. J. clin. Nutr.* **25**, 582.

Osuji, P. O. (1973) Ph.D. thesis,University of Aberdeen.

Osuji, P. O.; Gordon, J. G.; Webster, A. J. F. (1975) *Br. J. Nutr.* **34**, 59.

Ott, E. A.; Smith, W. H.; Stob, M.; Beeson, W. M. (1964) *J. Nutr.* **82**, 41.

Ott, E. A.; Smith, W. H.; Stob, M.; Parker, H. E.; Harrington, R. B.; Beeson, W. M. (1965) *J. Nutr.* **87**, 459.

Otto, J. S. (1938) *Onderstepoort J. vet. Sci.* **10**, 281.

Overman, O. R.; Gaines, W. L. (1933) *J. agric. Res.* **46**, 1109.

Overman, O. R.; Gaines, W. L. (1948) *J. Anim. Sci.* **7**, 55.

Overman, O. R.; Sanmann, F. P.; Wright, K. E. (1929) *Bull. Ill. agric. Exp. Stn* No. 325.

Owen, E. C. (1948) *Biochem. J.* **43**, 243.

Owen, J. B.; Ingleton, J. W. (1963) *J. agric. Sci., Camb.* **61**, 329.

Owen, J. B.; Miller, E. L.; Bridge, P. S. (1968) *J. agric. Sci., Camb.* **70**, 223.

Owen, J. B.; Miller, E. L.; Bridge, P. S. (1971) *J. agric. Sci., Camb.* **77**, 195.

Page, H. M.; Erwin, E. S.; Varnell, T. R.; Roubicek, C. B. (1958) *Am. J. Physiol.* **194**, 313.

Paladines, O. L.; Reid, J. T.; Bensadoun, A.; Van Niekerk, B. D. H. (1964) *J. Nutr.* **82**, 145.

Palmquist, D. L.; Mattos, W.; Stone, R. L. (1977) *Lipids* **12**, 235.

Paloheimo, L. (1944) *Suom. maatal. Seur. Julk.* No. 56, 5.

Paloheimo, L.; Mäkelä, A. (1959) *Suom. maatal. Seur. Julk.* No. 94, 15.

Panaretto, B. A. (1964) *Aust. J. agric. Res.* **15**, 771.

Papasteriadis, A. A. (1973) *Epistemonike Epeteris Kteniatrikes Skholes (Scient. Yb. vet. Fac.) Thessaloniki* **14**, 167.

Paquay, R.; Lomba, F.; Lousse, A.; Bienfet, V. (1968) *J. agric. Sci., Camb.* **71**, 173.

Paquay, R.; Lomba, F.; Lousse, A.; Bienfet, V. (1969a) *J. agric. Sci., Camb.* **73**, 223.

Paquay, R.; Lomba, F.; Lousse, A.; Bienfet, V. (1969b) *J. agric. Sci., Camb.* **73**, 445.

Paquay, R.; de Baere, R.; Lousse, A. (1970a) *J. agric. Sci., Camb.* **74**, 423.

Paquay, R.; de Baere, R.; Lousse, A. (1970b) *J. agric. Sci., Camb.* **75**, 251.

(S) Paquay, R.; de Baere, R.; Lousse, A. (1972) *Br. J. Nutr.* **27**, 27.

(S) Paquay, R.; Godeau, J. M.; de Baere, R.; Lousse, A. (1973) *J. Dairy Res.* **40**, 93.

(S) Parkins, J. J.; Fraser, J.; Ritchie, N. S.; Hemingway, R. G. (1974) *Anim. Prod.* **19**, 321.

Patle, B. R.; Mudgal, V. D. (1976) *Indian J. Anim. Sci.* **46**, 215.

Patle, B. R.; Mudgal, V. D. (1977) *Br. J. Nutr.* **37**, 23.

Paulson, G. D.; Baumann, C. A.; Pope, A. L. (1966) *J. Anim. Sci.* **25**, 1054.

Paulson, G. D.; Broderick, G. A.; Baumann, C. A.; Pope, A. L. (1968) *J. Anim. Sci.* **27**, 195.

Payne, J. M. (1963) *Vet. Rec.* **75**, 848.

Payne, J. M.; Manston, R. (1967) *Vet. Rec.* **81**, 214.

Peart, J. N. (1967) *J. agric. Sci., Camb.* **68**, 365.

Peart, J. N. (1968) *J. agric. Sci., Camb.* **70**, 87.

Peart, J. N. (1970) *J. agric. Sci., Camb.* **75**, 459.

Peart, J. N.; Edwards, R. A.; Donaldson, E. (1972) *J. agric. Sci., Camb.* **79**, 303.

Pedersen, N. D.; Whanger, P. D.; Weswig, P. H.; Muth, O. H. (1972) *Bioinorg. Chem.* **2**, 33.

Peirce, A. W. (1934) *Aust. J. exp. Biol. med. Sci.* **12**, 7.

Peirce, A. W. (1936) *Aust. J. exp. Biol. med. Sci.* **14**, 187.

Peirce, A. W. (1945) *Aust. J. exp. Biol. med. Sci.* **23**, 295.

Peirce, A. W. (1954) *Aust. J. agric. Res.* **5**, 470.

Peirce, A. W. (1959) *Aust. J. agric. Res.* **10**, 725.

Peirce, A. W. (1960) *Aust. J. agric. Res.* **11**, 548.

Peirce, A. W. (1962) *Aust. J. agric. Res.* **13**, 479.

Penning, P. D.; Bradfield, P. G. E.; Treacher, T. T. (1971) *Anim. Prod.* **13**, 365.

Penning, P. D.; Penning, I. M.; Treacher, T. T. (1977) *J. Agric. Sci., Camb.* **88**, 579.

Perrin, D. R. (1958) *J. Dairy Res.* **25**, 215.

Perry, T. W.; Beeson, W. M.; Smith, W. H.; Mohler, M. T. (1976) *J. Anim. Sci.* **42**, 192.

Pettyjohn, J. D.; Everett, J. P., Jr.; Mochrie, R. D. (1963) *J. Dairy Sci.* **46**, 710.

Pfeffer, E.; Kaufmann, W.; Dirksen, G. (1972) *Z. Tierphysiol. Tierernähr. Futtermittelk.* Suppl. 1, p. 22.

Phillips, D. S. M. (1968) *N.Z. Jl agric. Res.* **11**, 267.

Phillips, G. D.; Sundaram, S. K. (1966) *J. Physiol., Lond.* **184**, 889.

Phillips, P. H.; Greenwood, D. A.; Hobbs, C. S.; Huffman, C. F.; Spencer, G. R. (1960) *Publs. natn. Res. Counc., Wash.* No. 824.

Phillips, W. E. J.; Evans, E. V.; Murray, T. K.; Campbell, J. A.; Branion, H. D.; Emslie, A. R. G. (1966) *Publs. Dep. Agric. Can.* No. 1238.

Phipps, R. H. (1975) *J. Br. Grassld. Soc.* **30**, 45.

Piatkowski, B. (1966) *Arch. Tierz.* **9**, 375.

Pierson, R. E. (1966) *J. Am. vet. med. Ass.* **149**, 1279.

Pilgrim, A. F.; Gray, F. V.; Weller, R. A.; Belling, C. B. (1970) *Br. J. Nutr.* **24**, 589.

Pinot, R.; Teissier, J. H. (1965) *Annls Zootech.* **14**, 261.

Polan, C. E.; Miller, C. N.; McGilliard, M. L. (1976) *J. Dairy Sci.* **59**, 1910.

Popovici, D.; Jurenkova, G.; Raitaru, M.; Chirila, E. (1973) *Lucr. ştiinţ. Inst. Cerc. Crest. Taurin,* **1**, 69.

Potter, B. J. (1966) *J. Physiol., Lond.* **184**, 605.

Potter, B. J. (1968) *J. Physiol., Lond.* **194**, 435.

Poulton, S. G.; Ashton, W. M. (1970) *J. agric. Sci., Camb.* **75**, 245.

Pradhan, K.; Hemken, R. W. (1968) *J. Dairy Sci.* **51**, 1377.

Premachandra, B. N.; Pipes, G. W.; Turner, C. W. (1958) *J. Dairy Sci.* **41**, 1609.

Preston, R. L.; Pfander, W. H. (1961) *J. Anim. Sci.* **20**, 947.

Preston, R. L.; Pfander, W. H. (1963) *J. Anim. Sci.* **22**, 844.

Preston, R. L.; Pfander, W. H. (1964) *J. Nutr.* **83**, 369.

Preston, R. L.; Schnakenberg, D. D.; Pfander, W. H. (1965) *J. Nutr.* **86**, 281.

Preston, T. R.; Willis, M. B. (1974) *Intensive Beef Production*, 2nd ed. Oxford: Pergamon Press.

Preston, T. R.; Whitelaw, F. G.; MacLeod, N. A. (1963) *Anim. Prod.* **5**, 147.

Preston, T. R.; Whitelaw, F. G.; MacLeod, N. A.; Philip, E. B. (1965) *Anim. Prod.* **7**, 53.

Price, J.; Suttle, N. F. (1975) *Proc. Nutr. Soc.* **34**, 9A.

Pryor, W. J. (1964) *Res. vet. Sci.* **5**, 123.

Pullar, J. D. (1963) In *Progress in nutrition and allied sciences* p. 187. (Ed. D. P. Cuthbertson). Edinburgh: Oliver & Boyd.

Pullar, J. D.; Webster, A. J. F. (1974) *Br. J. Nutr.* **31**, 377.

Purcell, D. A.; Raven, A. M.; Thompson, R. H. (1971) *Res. vet. Sci.* **12**, 598.

Purser, D. B.; Buechler, S. M. (1966) *J. Dairy Sci.* **49**, 81.

Purves, H. D. (1964) In *The Thyroid Gland* vol. 2, p. 1. (Eds. R. Pitt-Rivers & W. R. Trotter). London: Butterworths.

Quarterman, J. (1966) *Vet. Rec.* **78**, 855.

Quarterman, J.; Mills, C. F. (1964) *Proc. Nutr. Soc.* **23**, x.

Quarterman, J.; Dalgarno, A. C.; McDonald, I. (1961) *Proc. Nutr. Soc.* **20**, xxviii.

Quarterman, J.; Dalgarno, A. C.; Adams, A. (1964) *Br. J. Nutr.* **18**, 79.

Quarterman, J.; Morrison, J. N.; Humphries, W. R.; Mills, C. F. (1977) *J. comp. Path.* **87**, 405.

Ragsdale, A. C.; Thompson, H. J.; Worstell, D. M.; Brody, S. (1950) *Res. Bull. Mo. agric. Exp. Stn* No. 460.

Ragsdale, A. C.; Thompson, H. J.; Worstell, D. M.; Brody, S. (1951) *Res. Bull. Mo. agric. Exp. Stn* No. 471.

Ragsdale, A. C.; Thompson, H. J.; Worstell, D. M.; Brody, S. (1953) *Res. Bull. Mo. agric. Exp. Stn* No. 521.

Ramberg, C. F., Jr; Mayer, G. P.; Kronfeld, D. S.; Phang, J. M.; Berman, M. (1970a) *Am. J. Physiol.* **219**, 1166.

Ramberg, C. F., Jr; Phang, J. M.; Mayer, G. P.; Norberg, A. I.; Kronfeld, D. S. (1970b) *J. Nutr.* **100**, 981.

Rattray, P. V.; Garrett, W. N.; East, N. E.; Hinman, N. (1973a). *J. Anim. Sci.* **37**, 853.

Rattray, P. V.; Garrett, W. N.; Hinman, N.; Garcia, I.; Castillo, J. (1973b) *J. Anim. Sci.* **36**, 115.

Rattray, P. V.; Garrett, W. N.; Meyer, H. H.; Bradford, G. E.; East, N. E.; Hinman, N. (1973c) *J. Anim. Sci.* **37**, 892.

Rattray, P. V.; Garrett, W. N.; East, N. E.; Hinman, N. (1974) *J. Anim. Sci.* **38**, 613.

Rattray, P. V.; Joyce, J. P. (1976) *N.Z. Jl agric. Res.* **19**, 299.

Rattray, P. V.; Garrett, W. N.; East, N. E.; Hinman, N. (1974a) *J. Anim. Sci.* **38**, 383.

Rattray, P. V.; Garrett, W. N.; Hinman, N.; East, N. E. (1974b) *J. Anim. Sci.* **38**, 378.

Raven, A. M. (1972) *J. agric. Sci., Camb.* **79**, 99.

Raven, A. M.; Robinson, K. L. (1964) *Rec. agric. Res.* **13**, 81.

Ray, S. N. (1942) *Indian J. vet. Sci.* **12**, 204.

Reid, G. W.; Greenhalgh, J. F. D.; Aitken, J. N. (1972) *J. agric. Sci., Camb.* **78**, 491.

Reid, J. T.; Robb, J. (1971) *J. Dairy Sci.* **54**, 553.

Reid, J. T.; Wellington, G. H.; Dunn, H. O. (1955) *J. Dairy Sci.* **38**, 1344.

Reid, J. T.; Bensadoun, A.; Bull, L. S.; Burton, J. H.; Gleeson, P. A.; Han, I. K.; Joo, Y. D.; Johnson, D. E.; McManus, W. R.; Paladines, O. L.; Stroud, J. W.; Tyrrell, H. F.; Van Niekerk, B. D. H.; Wellington, G. W. (1968) *Publs natn. Res. Coun., Wash.* No. 1598, p. 19.

Reinart, A.; Nesbitt, J. M. (1956) *Int. Dairy Congr. 14*, Rome, I, part 2, p. 911.

Reis, P. J. (1970) *Aust. J. biol. Sci.* **23**, 441.

Rendig, V. V.; Weir, W. C. (1957) *J. Anim. Sci.* **16**, 451.

Renkema, J. A.; Senshu, T.; Gaillard, B. D. E.; Brouwer, E. (1962) *Nature, Lond.* **195**, 389.

Ribeiro, J. M. de C. R. (1976) Ph.D. thesis, University of Aberdeen.

Ribeiro, J. M. de C. R.; Brockway, J. M.; Webster, A. J. F. (1977) *Anim. Prod.* **25**, 107.

Richards, J. I.; Bridge, P. S.; Spratling, F. R.; Abrams, J. T. (1970) *Int. J. Vitam. Nutr. Res.* **40**, 567.

Ricketts, R. E.; Weinman, D. E.; Campbell, J. R.; Tumbleson, M. E. (1970) *Am. J. vet. Res.* **31**, 1023.

Riddell, W. H.; Hughes, J. S.; Fitch, J. B. (1934) *Tech. Bull. Kans. St. Coll. Agric. appl. Sci.* No. 36.

Ritchie, N. S.; Hemingway, R. G. (1963) *J. agric. Sci., Camb.* **60**, 305.

Ritzman, E. G.; Benedict, F. G. (1938) *Publs Carnegie Instn* No. 494.

Ritzman, E. G.; Colovos, N. F. (1943) *Tech. Bull. New Hamps. agric. Exp. Stn* No. 80.

Rizaev, Z. N. (1965a) *Ovtsevodstvo* No. 8, p. 28.

Rizaev, Z. N. (1965b) *Ovtsevodstvo* No. 11, p. 33.

Robelin, J.; Geay, Y. (1976) In *Energy metabolism of farm animals*: 7th symp., Vichy, p. 213. (Ed. M. Vermorel). Clermont-Ferrand: de Bussac. (*Publs Eur. Ass. Anim. Prod.* **19**)

Roberts, W. K.; St. Omer, V. V. E. (1965) *J. Anim. Sci.* **24**, 902.

Roberts, W. K.; Findlay, G. M.; Stringham, E. W. (1965) *Can. J. comp. Med.* **29**, 43.

Robertson, H. A.; Falconer, I. R. (1961a) *J. Endocr.* **21**, 411.

Robertson, H. A.; Falconer, I. R. (1961b) *J. Endocr.* **22**, 133.

(S) Robertson, I. S.; Paver, H.; Wilson, J. C. (1970) *J. agric. Sci., Camb.* **74**, 299.

(S) Robinson, J. J.; Forbes, T. J. (1966) *Br. J. Nutr.* **20**, 263.

(S) Robinson, J. J.; Forbes, T. J. (1967) *Br. J. Nutr.* **21**, 879.

(S) Robinson, J. J.; Forbes, T. J. (1968) *Anim. Prod.* **10**, 297.

(S) Robinson, J. J.; Forbes, T. J. (1970a) *Anim. Prod.* **12**, 95.

(S) Robinson, J. J.; Forbes, T. J. (1970b) *Anim. Prod.* **12**, 601.

(S) Robinson, J. J.; Forbes, T. J. (1970c) *J. agric. Sci., Camb.* **74**, 415.

(S) Robinson, J. J.; Fraser, C.; Corse, E. L.; Gill, J. C. (1970) *J. agric. Sci., Camb.* **75**, 403.

(S) Robinson, J. J.; Fraser, C.; Gill, J. C.; McHattie, I. (1974) *Anim. Prod.* **19**, 331.

Rodel, M. G. W. (1971) *Rhodesia agric. J.* **68**, 109.

Rodel, M. G. W. (1972) *Rhodesia agric. J.* **69**, 59.

Roffler, R. E.; Satter, L. D. (1975) *J. Dairy Sci.* **58**, 1889.

Rojas, M. A.; Dyer, I. A.; Cassatt, W. A. (1965) *J. Anim. Sci.* **24**, 664.

Ronning, M.; Laben, R. C. (1966) *J. Dairy Sci.* **49**, 1080.

Ronning, M.; Berousek, E. R.; Gallup, W. D.; Griffith, J. L. (1959) *Tech. Bull. Okla. St. Univ.* No. T-16.

Rook, J. A. F.; Balch, C. C. (1958) *J. agric. Sci., Camb.* **51**, 199.

Rook, J. A. F.; Balch, C. C. (1962) *J. agric. Sci., Camb.* **58**, 103.

Rook, J. A. F.; Campling, R. C. (1962) *J. agric. Sci., Camb.* **59**, 225.

Rook, J. A. F.; Line, C. (1962) *Int. Dairy Congr. 16, Copenhagen* A, 57.

Rook, J. A. F.; Storry, J. E. (1962) *Proc. Nutr. Soc.* **21**, xl.

Rook, J. A. F.; Balch, C. C.; Line, C. (1958) *J. agric. Sci., Camb.* **51**, 189.

Rook, J. A. F.; Campling, R. C.; Johnson, V. W. (1964) *J. agric. Sci., Camb.* **62**, 273.

Ross, D. B. (1964) *Vet. Rec.* **76**, 875.

Ross, D. B. (1966) *Br. vet. J.* **122**, 279.

Roubicek, C. B. (1969) In *Animal Growth & Nutrition*, p. 353. (Eds. E. S. E. Hafez and I. A. Dyer). London: Baillière & Tindall.

Rousseau, J. E., Jr; Eaton, H. D.; Helmboldt, C. F.; Jungherr, E. L.; Robrish, S. A.; Beall, G.; Moore, L. A. (1954) *J. Dairy Sci.* **37**, 889.

Rowland, A. C. (1970) *Anim. Prod.* **12**, 291.

Rowland, S. J.; Rook, J. A. F. (1949) *Rep. natn. Inst. Res. Dairy* p. 42.

Roy, J. H. B. (1967) *Sb. vys. Šk. zeměd. Brne* **36**, 325.

Roy, J. H. B. (1969) *Proc. Nutr. Soc.* **28**, 160.

Roy, J. H. B. (1975) In *Perinatal ill-health in calves*: seminar, Compton, Berks, p. 125. (Ed. J. M. Rutter). Commission of European Communities.

Roy, J. H. B.; Gaston, H. J.; Shillam, K. W. G.; Thompson, S. Y.; Stobo, I. J. F.; Greatorex, J. C. (1964) *Br. J. Nutr.* **18**, 467,

Roy, J. H. B.; Stobo, I. J. F.; Gaston, H. J. (1970a) *Br. J. Nutr.* **24**, 459.

Roy, J. H. B.; Stobo, I. J. F.; Gaston, H. J.; Greatorex, J. C. (1970b) *Br. J. Nutr.* **24**, 441.

Roy, J. H. B.; Stobo, I. J. F.; Gaston, H. J.; Ganderton, P.; Shotton, S. M.; Ostler, D. C. (1971) *Br. J. Nutr.* **26**, 363.

Roy, J. H. B.; Stobo, I. J. F.; Ganderton, P.; Shotton, S. M. (1973a) *Anim. Prod.* **16**, 215.

Roy, J. H. B.; Stobo, I. J. F.; Gaston, H. J.; Shotton, S. M.; Ganderton, P. (1973b) *Anim. Prod.* **17**, 97.

Roy, J. H. B.; Stobo, I. J. F.; Gaston, H. J.; Shotton, S. M.; Ganderton, P. (1973c) *Anim. Prod.* **17**, 109.

Roy, J. H. B.; Stobo, I. J. F.; Shotton, S. M.; Ganderton, P.; Gillies, C. M. *Br. J. Nutr.* **38**, 167.

Rubner, M. (1902) *Die Gesetze des Energieverbrauchs bei der Ernährung.* Berlin: Paul Parey.

Rupel, I. W.; Bohstedt, G.; Hart, E. B. (1933) *Res. Bull. Wis. agric. Exp. Stn* No. 115.

Russel, A. J. F.; Gunn, R. G.; Doney, J. M. (1968) *Anim. Prod.* **10**, 43.

Ryder, M. L.; Stephenson, S. K. (1968) *Wool growth*, p. 33. New York: Academic Press.

Salem, M. A. I. (1970) *SchrReihe Max-Planck-Inst. Tierzucht* No. 48.

Salter, D. N.; Smith. R. H. (1977) *Proc. Nutr. Soc.* **36**, 54A.

Sansom, B. F.; Allen, W. M.; Davies, D. C.; Hoare, M. N.; Stenton, J. R.; Vagg, M. J. (1976a) *Vet. Rec.* **99**, 310.

Sansom, B. F.; Gibbons, R. A.; Dixon, S. N.; Russell, A. M.; Symonds, H. W. (1976b) In *Nuclear techniques in animal production and health:* int. symp. Vienna, p. 179. Vienna: IAEA.

Saraswat, R. C.; Arora, S. P. (1972) *Indian J. Anim. Sci.* **42**, 358.

Sasser, L. B.; Ward, G. M.; Johnson, J. E. (1966) *J. Dairy Sci.* **49**, 893.

Scheunert, A.; Klein, W.; Steuber, M. (1922) *Biochem. Z.* **133**, 137.

Schiemann, R. (1970) *Wiss. Z. Humboldt-Univ. Berl.* **19**, 35.

Schiemann, R.; Jentsch, W.; Wittenburg, H. (1971) *Arch. Tierernähr.* **21**, 223.

Schiemann, R.; Henseler, G.; Jentsch, W.; Wittenburg, H. (1974) *Arch. Tierernähr.* **24**, 105.

Schoeman, E. A.; de Wet, P. J.; Burger, W. J. (1972) *Agroanimalia* **4**, 35.

Schroder, J. D.; Hansard, S. L. (1958) *J. Anim. Sci.* **17**, 343.

Schubert, J. R.; Muth, O. H.; Oldfield, J. E.; Remmert, L. F. (1961) *Fedn Proc. Fedn Am. Socs exp. Biol.* **20**, 689.

Schulz, E.; Oslage, H. J.; Daenicke, R. (1974) *Z. Tierphysiol. Tierernähr. Futtermittelk.* Suppl. No. 4.

Schwab, C. G.; Broderick, G. A.; Satter, L. D. (1971) *J. Dairy Sci.* **54**, 788.

Schwark, H. J.; Kunert, G. (1971) *Arch. Tierz.* **14**, 359.

Schwarz, K. (1974) In *Trace element metabolism in animals–2:* 2nd int. symp., Madison, Wis., 1973, p. 355. (Eds. W. G. Hoekstra, J. W. Suttie, H. E. Ganther and W. Mertz). Baltimore, Md: University Park Press.

Schwarz, W. A.; Kirchgessner, M. (1975a) *Arch. Tierernähr.* **25**, 597.

Schwarz, W. A.; Kirchgessner, M. (1975b) *Z. Tierphysiol. Tierernähr. Futtermittelk.* **35**, 1.

Searle, T. W. (1970a) *J. agric. Sci., Camb.* **74**, 357.

Searle, T. W. (1970b) *J. agric. Sci., Camb.* **75**, 497.

Searle, T. W. (1972) Ph.D. thesis, University of Sydney.

Searle, T. W.; Graham, N. McC. (1972) *Aust. J. agric. Res.* **23**, 339.

Searle, T. W.; Graham, N. McC.; O'Callaghan, M. (1972) *J. agric. Sci., Camb.* **79**, 371.

Seawright, D. (1966) *Vet. Rec.* **78**, 320.

Šebela, F. (1972) *Sb. čsl. Akad. zeměd. Věd E* **17**, 243.

Sewell, R. F.; McDowell, L. J. (1966) *J. Nutr.* **89**, 64.

Shand, A.; Lewis, G. (1957) *Vet. Rec.* **69**, 618.

Sharma, H. R.; Ingalls, J. R.; Parker, R. J. (1974) *Can. J. Anim. Sci.* **54**, 305.

Shatalina, A. S.; Veretennikova, V. M. (1968) *Nauch. Trudy tashkent. gos. Univ.* No. 334, p. 43.

Sheppard, A. J.; Johnson, B. C. (1957) *J. Nutr.* **61**, 195.

Shillam, K. W. G.; Roy, J. H. B. (1963) *Br. J. Nutr.* **17**, 171.

Shreeve, J. E.; Edwin, E. E. (1974) *Vet. Rec.* **94**, 330.

Simesen, M. G.; Møller, T. (1969) *Medlemsbl. danske Dyrlaegeforen.* **52**, 5.

Simesen, M. G.; Lunaas, T.; Rogers, T. A.; Luick, J. R. (1962) *Acta vet. scand.* **3**, 175.

Sinclair, D. P.; Andrews, E. D. (1954) *N.Z. vet. J.* **2**, 72.

Sinclair, D. P.; Andrews, E. D. (1958) *N.Z. vet. J.* **6**, 87.

Sinclair, D.P.; Andrews, E. D. (1961) *N.Z. vet. J.* **9**, 96.

Singh, J. R.; Talapatra, S. K. (1961) *Indian J. vet. Sci.* **31**, 223.

Singh, M.; Mahadevan, V. (1968) *Indian vet. J.* **45**, 659.

Singh, M.; Mahadevan, V. (1970) *Anim. Prod.* **12**, 185.

Singh, O. N.; Henneman, H. A.; Reineke, E. P. (1956) *J. Anim. Sci.* **15**, 625.

Sissons, J. W.; Smith, R. H. (1976) *Br. J. Nutr.* **36**, 421.

Sklan, D.; Budowski, P. (1974) *J. Dairy Sci.* **57**, 56.

Sklan, D.; Volcani, R.; Budowski, P. (1971) *J. Dairy Sci.* **54**, 515.

Sklan, D.; Volcani, R.; Budowski, P. (1972) *Br. J. Nutr.* **27**, 365.

Slen, S. B.; Clark, R. D.; Hironoka, R. (1963) *Can. J. Anim. Sci.* **43**, 16.

Smith, B. S. W.; Field, A. C.; Suttle, N. F. (1968) *J. comp. Path.* **78**, 449.

Smith, G. S.; Durdle, W. M.; Zimmerman, J. E.; Neumann, A. L. (1964) *J. Anim. Sci.* **23**, 625.

Smith, J. C., Jr.; McDaniel, E. G.; Fan, F. F.; Halsted, J. A. (1973) *Science, N.Y.* **181**, 954.

Smith, R. C.; Moussa, N. M.; Hawkins, G. E. (1974) *Br. J. Nutr.* 32, 529.

Smith, R. H. (1957) *Biochem. J.* **67**, 472.

Smith, R. H. (1958) *Biochem. J.* **70**, 201.

Smith, R. H. (1959) *Biochem. J.* **71**, 306.

Smith, R. H. (1961) *J. agric. Sci., Camb.* **56**, 343.

Smith, R. H. (1975) In *Digestion and metabolism in the ruminant:* 4th int. symp. Ruminant Physiol., Sydney, 1974, p. 399. (Eds. I. W. McDonald and A. C. I. Warner). Armidale: University of New England Publishing Unit.

Smith, R. H.; McAllan, A. B. (1974) *Br. J. Nutr.* **31**, 27.

Smith, R. H.; Mohamed, O. E. (1977) *Proc. Nutr. Soc.* **36**, 153A.

Smith, R. H.; Sissons, J. W. (1975) *Br. J. Nutr.* **33**, 329.

Smith, R. H.; McAllan, A. B.; Hewitt, D., Lewis, P. E. (1978) *J. agric. Sci., Camb.* **90**, 557.

Smith, R. M.; Marston, H. R. (1970) *Br. J. Nutr.* **24**, 857.

Smith, R. M.; Marston, H. R. (1971) *Br. J. Nutr.* **26**, 41.

Smith, S. E.; Aines, P. D. (1959) *Bull. Cornell agric. Exp. Stn.* No. 938.

Smuts, D. B.; Marais, J. S. C. (1938) *Onderstepoort J. vet. Sci.* **11**, 399.

Smuts, D. B.; Marais, J. S. C. (1939) *Onderstepoort J. vet. Sci.* **13**, 219.

Soar, J. B.; Buttery, P. J.; Lewis, D. (1973) *Proc. Nutr. Soc.* **32**, 77A.

Somers, M.; Gawthorne, J. M. (1969) *Aust. J. exp. Biol. med. Sci.* **47**, 227.

Sørensen, P. H. (1958) *Beretn. Forsøgslab.* No. 302.

Sotola, J. (1930) *J. agric. Res.* **40**, 79.

Spas, A.; Papasteriadis, A.; Agiannidis, A.; Lazaridis, T. (1966) *Rep. vet. Fac. Univ. Thessaloniki* **7**, 167.

Spedding, C. R. W. (1970) *Sheep Production and Grazing Management* 2nd ed., p. 73. London: Ballière, Tindall & Cassell.

Sperling, G. (1951) *Futter Fütter.* No. 13, p. 94.

Spratling, F. R.; Bridge, P. S.; Barnett, K. C.; Abrams, J. T.; Palmer, A. C.; Sharman, I. M. (1965) *Vet. Rec.* **77**, 1532.

Stacy, B. D.; Brook, A. H.; Short, B. F. (1963) *Aust. J. agric. Res.* **14**, 340.

Standish, J. F.; Ammerman, C. B.; Simpson, C. F.; Neal, F. C.; Palmer, A. Z. (1969) *J. Anim. Sci.* **39**, 496.

Standish, J. F.; Ammerman, C. B.; Palmer, A. Z.; Simpson, C. F. (1971) *J. Anim. Sci.* **33**, 171.

Steenbock, H.; Nelson, V.; Hart, E. B. (1915) *Res. Bull. Wis. agric. Exp. Stn.* No. 36.

Stewart, W. L.; Allcroft, R. (1956) *Vet. Rec.* **68**, 723.

Stillings, B. R.; Bratzler, J. W.; Marriott, L. F.; Miller, R. C. (1964) *J. Anim. Sci.* **23**, 1148.

Stobo, I. J. F.; Roy, J. H. B. (1964) *Anim. Prod.* **6**, 253.

(S) Stobo, I. J. F.; Roy, J. H. B. (1973) *Br. J. Nutr.* **30**, 113.

(S) Stobo, I. J. F.; Roy, J. H. B.; Gaston, H. J. (1967a) *Anim. Prod.* **9**, 7.

(S) Stobo, I. J. F.; Roy, J. H. B.; Gaston, H. J. (1967b) *Anim. Prod.* **9**, 23.

(S) Stobo, I. J. F.; Roy, J. H. B.; Gaston, H. J. (1967c) *Anim. Prod.* **9**, 155.

Stocking, V. A.; Brew, J. D. (1920) *Dairyman's League News* Jan. 10.

St. Omer, V. V. E.; Roberts, W. K. (1967) *Can. J. Anim. Sci.* **47**, 39.

Storry, J. E.; Rook, J. A. F. (1963) *J. agric. Sci., Camb.* **61**, 167.

Suttie, J. W. (1969) *J. Air Pollutn Ass.* **19**, 239.

Suttle, N. F. (1968) *Proc. Nutr. Conf. Feed Mfrs Univ. Nott.* **2**, p. 150. (Eds. H. Swan and D. Lewis). London: Churchill.

Suttle, N. F. (1973) *Proc. Nutr. Soc.* **32**, 24A.

Suttle, N. F. (1974a) *Br. J. Nutr.* **32**, 395.

Suttle, N. F. (1974b) *Br. J. Nutr.* **32**, 559.

Suttle, N. F. (1975a) *J. agric. Sci., Camb.* **84**, 255.

Suttle, N. F. (1975b) *Br. J. Nutr.* **34**, 411.

Suttle, N. F. (1977) *Anim. Feed Sci. Technol.* **2**, 235.

Suttle, N. F.; Angus, K. W. (1978) *J. comp. Path.* **88**, 137.

Suttle, N. F.; Field, A. C. (1967) *Br. J. Nutr.* **21**, 819.

Suttle, N. F.; Field, A. C. (1968) *J. comp. Path.* **78**, 351.

Suttle, N. F.; Field, A. C. (1969) *Br. J. Nutr.* **23**, 81.

Suttle, N. F.; Field, A. C. (1974) *Vet. Rec.* **95**, 165.

Suttle, N. F.; McLauchlan, M. (1976) *Proc. Nutr. Soc.* **35**, 22A.

Suttle, N. F.; Price, J. (1976) *Anim. Prod.* **23**, 233.

Suttle, N. F.; Alloway, B. J.; Thornton, I. (1975) *J. agric. Sci., Camb.* **84**, 249.

Sutton, A. L.; Elliot, J. M. (1972) *J. Nutr.* **102**, 1341.

Sutton, J. D.; Smith, R. H.; McAllan, A. B.; Storry, J. E.; Corse, D. A. (1975) *J. agric. Sci., Camb.* **84**, 317.

Suveges, T.; Ratz, F.; Salyi, G. (1971) *Acta vet. hung.* **21**, 383.

Swan, J. B. (1952) *N.Z. vet. J.* **1**, 25.

Swanson, E. W. (1972) *J. Dairy Sci.* **55**, 1763.

Swanson, E. W.; Herman, H. A. (1943) *Res. Bull. Mo agric. Exp. Stn.* No. 372.

Swanson, E. W.; Martin, G. C.; Pardue, F. E.; Gorman, G. M. (1968) *J. Anim. Sci.* **27**, 541.

Swenerton, H.; Hurley, L. S. (1968) *J. Nutr.* **95**, 8.

Swett, W. W.; Matthews, C. A.; Fohrman, M. H. (1948) *Tech. Bull. U.S. Dep. Agric.* No. 964.

Sykes, A. R.; Dingwall, R. A. (1975) *J. agric. Sci., Camb.* **84**, 245.

Sykes, A. R.; Dingwall, R. A. (1976) *J. agric. Sci., Camb.* **86**, 587.

Sykes, A. R.; Field, A. C. (1972a) *J. agric. Sci., Camb.* **78**, 109.

Sykes, A. R.; Field, A. C. (1972b) *J. agric. Sci., Camb.* **78**, 119.

Sykes, A. R.; Field, A. C. (1972c) *J. agric. Sci., Camb.* **78**, 127.

Sykes, J. F. (1955) *Yb. Agric. U.S. Dep. Agric.* p. 14.

Symonds, H. W.; Manston, R.; Payne, J. M.; Sansom, B. F. (1966) *Br. vet. J.* **122**, 196.

Tagari, H. (1969) *Br. J. Nutr.* **23**, 455.

Tagari, H.; Roy, J. H. B. (1969) *Br. J. Nutr.* **23**, 763.

Tagari, H.; Ben Gedalya, D.; Shevach, Y.; Bondi, A. (1971) *J. agric. Sci., Camb.* **77**, 413.

Tait, R. M.; Krishnamurti, C. R.; Gilchrist, E. W.; MacDonald, K. (1971) *Can. vet. J.* **12**, 73.

Tamminga, S. (1975) *Neth. J. agric. Sci.* **23**, 89.

Tayler, J. C. (1959) *Nature, Lond.* **184**, 2021.

Tayler, J. C.; Wilkins, R. J. (1976) In *Principles of cattle production*, p. 343. (Eds. H. Swan & W. H. Broster). London: Butterworths. (*Proc. Easter Sch. agric. Sci. Univ. Nott.* **23**, 1975).

Telle, P. P.; Preston, R. L.; Kintner, L. D.; Pfander, W. H. (1964) *J. Anim. Sci.* **23**, 59.

Terlecki, S.; Markson, L. M. (1959). *Vet. Rec.* **71**, 508.

Ternouth, J. H.; Roy, J. H. B.; Siddons, R. C. (1974) *Br. J. Nutr.* **31**, 13.

Ternouth, J. H.; Roy, J. H. B.; Thompson, S. Y.; Toothill, J.; Gillies, C. M.; Edwards-Webb, J. D. (1975) *Br. J. Nutr.* **33**, 181.

Terry, R. A.; Spooner, M. C.; Osbourn, D. F. (1975) *J. agric. Sci., Camb.* **84**, 373.

Thafvelin, B.; Oksanen, H. E. (1966) *J. Dairy Sci.* **49**, 282.

Theiler, A. (1931) *Vet. Rec.* **11**, 1143.

Theiler, A. (1934) *Vet. J.* **90**, 143.

Theiler, A.; Green, H. H.; Du Toit, P. J. (1927) *J. agric. Sci., Camb.* **17**, 291.

Thickett, W. S.; Cuthbert, N. H.; Brigstocke, T. D. A.; Wilson, P. N.; Lindemann, M. A. (1980) *Anim. Prod.* (in Press).

Thomas, C.; Wilkinson, J. M.; Tayler, J. C. (1975) *J. agric. Sci., Camb.* **84**, 353.

Thomas, J. W. (1952) *J. Dairy Sci.* **35**, 1107.

Thomas, J. W. (1971) *J. Dairy Sci.* **54**, 1629.

Thomas, J. W.; Moore, L. A. (1951) *J. Dairy Sci.* **34**, 916.

Thomas, J. W.; Okamoto, M. (1956) *J. Dairy Sci.* **39**, 928.

Thomas, T. P. (1971) *Anim. Prod.* **13**, 399.

Thompson, A.; Hansard, S. L.; Bell, M. C. (1959) *J. Anim. Sci.* **18**, 187.

Thompson, H. J.; Worstell, D. M.; Brody, S. (1949) *Res. Bull. Mo. agric. Exp. Stn.* No. 436.

Thompson, S. Y. (1959) *Proc. 15th Int. Dairy Congr.* p. 247.

Thompson, S. Y. (1970) *Proc. 18th int. Dairy Congr.* 1E, 569.

Thompson, S. Y.; Henry, K. M.; Kon, S. K. (1964) *J. Dairy Res.* **31**, 1.

Thomson, D. J. (1972) *Proc. Nutr. Soc.* **31**, 127.

Thomson, D. J.; Beever, D. E. (1972) *Proc. Nutr. Soc.* **31**, 66A.

Thomson, D. J.; Beever, D. E.; Coelho da Silva, J. F.; Armstrong, D. G. (1972) *Br.J. Nutr.* **28**, 31.

Thorbek, G. (1970) In *Energy metabolism of farm animals:* 5th symp., Vitznau, p. 129. (Eds. A. Schürch & C. Wenk). Zurich: Juris Druck. (*Publs. Eur. Ass. Anim. Prod.* **13**).

Thornton, I. (1974) In *Trace element metabolism in animals—2:* 2nd int. symp., Madison, Wis., 1973, p. 451. (Eds. W. G. Hoekstra, J. W. Suttie, H. E. Ganther & W. Mertz). Baltimore, Md: University Park Press.

Thornton, I.; Kershaw, G. F.; Davies, M. K. (1972) *J. agric. Sci., Camb.* **78**, 157.

Tillman, A. D.; Brethour, J. R. (1958a) *J. Anim. Sci.* **17**, 100.

Tillman, A. D.; Brethour, J. R. (1958b) *J. Anim. Sci.* **17**, 104.

Tillman, A. D.; Brethour, J. R. (1958c) *J. Anim. Sci.* **17**, 782.

Tillman, A. D.; Brethour, J. R. (1958d) *J. Anim. Sci.* **17**, 792.

Tillman, A. D.; Brethour, J. R.; Hansard, S. L. (1959) *J. Anim. Sci.* **18**, 249.

Tisdall, F. F.; Brown, A. (1949) *J. Am. med. Ass.* **92**, 860.

Todd, G. C.; Krook, L. (1966) *Pathologia vet.* **3**, 379.

Todd, J. R. (1969) *Proc. Nutr. Soc.* **28**, 189.

Todd, J. R. (1972) *Proc. Br. Soc. Anim. Prod.* **1**, 93.

Todd, J. R.; Gribben, H. J. (1965) *Vet. Rec.* **77**, 498.

Todd, J. R.; Gracey, J. F.; Thompson, R. H. (1962) *Br. vet. J.* **118**, 482.

Todd, J. R.; Milne, A. A.; How, P. F. (1967) *Vet. Rec.* **81**, 653.

Tomas, F. M.; Somers, M. (1974) *Aust. J. agric. Res.* **25**, 475.

Topps, J. H.; Elliott, R. C.; Johnson, P. D.; Reed, W. D. C. (1966) *Proc. Nutr. Soc.* **25**, xxxv.

Topps, J. H.; Kay, R. N. B.; Goodall, E. D. (1968) *Br. J. Nutr.* **22**, 261.

Toutain, P. L.; Webster, A. J. F. (1975) *C. r. hebd. Séanc. Acad. Sci., Paris* **281D**, 1605.

(S) Treacher, R. J. (1977) *Proc. Nutr. Soc.* **36**, 39A.

(S) Treacher, R. J.; Little, W.; Collis, K. A.; Stark, A. J. (1976) *J. Dairy Res.* **43**, 357.

Trigg, T. E. (1974) Ph.D thesis, University of Aberdeen.

Trinder, N.; Woodhouse, C. D.; Renton, C. P. (1969) *Vet. Rec.* **85**, 550.

Tulloh, N. M. (1966) *N.Z. Jl agric. Res.* **9**, 252.

Turk, K. L.; Morrison, F. B.; Maynard, L. A. (1934) *J. agric. Res.* **48**, 555.

Turk, K. L.; Morrison, F. B.; Maynard, L. A. (1935) *J. agric. Res.* **51**, 401.

Turner, W. A.; Hartman, A. M. (1929) *J. Nutr.* **1**, 445.

Turner, W. A.; Harding, T. S.; Hartman, A. M. (1927) *J. agric. Res.* **35**, 625.

Tyrrell, H. F.; Reid, J. T. (1965) *J. Dairy Sci.* **48**, 1215.

Tyrrell, H. F.; Reid, J. T.; Moe, P. W. (1966) *J. Dairy Sci.* **49**, 739.

Tyrrell, H. F.; Reynolds, P. J.; Moe, P. W. (1975) In *Energy Metabolism of Farm Animals,* p. 57 (Ed. M. Vermorel) (*Publs. Europ. Assoc. Anim. Prod.* **19**).

Ulyatt, M. J.; MacRae, J. C. (1974) *J. agric. Sci., Camb.* **82**, 295.

Ulyatt, M. J.; MacRae, J. C.; Clarke, R. T. J.; Pearce, P. D. (1975) *J. agric. Sci., Camb.* **84**, 453.

Underwood, E. J. (1977) *Trace Elements in human and animal nutrition,* 4th edn. New York: Academic Press.

Underwood, E. J.; Somers, M. (1969) *Aust. J. agric. Res.* **20**, 889.

Ursescu, A.; Timariu, S.; Stavăr, P.; Mateescu, V. (1972) *Lucr. ştiinţ. Inst. Cerc. Nutr. anim.* **1**, 331.

Utley, P. R.; Bradley, N. W.; Boling, J. A. (1970) *J. Anim. Sci.* **31**, 130.

Van Adrichem, P. W. M. (1965) *Tijdschr. Diergeneesk.* **90**, 1371.

Van Adrichem, P. W. M.; Van Leeuwen, J. M.; Van Kluijve, J. J. (1970) *Tijdschr. Diergeneesk.* **95**, 1170.

Van Der Grift, J. (1955) *Versl. landbouwk. Onderz.* No. 61.10.

Van der Honing, Y.; Van Es, A. J. H. (1974) In *Energy metabolism of farm animals:* 6th symp., Hohenheim, 1973, p. 209. (Eds. K. H. Menke, H.-J. Lantzsch and J. R. Reichl.) Hohenheim: Universitäts Dokumentationsstelle. (*Publs. Eur. Ass. Anim. Prod.* **14**).

Van der Honing, Y.; Steg, A.; Van Es, A. J. H. (1977) *Livestk. Prod. Sci.* **4**, 57.

Van der Horst, C. J. G. (1960) *Tijdschr. Diergeneesk.* **85**, 1060.

Vanderveen, J. E.; Keener, H. A. (1964) *J. Dairy Sci.* **47**, 1224.

Van Dijk. J. B. (1966) *Tijdschr. Diergeneesk.* **91**, 733.

Van Es, A. J. H. (1961) *Versl. landb. Onderz.* No. 67.5.

Van Es, A. J. H. (1969) *Proc. 3rd gen. Mtg. Eur. Grassld. Fedn., Braunschweig,* p. 275.

Van Es, A. J. H. (1970) Cited by Flatt *et al.* (1972).

Van Es, A. J. H. (1972) *Handbuch der Tierernährung, vol. 2*, p. 1 (Eds. W. Lenkeit, K. Breirem and E. Crasemann). Berlin: Paul Parey.

Van Es, A. J. H. (1975) *Livest. Prod. Sci.* **2**, 95.

Van Es, A. J. H. (1976) In *Principles of cattle production,* p. 237. (Eds. H. Swan & W. H. Broster). London: Butterworths. (*Proc. Easter Sch. agric. Sci. Univ. Nott.* 1975, **23**.).

Van Es, A. J. H.; Nijkamp, H. J. (1969) In *Energy metabolism of farm animals:* 4th symp., Warsaw, 1967, p. 203. (Eds. K. L. Blaxter, J. Kielanowski and G. Thorbek). Newcastle upon Tyne: Oriel Press. (*Publs. Eur. Ass. Anim. Prod.* **12**).

Van Es, A. J. H.; Nijkamp, H. J.; Van Weerden, E. J.; Van Hellemond, K. K. (1969) In *Energy metabolism of farm animals:* 4th symp., Warsaw, 1967, p. 197. (Eds. K. L. Blaxter, J. Kielanowski & G. Thorbek). Newcastle upon Tyne: Oriel Press. (*Publs. Eur. Ass. Anim. Prod.* **12**).

Van Es, A. J. H.; Nijkamp, H. J.; Vogt, J. E. (1970) In *Energy metabolism of farm animals:* 5th symp., Vitznau, p. 61. (Eds. A. Schürch & C. Wenk). Zurich: Juris Druck. (*Publs. Eur. Ass. Anim. Prod.* **13**).

Van Gils, J. H. J.; Zayed, I. (1966) *Tijdschr. Diergeneesk.* **91**, 1375.

Van Landingham, A. H.; Henderson, H. O.; Bowling, G. A. (1935) *J. Dairy Sci.* **18**, 557.

Van Landingham, A. H.; Henderson, H. O.; Bowling, G. A. (1936) *J. Dairy Sci.* **19**, 597.

Van Leeuwen, J. M. (1972) *Neth. J. agric. Sci.* **20**, 35.

Van Leeuwen, J. M.; Van der Grift, J. (1969) *Versl. landbouwk. Onderz.* No. 731, p. 18.

Van Leeuwen, J. M.; Van Kluijve, J. J. (1971) *Neth. J. agric. Sci.* **19**, 189.

Vanschoubroek, F. X. (1958) *Neth. Milk Dairy J.* **12**, 12.

Van't Klooster, A. Th.; Rogers, P. A. M. (1969) *Meded. LandbHogesch. Wageningen* 69–11.

Van Ulsen, F. W. (1972) *Tijdschr. Diergeneesk.* **97**, 735.

Varela-Alvarez, H.; Wilson, L. L.; Rugh, M. C.; Gracia-Garza, E.; Simpson, M. J. (1970) *J. Dairy Sci.* **53**, 1783.

Vercoe, J. E. (1969) *Aust. J. agric. Res.* **20**, 607.

Vercoe, J. E. (1970) In *Energy metaoblism of farm animals:* 5th symp., Vitznau, p. 85 (Eds. A. Schürch & C. Wenk). Zurich: Juris Druck. (*Publs. Eur. Ass. Anim. Prod.* **13**.)

Vercoe, J. E.; Frisch, J. E. (1974) In *Energy Metabolism of Farm Animals,* p. 131 (Eds. K. H. Menke *et al*). (*Publs. Europ. Assoc. Anim. Prod.* **14**.)

Verdura, T.; Zamora, I. (1970) *Rev. cubana Cienc. agric.* (*Eng. Ed.*) **4**, 209.

Vermorel, M.; Bouvier, J. C.; Thivend, P.; Toullec, R. (1974) In *Energy metabolism of farm animals:* 6th symp., Hohenheim, 1973, p. 143 (Eds. K. H. Menke, H.-J. Lantzsch and J. R. Reichl). Hohenheim: Universitäts Dokumentationsstelle. (*Publs. Eur. Ass. Anim. Prod.* **14**).

Visek, W. J. (1973) *Proc. Cornell Nutr. Conf. Feed Mfrs.* p. 89.

Visek, W. J.; Munroe, R. A.; Swanson, E. W.; Comar, C. L. (1953) *J. Nutr.* **50**, 23.

Völtz, W. (1920) *Biochem. Z.* **102**, 151.

Wagner, D. G.; Loosli, J. K. (1967) *Mem. Cornell agric. Exp. Stn* No. 400.

Wainman, F. W.; Smith, J. S.; Dewey, P. J. S. (1975) *J. agric. Sci., Camb.* **84**, 109.

Waldo, D. R. (1973) *J. Anim. Sci.* **37**, 1062.

Walker, D. J.; Nader, C. J. (1968) *Appl. Microbiol.* **16**, 1124.

Walker, D. M. (1972) *J. agric. Sci., Camb.* **79**, 171.

Walker, D. M.; Cook, L. J. (1967) *Br. J. Nutr.* **21**, 237.

Walker, D. M.; Faichney, G. J. (1964a) *Br. J. Nutr.* **18**, 187.

Walker, D. M.; Faichney, G. J. (1964b) *Br. J. Nutr.* **18**, 295.

Walker, D. M.; Jagusch, K. T. (1969) In *Energy metabolism of farm animals:* 4th symp., Warsaw, 1967, p. 187. (Eds. K. L. Blaxter, J. Kielanowski & G. Thorbek). Newcastle upon Tyne: Oriel Press. (*Publs. Eur. Ass. Animl Prod.* **12**.)

Walker, D. M.; Norton, B. W. (1971a) *J. agric. Sci., Camb.* **77**, 363.

Walker, D. M.; Norton, B. W. (1971b) *Br. J. Nutr.* **26**, 15.

Walker, D. M.; Thompson, S. Y.; Bartlett, S.; Kon, S. K. (1949) *Int. Dairy Congr. 12, Stockholm* **1**, 83.

Walker, D. M.; Cook, L. J.; Jagusch, K. T. (1967) *Br. J. Nutr.* **21**, 275.

Wallace, L. R. (1948) *J. agric. Sci., Camb.* **38**, 93.

Wallis, G. C. (1938) *J. Dairy Sci.* **21**, 315.

Wallis, G. C. (1941) *J. Dairy Sci.* **24**, 517.

Wallis, G. C. (1944) *Bull. S. Dak. agric. Exp. Stn* No. 372.

Wallis, G. C. (1952) *J. Am. vet. med. Ass.* **120**, 392.

Wallis, G. C.; Palmer, L. S.; Gullickson, T. W. (1935) *J. Dairy Sci.* **18**, 213.

Ward, G.; Marion, G. B.; Campbell, C. W.; Dunham, J. R. (1971) *J. Dairy Sci.* **54**, 204.

Warner, A. C. I.; Stacy, B. D. (1972) *Q. Jl exp. Physiol.* **57**, 89.

Watson, M. J.; Savage, G. P.; Armstrong, D. G. (1972) *Proc. Nutr. Soc.* **31**, 98A.

Webster, A. J. F. (1967) *Br. J. Nutr.* **21**, 769.

Webster, A. J. F. (1971) In *Bioenergetics*, p. 42. (Ed. R. M. Smith). Washington, D.C.: Federation of American Societies for Experimental Biology.

Webster, A. J. F. (1975) In *Progress in Animal Biometeorology* vol. 1, p. 278 (Ed. H. D. Johnson). Amsterdam: Swets & Zeitlinger.

Webster, A. J. F. (1976) In *Meat animals: growth and productivity* (NATO adv. Study Inst., Prestbury, 1974) p. 89. (Eds. D. Lister, D. N. Rhodes, V. R. Fowler & M. F. Fuller). New York: Plenum Press.

Webster, A. J. F.; Blaxter, K. L. (1966) *Res. vet. Sci.* **7**, 466.

Webster, A. J. F.; Brockway, J.; Smith, J. S. (1974) *Anim. Prod.* **19**, 127.

Webster, A. J. F.; Chlumecky, J.; Young, B. A. (1970) *Can. J. Anim. Sci.* **50**, 89.

Webster, A. J. F.; Hays, F. L. (1968) *Can. J. Physiol. Pharmac.* **46**, 577.

Webster, A. J. F.; Hicks, A. M.; Hays, F. L. (1969) *Can. J. Physiol. Pharmac.* **47**, 553.

Webster, A. J. F.; Osuji, P. O.; White, F.; Ingram, J. F. (1975) *Br. J. Nutr.* **34**, 125.

Webster, A. J. F.; Smith, J. S.; Brockway, J. M. (1972) *Anim. Prod.* **15**, 189.

Webster, A. J. F.; Smith, J. S.; Crabtree, R. M.; Mollison, G. S. (1976) *Anim. Prod.* **23**, 329.

Webster, A. J. F.; Valks, D. (1966) *Proc. Nutr. Soc.* **25**, xxii.

Webster, W. M. (1932) *Aust. vet. J.* **8**, 199.

Weeth, H. J.; Haverland, L. H. (1961) *J. Anim. Sci.* **20**, 518.

Weeth, H. J.; Lesperance, A. L. (1965) *J. Anim. Sci.* **24**, 441.

Weeth, H. J.; Haverland, L. H.; Cassard, D. W. (1960) *J. Anim. Sci.* **19**, 845.

Weeth, H. J.; Sawhney, D. S.; Lesperance, A. L. (1967) *J. Anim. Sci.* **26**, 418.

Weeth, H. J.; Lesperance, A. L.; Bohman, V. R. (1968) *J. Anim. Sci.* **27**, 739.

Weichenthal, B. A.; Embry, L. B.; Emerick, R. J.; Whetzal, F. W. (1963) *J. Anim. Sci.* **22**, 979.

Weiss, E.; Bauer, P. (1968) *Zentralbl. Vet Med.* **15A**, 156.

Weller, R. A. (1957) *Aust. J. biol. Sci.* **10**, 384.

Weniger, J. H.; Funk, K.; König, K. H. (1955) *Arch. Tierernähr.* **5**, 216.

Wernery, H.; Heeschen, W.; Reichmuth, J.; Tolle, A. (1973) *Wien. tierärztl. Mschr.* **60**, 55.

Westerhuis, J. H. (1974) *Agric. Res. Rep. Wageningen* No. 814.

Weston, R. H.; Hogan, J. P. (1968) *Aust. J. agric. Res.* **19**, 567.

Weston, R. H.; Hogan, J. P. (1973) *Proc. Eur. Ass. Anim. Prod., Moscow*, p. 778.

Whanger, P. D.; Muth, O. H.; Oldfield, J. E.; Weswig, P. H. (1969) *J. Nutr.* **97**, 553.

Whanger, P. D.; Weswig, P. H.; Muth, O. H.; Oldfield, J. E. (1970) *Am. J. vet. Res.* **31**, 965.

Whanger, P. D.; Weswig, P. H.; Oldfield, J. E.; Cheeke, P. R.; Muth, O. H. (1972) *Nutr. Rep. int.* **6**, 21.

Whanger, P. D.; Weswig, P. H.; Oldfield, J. E.; Cheeke, P. R.; Schmitz, J. A. (1976) *Nutr. Rep. int.* **13**, 159.

Whiting, F.; Slen, S. B.; Bezeau, L. M. (1952) *Scient. Agric.* **32**, 365.

Whitlock, R. H.; Kessler, M. J.; Tasker, J. B. (1975) *Cornell Vet.* **65**, 512.

Wiegner, G.; Ghoneim, A. (1930) *Tierernährung* **2**, 193.

Wiener, G. (1967) *Anim. Prod.* **9**, 177.

Wiener, G.; Field, A. C. (1970) In *Trace element metabolism in animals:* WAAP/IBP int. symp., Aberdeen, 1969, p. 92. (Ed. C. F. Mills). Edinburgh: Livingstone.

Wiese, A. C.; Johnson, B. C.; Nevens, W. B. (1946) *Proc. Soc. exp. Biol. Med.* **63**, 521.

Wiese, A. C.; Johnson, B. C.; Mitchell, H. H.; Nevens, W. B. (1947) *J. Nutr.* **33**, 263.

Wilkins, R. J. (1974*a*) In *Proc. Nutr. Conf. Feed Mfrs. Univ. Nott.* **8**, 167. (Eds. H. Swan and D. Lewis). London: Butterworths.

Wilkins, R. J. (1974*b*) *Anim. Prod.* **19**, 87.

Williams, A. P.; Smith, R. H. (1974) *Br. J. Nutr.* **32**, 421.

Williams, A. P.; Smith, R. H. (1976) *Br. J. Nutr.* **36**, 199.

Williams, A. P.; Bishop, D. R.; Cockburn, J. E.; Scott, K. J. (1976) *J. Dairy Res.* **43**, 325.

Williams, R. B.; Bremner, I. (1976) *Proc. Nutr. Soc.* **35**, 86A.

Williams, V. J.; Roy, J. H. B.; Gillies, C. M. (1976) *Br. J. Nutr.* **36**, 317.

Wilman, D. (1965) *J. Br. Grassld. Soc.* **20**, 248.

Wilson, A. D. (1966) *Aus. J. agric. Res.* **17**, 503.

Wilson, A. D.; Hindley, N. L. (1968) *Aust. J. agric. Res.* **19**, 597.

Wilson, G. F. (1964) *N.Z. Jl agric. Res.* **7**, 432.

Wilson, J. G. (1966) *Vet. Rec.* **79**, 562.

Wilson, P. N. (1961) *J. agric. Sci., Camb.* **56**, 351.

Winchester, C. F.;Morris, M. J. (1956) *J. Anim. Sci.* **15**, 722.

Wise, M. B.; Ordoveza, A. L.; Barrick, E. R. (1963) *J. Nutr.* **79**, 79.

Witt, M.; Andreae, U.; Kallweit, E.; Huth, F. W.; Werhann, E.; Sekhausen, D.; Röseler, W. (1971) *SchrReihe. Max-Planck-Inst. Tierzucht* **53**, 10, 18.

Wohlt, J. E.; Sniffen, C. J.; Hoover, W. H. (1973) *J. Dairy Sci.* **56**, 1052.

Wood, P. D. P. (1969) *Anim. Prod.* **11**, 307.

Wood, T. B. (1927) *J. Minist. Agric.* **34**, 295.

Woodside, J. D. (1973) *Agriculture North. Ire.* **48**, 52.

Wright, E. (1955) *N.Z. Jl Sci. Technol. A* **37**, 332.

Wright, E.; Andrews, E. D. (1955) *N.Z. Jl Sci. Technol. A* **37**, 83.

Wright, M. J.; Davison, K. L. (1964) *Adv. Agron.* **16**, 197.

Wynne, K. N.; McClymont, G. L. (1956) *Aust. J. agric. Res.* **7**, 45.

Yadava, R. K.; Conrad, H. R.; Gilmore, L. O. (1973) *Indian J. Anim. Sci.* **43**, 16.

Young, B. A. (1966) *Aust. J. agric. Res.* **17**, 355.

Young, N. E.; Newton, J. E. (1974) *J. Br. Grassld Soc.* **29**, 117.

Young, N. E.; Newton, J. E. (1975) *Anim. Prod.* **21**, 203.

Young, V. R.; Lofgreen, G. P.; Luick, J. (1966) *Br. J. Nutr.* **20**, 795.

Yousef, M. K.; Johnson, H. D. (1966) *Life Sci.* **5**, 1349.

Zak, V. I. (1968) *Byull. èksp. Biol. Med.* No. 3, p. 51.

Ždanovski, N.; Dozet, N.; Stanišić, M. (1970) *Mljekarstvo* **20**, 266.

Zuntz, N.; Hagemann, O. (1898) *Untersuchungen über den Stoffwechsel des Pferdes bei Ruhe und Arbeit. Landw. Jb* **27**. Berlin: Paul Parey.

Index

Adnexa 6, 9, 26
Amino acids, utilization 130
Ammonia, toxicity 168

Biotin, requirement 291
Body composition 1–58

Calcium, absorption 184, 186, 217
  body 5, 17, 23, 27, 28, 34, 37
  effect on copper retention 229
  endogenous loss 185, 217
  relation to vitamin D requirements 278
  requirements 188
Calf, milk-fed, copper absorption 226
  energy requirements 97
  energy utilization 87
  feed intake 71
  iron requirements 237
  magnesium requirements 203, 205, 206
  protein requirements 146
  vitamin A requirements 274, 276, 281
  water intake 296
  zinc requirements 259
  newborn, composition 24, 37
Cattle, beef, nutrient requirements 311
  growth, body composition 29, 37, 56, 105
  calcium requirements 188, 190, 311
  copper requirements 225, 232
  feed intake 60
  iodine requirements 252
  iron requirements 239
  magnesium requirements 206, 311
  manganese requirements 264
  phosphorus requirements 196, 311
  potassium requirements 213
  sodium and chlorine requirements 214, 311
  vitamin D requirements 283, 288
  zinc requirements 259
  maintenance, energy requirements 97
  maintenance and growth, energy requirements 107, 118
  protein requirements 142, 148, 151, 164, 311
  vitamin A requirements 278
Cerebrocortical necrosis 292
Chlorine, body 21, 28, 36, 37
  losses 213
  requirements 215
Choline, requirements 291
Cobalt, deficiency 241
  requirements 242
  toxicity 242
Concentrate diets 41, 59, 63, 227
Copper, availability, effects of diet 227
  requirements, cattle 223, 232
  sheep 222, 230
  toxicity 233
Cow, lactation, body composition 37
  calcium absorption 186, 218
  calcium requirements 189, 313
  chlorine requirements 215
  copper requirements 225, 232
  energy requirements 111, 118
  energy utilization 89
  feed intake 66, 313
  gut fill 41
  iodine requirements 252

magnesium absorption 204, 205
magnesium requirements 207, 313
nutrient requirements 313
phosphorus absorption 220
phosphorus requirements 197, 198, 200, 313
protein requirements 138, 149, 155, 161, 163, 166, 313
ration formulation 141, 161
sodium requirements 215, 313
vitamin A requirements 275, 278, 313
water intake 297, 306
zinc requirements 259
pregnancy, calcium requirements 188, 190, 312
chlorine requirements 215
copper requirements 225, 232
energy deposition 106
energy requirements 113, 118
energy utilization 87
magnesium requirements 207, 312
nutrient deposition 26, 37, 58
nutrient requirements 312
phosphorus requirements 196, 198, 312
protein requirements 149, 156, 312
sodium requirements 215, 312
vitamin A requirements 275, 278, 312
vitamin D requirements 283, 312
water intake 297
zinc requirements 259

Drinking, frequency 304
Dry matter intake 61, 307, 308

Eating, energy cost 101
Energy requirements 73–119
Energy units 73
Energy value, body 2, 26, 29, 37, 105
Ewe, lactation, body composition 23
  calcium requirements 190, 191, 314
  copper requirements 223, 231
  energy requirements 116, 119
  feed intake 70
  gut fill 40
  iodine requirements 253
  magnesium requirements 207, 314
  nutrient requirements 314
  phosphorus requirements 197, 199, 314
  protein requirements 150, 158, 162, 314
  vitamin A requirements 278, 314
  water intake 300, 306
  zinc requirements 259
  pregnancy, body composition 22
  calcium requirements 190, 191, 314
  copper requirements 222, 231
  energy requirements 107, 115, 119
  energy utilization 87
  gut fill 40
  iodine requirements 253
  iron requirements 239
  magnesium requirements 207, 314
  nutrient deposition 58
  nutrient requirements 314
  phosphorus requirements 197, 199, 314
  protein requirements 150, 157, 314
  vitamin A requirements 278, 314
  water intake 300, 306
  zinc requirements 259

Fasting metabolism 95
Fat, body 5, 10, 22, 26, 29, 37, 54, 55, 57
Fat deposition 86
Fattening 83, 93, 104, 111
Fatty acids, essential 289
   polyunsaturated, myopathies 249
   volatile 78, 85, 293
Feed intake 59, 75, 103, 302
Feedingstuffs, digestibility in rumen 170
Feeds, energy values, calculation 117
Fertility, cow, role of vitamin D 285
Fleece 10, 16, 17, 49, 223, 258
Fluorine, toxicity 265
Foetus, analysis 1
   calf 24, 56, 58, 226
   lamb 2, 3, 51, 58
Folic acid, requirements 291

Goat, iodine requirements 253
Goitrogens, in feeds 254
Grass, magnesium absorption 205, 206
Growth, compensatory 16, 34, 40
   efficiency 83
   rate, cattle 34
      sheep 16
Gut contents 38–45

Hypomagnesaemia 208, 210

Iodine, deficiency 251
   requirements 252
   toxicity 256
Iron, availability 237
   deficiency 234
   effect on copper retention 230
   requirements, cattle 237
      sheep 239
   toxicity 239

Lactation see Cow, lactation; Ewe, lactation
Lamb, milk-fed, body composition 13
      copper absorption 226
      energy requirements 115, 119
      energy utilization 87
      feed intake 71
      magnesium requirements 203, 205
      protein requirements 147
   newborn, composition 3, 21
Lead, toxicity 266

Magnesium, absorption 184, 203
   body 6, 19, 23, 28, 36, 37
   endogenous loss 201
   requirements 206, 309–314
Maintenance, energy requirements 80
Manganese, deficiency 263
   requirements 264
   toxicity 265
Metabolizability of energy 59, 60, 68, 75, 78, 161, 307
Methane production 77
Milk, cow 44, 46, 47, 48, 291
   ewe 43, 45, 47, 48, 254, 291
Milk diets 41, 145
Milk fever 189, 286

Milk substitutes 71, 87, 97, 233, 239, 292, 296
Molasses, toxicity 292
Molybdenum, effect on copper availability 227

Nicotinic acid, requirements 291
Nitrate, toxicity 169
Nitrite, toxicity 169
Nitrogen, absorption 129, 176
   endogenous, in urine 178
   microbial 126, 175
   requirements, calculation 138
Nutrient requirements, as concentrations 307–314

Pelleted feeds 41, 84, 126
Phosphorus, absorption 195, 218, 220
   body 5, 18, 19, 23, 27, 28, 36, 37
   deficiency 199
   endogenous loss 192, 218, 219
   relation to vitamin D requirement 278
   requirements 196, 309–314
Potassium, body 6, 19, 21, 23, 28, 36, 37
   losses 211
   requirements 212
Pregnancy see Cow, pregnancy; Ewe, pregnancy
Protein, body 5, 10, 11, 22, 26, 29, 37, 54, 55, 57
   degradability 122, 134, 161, 179, 309, 310
   deposition 86
   requirements 121–181, 308, 310–314

Riboflavin, requirements 291
Rickets 281
Roughage diets 41, 59, 60, 126, 203
Rumen, digestion 74, 122, 125, 134, 170

Salt, tolerance 216, 303
Selenium, deficiency 244
   relation to vitamin E 243
   toxicity 250
   utilization 248
Sheep, fasting metabolism 99
   growth, body composition 9, 13, 21, 52, 106
      calcium requirements 189, 191, 313
      copper requirements 222, 231
      feed intake 60
      magnesium requirements 207, 313
      nutrient requirements 313
      phosphorus requirements 197, 313
      potassium requirements 213
      sodium and chlorine requirements 214, 216
      zinc requirements 259
   maintenance and growth, energy requirements 114, 118
      protein requirements 149, 157, 164, 313
      vitamin A requirements 278, 313
   vitamin D requirements 284, 288, 313
   water intakes 301, 306
Silage 41, 60, 62, 64, 126, 163
Sodium, body 6, 21, 23, 28, 36, 37
   losses 213
   requirements 215, 309–314
Soil, effect on copper availability 229
Sulphur, effect on copper availability 227
   relation to nitrogen in feeds 181
   relation to selenium utilization 248
   requirements 166
Sunlight, relation to vitamin D requirement 287

Temperature, environmental   102, 298, 300, 302
Thiamin, requirements   291, 292
Trace elements   221–267

Urea, utilization   128, 140, 163
Uterus   3, 6, 8, 9, 26, 58

Vitamin A, deficiency   269
  requirements, cattle   274, 278, 311–313
    sheep   276, 278
  toxicity   273
  utilization   271
Vitamin B$_6$, requirements   291
Vitamin B$_{12}$, requirements   291
  synthesis   240, 293
Vitamin D, activity   278
  deficiency   280, 285

requirements, cattle   283, 288, 311–313
  sheep   284, 288
  toxicity   286
Vitamin E, activity   243
  deficiency   244
  utilization   248

Walking, energy cost   101
Water, requirements   295–306
Weight gains, energy value   105
Wool, see Fleece
Work, muscular   82, 100

Zinc, deficiency   256, 261
  effect on copper retention   230, 262
  requirements   256
  toxicity   262